Molecular Approaches to Genetic Diversity

Molecular Approaches to Genetic Diversity

Editor:

Dr. Q. Tian Wang, PhD

Delve Publishing
www.delvepublishing.com

Molecular Approaches to Genetic Diversity

Editor: Dr. Q. Tian Wang, PhD

708 3rd Avenue, 6th Floor
New York, NY 10017
United States of America
www.delvepublishing.com

ISBN: 978-1-68095-758-7

Library of Congress Control Number: 2016951514

About the Editor

Dr. Q. Tian Wang, PhD

Dr. Q. Tian Wang holds a PhD in Biochemistry, Molecular Biology and Cell Biology from Northwestern University and has received postdoctoral training at Stanford University. She has extensive biomedical research experience and has directed multiple award-winning research programs studying the molecular basis of embryonic development and disease. She is currently a Science Officer in support of Congressionally Directed Medical Research Programs.

Contents

List of Contributors

J. Gustav Smith
Department of Cardiology, Department of Clinical Sciences, Lund University, Lund, Sweden
Department of Heart Failure and Valvular Disease, Skåne University Hospital, Lund, Sweden
Program in Medical and Population Genetics, Broad Institute of Harvard and MIT, Cambridge, Massachusetts, United States of America
Center for Human Genetic Research and Cardiovascular Research Center, Harvard Medical School and Massachusetts General Hospital, Boston, Massachusetts, United States of America
Department of Clinical Sciences, Lund University, Malmö, Sweden

Janine F. Felix
Department of Epidemiology, Erasmus MC, University Medical Center Rotterdam, Rotterdam, the Netherlands
Netherlands Consortium for Healthy Aging (NGI-NCHA), The Netherlands Genomics Initiative, Leiden, the Netherlands

Alanna C. Morrison
Human Genetics Center, University of Texas Health Science Center at Houston, Houston, Texas, United States of America

Andreas Kalogeropoulos
Emory Clinical Cardiovascular Research Institute, Emory University, Atlanta, Georgia, United States of America

Stella Trompet
Department of Cardiology, Leiden University Medical Center, Leiden, the Netherlands
Department of Gerontology and Geriatrics, Leiden University Medical Center, Leiden, the Netherlands

Jemma B. Wilk
Brigham and Women's Hospital and Harvard Medical School, Boston, Massachusetts, United States of America

Olof Gidlöf
Department of Cardiology, Department of Clinical Sciences, Lund University, Lund, Sweden

Xinchen Wang
Program in Medical and Population Genetics, Broad Institute of Harvard and MIT, Cambridge, Massachusetts, United States of America
Department of Biology, Massachusetts Institute of Technology, Cambridge, Massachusetts, United States of America

Michael Morley
Perelman School of Medicine, University of Pennsylvania, Philadelphia, Pennsylvania, United States of America

Michael Mendelson
The Framingham Heart Study, Framingham, Massachusetts, United States of America
The Population Sciences Branch, National Heart, Lund and Blood Institute, Bethesda, Maryland, United States of America
Department of Cardiology, Boston Children's Hospital, Boston, Massachusetts, United States of America

Roby Joehanes
The Framingham Heart Study, Framingham, Massachusetts, United States of America
The Population Sciences Branch, National Heart, Lund and Blood Institute, Bethesda, Maryland, United States of America

Symen Ligthart
Department of Epidemiology, Erasmus MC, University Medical Center Rotterdam, Rotterdam, the Netherlands

Xiaoyin Shan
Perelman School of Medicine, University of Pennsylvania, Philadelphia, Pennsylvania, United States of America

Joshua C. Bis
Department of Epidemiology, University of Washington, Seattle, Washington, United States of America

Ying A. Wang
Novartis Institutes for BioMedical Research, Cambridge, Massachusetts, United States of America

Marketa Sjögren
Department of Clinical Sciences, Lund University, Malmö, Sweden

Julius Ngwa
Department of Biostatistics, Boston University School of Public Health, Boston, Massachusetts, United States of America

Jeffrey Brandimarto
Perelman School of Medicine, University of Pennsylvania, Philadelphia, Pennsylvania, United States of America

David J. Stott
Academic Section of Geriatric Medicine, Institute of Cardiovascular and Medical Sciences, Faculty of Medicine, University of Glasgow, Glasgow, United Kingdom

David Aguilar
Baylor College of Medicine, Houston, Texas, United States of America

Kenneth M. Rice
Department of Biostatistics, University of Washington, Seattle, Washington, United States of America

Howard D. Sesso
Brigham and Women's Hospital and Harvard Medical School, Boston, Massachusetts, United States of America

Serkalem Demissie
Department of Biostatistics, Boston University School of Public Health, Boston, Massachusetts, United States of America

Brendan M. Buckley
Department of Pharmacology and Therapeutics, University College Cork, Cork, Ireland

Kent D. Taylor
Institute for Translational Genomics and Population Sciences, Los Angeles Biomedical Research Institute and Department of Pediatrics, Harbor-UCLA Medical Center, Torrance, California, United States of America

Ian Ford
Robertson Center for Biostatistics, University of Glasgow, Glasgow, United Kingdom

Chen Yao
The Framingham Heart Study, Framingham, Massachusetts, United States of America
The Population Sciences Branch, National Heart, Lund and Blood Institute, Bethesda, Maryland, United States of America

Chunyu Liu
The Framingham Heart Study, Framingham, Massachusetts, United States of America
The Population Sciences Branch, National Heart, Lund and Blood Institute, Bethesda, Maryland, United States of America

Nona Sotoodehnia
Cardiovascular Health Research Unit, Department of Medicine, University of Washington, Seattle, Washington, United States of America

Pim van der Harst
Department of Cardiology, University of Groningen, University Medical Center Groningen, Groningen, the Netherlands

Bruno H. Ch. Stricker
Department of Epidemiology, Erasmus MC, University Medical Center Rotterdam, Rotterdam, the Netherlands
Department of Internal Medicine, Erasmus MC, University Medical Center Rotterdam, Rotterdam, the Netherlands
Inspectorate for Health Care, The Hague, the Netherlands
Department of Medical Informatics, Erasmus MC, University Medical Center Rotterdam, Rotterdam, the Netherlands

Stephen B. Kritchevsky
Department of Internal Medicine, Section on Geronotology and Geriatric Medicine, Wake Forest School of Medicine, Winston-Salem, North Carolina, United States of America

Yongmei Liu
Department of Epidemiology and Prevention, Division of Public Health Sciences, Wake Forest University Health Sciences, Winston-Salem, North Carolina, United States of America

J. Michael Gaziano
Brigham and Women's Hospital and Harvard Medical School, Boston, Massachusetts, United States of America

Albert Hofman
Department of Epidemiology, Erasmus MC, University Medical Center Rotterdam, Rotterdam, the Netherlands

Christine S. Moravec
Department of Cardiovascular Medicine, Cleveland Clinic Foundation, Cleveland, Ohio, United States of America

André G. Uitterlinden
Department of Epidemiology, Erasmus MC, University Medical Center Rotterdam, Rotterdam, the Netherlands
Netherlands Consortium for Healthy Aging (NGI-NCHA), The Netherlands Genomics Initiative, Leiden, the Netherlands
Department of Internal Medicine, Erasmus MC, University Medical Center Rotterdam, Rotterdam, the Netherlands

Manolis Kellis
Program in Medical and Population Genetics, Broad Institute of Harvard and MIT, Cambridge, Massachusetts, United States of America
Department of Biology, Massachusetts Institute of Technology, Cambridge, Massachusetts, United States of America

Joyce B. van Meurs
Department of Internal Medicine, Erasmus MC, University Medical Center Rotterdam, Rotterdam, the Netherlands

Kenneth B. Margulies
Perelman School of Medicine, University of Pennsylvania, Philadelphia, Pennsylvania, United States of America

Abbas Dehghan
Department of Epidemiology, Erasmus MC, University Medical Center Rotterdam, Rotterdam, the Netherlands

Daniel Levy
The Framingham Heart Study, Framingham, Massachusetts, United States of America
The Population Sciences Branch, National Heart, Lund and Blood Institute, Bethesda, Maryland, United States of America

Björn Olde
Department of Cardiology, Department of Clinical Sciences, Lund University, Lund, Sweden

Bruce M. Psaty
Department of Epidemiology, University of Washington, Seattle, Washington, United States of America
Cardiovascular Health Research Unit, Department of Medicine, University of Washington, Seattle, Washington, United States of America
Department of Health Services, University of Washington, Seattle, Washington, United States of America
Group Health Research Institute, Group Health Cooperative, Seattle, Washington, United States of America

L. Adrienne Cupples
Department of Biostatistics, Boston University School of Public Health, Boston, Massachusetts, United States of America

J. Wouter Jukema
Department of Cardiology, Leiden University Medical Center, Leiden, the Netherlands
Durrer Center for Cardiogenetic Research, Amsterdam, the Netherlands
Interuniversity Cardiology Institute of the Netherlands, Utrecht, the Netherlands

Luc Djousse
Brigham and Women's Hospital and Harvard Medical School, Boston, Massachusetts, United States of America

Oscar H. Franco
Department of Epidemiology, Erasmus MC, University Medical Center Rotterdam, Rotterdam, the Netherlands
Netherlands Consortium for Healthy Aging (NGI-NCHA), The Netherlands Genomics Initiative, Leiden, the Netherlands

Eric Boerwinkle
Human Genetics Center, University of Texas Health Science Center at Houston, Houston, Texas, United States of America
Human Genome Sequencing Center, Baylor College of Medicine, Houston, Texas, United States of America

Laurie A. Boyer
Department of Biology, Massachusetts Institute of Technology, Cambridge, Massachusetts, United States of America

Christopher Newton-Cheh
Program in Medical and Population Genetics, Broad Institute of Harvard and MIT, Cambridge, Massachusetts, United States of America
Center for Human Genetic Research and Cardiovascular Research Center, Harvard Medical School and Massachusetts General Hospital, Boston, Massachusetts, United States of America

Javed Butler
Emory Clinical Cardiovascular Research Institute, Emory University, Atlanta, Georgia, United States of America

Ramachandran S. Vasan
Departments of Medicine and Preventive Medicine, Boston University School of Medicine, Boston, Massachusetts, United States of America

Thomas P. Cappola
Perelman School of Medicine, University of Pennsylvania, Philadelphia, Pennsylvania, United States of America

Nicholas L. Smith
Department of Epidemiology, University of Washington, Seattle, Washington, United States of America
Department of Health Services, University of Washington, Seattle, Washington, United States of America
Seattle Epidemiologic Research and Information Center, Department of Veteran Affairs Office of Research and Development, Seattle, Washington, United States of America

Thalia Antoniadi
Bristol Genetics Laboratory, North Bristol NHS Trust, Southmead Hospital, Bristol BS10 5NB, UK

Chris Buxton
Bristol Genetics Laboratory, North Bristol NHS Trust, Southmead Hospital, Bristol BS10 5NB, UK

Gemma Dennis
Bristol Genetics Laboratory, North Bristol NHS Trust, Southmead Hospital, Bristol BS10 5NB, UK

Natalie Forrester
Bristol Genetics Laboratory, North Bristol NHS Trust, Southmead Hospital, Bristol BS10 5NB, UK

Debbie Smith
Bristol Genetics Laboratory, North Bristol NHS Trust, Southmead Hospital, Bristol BS10 5NB, UK

Peter Lunt
Department of Social & Community Medicine, University of Bristol, Oakfield House, Bristol BS8 2BN, UK.

Sarah Burton-Jones
Bristol Genetics Laboratory, North Bristol NHS Trust, Southmead Hospital, Bristol BS10 5NB, UK

Celeste Sassi
Reta Lila, Weston Research Laboratories, Department of Molecular Neuroscience, UCL Institute of Neurology, London, UK
Laboratory of Neurogenetics, National Institute on Aging, National Institutes of Health, Bethesda, MD, USA
Department of Experimental Neurology, Center for Stroke Research Berlin (CSB), Charite' Universitätmedizin, Berlin, Germany
German Center for Neurodegenerative Diseases (DZNE), Berlin site, Germany

Michael A. Nalls
Laboratory of Neurogenetics, National Institute on Aging, National Institutes of Health, Bethesda, MD, USA

Perry G. Ridge
Department of Biology, Brigham Young University, Provo, UT, USA

Jesse R. Gibbs
Laboratory of Neurogenetics, National Institute on Aging, National Institutes of Health, Bethesda, MD, USA

Jinhui Ding
Laboratory of Neurogenetics, National Institute on Aging, National Institutes of Health, Bethesda, MD, USA

Michelle K. Lupton
King's College London, Institute of Psychiatry, Psychology and Neuroscience, London, UK
QIMR Berghofer Medical Research Institute, Brisbane, Queensland, Australia

Claire Troakes
King's College London, Institute of Psychiatry, Psychology and Neuroscience, London, UK

Katie Lunnon
King's College London, Institute of Psychiatry, Psychology and Neuroscience, London, UK
Institute of Biomedical and Clinical Science, University of Exeter Medical School, Exeter, Devon, UK

Safa Al-Sarraj
King's College London, Institute of Psychiatry, Psychology and Neuroscience, London, UK

Kristelle S. Brown
Translation Cell Sciences-Human Genetics, School of Life Sciences, Queens Medical Centre, University of Nottingham, Nottingham, UK

Christopher Medway
Translation Cell Sciences-Human Genetics, School of Life Sciences, Queens Medical Centre, University of Nottingham, Nottingham, UK

Naomi Clement
Translation Cell Sciences-Human Genetics, School of Life Sciences, Queens Medical Centre, University of Nottingham, Nottingham, UK

Jenny Lord
Translation Cell Sciences-Human Genetics, School of Life Sciences, Queens Medical Centre, University of Nottingham, Nottingham, UK

James Turton
Translation Cell Sciences-Human Genetics, School of Life Sciences, Queens Medical Centre, University of Nottingham, Nottingham, UK

Jose Bras
Reta Lila, Weston Research Laboratories, Department of Molecular Neuroscience, UCL Institute of Neurology, London, UK

Maria R. Almeida
Neurogenetics Laboratory, Center for Neuroscience and Cell Biology, University of Coimbra, Coimbra, Portugal

Henne Holstege
Department of Neurology, Alzheimer Center, VU University Medical Center, Neuroscience Campus Amsterdam, Amsterdam, the Netherlands

Eva Louwersheimer
Department of Neurology, Alzheimer Center, VU University Medical Center, Neuroscience Campus Amsterdam, Amsterdam, the Netherlands

Wiesje M. van der Flier
Department of Neurology, Alzheimer Center, VU University Medical Center, Neuroscience Campus Amsterdam, Amsterdam, the Netherlands

Philip Scheltens
Department of Neurology, Alzheimer Center, VU University Medical Center, Neuroscience Campus Amsterdam, Amsterdam, the Netherlands

John C. Van Swieten
Department of Neurology, Alzheimer Center, VU University Medical Center, Neuroscience Campus Amsterdam, Amsterdam, the Netherlands
Department of Neurology, Erasmus Medical Centre, Rotterdam, the Netherlands

Isabel Santana
Neurology Department, Centro Hospitalar e Universitário de Coimbra, Coimbra, Portugal
Faculty of Medicine, Coimbra University, Coimbra, Portugal

Catarina Oliveira
CNC—Center for Neuroscience and Cell Biology, University of Coimbra, Coimbra, Portugal
Laboratory of Biochemistry, Faculty of Medicine, University of Coimbra, Coimbra, Portugal

Kevin Morgan
Translation Cell Sciences-Human Genetics, School of Life Sciences, Queens Medical Centre, University of Nottingham, Nottingham, UK

John F. Powell
King's College London, Institute of Psychiatry, Psychology and Neuroscience, London, UK

John S. Kauwe
Department of Biology, Brigham Young University, Provo, UT, USA
Department of Neuroscience, Brigham Young University, Provo, UT, USA

Carlos Cruchaga
Division of Biology and Biomedical Sciences, Washington University, St. Louis, MO, USA

Alison M. Goate
Icahn School of Medicine at Mount Sinai, Icahn Medical Institute, New York, NY, USA

Andrew B. Singleton
Laboratory of Neurogenetics, National Institute on Aging, National Institutes of Health, Bethesda, MD, USA

Rita Guerreiro
Reta Lila, Weston Research Laboratories, Department of Molecular Neuroscience, UCL Institute of Neurology, London, UK

John Hardy
Reta Lila, Weston Research Laboratories, Department of Molecular Neuroscience, UCL Institute of Neurology, London, UK

Benjamin R. Lin
Stein Eye Institute, David Geffen School of Medicine at UCLA, Los Angeles, California, United States of America

Derek J. Le
Stein Eye Institute, David Geffen School of Medicine at UCLA, Los Angeles, California, United States of America

Yabin Chen
Ophthalmic Genetics and Visual Function Branch, National Eye Institute, National Institutes of Health, Bethesda, Maryland, United States of America

Qiwei Wang
Ophthalmic Genetics and Visual Function Branch, National Eye Institute, National Institutes of Health, Bethesda, Maryland, United States of America

D. Doug Chung
Stein Eye Institute, David Geffen School of Medicine at UCLA, Los Angeles, California, United States of America

Ricardo F. Frausto
Stein Eye Institute, David Geffen School of Medicine at UCLA, Los Angeles, California, United States of America

Christopher Croasdale
Davis Duehr Dean Clinic, Madison, Wisconsin, United States of America

Richard W. Yee
Cross Ophthalmology Associates, Houston, Texas, United States of America

Fielding J. Hejtmancik
Ophthalmic Genetics and Visual Function Branch, National Eye Institute, National Institutes of Health, Bethesda, Maryland, United States of America

Anthony J. Aldave
Stein Eye Institute, David Geffen School of Medicine at UCLA, Los Angeles, California, United States of America

Agnieszka K. Witkiewicz
Simmons Cancer Center, UT Southwestern Medical Center, Dallas, Texas 75390, USA
Department of Pathology, UT Southwestern Medical Center, Dallas, Texas 75390, USA

Elizabeth A. McMillan
Department of Cell Biology, UT Southwestern Medical Center, Dallas, Texas 75390, USA

Uthra Balaji
Department of Pathology, UT Southwestern Medical Center, Dallas, Texas 75390, USA

GuemHee Baek
Department of Pathology, UT Southwestern Medical Center, Dallas, Texas 75390, USA

Wan-Chi Lin
Eppley Institute for Research in Cancer and Allied Diseases, University of Nebraska Medical Center, Omaha Nebraska 68198, USA

John Mansour
Department of Surgery, UT Southwestern Medical Center, Dallas, Texas 75390, USA

Mehri Mollaee
Department of Pathology, Thomas Jefferson University Philadelphia Pennsylvania 19107, USA. 7Department of Surgery, Thomas Jefferson University, Philadelphia Pennsylvania 19107, USA

Kay-Uwe Wagner
Eppley Institute for Research in Cancer and Allied Diseases, University of Nebraska Medical Center, Omaha Nebraska 68198, USA

Prasad Koduru
Department of Pathology, UT Southwestern Medical Center, Dallas, Texas 75390, USA

Adam Yopp
Department of Surgery, UT Southwestern Medical Center, Dallas, Texas 75390, USA

Michael A. Choti
Department of Surgery, UT Southwestern Medical Center, Dallas, Texas 75390, USA

Charles J. Yeo
Department of Surgery, Thomas Jefferson University, Philadelphia Pennsylvania 19107, USA

Peter McCue
Department of Pathology, Thomas Jefferson University Philadelphia Pennsylvania 19107, USA

Michael A. White
Simmons Cancer Center, UT Southwestern Medical Center, Dallas, Texas 75390, USA
Department of Cell Biology, UT Southwestern Medical Center, Dallas, Texas 75390, USA

Erik S. Knudsen
Simmons Cancer Center, UT Southwestern Medical Center, Dallas, Texas 75390, USA
Department of Pathology, UT Southwestern Medical Center, Dallas, Texas 75390, USA

Cristina Valente
IPATIMUP, Institute of Molecular Pathology and Immunology of the University of Porto, Porto, Portugal
Faculty of Sciences, University of Porto, Porto, Portugal

Luis Alvarez
IPATIMUP, Institute of Molecular Pathology and Immunology of the University of Porto, Porto, Portugal

Sarah J. Marks
Department of Zoology, University of Oxford, Oxford, UK

Ana M. Lopez-Parra
Departamento de Toxicología y Legislación Sanitaria, Facultad de Medicina, Universidad Complutense de Madrid, Madrid, Spain

Walther Parson
Institute of Legal Medicine, Innsbruck Medical University, Innsbruck, Austria
Penn State Eberly College of Science, University Park, Pennsylvania, USA

Ockie Oosthuizen
School of Medicine, University of Namibia, Windhoek, Namibia

Erica Oosthuizen
School of Medicine, University of Namibia, Windhoek, Namibia

António Amorim
IPATIMUP, Institute of Molecular Pathology and Immunology of the University of Porto, Porto, Portugal
Faculty of Sciences, University of Porto, Porto, Portugal

Cristian Capelli
Department of Zoology, University of Oxford, Oxford, UK

Eduardo Arroyo-Pardo
Departamento de Toxicología y Legislación Sanitaria, Facultad de Medicina, Universidad Complutense de Madrid, Madrid, Spain

Leonor Gusmão
IPATIMUP, Institute of Molecular Pathology and Immunology of the University of Porto, Porto, Portugal
DNA Diagnostic Laboratory (LDD), State University of Rio de Janeiro (UERJ), Rio de Janeiro, Brazil.

Maria J. Prata
IPATIMUP, Institute of Molecular Pathology and Immunology of the University of Porto, Porto, Portugal
Faculty of Sciences, University of Porto, Porto, Portugal

Vinicius Henrique da Silva
Animal Biotechnology Laboratory, Animal Science Department, University of São Paulo (USP)/Luiz de Queiroz College of Agriculture (ES-ALQ), Piracicaba, São Paulo, Brazil

Luciana Correia de Almeida Regitano
Embrapa Pecuária Sudeste, São Carlos, São Paulo, Brazil

Ludwig Geistlinger
Institute of Bioinformatics, Department of Informatics, Ludwig-MaximiliansUniversität München (LMU), Amalienstrasse 17, 80333, München, Germany

Fábio Pértille
Animal Biotechnology Laboratory, Animal Science Department, University of São Paulo (USP)/Luiz de Queiroz College of Agriculture (ES-ALQ), Piracicaba, São Paulo, Brazil

Poliana Fernanda Giachetto
Embrapa Informática Agropecuaria, Campinas, São Paulo, Brazil

Ricardo Augusto Brassaloti
Animal Biotechnology Laboratory, Animal Science Department, University of São Paulo (USP)/Luiz de Queiroz College of Agriculture (ES-ALQ), Piracicaba, São Paulo, Brazil

Natália Silva Morosini
Animal Biotechnology Laboratory, Animal Science Department, University of São Paulo (USP)/Luiz de Queiroz College of Agriculture (ES-ALQ), Piracicaba, São Paulo, Brazil

Ralf Zimmer
Institute of Bioinformatics, Department of Informatics, Ludwig-MaximiliansUniversität München (LMU), Amalienstrasse 17, 80333, München, Germany

Luiz Lehmann Coutinho
Animal Biotechnology Laboratory, Animal Science Department, University of São Paulo (USP)/Luiz de Queiroz College of Agriculture (ESALQ), Piracicaba, São Paulo, Brazil

Shadi Nayeri
Department of Agricultural, Food and Nutritional Science, University of Alberta, Edmonton, Alberta T6G 2P5, Canada

Mehdi Sargolzaei
Department of Animal Bioscience, Centre for the Genetic Improvement of Livestock, University of Guelph, Ontario, Canada
The Semex Alliance, Guelph, Canada

Mohammed K. Abo-Ismail
Department of Animal Bioscience, Centre for the Genetic Improvement of Livestock, University of Guelph, Ontario, Canada
Department of Animal and Poultry Production, Damanhour University, Damanhour, Egypt

Natalie May
Department of Agricultural, Food and Nutritional Science, University of Alberta, Edmonton, Alberta T6G 2P5, Canada

Stephen P. Miller
Department of Animal Bioscience, Centre for the Genetic Improvement of Livestock, University of Guelph, Ontario, Canada
Invermay Agricultural Centre, AgResearch Limited, Mosgiel, New Zealand

Flavio Schenkel
Department of Animal Bioscience, Centre for the Genetic Improvement of Livestock, University of Guelph, Ontario, Canada

Stephen S. Moore
Department of Agricultural, Food and Nutritional Science, University of Alberta, Edmonton, Alberta T6G 2P5, Canada
Agriculture and Food Innovation, Centre for Animal Science, University of Queensland, Queensland, Australia

Paul Stothard
Department of Agricultural, Food and Nutritional Science, University of Alberta, Edmonton, Alberta T6G 2P5, Canada

Hasina Begum
1 International Rice Research Institute, Los Baños, Philippines

Jennifer E. Spindel
Department of Plant Breeding and Genetics, Cornell University, Ithaca, NY, United States of America

Antonio Lalusin
Crop Science Cluster, University of the Philippines Los Baños, Los Baños, Philippines

Teresita Borromeo
Crop Science Cluster, University of the Philippines Los Baños, Los Baños, Philippines

Glenn Gregorio
International Rice Research Institute, Los Baños, Philippines

Jose Hernandez
Crop Science Cluster, University of the Philippines Los Baños, Los Baños, Philippines

Parminder Virk
International Center for Tropical Agriculture, Cali, Colombia

Bertrand Collard
International Rice Research Institute, Los Baños, Philippines

Susan R. McCouch
Department of Plant Breeding and Genetics, Cornell University, Ithaca, NY, United States of America

Jorge Urrestarazu
IRHS, INRA, AGROCAMPUS-Ouest, Université d'Angers, SFR 4207 QUASAV, 42 rue Georges Morel, 49071 Beaucouzé cedex, France
Department of Agricultural Sciences, University of Bologna, Viale Giuseppe Fanin 44, 40127 Bologna, Italy
Public University of Navarre (UPNA), Campus Arrosadia, 31006 Pamplona, Spain

Caroline Denancé
IRHS, INRA, AGROCAMPUS-Ouest, Université d'Angers, SFR 4207 QUASAV, 42 rue Georges Morel, 49071 Beaucouzé cedex, France

Elisa Ravon
IRHS, INRA, AGROCAMPUS-Ouest, Université d'Angers, SFR 4207 QUASAV, 42 rue Georges Morel, 49071 Beaucouzé cedex, France

Arnaud Guyader
IRHS, INRA, AGROCAMPUS-Ouest, Université d'Angers, SFR 4207 QUASAV, 42 rue Georges Morel, 49071 Beaucouzé cedex, France

Rémi Guisnel
IRHS, INRA, AGROCAMPUS-Ouest, Université d'Angers, SFR 4207 QUASAV, 42 rue Georges Morel, 49071 Beaucouzé cedex, France

Laurence Feugey
IRHS, INRA, AGROCAMPUS-Ouest, Université d'Angers, SFR 4207 QUASAV, 42 rue Georges Morel, 49071 Beaucouzé cedex, France

Charles Poncet
Plateforme Gentyane, INRA UMR1095 Genetics, Diversity and Ecophysiology of Cereals, 63100 Clermont-Ferrand, France

Marc Lateur
CRA-W, Centre Wallon de Recherches Agronomiques, Plant Breeding & Biodiversity, Bâtiment Emile Marchal, Rue de Liroux, 4 - 5030 Gembloux, Belgium

Patrick Houben
CRA-W, Centre Wallon de Recherches Agronomiques, Plant Breeding & Biodiversity, Bâtiment Emile Marchal, Rue de Liroux, 4 - 5030 Gembloux, Belgium

Matthew Ordidge
School of Agriculture, Policy and Development, University of Reading, Whiteknights, Reading RG6 6AR, UK

Felicidad Fernandez-Fernandez
NIAB EMR, East Malling Research, East Malling, Kent ME19 6BJ, United Kingdom

Kate M. Evans
Washington State University Tree Fruit Research and Extension Center, 1100 N Western Ave, Wenatchee, WA 98801, USA.

Frantisek Paprstein
RBIPH, Research and Breeding Institute of Pomology Holovousy Ltd., 508 01 Horice, Czech Republic

Jiri Sedlak
RBIPH, Research and Breeding Institute of Pomology Holovousy Ltd., 508 01 Horice, Czech Republic

Hilde Nybom
Department of Plant Breeding, Balsgård, Fjälkestadsvägen 459, Swedish University of Agricultural Sciences, 291 94 Kristianstad, Sweden

Larisa Garkava-Gustavsson
Department of Plant Breeding, Swedish University of Agricultural Sciences, Box 101230 53 Alnarp, Sweden

Carlos Miranda
Public University of Navarre (UPNA), Campus Arrosadia, 31006 Pamplona, Spain

Jennifer Gassmann
Agroscope, Institute for Plant Production Sciences IPS, Schloss 1, Wädenswil, Switzerland

Markus Kellerhals
Agroscope, Institute for Plant Production Sciences IPS, Schloss 1, Wädenswil, Switzerland

Ivan Suprun
NCRRIH&V, North Caucasian Regional Research Institute of Horticulture and Viticulture, 39, 40-letiya Pobedy street, Krasnodar 350901, Russian Federation

Anna V. Pikunova
VNIISPK, The All Russian Research Institute of Fruit Crop Breeding, 302530, p/o Zhilina, Orel district, Russian Federation

Nina G. Krasova
VNIISPK, The All Russian Research Institute of Fruit Crop Breeding, 302530, p/o Zhilina, Orel district, Russian Federation

Elnura Torutaeva
Kyrgyz National Agrarian University, 68 Mederova Street, 720005, Bishkek, Kyrgyzstan.

Luca Dondini
Department of Agricultural Sciences, University of Bologna, Viale Giuseppe Fanin 44, 40127 Bologna, Italy

Stefano Tartarini
Department of Agricultural Sciences, University of Bologna, Viale Giuseppe Fanin 44, 40127 Bologna, Italy

François Laurens
IRHS, INRA, AGROCAMPUS-Ouest, Université d'Angers, SFR 4207 QUASAV, 42 rue Georges Morel, 49071 Beaucouzé cedex, France

Charles-Eric Durel
RHS, INRA, AGROCAMPUS-Ouest, Université d'Angers, SFR 4207 QUASAV, 42 rue Georges Morel, 49071 Beaucouzé cedex, France

Fabien Nogué
INRA AgroParisTech, IJPB, UMR 1318, INRA Centre de Versailles, Route de Saint Cyr, 78026 Versailles Cedex, France

Kostlend Mara
INRA AgroParisTech, IJPB, UMR 1318, INRA Centre de Versailles, Route de Saint Cyr, 78026 Versailles Cedex, France

Cécile Collonnier
INRA AgroParisTech, IJPB, UMR 1318, INRA Centre de Versailles, Route de Saint Cyr, 78026 Versailles Cedex, France

Josep M. Casacuberta
Center for Research in Agricultural Genomics, CRAG (CSIC-IRTA-UAB-UB), Campus UAB, Cerdanyola del Valle`s, 08193 Barcelona, Spain

Yiping Qi
Department of Genetics, Cell Biology & Development and Center for Genome Engineering, University of Minnesota, Minneapolis, Minnesota 55455

Xiaohong Li
Department of Genetics, Cell Biology & Development and Center for Genome Engineering, University of Minnesota, Minneapolis, Minnesota 55455

Yong Zhang
Department of Genetics, Cell Biology & Development and Center for Genome Engineering, University of Minnesota, Minneapolis, Minnesota 55455
Department of Biotechnology, School of Life Sciences and Technology, University of Electronic Science and Technology of China, Chendu 610054, China

Colby G. Starker
Department of Genetics, Cell Biology & Development and Center for Genome Engineering, University of Minnesota, Minneapolis, Minnesota 55455

Nicholas J. Baltes
Department of Genetics, Cell Biology & Development and Center for Genome Engineering, University of Minnesota, Minneapolis, Minnesota 55455

Feng Zhang
Department of Genetics, Cell Biology & Development and Center for Genome Engineering, University of Minnesota, Minneapolis, Minnesota 55455

Jeffry D. Sander
Molecular Pathology Unit, Center for Computational and Integrative Biology, and Center for Cancer Research, Massachusetts General Hospital, Charlestown, Massachusetts 02129
Department of Pathology, Harvard Medical School, Boston, Massachusetts 02115

Deepak Reyon
Molecular Pathology Unit, Center for Computational and Integrative Biology, and Center for Cancer Research, Massachusetts General Hospital, Charlestown, Massachusetts 02129
Department of Pathology, Harvard Medical School, Boston, Massachusetts 02115
Department of Genetics, Development and Cell Biology, Interdepartmental Graduate Program in Bioinformatics & Computational Biology, Iowa State University, Ames, Iowa 50011

J. Keith Joung
Molecular Pathology Unit, Center for Computational and Integrative Biology, and Center for Cancer Research, Massachusetts General Hospital, Charlestown, Massachusetts 02129
Department of Pathology, Harvard Medical School, Boston, Massachusetts 02115

Daniel F. Voytas
Department of Genetics, Cell Biology & Development and Center for Genome Engineering, University of Minnesota, Minneapolis, Minnesota 55455

Antonio R. Castilla
Department of Integrative Biology, 401 Biological Laboratories, University of Texas, Austin, TX, 78712, United States of America

Nathaniel Pope
Department of Integrative Biology, 401 Biological Laboratories, University of Texas, Austin, TX, 78712, United States of America

Rodolfo Jaffé
Vale Institute of Technology–Sustainable Development, Rua Boaventura da Silva 955, Belém, PA 66055–090, Brazil
Department of Ecology, University of São Paulo, Rua do Matão 321, São Paulo, SP 05508–090, Brazil

Shalene Jha
Department of Integrative Biology, 401 Biological Laboratories, University of Texas, Austin, TX, 78712, United States of America

Michael R. Gillings
Department of Biological Sciences, Macquarie University, Sydney, NSW 2109, Australia

Ian T. Paulsen
Department of Chemistry and Biomolecular Sciences, Macquarie University, Sydney, NSW 2109, Australia

Sasha G. Tetu
Department of Chemistry and Biomolecular Sciences, Macquarie University, Sydney, NSW 2109, Australia

Selva Perumal Gunaletchumy
Department of Medical Microbiology, Faculty of Medicine, University of Malaya, 50603 Kuala Lumpur Malaysia,

Indran Seevasant
Department of Medical Microbiology, Faculty of Medicine, University of Malaya, 50603 Kuala Lumpur Malaysia,

Mun Hua Tan
Malaysian Genomics Resource Centre Berhad, 59200 Kuala Lumpur Malaysia

Laurence J. Croft
Malaysian Genomics Resource Centre Berhad, 59200 Kuala Lumpur Malaysia

Hazel M. Mitchell
School of Biotechnology and Biomolecular Sciences, University of New South Wales, Sydney NSW 2052 Australia

Khean Lee Goh
Department of Medicine, Faculty of Medicine, University of Malaya, 50603 Kuala Lumpur Malaysia

Mun Fai Loke
Department of Medical Microbiology, Faculty of Medicine, University of Malaya, 50603 Kuala Lumpur Malaysia,

Jamuna Vadivelu
Department of Medical Microbiology, Faculty of Medicine, University of Malaya, 50603 Kuala Lumpur Malaysia,

Craig Winstanley
Department of Clinical Infection, Microbiology and Immunology, Institute of Infection and Global Health, Ronald Ross Building, University of Liverpool, 8 West Derby Street, Liverpool, L69 7BE, UK

Siobhan O'Brien
Department of Biology, University of York, Wentworth Way, York, YO10 5DD, UK

Michael A. Brockhurst
Department of Biology, University of York, Wentworth Way, York, YO10 5DD, UK

Elizabeth J. Tanner
Gladstone Institutes (Virology and Immunology), San Francisco, California, United States of America

Karla A. Kirkegaard
Department of Genetics, Stanford University School of Medicine, Stanford, California, United States of America

Leor S. Weinberger
Gladstone Institutes (Virology and Immunology), San Francisco, California, United States of America
Department of Biochemistry and Biophysics, University of California, San Francisco, San Francisco, California, United States of America

Preface

Genetic diversity, or the variety of genes within a species, plays important roles in phenotypic diversity and the species' ability to adapt to environments. Studies of genetic diversity mainly focus on two areas. The first area is concerned with understanding and characterizing genetic diversity in humans, animals, plants, and microbes. The second focus area is exploiting genetic diversity for the purposes of medicine, agriculture, and conservation. To help readers appreciate the wide range of research questions on genetic diversity and how molecular approaches are utilized to address these questions, this book is organized into three chapters to showcase selected studies of genetic diversity in humans, animals and plants, and microbes, respectively. The post-genomic era provides powerful tools for the studies of genetic diversity. For example, as of 2016, the single nucleotide polymorphism (SNP) database maintained by the National Center for Biotechnology Information (NCBI) lists ~150 million SNPs. The huge amount of available genomic information greatly facilitates genome-wide association studies (GWAS). GWAS has been applied to the studies of various human diseases including cancer, heart diseases and diabetes. In agriculture, GWAS is used to identify genomic regions associated with desirable breeding traits in cattle or agronomic traits in plants. Thanks to the advancement in sequencing technology, it has become feasible and increasingly routine to sequence multi-gene panels, exomes, and whole genomes. These approaches are being widely used for the investigation of genetic variants-of-interest as well as genetic diagnosis. Gene panels offer a quick way to interrogate a panel of genes that are known to be associated with a specific disease or condition. Compared to sequencing of gene panels, whole exome sequencing (WES) and whole genome sequencing (WGS) cost more for each sample but generate much larger amount of data such that the cost per variant is significantly less. By interrogating millions of variants per genome, WES/WGS can be very powerful in identify new variants that are associated with diseases, disease risks, or prognosis. Another important technological development is gene editing technology. Genetic diversity in plants and animals provide the raw material for crop and livestock breeders, respectively. Gene editing technology greatly enhances the efficiency of breeding by allowing breeders to create transgenic elite lines and manipulate multiple loci simultaneously.

Editor
Dr. Q. Tian Wang, PhD

INTRODUCTION

Genetics is the study of biological inheritance. Geneticists examine a wide variety of inherited traits, from the ability to bear large numbers of fruit in trees, to eye color in mammals. This discipline is a major cornerstone of the larger field of biology. People who study it do so in a wide variety of situations, from research laboratories to pharmaceutical companies, and new breakthroughs are consistently being made.

Humans have understood genetics on some level for thousands of years. They have been improving crops and animals through breeding for quite some time, selecting desirable traits and attempting to propagate them. Casual observations have also been an important part of human life for thousands of years, with people noting that members of certain families tend to have distinctive anatomical features. The Hapsburgs of Austria, for example, are famous for their prognathic jaws.

In the 1800s, a monk named Gregor Mendel began doing controlled experiments with peas which were designed to unravel the mysteries of gene inheritance. He is often regarded as the father of modern genetics, because although he was not able to identify individual genes, he did demonstrate that a combination of dominant and recessive traits determine the physical appearance of an organism. In the 1950s, research finally revealed that DNA in the chromosomes carried the genetic code of the host organism.

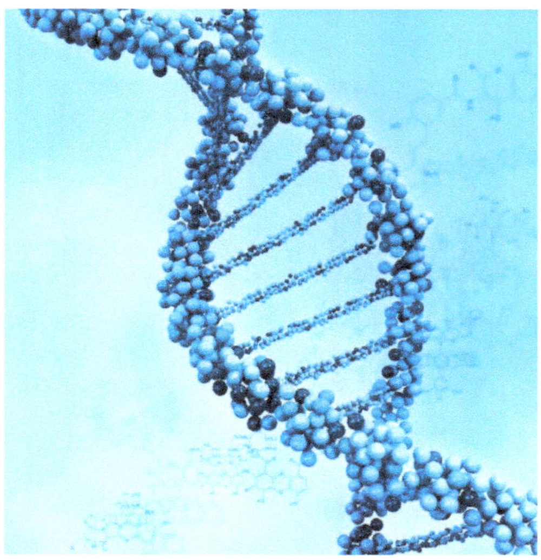

With the confirmation that DNA held the secret to inheritance, people also started charting the genetic code, finding the specific locations on the chromosomes where certain genes appeared. Along the way, they learned a great deal about DNA and inheritance, and the DNA code is continually being studied to identify new genes. Understanding this coding has allowed researchers to manipulate it, removing or altering sections to create a specific desired outcome, such as a crop that is resistant to pests. Knowledge of the human genetic code has also led to gene therapy, in which specific problems are addressed or prevented by manipulating the code of an individual.

The study of this field doesn't just explain the manifestation of physical traits like hair color. Genetics is also used to explore inherited disease, and to determine whether or not people are more susceptible to certain diseases because of their genes. The discipline also involves what happens when things go wrong, as in the case of an organism that inherits an extra chromosome. Some people have also theorized that genes may play a role in behavior, with certain people being predisposed to depression, addiction, and other behavioral issues.

What is Genetic diversity?

Genetic diversity is the variation of heritable characteristics present in a population of the same species. It serves an important role in evolution by allowing a species to adapt to a new environment and to fight off parasites. It is applicable to domesticated species, which typically have low levels of diversity. Studying genetic diversity in humans can help researchers form theories on human origins.

Living things contain in their cells the basic instructions, or blueprints, for their own development. Many of these instructions, which are called genes, result in physical characteristics that affect the way organisms interact with their environment. Variations in such characteristics within the same species give rise to genetic diversity. For a species to adapt to an ever-changing ecosystem, a significant level of variation must be present. Those individuals that possess favorable characteristics will go on to reproduce, while those that don't do not tend to pass on their characteristics to many offspring.

Domesticated species often have low levels of genetic diversity. This is caused by the artificial selection, or preferential breeding, of crops and animals for traits that humans find preferable. While this can have positive short-term results, such as a richer harvest, low diversity among domesticated species poses risks. A newly-evolved virus or bacteria strand can invade a population of nearly identical organisms very rapidly. The protection that diversity generally offers in wild populations is lost in this scenario.

The Irish Potato Famine between 1845 and 1852 was caused by a parasite invading a large population of nearly identical potatoes. The parasite was a water mold named Phytophthora infesting. This famine caused Ireland's population, which was widely dependent on potatoes for food, to decrease by 20 to 25 percent.

Human genetic diversity typically varies based on the geographic location of a population. This has led biologists and anthropologists to study these diversity levels in the quest to understand human origins. Genetic diversity levels in Africa, for example, have been found to be higher than in many of areas of the world. Researchers have developed models of human origins from this evidence. The recent Out-of-Africa model, which suggests modern humans have a common origin in Africa, is one such example.

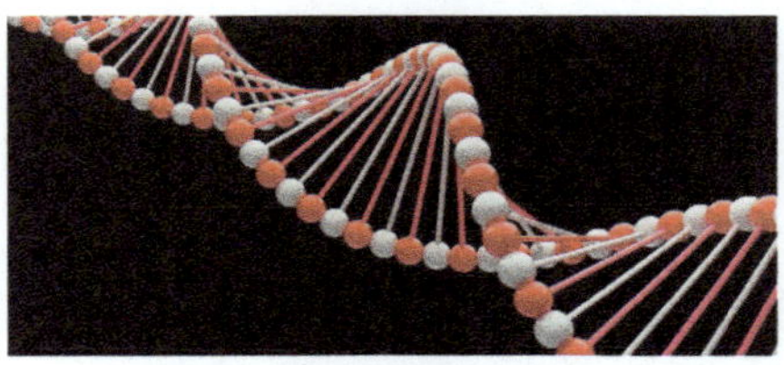

Biodiversity refers to the level of variation of all living things within an ecosystem. The importance of biodiversity to an ecosystem is analogous to the importance of genetic diversity to a population. Both forms of diversity contribute to the health and robustness of the larger system. When these levels of diversity fall, both systems are less able to adapt to a changing environment.

The Importance of Genetic Diversity

Most higher organisms (both plants and animals) reproduce sexually—that is, they produce offspring through the union of reproductive cells from two different parents. Think back to our example of the violet. Violets produce showy flowers that attract insects that carry pollen from one plant to the next. Offspring resulting from this cross-pollination are genetically distinct from either parent. Violets also produce flowers that never open, and are self-pollinated. The resulting offspring are genetically similar, though not identical, to the parent. And, finally, violets send out creeping stems. Plants sprouting from these runners are genetically identical to the parent plant. We've talked about genetic diversity—but why is it important, and how does it fit in with our general topic of plant sexual reproduction? Sexual reproduction is critical for maintaining genetic diversity within a species because it combines the parents' genetic material, resulting in offspring with unique genetic blueprints—different from either parent.

Genetic diversity is important for two reasons

First of all, when a population of an organism contains a large gene pool—that is, if the genetic blueprints of individuals in the population vary significantly—the group has a greater chance of surviving and flourishing than a population with limited genetic variability.

Why is this so? Because some of the individuals may have inherited traits making them particularly resistant to disease or tolerant of cold, for example. Or they may possess other traits that increase their chance for survival. In nature, the "fittest" individuals succeed and go on to reproduce—Darwin termed this process "natural selection." Suppose there's an outbreak of a disease that threatens to wipe out an entire species. The more genetic variability there is within that species, the higher the likelihood that at least some of the individuals will be resistant, and will survive.

In the lab, plant breeders take advantage of these genetic variants to improve existing plants and create new varieties. Through cross breeding they strive to breed in disease resistance, superior fruit production, increased cold tolerance, or other desirable traits.

Genetic diversity also reduces the incidence of unfavorable inherited traits. In a small, isolated population of organisms, individuals may be forced to breed with close relatives. When this happens, the genetic makeup of the individuals becomes more and more uniform, and genetic flaws become increasingly more common. This phenomenon is called inbreeding.

When closely related organisms (siblings or cousins, for example) interbreed, any genetic weaknesses that are hidden in the parents can be multiplied in the offspring. For example, animals can be carriers of a gene for an inherited disease, but not show any symptoms. If they mate with a partner who is also a carrier, then the offspring may exhibit symptoms of the disease. (We'll talk more about inherited traits next week.) In an inbred population, chances are greater that carriers will interbreed. Over time, the entire population is weakened.

Genetic diversity in Plant

Genetic diversity within a population refers to the number of different alleles (the alternate forms of genes) of all genes and the frequency with which they appear. Variation is high when there are many different alleles of all genes and many different combinations of those alleles. A gene pool is the collective set of alleles found in a population. Usually gene pools refer to a particular species (individuals can interbreed freely). Plant breeders can also make use of the gene pool of closely related species that can cross and produce fertile hybrids (called interspecific hybrids).

Maize ears from CIMMYT's collection, showing a wide variety of colors and shapes. CIMMYT's germplasm bank contains about 28,000 unique samples of cultivated maize and its wild relatives.

The level of genetic variation in a population is constantly changing: different alleles of a single gene can appear and disappear from time to time within a population. This means that the gene pool of a population is always dynamic.

For some characteristics, such as maize or sorghum grain color, there is a great level of genetic variation in nature.

The term "bean" is used to denote the seed of a large number of related plants: broad or fava beans, common beans, runner beans, soybeans, peas, chickpeas, vetches, and lupines. These are in fact different species.

In biology, a species is one of the basic units of biological classification, defined as a group of organisms capable of interbreeding and producing fertile offspring, or as a separately evolving lineage with a single gene pool.

In practice defining a species is not simple (this is known as the "species problem". Some separately evolving groups (i.e. "species") may continue to interbreed to some extent, and human intervention has resulted in extraordinary changes in the appearance of plants and animals during their domestication.

Plant breeders define species as populations of organisms that have a high level of genetic similarity.

Genetic biodiversity

All forms of life on earth, whether microbes, plants, animals, or human beings, contain genes. Genetic diversity is the sum of genetic information contained in the genes of individual plants, animals, and micro-organisms. Each species is the storehouse of an immense amount of genetic information in the form of traits, characteristics, etc. The number of genes ranges from about 1000 in bacteria to more than 400 000 in many flowering plants. Each species consists of many organisms and virtually no two members of the same species are genetically identical.

An important conservation consequence of this is that even if an endangered species is saved from extinction it has probably lost some of its internal diversity. Consequently, when populations expand again, they become more genetically uniform than their ancestors. There are mathematical formulas to express a genetically effective population size that explain the genetic effects on populations that have gone through a bottleneck before expanding again such as the African cheetah or the North American bison. Subsequent inbreeding in small populations may result in:

- Reduced fertility and
- Increased susceptibility to disease.

Genetic differentiation within species occurs as a result of sexual reproduction, in which genetic differences between individuals are combined in their offspring to produce new combinations of genes or from mutations causing changes in the DNA. Genetic diversity is usually mentioned with reference to agriculture and maintaining food security. This is because genetic erosion of several crops has already occurred leading to the world's dependence for food on just a few species. Currently, a mere 100-odd species account for 90% of the supply of food crops, and three crops – rice, maize, and wheat – account for 69% of the calories and 56% of the proteins that people derive from plants.

Significance of Genetic Diversity

While some individuals might be able to tolerate an increased load of pollutants in their environment, others, carrying different genes, might suffer from infertility or even die under the exact same environmental conditions. Whilst the former will continue to live in the environment the latter will either have to leave it or die. This process is called natural selection and it leads to the loss of genetic diversity in certain habitats. However, the individuals that are no longer present might have carried genes for faster growth or for the ability to cope better with other stress factors.

How do human activities affect genetic diversity?

Any change in the environment - natural or human induced causes a selection of events that only the fittest survive.

Anthropogenic impact is particularly apparent in the coastal zone and increases the number of changes occurring to individual and populations. Such pressure is exerted by

- artificial selection (harvesting, aquaculture)
- degradation of habitats (leading to a reduction of total stocks and thus increasing the likeliness of inbreeding) and
- the release of farmed fish into the wild. These activities reduce the sum of genes available, thus leaving behind a population that is less capable of tolerating any further natural or anthropogenically caused changes in environment.

These activities reduce the sum of genes available, thus leaving behind a population that is less capable of tolerating any further natural or human disturbances in environment.

Why prevent the loss of genetic diversity?

The loss of genetic diversity is difficult to see or measure. In contrast, the reduction and extinction of populations is far easier to see. Extinction is not only the loss of whole species, but is also preceded by a loss of genetic diversity within the species.

This loss reduces the species ability to perform its inherent role in the whole ecosystem. The loss of genetic diversity within a species can result in the loss of useful and desirable traits (e.g. resistance to parasites). Reduced diversity may eliminate options to use untapped resources for food production, industry and medicine.

SECTION I: GENETIC DIVERSITY IN HUMANS: ROLES IN COMPLEX DISEASES

CHAPTER 1 GENETIC VARIATION ON CHROMOSOME 5Q22

J. Gustav Smith, Janine F. Felix, Alanna C. Morrison, Andreas Kalogeropoulos, Stella Trompet Jemma B. Wilk, Olof Gidlöf, Xinchen Wang, Michael Morley, Michael Mendelson, Roby Joehanes, Symen Ligthart, Xiaoyin Shan, Joshua C. Bis, Ying A. Wang, Marketa Sjögren, Julius Ngwa, Jeffrey Brandimarto, David J. Stott, David Aguilar, Kenneth M. Rice, Howard D. Sesso, Serkalem Demissie, Brendan M. Buckley, Kent D. Taylor, Ian Ford, Chen Yao, Chunyu Liu, Nona Sotoodehnia, Pim van der Harst, Bruno H. Ch. Stricker, Stephen B. Kritchevsky, Yongmei Liu, J. Michael Gaziano, Albert Hofman , Christine S. Moravec, André G. Uitterlinden, Manolis Kellis, Joyce B. van Meurs, Kenneth B. Margulies, Abbas Dehghan , Daniel Levy, Björn Olde, Bruce M. Psaty, L. Adrienne Cupples, J. Wouter Jukema, Luc Djouss, Oscar H. Franco, Eric Boerwinkle, Laurie A. Boyer, Christopher Newton-Cheh, Javed Butler, Ramachandran S. Vasan, Thomas P. Cappola, Nicholas L. Smith

INTRODUCTION

Heart failure (HF) is a common clinical condition in which the heart fails to maintain blood circulation adequate to meet the metabolic demands of the body without increased cardiac filling pressures. HF is the result of chronic ventricular remodelling initiated by myocardial injury, volume/pressure overload, or intrinsic cardiomyopathic processes. Progression of HF is a complex process involving many tissues, driven by activation of neurohormonal pathways, which induce gradual myocardial hypertrophy, ventricular dilation, and deterioration of cardiac function, often resulting in death from low cardiac output, arrhythmia, or thromboembolic complications [1]. Activation of such neurohormonal pathways in the short term increases cardiac output when necessary. However, long-term activation results in accelerated ventricular remodelling and myocyte death. Inhibitors of deleterious neurohormonal pathways, including adrenergic [2–4] and renin-angiotensin-aldosterone (RAAS) [5–8] pathways have been shown to improve ventricular function and survival in patients with HF and are the mainstay of current pharmacological treatment of HF [9–10]. Despite advances in therapy with neurohormonal antagonists, mortality after onset of HF remains high [9–13] and continued progress to identify additional therapeutic targets is needed.

Genome-wide association (GWA) studies have the potential to identify in an agnostic manner genetic variants related to clinical outcomes in humans and has led to the identification of novel pathways [14] and potential treatments [15] for cardiovascular traits. Heritable factors have been shown to be predictive of mortality in certain heart failure patients [16]. We therefore implemented a genome-wide association approach to identify novel molecular determinants of mortality in patients with new-onset HF.

RESULTS

Two-Stage GWA Study

We expanded our previously published GWA study [17] of HF mortality with additional samples and extended follow-up in Stage 1. Stage 1 included 2,828 new-onset HF patients from five community-based cohorts, thus representative of the general population of HF patients, as part of the Cohorts for Heart and Aging Research in Genomic Epidemiology (CHARGE) consortium [18]: the Atherosclerosis Risk in Communities (ARIC) Study, the Cardiovascular Health Study (CHS), the Framingham Subjects in Stage 1 cohorts were of European ancestry, predominantly male, and approximately 20–30% had a history of myocardial infarction at the time of HF diagnosis. Additional characteristics are shown in **Table 1**. During an average follow-up time of 3.5 years, 1,798 deaths occurred. The sample-size weighted average 1-year mortality rate was 28%. Among deaths, 51% were classified as cardiovascular, 19% were due to neoplasms, 10% were respiratory deaths, and the remaining were due to other miscellaneous causes.

Citation: Smith JG, Felix JF, Morrison AC, Kalogeropoulos A, Trompet S, Wilk JB, et al. (2016) Genetic Variation on Chromosome 5q22. PLoS Genet 12(5): e1006034. doi:10.1371/journal.pgen.1006034.

Genotyping using high-density Illumina or Affymetrix single nucleotide polymorphism (SNP) arrays, followed by imputation to the HapMap CEU release 22 imputation panel was performed in each cohort. Population stratification was assessed and corrected in each cohort. Association with time to death following HF diagnosis was examined in each cohort using Cox proportional hazards models with censoring at loss to follow-up. Mild inflation of test statistics was observed only in the Framingham Heart Study (FHS) as shown in **S1 Fig** ($\lambda_{GC} = 1.07$, other cohorts ≤ 1.03), and genomic control was applied in each individual study. In the meta-analysis of all cohorts, there was no evidence of inflated test statistics overall ($\lambda_{GC} = 1.00$) as shown in **S2 Fig**, so no further genomic control was needed. Results for all SNPs across the genome are plotted in **S3 Fig**. Single nucleotide polymorphisms (SNPs) passing a significance threshold specified *a priori* as$P < 5.0x10^{-7}$, as used in our previous article [17], were carried forward to a second stage of genotyping in independent cohorts.

Table 1. Characteristics of cohorts in stage 1

	ARIC	ARIC2	CHS	FHS	Health ABC	RS	RS2
HF sample size	691	84	838	249	173	748	45
Age at HF (years)	67.3 (6.6)	68.8 (6.5)	82.4 (6.1)	80.7 (9.6)	79.1 (3.8)	79.6 (8.0)	73.7 (8.6)
Male sex (%)	60.1	58.3	56.3	47.8	61.8	48.8	68.9
Body mass index (kg/m^2)	29.5 (5.9)	28.7 (6.3)	27.0 (4.6)	28.5 (5.4)	26.5 (4.3)	27.2 (4.1)	27.1 (4.0)
Diabetes (%)	28.7	29.8	18.1	17.7	20.8	8.6	11.1
Hypertension (%)	58.5	53.7	52.3	76.3	79.8	81.0	73.3
Current smoking (%)	28.6	33.3	8.4	10.4	7.5	18.6	26.7
History of MI (%)	25.2	26.2	5.4	27.7	24.9	28.1	28.9
Follow-up time (years)	3.9 (3.9)	4.1 (4.0)	3.5 (2.4)	1.8 (1.9)	3.0 (2.7)	3.9 (3.7)	3.6 (2.6)
Mortality rate, 1 year	0.25	0.22	0.28	0.34	0.27	0.30	0.25
All-cause mortality (n, %)	327 (47.3)	42 (50.0)	606 (72.3)	156 (62.7)	87 (50.3)	556 (74)	24 (53)

Age, body mass index and follow-up time are presented as mean (standard deviation). Categorical variables are presented as percentages. Body mass index, diabetes, hypertension and smoking refer to the nearest study exam prior to heart failure diagnosis whereas age, follow-up time, mortality rate and all-cause death refer to the time of HF diagnosis. Mortality rate refers to the 1-year Kaplan-Meier estimate, with censoring at end of or loss to follow-up. The prevalence of myocardial infarction in CHS is low, because this was an exclusion criterion at the study baseline. MI, Myocardial infarction.

doi:10.1371/journal.pgen.1006034.t001

Five SNPs on chromosome 5q22 and one SNP on chromosome 3p22 passed the pre-specified$P$-value threshold. Results for all six SNPs are shown in **Table 2** and **S3 Table**. The five SNPs on chromosome 5q22 were highly correlated (pairwise $r^2 > 0.9$). Two sentinel SNPs, rs9885413 and rs12638540, on chromosomes 5q22 and 3p22, respectively, were next genotyped in 1,870 European-ancestry subjects with new-onset HF from four independent cohorts in Stage 2: Malmö Diet and Cancer, Malmö Preventive Project, Physicians' Health Study, and the PROSPER trial. Characteristics of populations in Stage 2 are shown in **S2 Table**. During an average sample-size weighted follow-up of 4.3 years in Stage 2 samples, 889 patients died. We observed evidence of association with mortality for rs9885413 on chromosome 5q22 (P = 0.006) but not for the SNP rs12638540 (P = 0.18) which reached nominal significance in our previous analysis [17]. Results for both SNPs are shown in **Table 2**. In the combined results from Stages 1 and 2, rs9885413 was associated with a 36% relative increase in mortality per minor allele ($P = 2.7x10^{-9}$). There was no evidence for effect heterogeneity across cohorts in the two stages (P for heterogeneity = 0.39) as shown in **S4 Table**. The SNP had a similar minor allele frequency (MAF = 0.07) across cohorts. Information on cause-specific mortality was available from death certificates in a subset of cohorts (**S5 Table**) and was explored descriptively due well-known problems with substantial misclassification in death certificate data and low power for agnostic GWAS of individual causes. The minor allele frequency was slightly higher for several causes of death associated with heart failure, including renal, pulmonary and endocrine mortality and death from ischemic heart disease.

Table 2. Association of genetic polymorphisms with HF mortality

SNP				Stage 1		Stage 2		Combined		Heterogeneity	
	Chr	*Alleles*	*MAF*	*HR (95% CI)*	*P*	*HR (95% CI)*	*P*	*HR (95% CI)*	*P*	I^2	*P*
rs12638540	3p22	G / A	0.04	1.50 (1.28–1.75)	4x10^{-7}	1.10 (0.89–1.37)	0.18	1.35 (1.19–1.54)	3.08x10^{-6}	0	0.53
rs9885413	5q22	T / G	0.07	1.40 (1.23–1.58)	1x10^{-7}	1.25 (1.05–1.49)	0.006	1.36 (1.23–1.51)	2.65x10^{-9}	0.15	0.39
rs12658193	5q22	T / G	0.07	1.39 (1.23–1.58)	2x10^{-7}	-		-		-	-
rs11956079	5q22	T / C	0.07	1.39 (1.23–1.58)	2x10^{-7}	-		-		-	-
rs10069077	5q22	T / C	0.07	1.39 (1.23–1.58)	2x10^{-7}	-		-		-	-
rs10068260	5q22	T / A	0.06	1.42 (1.24–1.62)	2x10^{-7}	-		-		-	-

Results for SNPs with P-value < 5x10^{-7} in stage 1 are presented with minor/major alleles, minor allele frequency (MAF) and effect estimate (HR, hazard ratio) with 95% confidence intervals (95% CI) per minor allele from multivariable models adjusting for sex and age at the time of diagnosis. The index SNP at each locus was carried forward to stage 2. Effect heterogeneity was evaluated across the combined stage 1 and 2 samples. Chr, Chromosome.

doi:10.1371/journal.pgen.1006034.t002

Lack of Association with Myocardial Traits

We next examined whether rs9885413 on chromosome 5q22 that was associated with HF mortality was also associated with differences in myocardial structure and function, which could potentially mediate the association (**S6 Table**). In 12,612 individuals from the EchoGen Consortium [19], the SNP was not associated with major echocardiographic characteristics. The SNP rs9885413 was not associated with incident HF in 20,926 individuals from the general population in the CHARGE-HF study [20], or with cardiac endocrine function, as determined by plasma levels of atrial and B-type natriuretic peptides (all $P > 0.05$), in a GWA study of 5,453 individuals from the population-based Malmö Diet and Cancer study [21]. No association was observed with electrocardiographic measures of cardiac conduction (n = 39,222) [22] or repolarization (n = 74,149) [23], which confer risk of ventricular arrhythmia, or with sudden cardiac death in 4,496 sudden death cases and over 25,000 controls from the general population.

Functional Enhancer Annotation

The lead SNP rs9885413 on chromosome 5q22 that was associated with mortality is located in an intergenic region, 100 kb downstream of the gene *SLC25A46*, 114 kb upstream of*TMEM232*, and 230 kb upstream of *TSLP* as shown in **Fig 1**. The SNP is not in linkage disequilibrium with any known coding SNP in the 1000 Genomes Project database (no coding SNP with $r^2 > 0.01$ to the sentinel SNP). We therefore sought to evaluate gene regulatory functions of this SNP.

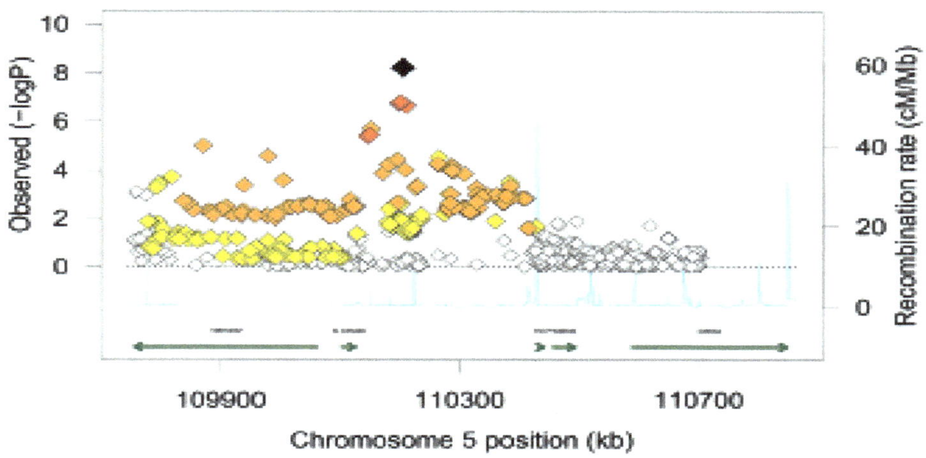

Figure 1. Regional association plot for HF mortality at the 5q22 locus. The plot covers the genomic region from 450 kb upstream of the SNP to 650 kb downstream. Diamonds represent SNPs. The large black diamond represents the SNP with the lowest *P*-value (rs9885413), with the *P*-value from the combined meta-analysis presented. Diamond color represents strength of pairwise correlation with the strongest SNP. Red represents $r^2 \geq 0.8$, orange represents r^2 0.5–0.8, yellow represents r^2 0.2–0.5, and white represents $r^2 < 0.2$. Recombination rate is plotted in the background and known genes are represented in the bottom of the plot. Positions refer to NCBI build 36. SNP correlations and recombination rates were obtained from HapMap release 22.

In 129 human tissues from the ROADMAP Epigenomics project [24], we studied whether rs9885413 or strongly correlated SNPs (a total of 9 at $r^2 > 0.8$) are located in regulatory regions, as determined by histone modification patterns. None of the 9 SNPs was located in an active regulatory region in cardiac tissues (**S7 Table**), but rs9885413 was located in a predicted enhancer in several epithelial or mesenchymal tissues, including keratinocytes, gastrointestinal cell types and adipose cells (**Fig 2** and **S7 Table**). Regulatory motif annotations in HaploReg indicate that the SNP causes a change in a regulatory motif predicted to bind the transcription factor NHLH1 as shown in **S8 Table**. Interestingly, NHLH1-null mice have been shown to be predisposed to premature, adult-onset unexpected death in the absence of signs of cardiac structural or conduction abnormalities, in particular when mice were exposed to stress [25]. Little is known about the function of NHLH1, but it is widely expressed in human tissues and has been shown to regulate expression of key inflammatory genes [26].

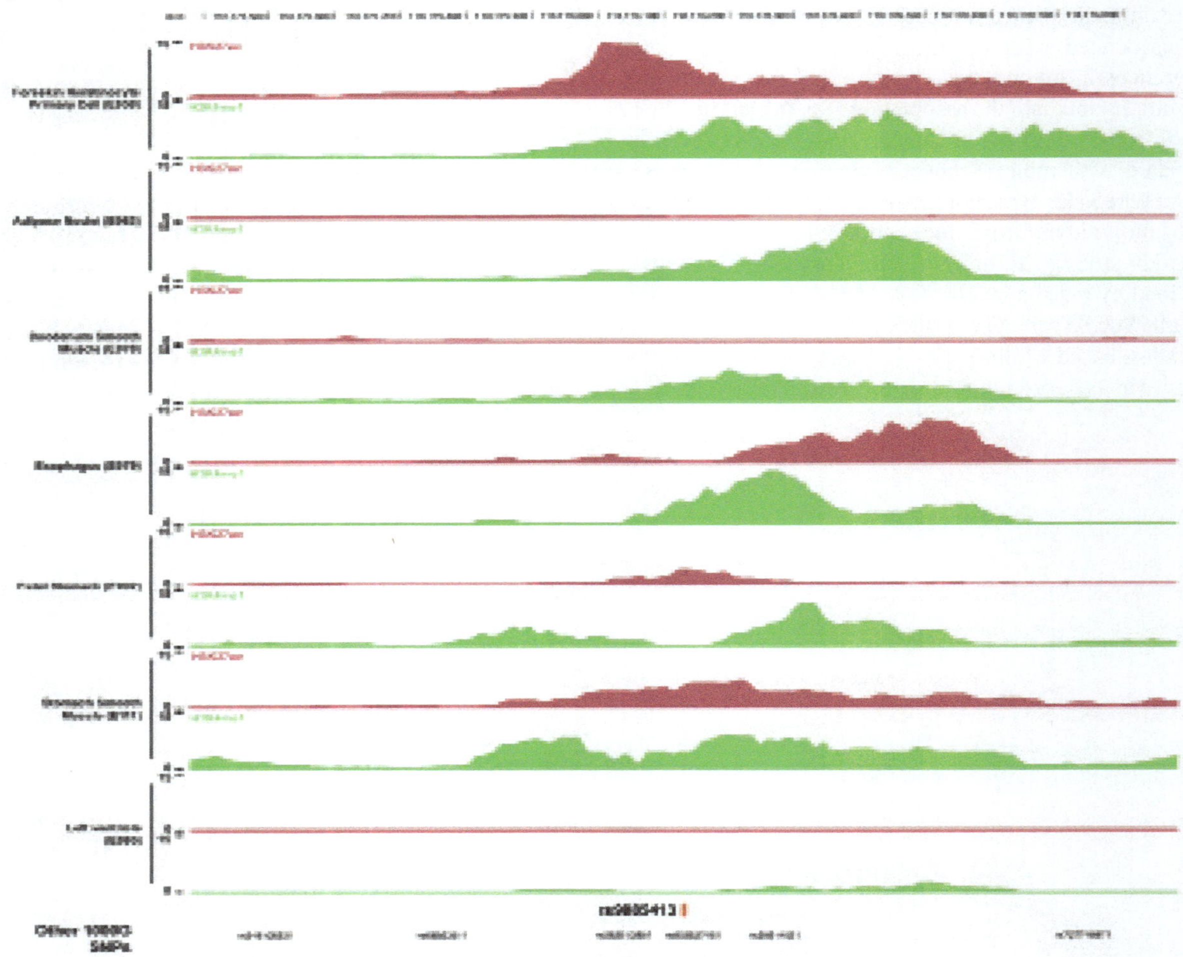

Figure 2. Epigenomic marks indicating enhancer activity at the 5q22 locus. Histone marks in samples from the ROADMAP Epigenomics Project at the enhancer overlapping the lead SNP rs9885413 in heart tissue (left ventricle) and in tissues with evidence of an active enhancer at the locus. Histone marks are monomethylation (H3K4Me1) of the fourth residue (lysine) and acetylation of the 27th residue (H3K27Ac) of histone H3. Positions refer to NCBI build 36.

To experimentally test the effect of rs9885413 on enhancer activity, the 100 bp region flanking the SNP (50 bp on either side) was cloned into a reporter vector and transfected into HEK293 cells expressing NHLH1. Luciferase activity measured after 24 hours was 4-fold higher with a construct corresponding to the risk allele as compared to the wild-type allele (**S4 Fig**, $P < 0.001$), indicating that the risk allele of rs9885413 substantially increases enhancer activity.

Effects on DNA Methylation

We next explored the association of rs9885413 with DNA methylation at the locus, providing functional evidence of epigenetic association and regulation of gene expression. DNA methylation was determined by a microarray assaying in total over 480 000 CpG methylation sites in whole blood samples from 2408 participants of the FHS. Of the 84 CpG methylation sites on the microarray within +/- 500 kb of the SNP, two were significantly associated with rs9885413: cg21070081 (beta 0.017 per T allele, $P = 9.0\text{x}10^{-69}$) and cg02061660 (beta -0.015 per T allele, $P = 4.5\text{x}10^{-40}$), thus constituting strong methylation quantitative trait loci (mQTLs) at the locus. Other, correlated SNPs at the locus were more strongly associated with each of these mQTLs as shown in **S5 Fig**: rs244431 for cg21070081 ($P = 6.7\text{x}10^{-369}$) and rs72774805 for cg02061660 ($P = 7.0\text{x}10^{-85}$). The SNP rs72774805 (perfect proxy SNP rs3844597 used) but not rs244431 was associated with heart failure mortality ($P = 3.3\text{x}10^{-3}$ and 0.08, respectively), indicating that the methylation site cg02061660 is more strongly related to the underlying signal for heart failure mortality. The association of rs9885413 with lower probability of methylation at cg02061660 was replicated in 731 participants from the Rotterdam study (beta -0.029 per T allele, $P = 1.7\text{x}10^{-11}$). Adjustment for cell types from direct measurement instead of estimates from methylation patterns did not abolish the association (beta -0.029 per T, $P = 1.2\text{x}10^{-6}$). Interestingly, differential methylation at this CpG site was also correlated with a SNP at the locus previously associated with allergic sensitization [27] (rs10056340, $P = 4.7\text{x}10^{-29}$ for mQTL), suggesting a link to inflammatory disease. This SNP was also modestly correlated with rs9885413 ($r^2 = 0.28$) and associated with heart failure mortality ($P = 0.01$). The association of cg02061660 with rs9885413 ($P = 0.52$) and rs10056430 ($P = 0.87$) was abolished in analyses conditioning for rs72774805, for which the association was also markedly attenuated ($P = 7.0 \times 10^{-33}$ and 2.1×10^{-46}, respectively) indicating that these correlated SNPs may reflect the same underlying signal.

Effects on Gene Expression

We further assessed the association of rs9885413 with gene expression. No gene was significantly associated with rs9885413 in the diverse tissues from the Gene-Tissue Expression (GTEx) project [28] after correction for multiple tests (**S9 Table**), although conclusions were limited by a small sample size.

We next assessed association of the SNP with gene expression in two large datasets with each of the tissues most relevant for the phenotype under study: heart tissue and whole blood. We observed no convincing evidence of association (**S10 Table**) with gene expression in 247 left ventricular samples from patients with advanced heart failure (n = 116) undergoing transplantation and from unused donors (n = 131). Finally, we tested the association of rs9885413 with the expression of genes at the locus in whole blood from 5257 FHS participants [29], and with DNA methylation at cg02061660 among 2262 FHS participants. All five genes at the locus (**Fig 1**) except *TMEM232* were expressed in blood. We did not observe association of the SNP rs9885413 with any transcript, but expression of one gene (*TSLP*) was significantly associated with the methylation status of cg02061660 ($P = 1.1 \times 10^{-4}$).

The *TSLP* gene encodes a cytokine released from epithelial cells that induces release of T cell-attracting chemokines from monocytes, promotes T helper type 2 cell responses, enhances maturation of dendritic cells and activates mast cells. It has also been linked to angiogenesis and fibrosis. A monoclonal antibody targeting and inhibiting *TSLP* is currently in clinical phase III trials for asthma and allergic inflammation after a promising phase II trial [30–32]. In the myocardium, the *TSLP* gene has very low expression (**S10 Table**) but expression has been described in mature myocardial fibroblasts, which are abundant in the myocardium but of substantially smaller volume than cardiomyocytes and likely contribute little to the overall myocardial RNA pool [31,32].

To examine whether the transcription factor NHLH1 affects the expression of any of the five genes in the locus (**Fig 1**), we knocked down NHLH1 in HEK293 cells using siRNAs. A 50% decrease in NHLH1 mRNA levels was seen 48 hours after transfection, confirming efficient knock down ($p<0.05$, **S6A Fig**). TSLP was the only gene at the locus affected by NHLH1 knock down, showing a 30% decrease compared to cells transfected with negative control siRNA ($p<0.05$, **S6A Fig**). Moreover, we observed a dose-response relation between level of NHLH1 knockdown and expression of TSLP in HEK293 cells ($r^2 = 0.74$, $p<0.0001$, **S6B Fig**).

Population Genetic Analysis

Finally, distribution of the risk allele of rs9885413 in human populations was assessed using data from HapMap phase II. The derived (non-ancestral) T allele (risk allele for mortality) was highly differentiated among human populations (**S7 Fig**) having risen to an allele frequency of 0.59 in a Nigerian population (HapMap YRI sample) but only 0.06 in a European population (CEU sample). The fixation index (F_{st}), a measure of population differentiation in allele frequencies, for comparison of YRI and CEU was 0.48 and more extreme F_{st} was observed in only 2.4% of SNPs in the HapMap phase 2 dataset. Consistent results were observed for another signature of recent positive selection, based on longer runs of haplotype homozygosity in carriers of the derived allele (standardized integrated haplotype score -0.766 in YRI, where negative score values indicate longer haplotypes on the background of the derived allele) [33]. These observations are consistent with positive selection in recent human history, with a selective sweep resulting in high frequency of the derived allele in western African populations. These findings are of particular interest as HF mortality is well known to be higher in populations of African ancestry, although the current study has not tested for the association with HF mortality in such populations [34].

DISCUSSION

We identified a SNP on chromosome 5q22 associated with increased mortality in subjects with HF. Although previous genome-wide association studies have described hundreds of loci associated with risk of disease onset, few have examined prognosis in subjects with manifest disease. This approach has the potential to generate targets for novel disease-modifying medications. Through a series of analyses *in silico* and *in vitro* we show that the SNP is located in an enhancer region, and confers increased activity of this enhancer.

Interestingly, mice deficient in the transcription factor NHLH1 predicted to bind a motif in this enhancer region have been reported to be predisposed to premature, adult-onset unexpected death in the absence of signs of cardiac structural or conduction abnormalities. NHLH1 has also been shown to regulate expression of key inflammatory cytokines such as interleukin-6 and tumor necrosis factor α. The SNP was not associated with any electrocardiographic, endocrine, or echocardiographic marker of increased risk in the general population, suggesting a mechanism specific to heart failure, an extracardiac pathway of importance in cardiac pathophysiology, or interaction with therapy for heart failure which we were unable to further test given the inception cohort design of this study. We also did not observe any robust eQTL associations for the SNP in heart. The SNP was however associated with a DNA methylation signature in whole blood that was also associated with a SNP previously associated with allergy, and with expression of the cytokine TSLP in blood. Knockdown of NHLH1 also resulted in lower expression of TSLP in HEK293 cells. This non-coding SNP may thus exert an influence on *TSLP* expression via altered NHLH1 enhancer function and DNA methylation at the methylation site cg02061660. Detailed characterization of causal variants and different association signals at the locus would however require finemapping and sequence data.

The TSLP cytokine is released from epithelial cells and fibroblasts and is considered important in initiation of inflammatory responses to tissue damage, particularly in the type 2 T-helper (Th2) pathways. Th2 pathways are central in the response to extracellular parasites but also play a key role in the pathophysiology of allergies and hypersensitivity reactions. A small subset of HF is known to be caused by Th2-mediated inflammation (eosinophilic cardiomyopathy), yet Th2 cells have received limited attention in HF pathophysiology. Recent experimental work implicates an important role of T-helper cells in HF progression for both systolic and diastolic heart failure, but has mainly focused on type 1 T-helper pathways [35,36]. It remains unclear if the mechanism for rs9885413 is through a specific etiology characterized by high mortality such as eosinophilic cardiomyopathy or a pathway involved in outcomes with manifest disease. The lack of association with HF incidence suggests that it may not act through incidence of a specific etiology, although firm conclusions are limited by sample size. We did not observe significant associations of the SNP with gene expression in any tissue. It is possible that adequately powered samples with a specific cell subtype in a specific context is needed to detect such associations, as illustrated by a recent study which only observed certain eQTLs with single-cell but not across averaged cells [37]. Indeed, baseline expression of *TSLP* was low in our samples, and is induced by tissue injury, microbes, viruses and proinflammatory cytokines [38].

Evidence of recent positive selection in individuals of African descent suggests that the HF risk allele may have been beneficial in some environments in recent human history. Inflammatory pathways are enriched for signals of recent positive selection, reflecting that infectious disease has been an important cause of mortality throughout recent evolution. Genes such as *HBB* and*APOL1* have also been reported to have been subject to recent positive selection in Africa by conferring protection against infectious diseases such as Malaria and Trypanosomiasis (sleeping sickness) [39], and *APOL1* alleles have also been linked to cardiovascular disease [40]. As cardiovascular disease and heart failure often presents after reproductive age, increased mortality in such patients would not be expected to exert purifying (negative) selective pressure. Whether SNPs at 5q22 contribute to higher mortality in subjects of African ancestry remains to be shown.

Thus, although additional work is needed to further clarify the tissues and pathways perturbed by this genetic variant and the mechanisms linking it to mortality in HF patients, the current findings implicate rs9885413 as a novel marker of increased risk among patients with HF. Complementary epigenomic evidence demonstrated candidate regions and genes, which may be mediators in cardiac pathophysiology and potential therapeutic targets to improve prognosis in patients with HF.

MATERIALS AND METHODS

Genome-Wide Association Study Stage 1

A genome-wide association (GWA) study was performed in a total of 2,828 subjects of European ancestry with HF from seven samples collected within five large community-based prospective cohort studies including the Atherosclerosis Risk in Communities (ARIC and ARIC2) Study, the Cardiovascular Health Study (CHS), the Framingham Study (FHS), the Health ABC (Health ABC) study and the Rotterdam Study (RS and RS2). Sample characteristics, data collection and clinical definitions have been described previously. First diagnosis of heart failure (new-onset) was ascertained using a variety of methods based on international published criteria, as detailed in **S1 Table**. Mortality was ascertained from telephone contacts with relatives and from medical records, death certificates and/ or municipal records.

Genotyping was performed using commercially available assays for genome-wide SNP detection. Imputation of non-genotyped SNPs was performed using CEU reference panels of SNP correlations from the HapMap project phase II, to characterize a total of 2.5 million SNPs. Imputation quality was assessed for each SNP from the ratio of observed over expected variance of allele dosage.

All-cause mortality following initial HF diagnosis was examined for association with additive allele dosage of each genotyped or imputed SNP using Cox proportional hazards models, with censoring at the end of or loss to follow-up. Models were adjusted for age at diagnosis, sex, and recruitment site in multicenter cohorts. In the family-based FHS, Cox models were implemented with clustering on pedigree to account for relatedness. Genomic control was applied to results from each cohort. Cohort-specific GWA results were combined using fixed effects meta-analysis with inverse variance weights. SNPs were excluded from cohort-level analyses if exhibiting implausible beta coefficients (< -5 or > 5) and from the meta-analysis for low heterozygosity (sample size-weighted minor allele frequency ≤ 0.03, corresponding to < 100 minor allele carriers with an endpoint).

Stage 2 Genotyping in Independent Samples and Combined Analysis

SNPs passing a *P*-value threshold defined *a priori* as $P < 5.0\times10^{-7}$ in the genome-wide stage 1 were carried forward to the second stage with targeted genotyping in 1,870 HF patients from four independent cohorts. For 2.5 million tests, this threshold limits the expected number of genome-wide false positives to approximately 1, assuming statistical independence of tests. The second stage included four independent cohorts; the Malmö Diet and Cancer Study (MDCS), the Malmö Preventive Project (MPP), the Physicians' Health Study (PHS) and the Prospective Study of Pravastatin in the Elderly at Risk (PROSPER) [47–50]. Heart failure ascertainment and time of death in these cohorts was similar to in stage 1 cohorts, as shown in **S1 Table**. Association analyses and meta-analysis of results were performed as in the first stage. Meta-analysis of stage 1 and 2 was performed, and a combined *P*-value $< 5.0\times10^{-8}$ was considered significant. Heterogeneity was assessed across the combined stage 1 and 2 cohorts using Cochran's Q test for heterogeneity, computed as the sum of the squared deviations of each study's effect from the weighted mean over the study variance, and the I^2 test, the percentage of total variation across studies that is due to heterogeneity rather than chance ($I^2 = [Q—df] / Q$) [51,52].

In Silico Studies of Effect on Cardiac Structure and Function

The association of replicated SNPs with measures of cardiac structure and function was evaluated from summary results of the following GWA consortia: EchoGen [19], CHARGE-HF [20], CHARGE-QRS [22], natriuretic peptides in 5453 subjects from the Malmö Diet and Cancer study [21], QT-IGC [23], and the CHARGE Sudden Cardiac Death consortium (manuscript in preparation).

Interrogation of Functional Motifs

The correlation of replicated SNPs with known coding SNPs was examined in the databases for the 1000 Genomes Project and phase III of the HapMap project, using SNAP [53]. The location of SNPs in relation to regulatory motifs was explored using histone methylation patterns generated as part of the ROADMAP Epigenomics project [24]. Enhancers were identified in each of the 129 ROADMAP tissues using the ChromHMM algorithm [54] from patterns of monomethylation (H3K4Me1) of the fourth residue (lysine) and acetylation of the 27th residue (H3K27Ac) of histone H3. The location of SNPs in relation to transcription factor binding sites was assessed in silico using HaploReg version 4.1 (http://www.broadinstitute.org/mammals/haploreg/haploreg.php) [55] and the UCSC Genome Browser (http://genome.ucsc.edu). In HaploReg, position weight matrices (PWMs; probabilistic representations of DNA sequence) were computed with p-values based on literature sources and ENCODE ChIP-Seq experiments as previously described [55], and only instances where a motif in the sequence passed a threshold of $P < 4^{-7}$ were considered. The NHLH1-binding motif was retrieved into HaploReg from the manually curated TRANSFAC database.

In Vitro Assessment of Enhancer Activity

Complementary DNA oligonucleotides corresponding to the 100 bp genomic region flanking rs9885413 (50 bp on either side of the SNP) were cloned into the luciferase reporter vector pGL3-Promoter (Promega, Madison, WI) using the MluI and BglII sites. Two different sets of oligos were cloned, one corresponding to the major allele of rs9885413 (pGL3P-G) and one to the minor allele (pGL3P-T). Oligonucleotide sequences were as following: major allele sense: CGC-GTCCTGCCTCACATAATCTTTTTGTTTGTCCCCCT-GAAATGGATTCTCAGCTGTTGCCCAA ACATTTCATCT-TAGCGTTCCAGGTTTGAACTCGCCCTCACGA, minor allele sense: CGCGTCCTGCCTCACATAATCTTTTTGTTT-GTCCCCCTGAAATGTATTC TCAGCTGTTGCCCAAA-CATTTCATCTTAGCGTTCCAGGTTTGAACTCGCCCT-CACGA, and the corresponding antisense sequences. The reporter vectors were co-transfected with the pRL-null vector at a ratio of 10:1 into HEK293 cells using Lipofectamine LTX (Life Technologies) according to the manufacturer's instructions. 24 hours post-transfection, luciferase activity was assayed using the Dual-Luciferase Reporter Assay System (Promega) and Glomax 20/20 Luminometer (Promega). The signal from the reporter vector was normalized to the signal from the pRL-null vector.

Gene Expression in Heart

Samples of left ventricular cardiac tissue from patients undergoing cardiac surgery were genotyped for the SNP rs9885413 and for all five transcripts within +/- 500 kb of the SNP. Samples of cardiac tissue were acquired from patients from the MAGNet consortium (http://www.med.upenn.edu/magnet/). Gene expression levels were determined using the Affymetrix ST1.1 gene expression array (Affymetrix, Santa Clara, CA, USA) in a cohort including 247 heart samples. Genotyping was performed using the Illumina OmniExpress array. Left ventricular free-wall tissue was harvested at time of cardiac surgery from subjects with heart failure undergoing transplantation or from unused transplant donors. In all cases, the heart was perfused with cold cardioplegia prior to cardiectomy to arrest contraction and prevent ischemic damage. Tissue specimens were then obtained and frozen in liquid nitrogen. Genomic DNA from left ventricle was extracted using the Gentra Puregene Tissue Kit (Qiagen) according to manufacturer's instruction.

Total RNA was extracted from left ventricle using the miRNeasy Kit (Qiagen) including DNAse treatment on column. RNA concentration and quality was determined using the NanoVue Plus spectrophotometer (GE Healthcare) and the Agilent 2100 RNA Nano Chip (Agilent). For all samples, genome-wide SNP genotypes were generated using the Illumina OmniExpress Array. Caucasian Ancestry was verified using multi-dimensional scaling of genotypes. For Gene expression array experiments, the Affymetrix ST1.1 Gene array was used. Data were normalized using the Robust Multi-array Average algorithm and batch effects were adjusted for using ComBat. Transcript expression levels were considered significantly higher than background noise if expression values from robust multiarray analysis in at least 10% of either cases or controls exceeded of the 80% quantile of expression of genes on the Y-chromosome in female hearts (5.24). Associations of expression levels for expressed genes with SNP genotypes were tested using a likelihood ratio test. Specifically, we fit a linear regression model $Y = \beta0 + \beta1*D + \beta2*g + \beta3*(g \times D)$ where Y is the $\log_2$ transformed expression level of a given probe, g is the genotype (coded as 0, 1, and 2) of the test SNP, and D is heart failure disease status (D = 1 for heart failure cases and D = 0 for unused donor controls). Association between the probe and test SNP was assessed by testing H0: $\beta2 = \beta3 = 0$ using a likelihood ratio test. Significance of the test statistic was evaluated by comparing with a Chi-squared distribution with two degrees of freedom. All models were additionally adjusted for age, gender, and study site.

DNA Methylation and Gene Expression in Whole Blood

The association of the SNP rs9885413 with DNA methylation was examined in 2408 participants from the FHS Offspring cohort. Methylation at cytosine-guanine dinucleotides (CpG) at the 5q22 locus (+/-500 kb from rs9885413) were ascertained from a gene-centric DNA methylation array (Infinium HumaMethylation450 BeadChip, Illumina, San Diego, CA, USA) which allows interrogation of 485,512 methylation sites across the genome. The array has coverage of at least one methylation site near 99% of RefSeq genes and 96% of CpG islands. Briefly, bisulfite-treated genomic DNA (1 µg) from peripheral blood samples underwent whole-genome amplification, array hybridization and scanning according to manufacturer instructions. Genotyping of rs9885413 was performed. Association of rs9885413 and the methylation probe cg02061660 with expression of the five genes at the locus (+/-500 kb from rs9885413) was examined from microarray data (Affymetrix Human Exon Array ST 1.0) in 5257 participants from the FHS Offspring cohort and Third Generation cohort. Procedures for RNA extraction, processing and analysis have been described previously (28). Linear mixed effect (LME) models were fit accounting for familial correlation, cell count heterogeneity and technical covariates to account for batch effects using the pedigreemm package in R [56]. Specifically, the mQTL model utilized a two-step approach: first, the DNA methylation beta-value (ratio of methylated probe intensity to total probe intensity) was residualized with adjustment for age, sex, cell count proportions (imputed using the Houseman method for granulocytes, monocytes, B-lymphocytes, CD4+ T lymphocytes, CD8+ T lymphocytes and NK cells) [57], measured technical covariates (row, chip, column), and the family structure covariance matrix. Second, DNA methylation residuals were specified as dependent variable, SNP genotype dosage as independent variable with additional adjustment for 558 SVAs (surrogate variable analysis) [58] and ten principal components from eigenstrat [59] to account for unmeasured batch effects. The eQTL models similarly residualized gene expression with adjustment for age, sex, imputed cell count proportions (imputed in Offspring Cohort participants utilizing gene expression markers of cell count proportions developed from the Third Generation participants with both gene expression and measured complete blood counts), and family structure covariance matrix. The residual of gene expression was specified as dependent variable and SNP dosage as independent variable adjusted for 20 PEER (probabilistic estimation of expression residuals) factors [60] to account for unmeasured technical and batch effects in the gene expression data. The eQTM models specified gene expression residual as dependent variable and DNA methylation residual as independent variable adjusted for 20 methylation SVAs and 20 expression SVAs to account for unmeasured technical and batch effects.

Replication of the association of rs9885413 with cg02061660 including the same covariates in the model as in FHS was attempted in blood samples from 750 randomly selected participants of the Rotterdam study (RS3) not included in the GWA stage, where information from the same DNA methylation array as FHS was available. DNA was extracted, bisulfite-treated using the Zymo EZ-96 DNA-methylation kit (Zymo Research, Irvine, CA, USA) and hybridized to arrays according to manufacturer instructions. During quality control samples showing incomplete bisulfite treatment were excluded (n = 5) as were samples with a low detection rate (<99%) (n = 7), or gender swaps (n = 4). Probes with a detection *P*-value>0.01 in >1% samples, were filtered out. A total number of 474,528 probes passed the quality control and the filtered β values were normalized with DASEN implemented in the *wateRmelon* package in R statistical software. Genotyping was performed using the Illumina 610quad array. Cell counts were estimated using the same method as in FHS and also directly measured on a Coulter AcT Diff II Hematology Analyzer (Beckman Coulter, Brea, CA) for granulocytes, monocytes, lymphocytes). Models including both estimated and directly measured cell counts were explored.

siRNA-Mediated Knock Down of NHLH1

HEK293 cells were seeded at 100,000 cells/well in a 6-well plate the day before transfection. Cells were transfected using Lipofectamine and 50 nM of siRNA designed to target human NHLH1 or negative control siRNA (Life Technologies, Carlsbad, CA, USA) according to the manufacturer's instructions. After 48 hours, cells were harvested and total RNA extracted using the miRNeasy Mini Kit (Qiagen, Hilden, Germany) according to the manufacturer's instructions. cDNA was synthesized using the RevertAid H- First Strand cDNA Synthesis Kit (Thermo Fischer Scientific, Waltham, MA, USA) using random hexamer primers and qPCR was performed with TaqMan assays for NHLH1, TMEM232, SLC25A4, WDR36, TSLP, CAMK4 and GAPDH on a StepOne Plus Real-Time PCR System (Life Technologies). Gene expression was normalized to GAPDH and expressed relative to cells transfected with negative control siRNA according to the ΔΔCt-method [61].

Population Genetic Analysis

The frequencies of ancestral and derived alleles of rs9885413 were examined in populations from the International HapMap Project (http://www.hapmap.org/) [62] and the Human Genome Diversity Project (HGDP, http://hagsc.org/hgdp/) [63]. The fixation index (F_{st}) was estimated as described by Weir and Cockerham [64], based on allele frequencies in HapMap stage II as also previously described [65]. The integrated haplotype score (iHS) was calculated from HapMap stage II data as described by Voight et al (http://haplotter.uchicago.edu/) [33]. Allele frequency distributions in HGDP populations were visualized using the HGDP selection browser (http://hgdp.uchicago.edu/) [66].

ACKNOWLEDGMENTS

A detailed list of acknowledgments is provided in the Supplementary Materials.

AUTHOR CONTRIBUTIONS

Conceived and designed the experiments: JGS JFF ACM RSV NLS. Performed the experiments: JGS JFF OG XW MMo RJ BO. Analyzed the data: JGS JFF OG XW MMe RJ BO. Wrote the paper: JGS. Data collection and analysis in the participating genome-wide association and functional studies: ARIC study: ACM DA EB. CHARGE-SCD/ CHARGE-QRS study: NS PvdH. CHS study: JCB KMR KDT BMP NLS. Functional enhancer annotation and siRNA knockdown: JGS OG XW MK BO LAB CNC. Health ABC: AK SBK YL JBu. FHS: MMe RJ YAW JN SD CY CL DL LAC RSV. MAGNet: MMo XS JBr CSM KBM TPC. MDC/MPP: JGS MS. PHS: JBW HDS JMG LD. PROSPER: ST DJS BMB IF JWJ. RS: JFF SL BHCS AH AGU JBvM AD OHF. QT-IGC: CNC.

SUPPORTING INFORMATION

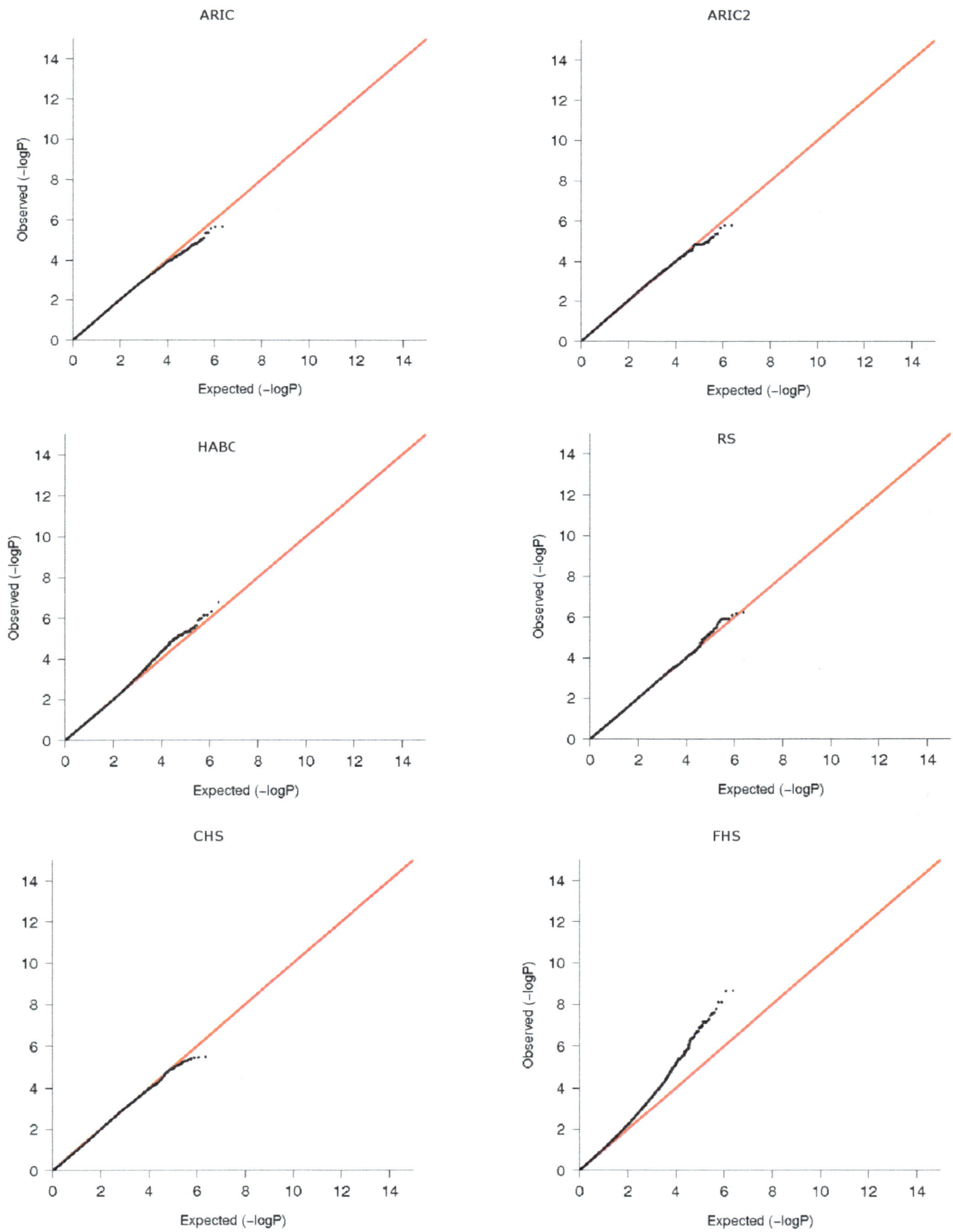

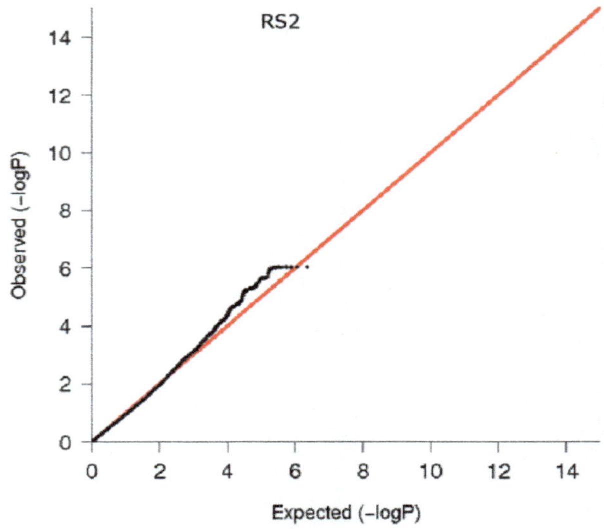

S1 Figure. Quantile-quantile plots for individual genome-wide association studies of HF mortality. Plotted on the x-axis are expected *P*-values under the null hypothesis and on the y-axis the observed *P*-values before genomic control has been applied.

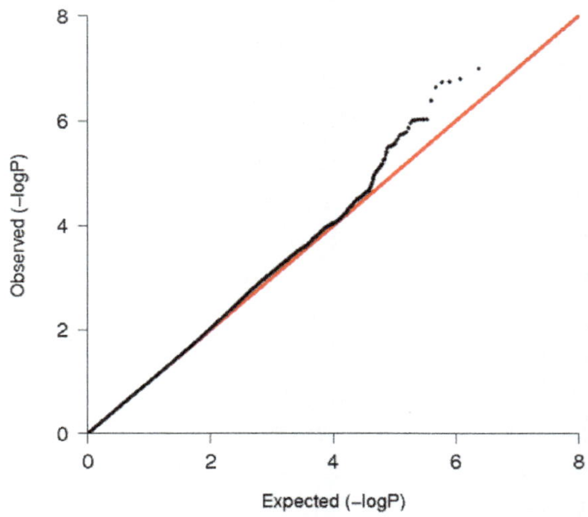

S2 Figure. Quantile-quantile plot for meta-analysis of genome-wide association studies of HF mortality. Plotted on the x-axis are expected *P*-values under the null hypothesis and on the y-axis the observed *P*-values.

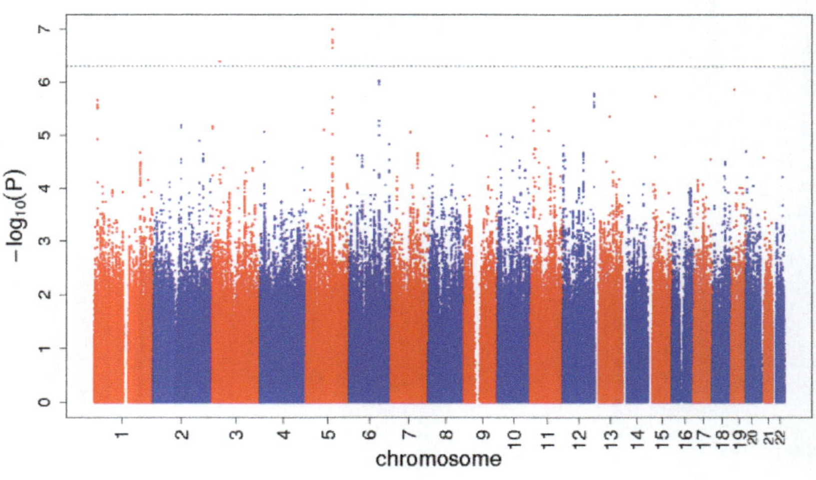

S3 Figure. HF mortality association results for 2.5 million SNPs across 22 autosomes in stage 1 cohorts. Each dot represents one SNP. The x-axis shows physical position by chromosome and the y-axis–$\log_{10}$ (*P*-value). The vertical dotted line indicates the a priori specified significance threshold for SNPs to be carried forward to stage 2 ($P < 5.0\text{x}10^{-7}$).

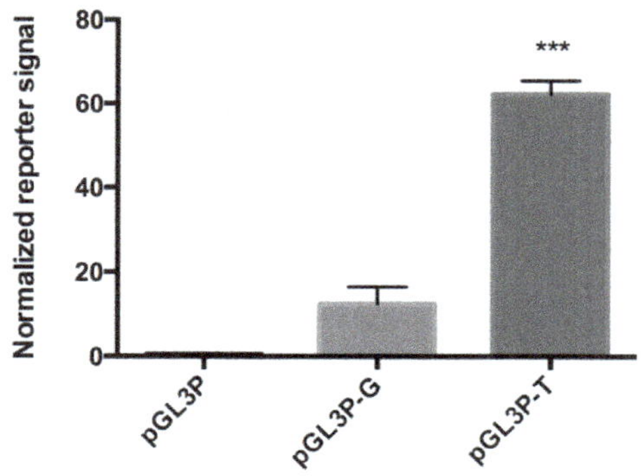

S4 Figure. Increased enhancer activity by the lead SNP. HEK293 cells were transfected with either empty vector (pGL3P) or vectors containing the major (pGL3P-G) or minor (pGL3P-T) allele of rs9885413 together with 100 bp flanking the SNP. Luciferase activity from the vectors was measured after 24 hours and normalized to that of a pRL-null vector. N = 3. ***$P < 0.001$.

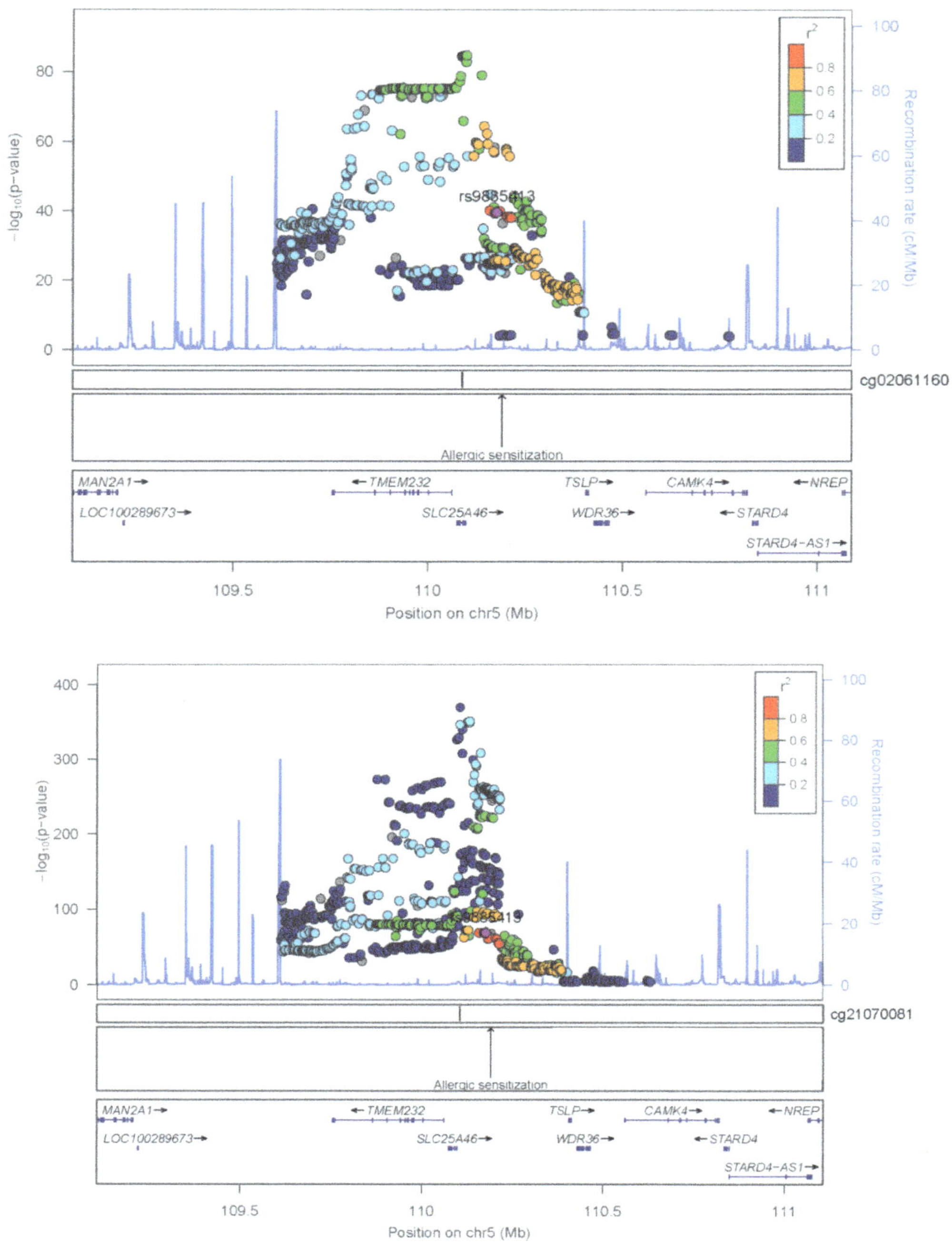

S5 Figure. Association of rs9885413 with DNA methylation. The plot covers the genomic region of +/-1 million bases from each CpG methylation. P-values refer to association of each SNP with two CpG methylation sites on chromosome 5q22: cg02061160 (top panel) and cg21070081 (bottom panel). Circles represent SNPs. The purple circle represents the SNP associated with heart failure mortality (rs9885413). Circle color represents strength of pairwise correlation with rs9885413, with r² according to the inset. CpG sites are illustrated below the regional plot, as is the location of the SNP associated with allergic sensitization (rs10056340). Recombination rate is plotted in the background and known genes are represented in the bottom of the plot. Positions refer to NCBI build 36. SNP correlations and recombination rates were obtained from the 1000 Genomes pilot and HapMap release 22, respectively. The plot was created using LocusZoom (http://locuszoom.sph.umich.edu/locuszoom/).

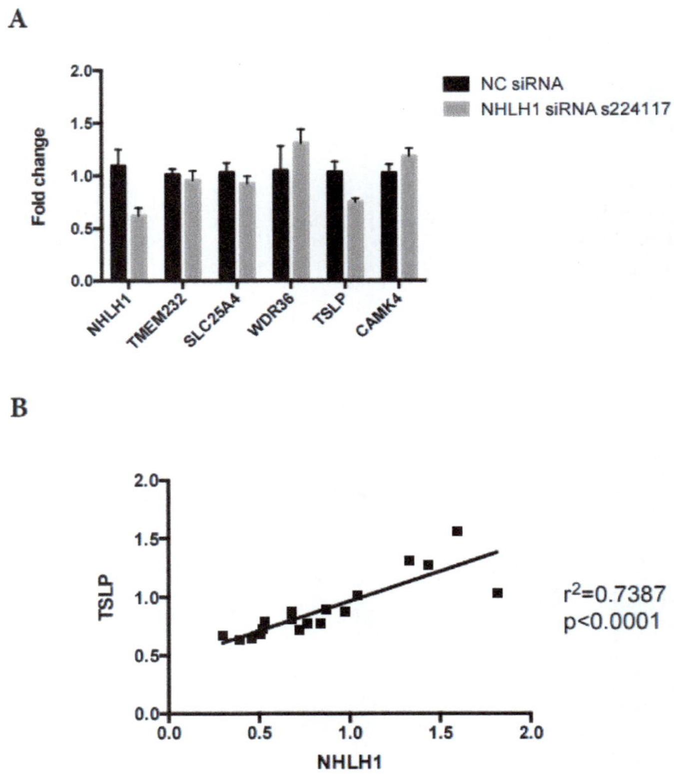

S6 Figure. siRNA-mediated knock down of NHLH1. HEK293 cells were transfected with siRNA targeted against NHLH1 and expression of the five genes surrounding rs9885413 were analyzed by qPCR. Expression levels were normalized to the housekeeping gene GAPDH and expressed relative to cells transfected with negative control (NC) siRNA. Results are from three separate experiments with triplicates in each sample group. *p<0.05 in two sample t-tests. A) Fold change for NHLH1 and individual genes at the locus. B) Dose-response relation of NHLH1 knockdown and TSLP expression.

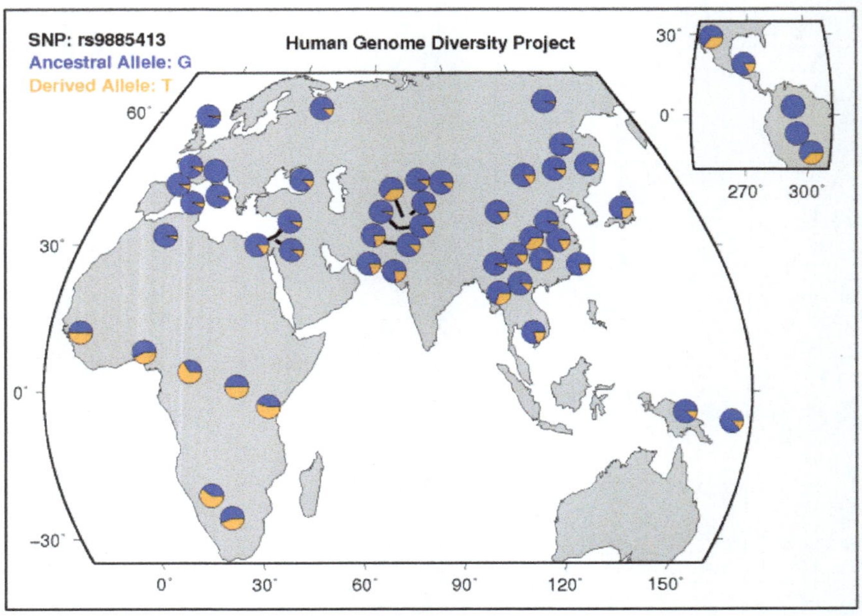

S7 Figure. Distribution of rs9885413 alleles in HGDP populations. Relative allele frequencies of the ancestral (G, blue) and derived (T, yellow) alleles of rs9885413 presented as pie slices across populations in the Human Genome Diversity Project (HGDP).

S1 Table. Methods of heart failure ascertainment and diagnosis across cohorts

	Stage 1					Stage 2		
	ARIC/ ARIC2	*CHS*	*FHS*	*RS/ RS2*	*Health ABC*	*MDCS/ MPP*	*PHS*	*PROS-PER*
Source of information								
Self-report, care-giver report	No	Yes	Yes	No	Yes	No	Yes	Yes
Administrative records	Yes	Yes	Yes	Yes	Yes	Yes	No	Yes
Inclusion of non-hospitalized events	Yes*	Yes	Yes	Yes	No	No	Yes	Yes
Review criteria applied	None	CHS	Framing-ham	ESC**	Health ABC	None	Framing-ham***	ESC
Event characteristics considered								
ICD 9/10 code	Yes	No	No	Yes	No	Yes	No	Yes
Signs and symptoms†	No	Yes	Yes	Yes	Yes	No	Yes***	Yes
Cardiac structure and function‡	No	Yes	Yes	Yes	Yes	No	Yes***	Yes
Physician diagnosis	No	Yes	Yes	Yes	Yes	Yes	Yes	Yes
Pharmaceutical treat-ment§	No	Yes	Yes	Yes	Yes	No	Yes***	Yes

*Included non-hospitalized deaths but not morbid events.

**European Society of Cardiology.

***In a validation sample as previously described [26].

†Signs and symptoms included in Framingham criteria.

‡Cardiac structure and function including cardiomegaly, dilated ventricle, decreased systolic function, or segmental wall-motion abnormalities.

§Pharmaceutical treatments include the prescribing of diuretics, digoxin, or vasodilator.

S2 Table. Characteristics of stage 2 cohorts. Age, body mass index and follow-up time are presented as mean (standard deviation). Categorical variables are presented as percentages. Body mass index, diabetes, hypertension and smoking refer to the nearest study exam prior to heart failure diagnosis whereas age, follow-up time, mortality rate and all-cause death refer to the time of HF diagnosis. Mortality rate refers to the 1-year Kaplan-Meier estimate, with censoring at loss to follow-up. MI: Myocardial infarction

	MDCS	MPP	PHS	PROSPER
HF sample size	856	203	600	211
Age at HF (years)	70.9 (7.9)	69.9 (8.7)	78.8 (8.0)	78.1 (3.7)
Male sex (%)	59.8	81.3	100	58.3
Body mass index (kg/m^2)	28.0 (4.8)	28.6 (5.3)	26.2 (3.9)	27.4 (4.5)
Diabetes (%)	12.1	25.4	19.5	14.2
Hypertension (%)	86.8	94.1	70.5	57.8
Current smoking (%)	30.7	25.6	3.9	26.1
History of MI (%)	29.4	21.2	17.7	28.0
Follow-up time (years)	4.6 (4.4)	6.0 (4.9)	4.1 (3.1)	1.8 (0.9)
Mortality rate, 1 year	0.19	0.09	0.12	0.23
All-cause death (n, %)	456 (53.0)	69 (34.0)	275 (45.8)	89 (42.2)

S3 Table. Results for all SNPs with $P < 1.0 \times 10^{-5}$

SNP	CHR	POS	STR	CA	A2	N	BETA	SE	P
rs9885413	5	110204027	+	G	T	2828	-0.333	0.062	9.97×10^{-8}
rs12658193	5	110200417	+	G	T	2828	-0.333	0.063	1.57×10^{-7}
rs11956079	5	110198295	+	C	T	2828	-0.332	0.064	1.77×10^{-7}
rs10069077	5	110197925	+	C	T	2828	-0.332	0.064	1.81×10^{-7}

rs10068260	5	110209995	+	T	A	2828	0.350	0.068	2.25×10^{-7}
rs12638540	3	32463538	+	G	A	2828	0.405	0.080	4.09×10^{-7}
rs2315841	6	122319575	+	C	A	2828	0.268	0.055	9.35×10^{-7}
rs9490362	6	122315940	+	C	G	2828	0.268	0.055	9.41×10^{-7}
rs9490363	6	122315962	+	C	T	2828	-0.268	0.055	9.42×10^{-7}
rs9375078	6	122307104	+	C	A	2828	0.268	0.055	9.50×10^{-7}
rs1402532	6	122304954	+	G	A	2828	-0.268	0.055	9.79×10^{-7}
rs1607934	6	122302456	+	G	T	2828	-0.268	0.055	9.94×10^{-7}
rs9375081	6	122329048	+	G	A	2828	-0.264	0.054	1.11×10^{-6}
rs4528684	19	14212574	+	C	T	2783	-0.331	0.068	1.36×10^{-6}
rs7965445	12	130428856	+	G	A	2828	-0.255	0.053	1.62×10^{-6}
rs12422521	12	130426315	+	C	T	2828	0.254	0.053	$1{,}70 \times 10^{-6}$
rs7299374	12	130428360	+	G	T	2828	-0.254	0.053	$1{,}77 \times 10^{-6}$
rs12422973	12	130430537	+	C	G	2828	0.253	0.053	$1{,}81 \times 10^{-6}$
rs2125623	15	29616999	+	C	T	2828	0.200	0.042	$1{,}85 \times 10^{-6}$
rs9326808	5	110150873	+	C	T	2828	0.393	0.083	$1{,}95 \times 10^{-6}$
rs10927603	1	15168271	+	C	T	2828	0.186	0.039	$2{,}21 \times 10^{-6}$
rs7977251	12	130436978	+	C	T	2828	-0.255	0.054	$2{,}35 \times 10^{-6}$
rs7976504	12	130437027	+	G	A	2828	-0.255	0.054	$2{,}69 \times 10^{-6}$
rs10754875	1	15166618	+	G	A	2828	0.185	0.039	$2{,}70 \times 10^{-6}$
rs6486658	12	130432243	+	C	T	2828	-0.247	0.053	$2{,}91 \times 10^{-6}$
rs7120489	11	12425973	+	G	A	2828	-0.301	0.064	$2{,}95 \times 10^{-6}$
rs2175773	1	15172259	+	G	A	2828	-0.182	0.039	$3{,}00 \times 10^{-6}$
rs4575078	1	15168869	+	G	T	2828	-0.184	0.039	$3{,}07 \times 10^{-6}$
rs7554572	1	15173226	+	C	T	2828	-0.184	0.039	$3{,}10 \times 10^{-6}$
rs880923	1	15168493	+	C	G	2828	-0.184	0.039	$3{,}13 \times 10^{-6}$
rs7724652	5	110149062	+	C	A	2828	-0.340	0.073	$3{,}35 \times 10^{-6}$
rs17447390	5	110144549	+	G	A	2828	0.340	0.074	$3{,}91 \times 10^{-6}$
rs656316	13	62874290	+	G	A	1990	-0.551	0.120	$4{,}45 \times 10^{-6}$
rs933934	11	12411657	+	G	T	2828	0.281	0.062	$5{,}25 \times 10^{-6}$
rs6569238	6	122330405	+	C	T	2828	0.252	0.055	$5{,}37 \times 10^{-6}$
rs11022357	11	12418898	+	C	T	2828	-0.281	0.062	$5{,}47 \times 10^{-6}$
rs10496499	2	116090407	+	C	A	2828	0.220	0.049	$6{,}56 \times 10^{-6}$
rs9372670	6	122331481	+	C	T	2828	-0.255	0.057	$6{,}71 \times 10^{-6}$
rs13075438	3	3316646	+	C	A	2828	0.236	0.053	$6{,}94 \times 10^{-6}$
rs2901337	2	116070814	+	G	A	2828	0.219	0.049	$7{,}27 \times 10^{-6}$
rs6773870	3	3314720	+	G	T	2828	-0.235	0.052	$7{,}66 \times 10^{-6}$
rs17670147	5	73940490	+	G	A	2828	-0.321	0.072	$8{,}03 \times 10^{-6}$
rs7129111	11	75543144	+	C	T	2828	-0.271	0.061	$8{,}26 \times 10^{-6}$
rs10741590	11	12419894	+	C	T	2828	-0.270	0.061	$8{,}31 \times 10^{-6}$
rs7683977	4	19096126	+	C	T	2828	-0.194	0.044	$8{,}70 \times 10^{-6}$
rs2190523	7	81748107	+	C	T	1990	0.639	0.144	$8{,}82 \times 10^{-6}$
rs10044330	5	109869920	+	G	A	2828	-0.379	0.086	$9{,}67 \times 10^{-6}$
rs10795987	10	12934587	+	G	A	2828	-0.152	0.034	$9{,}71 \times 10^{-6}$

Genome position (POS) refers to NCBI build 36. CHR, chromosome. STR, strand. CA, coded allele. A2, non-coded allele. N, sample size. BETA, beta estimate. SE, standard error of beta estimate. *P*, *P*-value.

S4 Table. Cohort-specific results for SNPs in combined stages 1 and 2

	rs9885413								rs12638540							
Cohort	**Strand**	**CA**	**A2**	**Caf**	**N**	**Beta**	**Se**	**P**	**Strand**	**Ca**	**A2**	**Caf**	**N**	**Beta**	**SE**	**P**
ARIC1	+	T	G	0.07	691	0.24	0.14	0.091	+	G	A	0.05	691	0.42	0.16	0.01
ARIC2	+	T	G	0.07	84	0.41	0.42	0.35	+	G	A	0.04	84	0.53	0.75	0.51
CHS	+	T	G	0.07	838	0.40	0.11	$1.7x10^{-4}$	+	G	A	0.04	838	0.34	0.14	0.01
FHS	+	T	G	0.05	249	0.36	0.19	0.05	+	G	A	0.03	249	-0.05	0.52	0.93
HABC	+	T	G	0.07	173	0.78	0.23	$8x10^{-4}$	+	G	A	0.04	173	0.14	0.39	0.72
Malmo	+	T	G	0.06	1013	0.36	0.12	0.004	+	G	A	0.05	1013	0.12	0.15	0.41
PHS	+	T	G	0.08	591	0.03	0.14	0.83	+	G	A	0.04	581	0.11	0.17	0.52
PROSPER	+	T	G	0.07	211	0.29	0.28	0.31	+	G	A	0.03	211	-0.57	0.72	0.42
RS1	+	T	G	0.06	748	0.15	0.13	0.24	+	G	A	0.04	748	0.54	0.14	$3x10^{-4}$
RS2	+	T	G	0.07	45	0.51	0.55	0.38	+	G	A	0.06	45	-0.23	0.80	0.77
Combined	+	T	G	0.07	3895	0.31	0.05	$2.7x10^{-9}$	+	G	A	0.04	3885	0.30	0.07	$3x10^{-6}$

CHR, chromosome. CA, coded allele. A2, non-coded allele. Caf, coded allele frequency. N, sample size. Beta, beta estimate. Se, standard error of beta estimate. P, p-value.

S5 Table. Allele frequency distribution across major causes of mortality

Cause of death	**Proportion of deaths**	**Frequency of minor (risk) homozygotes or heterozygotes**
Neoplasms	0.19	0.13
Cardiovascular	0.51	0.14
Ischemic heart disease	0.36	0.15
Cerebrovascular	0.07	0.14
Infectious	0.01	0.00
Injuries	0.01	0.09
Psychiatric	0.02	0.15
Respiratory	0.10	0.15
Renal	0.02	0.16
Neurological	0.03	0.13
Endocrine	0.03	0.19
Blood diseases	0.001	0.00
Digestive	0.02	0.15
Other	0.05	0.12
Total	1.0	0.14

S6 Table. In silico studies of associations of rs9885413 with cardiac structure and function

	***P*-value**
Echocardiography	
Systolic dysfunction	0.67
Fractional shortening	0.46

Left ventricular mass	0.96
Left ventricular wall thickness	0.72
Left ventricular internal diastolic dimension	0.34
Left atrial size	0.58
Electrocardiography	
QRS interval duration	0.37
QT interval duration	0.23
Natriuretic peptides	
Atrial natriuretic peptide	0.77
B-type natriuretic peptide	0.31
Heart failure, incident	0.30
Sudden cardiac death	0.77

S7 Table. Enhancer annotations in the ROADMAP Epigenomics Project. Enhancer regions in the 129 tissues from the ROADMAP Epigenomics Project, as determined from the ChromHMM algorithm from patterns of monomethylation of the fourth residue (lysine) of histone H3 (H3K4Me1). Enhancers overlapping the lead SNP on chromosome 5q22 (rs9885413) and strongly correlated SNPs ($r^2 > 0.8$), with enhancers overlapping the lead SNP shown in bold

Tissue	Chr	Start pos	End pos	SNP	ROADMAP tissue
E021	**5**	**110176000**	**110176200**	**rs9885413**	**iPS DF 6.9 Cell Line**
E058	**5**	**110176000**	**110177200**	**rs9885413**	**Foreskin Keratinocyte Primary Cells skin03**
E063	**5**	**110176000**	**110176600**	**rs9885413**	**Adipose Nuclei**
E078	**5**	**110176000**	**110176400**	**rs9885413**	**Duodenum Smooth Muscle**
E079	**5**	**110176000**	**110176600**	**rs9885413**	**Esophagus**
E092	**5**	**110175800**	**110176600**	**rs9885413**	**Fetal Stomach**
E111	**5**	**110174600**	**110177000**	**rs9885413**	**Stomach Smooth Muscle**
E127	**5**	**110175800**	**110177400**	**rs9885413**	**NHEK-Epidermal Keratinocyte Primary Cells**
E076	5	110172400	110173600	rs12658193	Colon Smooth Muscle
E078	5	110172200	110172800	rs12658193	Duodenum Smooth Muscle
E092	5	110172000	110174400	rs12658193	Fetal Stomach
E103	5	110172000	110172600	rs12658193	Rectal Smooth Muscle
E115	5	110170000	110171000	rs10069077	Dnd41 TCell Leukemia Cell Line
E023	5	110219800	110220600	rs12110223	Mesenchymal Stem Cell Derived Adipocyte Cultured Cells
E025	5	110219200	110220600	rs12110223	Adipose Derived Mesenchymal Stem Cell Cultured Cells
E052	5	110219000	110220600	rs12110223	Muscle Satellite Cultured Cells
E023	5	110232200	110233000	rs1423154	Mesenchymal Stem Cell Derived Adipocyte Cultured Cells
E025	5	110232200	110233200	rs1423154	Adipose Derived Mesenchymal Stem Cell Cultured Cells
E052	5	110231800	110233600	rs1423154	Muscle Satellite Cultured Cells
E125	5	110232200	110233000	rs1423154	NH-A Astrocytes Primary Cells
E126	5	110232200	110233400	rs1423154	NHDF-Ad Adult Dermal Fibroblast Primary Cells
E128	5	110232200	110233600	rs1423154	NHLF Lung Fibroblast Primary Cells
E003	5	110235400	110235600	rs7703957	H1 Cell Line
E010	5	110235200	110235600	rs7703957	H9 Derived Neuron Cultured Cells
E016	5	110234800	110236200	rs7703957	HUES64 Cell Line
E017	5	110235200	110235600	rs7703957	IMR90 fetal lung fibroblasts Cell Line
E018	5	110234800	110237800	rs7703957	iPS-15b Cell Line
E019	5	110235400	110237600	rs7703957	iPS-18 Cell Line

E023	5	110235400	110237200	rs7703957	Mesenchymal Stem Cell Derived Adipocyte Cultured Cells
E024	5	110235200	110236600	rs7703957	ES-UCSF4 Cell Line
E025	5	110235200	110236000	rs7703957	Adipose Derived Mesenchymal Stem Cell Cultured Cells
E026	5	110235400	110237600	rs7703957	Bone Marrow Derived Cultured Mesenchymal Stem Cells
E028	5	110234800	110236600	rs7703957	Breast variant Human Mammary Epithelial Cells (vH-MEC)
E055	5	110232800	110235600	rs7703957	Foreskin Fibroblast Primary Cells skin01
E057	5	110234800	110236400	rs7703957	Foreskin Keratinocyte Primary Cells skin02
E058	5	110234800	110236200	rs7703957	Foreskin Keratinocyte Primary Cells skin03
E116	5	110235200	110237000	rs7703957	GM12878 Lymphoblastoid Cell Line
E119	5	110235000	110237800	rs7703957	HMEC Mammary Epithelial Primary Cells
E126	5	110235400	110236800	rs7703957	NHDF-Ad Adult Dermal Fibroblast Primary Cells
E127	5	110235000	110237400	rs7703957	NHEK-Epidermal Keratinocyte Primary Cells
E128	5	110235400	110238400	rs7703957	NHLF Lung Fibroblast Primary Cells
E129	5	110235200	110237800	rs7703957	Osteoblast Primary Cells
E001	5	110235600	110238000	rs1862532	ES-I3 Cell Line
E002	5	110235600	110236000	rs1862532	ES-WA7 Cell Line
E014	5	110235800	110238800	rs1862532	HUES48 Cell Line
E015	5	110235600	110237000	rs1862532	HUES6 Cell Line
E016	5	110234800	110236200	rs1862532	HUES64 Cell Line
E018	5	110234800	110237800	rs1862532	iPS-15b Cell Line
E019	5	110235400	110237600	rs1862532	iPS-18 Cell Line
E020	5	110235800	110237000	rs1862532	iPS-20b Cell Line
E023	5	110235400	110237200	rs1862532	Mesenchymal Stem Cell Derived Adipocyte Cultured Cells
E024	5	110235200	110236600	rs1862532	ES-UCSF4 Cell Line
E025	5	110235200	110236000	rs1862532	Adipose Derived Mesenchymal Stem Cell Cultured Cells
E026	5	110235400	110237600	rs1862532	Bone Marrow Derived Cultured Mesenchymal Stem Cells
E028	5	110234800	110236600	rs1862532	Breast variant Human Mammary Epithelial Cells (vH-MEC)
E057	5	110234800	110236400	rs1862532	Foreskin Keratinocyte Primary Cells skin02
E058	5	110234800	110236200	rs1862532	Foreskin Keratinocyte Primary Cells skin03
E116	5	110235200	110237000	rs1862532	GM12878 Lymphoblastoid Cell Line
E119	5	110235000	110237800	rs1862532	HMEC Mammary Epithelial Primary Cells
E120	5	110235800	110237200	rs1862532	HSMM Skeletal Muscle Myoblasts Cell Line
E125	5	110235800	110236600	rs1862532	NH-A Astrocytes Primary Cells
E126	5	110235400	110236800	rs1862532	NHDF-Ad Adult Dermal Fibroblast Primary Cells
E127	5	110235000	110237400	rs1862532	NHEK-Epidermal Keratinocyte Primary Cells
E128	5	110235400	110238400	rs1862532	NHLF Lung Fibroblast Primary Cells
E129	5	110235200	110237800	rs1862532	Osteoblast Primary Cells
E001	5	110235600	110238000	rs1862530	ES-I3 Cell Line
E014	5	110235800	110238800	rs1862530	HUES48 Cell Line
E015	5	110235600	110237000	rs1862530	HUES6 Cell Line
E016	5	110234800	110236200	rs1862530	HUES64 Cell Line

E018	5	110234800	110237800	rs1862530	iPS-15b Cell Line
E019	5	110235400	110237600	rs1862530	iPS-18 Cell Line
E020	5	110235800	110237000	rs1862530	iPS-20b Cell Line
E023	5	110235400	110237200	rs1862530	Mesenchymal Stem Cell Derived Adipocyte Cultured Cells
E024	5	110235200	110236600	rs1862530	ES-UCSF4 Cell Line
E026	5	110235400	110237600	rs1862530	Bone Marrow Derived Cultured Mesenchymal Stem Cells
E028	5	110234800	110236600	rs1862530	Breast variant Human Mammary Epithelial Cells (vH-MEC)
E049	5	110236000	110236800	rs1862530	Mesenchymal Stem Cell Derived Chondrocyte Cultured Cells
E057	5	110234800	110236400	rs1862530	Foreskin Keratinocyte Primary Cells skin02
E058	5	110234800	110236200	rs1862530	Foreskin Keratinocyte Primary Cells skin03
E116	5	110235200	110237000	rs1862530	GM12878 Lymphoblastoid Cell Line
E119	5	110235000	110237800	rs1862530	HMEC Mammary Epithelial Primary Cells
E120	5	110235800	110237200	rs1862530	HSMM Skeletal Muscle Myoblasts Cell Line
E125	5	110235800	110236600	rs1862530	NH-A Astrocytes Primary Cells
E126	5	110235400	110236800	rs1862530	NHDF-Ad Adult Dermal Fibroblast Primary Cells
E127	5	110235000	110237400	rs1862530	NHEK-Epidermal Keratinocyte Primary Cells
E128	5	110235400	110238400	rs1862530	NHLF Lung Fibroblast Primary Cells
E129	5	110235200	110237800	rs1862530	Osteoblast Primary Cells
E001	5	110235600	110238000	rs17132484	ES-I3 Cell Line
E014	5	110235800	110238800	rs17132484	HUES48 Cell Line
E018	5	110234800	110237800	rs17132484	iPS-15b Cell Line
E019	5	110235400	110237600	rs17132484	iPS-18 Cell Line
E025	5	110236200	110237800	rs17132484	Adipose Derived Mesenchymal Stem Cell Cultured Cells
E026	5	110235400	110237600	rs17132484	Bone Marrow Derived Cultured Mesenchymal Stem Cells
E028	5	110237000	110237600	rs17132484	Breast variant Human Mammary Epithelial Cells (vH-MEC)
E052	5	110236400	110238600	rs17132484	Muscle Satellite Cultured Cells
E058	5	110237200	110237400	rs17132484	Foreskin Keratinocyte Primary Cells skin03
E119	5	110235000	110237800	rs17132484	HMEC Mammary Epithelial Primary Cells
E125	5	110237000	110238600	rs17132484	NH-A Astrocytes Primary Cells
E127	5	110235000	110237400	rs17132484	NHEK-Epidermal Keratinocyte Primary Cells
E128	5	110235400	110238400	rs17132484	NHLF Lung Fibroblast Primary Cells
E129	5	110235200	110237800	rs17132484	Osteoblast Primary Cells

S8 Table. Location of rs9885413 in the predicted NHLH1 binding site

T T **C/A*** A T T C T C A G C T G T T G C C C A A

* The two alleles of the SNP rs9885413.

S9 Table. Association of rs9885413 with gene expression in the GTEx Project. Association of rs9885413 with expression of the five genes at the chromosomal locus in 13 tissues from the Gene-Tissue Expression Project with data available in > 60 samples. Empirical*P*-values were computed using a permutation approach as described in documentation for the GTEx project (http://www.gtexportal.org). N/A, no results available in GTEx. Effect direction is presented for associations with $P \leq 0.1$

SNP	Transcript	GTEX tissue	Direction	*P* value
rs9885413	TMEM232	Adipose Subcutaneous	+	0.04

rs9885413	TMEM232	Aorta	+	0.06
rs9885413	TMEM232	Artery Tibial	+	0.03
rs9885413	TMEM232	Blood		N/A
rs9885413	TMEM232	Esophagus Mucosa	+	0.01
rs9885413	TMEM232	Esophagus Muscularis	+	0.02
rs9885413	TMEM232	Heart left ventricle		N/A
rs9885413	TMEM232	Lung		0.9
rs9885413	TMEM232	Muscle skeletal		N/A
rs9885413	TMEM232	Nerve Tibial	+	0.01
rs9885413	TMEM232	Skin sun exposed	+	0.1
rs9885413	TMEM232	Stomach		0.9
rs9885413	TMEM232	Thyroid	+	0.05
rs9885413	SLC25A46	Adipose Subcutaneous		0.2
rs9885413	SLC25A46	Aorta		0.9
rs9885413	SLC25A46	Artery Tibial		0.3
rs9885413	SLC25A46	Blood		0.8
rs9885413	SLC25A46	Esophagus Mucosa	-	0.008
rs9885413	SLC25A46	Esophagus Muscularis		0.9
rs9885413	SLC25A46	Heart left ventricle		1.0
rs9885413	SLC25A46	Lung		0.2
rs9885413	SLC25A46	Muscle skeletal		1.0
rs9885413	SLC25A46	Nerve Tibial		0.6
rs9885413	SLC25A46	Skin sun exposed	+	0.005
rs9885413	SLC25A46	Stomach		0.6
rs9885413	SLC25A46	Thyroid		0.8
rs9885413	TSLP	Adipose Subcutaneous	-	0.09
rs9885413	TSLP	Aorta		1.0
rs9885413	TSLP	Artery Tibial		0.8
rs9885413	TSLP	Blood		N/A
rs9885413	TSLP	Esophagus Mucosa		0.9
rs9885413	TSLP	Esophagus Muscularis		0.5
rs9885413	TSLP	Heart left ventricle		0.4
rs9885413	TSLP	Lung		0.7
rs9885413	TSLP	Muscle skeletal		0.5
rs9885413	TSLP	Nerve Tibial		0.8
rs9885413	TSLP	Skin sun exposed		0.7
rs9885413	TSLP	Stomach		0.8
rs9885413	TSLP	Thyroid		0.5
rs9885413	WDR36	Adipose Subcutaneous		0.6
rs9885413	WDR36	Aorta		0.8
rs9885413	WDR36	Artery Tibial		0.8
rs9885413	WDR36	Blood		0.5
rs9885413	WDR36	Esophagus Mucosa		0.2
rs9885413	WDR36	Esophagus Muscularis		0.4
rs9885413	WDR36	Heart left ventricle		0.9

rs9885413	WDR36	Lung	+	0.06
rs9885413	WDR36	Muscle skeletal		0.5
rs9885413	WDR36	Nerve Tibial	+	0.1
rs9885413	WDR36	Skin sun exposed		0.9
rs9885413	WDR36	Stomach		0.6
rs9885413	WDR36	Thyroid		0.8
rs9885413	CAMK4	Adipose Subcutaneous	-	0.1
rs9885413	CAMK4	Aorta		0.4
rs9885413	CAMK4	Artery Tibial		1.0
rs9885413	CAMK4	Blood		0.8
rs9885413	CAMK4	Esophagus Mucosa		0.2
rs9885413	CAMK4	Esophagus Muscularis	-	0.03
rs9885413	CAMK4	Heart left ventricle		N/A
rs9885413	CAMK4	Lung		0.2
rs9885413	CAMK4	Muscle skeletal		N/A
rs9885413	CAMK4	Nerve Tibial	+	0.02
rs9885413	CAMK4	Skin sun exposed	+	0.02
rs9885413	CAMK4	Stomach	-	0.1
rs9885413	CAMK4	Thyroid	+	0.1

S10 Table. Association of rs9885413 with gene expression in human heart. Association of rs9885413 with expression of the five genes at the locus on chromosome 5q22 in 247 human heart samples from 116 patients with advanced heart failure (HF) and 131 unused transplant donor hearts (controls). Transcript expression levels were considered significantly higher than background noise if expression values from robust multiarray analysis in at least 10% of either cases or controls exceeded of the 80% quantile of expression of genes on the Y-chromosome in female hearts (5.24). Positive direction indicates higher expression with the risk (minor) allele of rs9885413

Transcript	HF mean	Control mean	Direction	*P* value
TMEM232	5.30	4.85	+	2.2×10^{-6}
SLC25A46	8.70	8.84		0.36
TSLP	3.27	3.31		N/A
WDR36	6.74	6.79		0.27
CAMK4	3.88	3.97		N/A

REFERENCES

1. Mudd JO, Kass DA. Tackling heart failure in the twenty-first century. Nature 2008; 451: 919–928. doi: 10.1038/nature06798. pmid:18288181
2. Hjalmarson A, Goldstein S, Fagerberg B, Wedel H, Waagstein F, Kjekshus J, et al. Effects of controlled-release metoprolol on total mortality, hospitalizations, and well-being in patients with heart failure: the Metoprolol CR/XL Randomized Intervention Trial in congestive heart failure (MERIT-HF). MERIT-HF Study Group. JAMA 2000; 283: 1295–1302. pmid:10714728 doi: 10.1001/jama.283.10.1295
3. The Cardiac Insufficiency Bisoprolol Study II (CIBIS-II): a randomised trial. Lancet 1999; 353; 9–13. pmid:10023943 doi: 10.1016/s0140-6736(98)11181-9
4. Packer M, Bristow MR, Cohn JN, Colucci WS, Fowler MB, Gilbert EM, et al. The effect of carvedilol on morbidity and mortality in patients with chronic heart failure. U.S. Carvedilol Heart Failure Study Group. N Engl J Med 1996; 334: 1349–1355. pmid:8614419 doi: 10.1056/nejm199605233342101
5. The CONSENSUS Trial Study Group. Effects of enalapril on mortality in severe congestive heart failure. Results of the Cooperative North Scandinavian Enalapril Survival Study (CONSENSUS). N Engl J Med 1987; 316: 1429–1435. pmid:2883575 doi: 10.1056/nejm198706043162301
6. The SOLVD Investigators. Effect of enalapril on survival in patients with reduced left ventricular ejection fractions and congestive heart failure. N Engl J Med 1991; 325: 293–302. pmid:2057034 doi: 10.1056/nejm199108013250501
7. Pitt B, Zannad F, Remme WJ, Cody R, Castaigne A, Perez A, et al. The effect of spironolactone on morbidity and mortality in patients with severe heart failure. Randomized Aldactone Evaluation Study Investigators. N Engl J Med 1999; 341: 709–717. pmid:10471456 doi: 10.1056/nejm199909023411001

8. Pfeffer MA, Swedberg K, Granger CB, Held P, McMurray JJ, Michelson EL, et al. Effects of candesartan on mortality and morbidity in patients with chronic heart failure: the CHARM-Overall programme. Lancet 2003; 362: 759–766. pmid:13678868 doi: 10.1016/s0140-6736(03)14282-1
9. Levy D, Kenchaiah S, Larson MG, Benjamin EJ, Kupka MJ, Ho KK, et al. Long-term trends in the incidence of and survival with heart failure. N Engl J Med 2002; 347: 1397–1402. pmid:12409541 doi: 10.1056/nejmoa020265
10. Shafazand M, Schaufelberger M, Lappas G, Swedberg K, Rosengren A. Survival trends in men and women with heart failure of ischaemic and non-ischaemic origin: data for the period 1987–2003 from the Swedish Hospital Discharge Registry. Eur Heart J 2009; 30: 671–678. doi: 10.1093/eurheartj/ehn541. pmid:19109351
11. Chen J, Normand SL, Wang Y, Krumholz HM. National and regional trends in heart failure hospitalization and mortality rates for Medicare beneficiaries, 1998–2008. JAMA 2011; 306: 1669–1678. doi: 10.1001/jama.2011.1474. pmid:22009099
12. Owan TE, Hodge DO, Herges RM, Jacobsen SJ, Roger VL, Redfield MM. Trends in prevalence and outcome of heart failure with preserved ejection fraction. N Engl J Med 2006; 355: 251–259. pmid:16855265 doi: 10.1056/nejmoa052256
13. Bhatia RS, Tu JV, Lee DS, Austin PC, Fang J, Haouzi A, et al. Outcome of heart failure with preserved ejection fraction in a population-based study. N Engl J Med 2006; 355: 260–269. pmid:16855266 doi: 10.1056/nejmoa051530
14. Musunuru K, Strong A, Frank-Kamenetsky M, Lee NE, Ahfeldt T, Sachs KV, et al. From noncoding variant to phenotype via SORT1 at the 1p13 cholesterol locus. Nature 2010; 466: 714–719. doi: 10.1038/nature09266. pmid:20686566
15. Cohen JC. Emerging LDL therapies: Using human genetics to discover new therapeutic targets for plasma lipids. 2010; J Clin Lipidol 7: S1–5. doi: 10.1016/j.jacl.2013.03.005
16. O'Mahony C, Jichi F, Pavlou M, Monserrat L, Anastasakis A, Rapezzi C, et al. A novel clinical risk prediction model for sudden cardiac death in hypertrophic cardiomyopathy (HCM risk-SCD). Eur Heart J 2014; 35: 2010–2020. doi: 10.1093/eurheartj/eht439. pmid:24126876
17. Morrison AC, Felix JF, Cupples LA, Glazer NL, Loehr LR, Dehghan A, et al. Genomic variation associated with mortality among adults of European and African ancestry with heart failure: the cohorts for heart and aging research in genomic epidemiology consortium. Circ Cardiovasc Genet 2010; 3: 248–255. doi: 10.1161/CIRCGENETICS.109.895995. pmid:20400778
18. Psaty BM, O'Donnell CJ, Gudnason V, Lunetta KL, Folsom AR, Rotter JI, et al. Cohorts for Heart and Aging Research in Genomic Epidemiology (CHARGE) Consortium: Design of prospective meta-analyses of genome-wide association studies from 5 cohorts. Circ Cardiovasc Genet 2009; 2: 73–80. doi: 10.1161/CIRCGENETICS.108.829747. pmid:20031568
19. Vasan RS, Glazer NL, Felix JF, Lieb W, Wild PS, Felix SB, et al. Genetic variants associated with cardiac structure and function: a meta-analysis and replication of genome-wide association data. JAMA 2009; 302: 168–178. doi: 10.1001/jama.2009.978-a. pmid:19584346
20. Smith NL, Felix JF, Morrison AC, Demissie S, Glazer NL, Loehr LR, et al. Association of genome-wide variation with the risk of incident heart failure in adults of European and African ancestry: a prospective meta-analysis from the cohorts for heart and aging research in genomic epidemiology (CHARGE) consortium. Circ Cardiovasc Genet 2010; 3: 256–266. doi: 10.1161/CIRCGENETICS.109.895763. pmid:20445134
21. Smith JG, Newton-Cheh C, Almgren P, Struck J, Morgenthaler NG, Bergmann A, et al. Assessment of conventional cardiovascular risk factors and multiple biomarkers for the prediction of incident heart failure and atrial fibrillation. J Am Coll Cardiol 2010; 56: 1712–1719. doi: 10.1016/j.jacc.2010.05.049. pmid:21070922
22. Sotoodehnia N, Isaacs A, de Bakker PI, Dorr M, Newton-Cheh C, Nolte IM, et al. Common variants in 22 loci are associated with QRS duration and cardiac ventricular conduction. Nat Genet 2010; 42: 1068–1076. doi: 10.1038/ng.716. pmid:21076409
23. Arking DE, Pulit SL, Crotti L, van der Harst P, Munroe PB, Koopmann TT, et al. Genetic association study of QT interval highlights role for calcium signaling pathways in myocardial repolarization. Nat Genet 2014; 46: 826–836. doi: 10.1038/ng.3014. pmid:24952745
24. Roadmap Epigenomics Consortium. Integrative analysis of 111 reference human epigenomes. Nature 2015; 518: 317–330. doi: 10.1038/nature14248. pmid:25693563
25. Cogliati T, Good DJ, Haigney M, Delgado-Romero P, Eckhaus MA, Koch WJ, et al. Predisposition to arrhythmia and autonomic dysfunction in Nhlh1-deficient mice. Mol Cell Biol 2002; 22: 4977–4983. pmid:12077327 doi: 10.1128/mcb.22.14.4977-4983.2002
26. Yang IV, Alper S, Lackford B, Rutledge H, Warg LA, Burch LH, et al. Novel regulators of the systemic response to lipopolysaccharide. Am J Respir Cell Mol Biol 2011; 45: 393–402. doi: 10.1165/rcmb.2010-0342OC. pmid:21131441
27. Bønnelykke K, Matheson MC, Pers TH, Granell R, Strachan DP, Alves AC, et al. Meta-analysis of genome-wide association studies identifies ten loci influencing allergic sensitization. Nat Genet 2013; 45: 902–6. doi: 10.1038/ng.2694. pmid:23817571
28. GTEx consortium. The Genotype-Tissue Expression (GTEx) project. Nat Genet 2013; 45: 580–585. doi: 10.1038/ng.2653. pmid:23715323
29. Zhang X, Joehanes R, Chen BH, Huan T, Ying S, Munson PJ, et al. Identification of common genetic variants controlling transcript isoform variation in human whole blood. Nat Genet 2015; 47: 345–52. doi: 10.1038/ng.3220. pmid:25685889
30. Gauvreau GM, O'Byrne PM, Boulet LP, Wang Y, Cockcroft D, Bigler J, et al. Effects of an anti-TSLP antibody on allergen-induced asthmatic responses.

N Engl J Med 2014; 370: 2102–10. doi: 10.1056/NEJMoa1402895. pmid:24846652

31. Porter KE, Turner NA. Cardiac fibroblasts: at the heart of myocardial remodeling. Pharmacol Ther 2009; 123: 255–78. doi: 10.1016/j.pharmthera.2009.05.002. pmid:19460403

32. Ying G, Zhang Y, Tang G, Chen S. Functions of thymic stromal lymphopoietin in non-allergic diseases. Cell Immunol 2015; 295: 144–9. doi: 10.1016/j.cellimm.2015.03.006. pmid:25880106

33. Voight BF, Kudaravalli S, Wen X, Pritchard JK. A map of recent positive selection in the human genome. PLoS biology 2006; 4: e72. pmid:16494531 doi: 10.1371/journal.pbio.0040072

34. Shah S. Heart failure with preserved ejection fraction in African Americans. Ethn Dis 2012; 22: 432–438. pmid:23140073

35. Hofmann U, Frantz S. Role of lymphocytes in myocardial injury, healing, and remodeling after myocardial infarction. Circ Res 2015; 116: 354–67. doi: 10.1161/CIRCRESAHA.116.304072. pmid:25593279

36. Laroumanie F, Douin-Echinard V, Pozzo J, Lairez O, Vinel C, Delage C, et al. CD4+ T cells promote the transition from hypertrophy to heart failure during chronic pressure overload. Circulation 2014; 129: 2111–24. doi: 10.1161/CIRCULATIONAHA.113.007101. pmid:24657994

37. Wills QF, Livak KJ, Tipping AJ, Enver T, Goldson AJ, Sexton DW, et al. Single-cell gene expression analysis reveals genetic associations masked in whole-tissue experiments. Nat Biotechnol 2013; 31: 748–52. doi: 10.1038/nbt.2642. pmid:23873083

38. Takai T. TSLP expression: cellular sources, triggers, and regulatory mechanisms. Allergol Int 2012; 61: 3–17. doi: 10.2332/allergolint.11-RAI-0395. pmid:22270071

39. Karlsson EK, Kwiatkowski DP, Sabeti PC. Natural selection and infectious disease in human populations. Nat Rev Genet 2014; 15: 379–93. doi: 10.1038/nrg3734. pmid:24776769

40. Ito K, Bick AG, Flannick J, Friedman DJ, Genovese G, Parfenov MG, et al. Increased burden of cardiovascular disease in carriers of APOL1 genetic variants. Circ Res 2014; 114: 845–50. doi: 10.1161/CIRCRESAHA.114.302347. pmid:24379297

41. Investigators ARIC. The Atherosclerosis Risk in Communities (ARIC) Study: design and objectives. Am J Epidemiol 1989; 129: 687–702. pmid:2646917

42. Fried LP, Borhani NO, Enright P, Furberg CD, Gardin JM, Kronmal RA, et al. The Cardiovascular Health Study: design and rationale. Ann Epidemiol 1991; 1: 263–276. pmid:1669507 doi: 10.1016/1047-2797(91)90005-w

43. Dawber T, Meadors GF, Moore FE, Epidemiologic approaches to heart disease: the Framingham study. American Journal of Public Health 1951; 41: 279–286. pmid:14819398 doi: 10.2105/ajph.41.3.279

44. Feinleib M, Kannel WB, Garrison RJ, The Framingham Offspring Study. Design and preliminary data. Preventive Medicine 1975; 4: 518–525. pmid:1208363 doi: 10.1016/0091-7435(75)90037-7

45. Hofman A, Darwish Murad S, van Duijn CM, Franco OH, Goedegebure A, Ikram MA, et al. The Rotterdam Study: 2014 objectives and design update. Eur J Epidemiol 2013; 28: 889–926. doi: 10.1007/s10654-013-9866-z. pmid:24258680

46. Kalogeropoulos A, Georgiopoulou V, Kritchevsky SB, Psaty BM, Smith NL, Newman AB, et al. Epidemiology of incident heart failure in a contemporary elderly cohort: the health, aging, and body composition study. Arch Intern Med 2009; 169: 708–715. doi: 10.1001/archinternmed.2009.40. pmid:19365001

47. Smith JG, Platonov PG, Hedblad B, Engstrom G, Melander O. Atrial fibrillation in the Malmo Diet and Cancer study: a study of occurrence, risk factors and diagnostic validity. Eur J Epidemiol 2010; 25: 95–102. doi: 10.1007/s10654-009-9404-1. pmid:19936945

48. Lyssenko V, Jonsson A, Almgren P, Pulizzi N, Isomaa B, Tuomi T, et al. Clinical risk factors, DNA variants, and the development of type 2 diabetes. N Engl J Med 2008; 359: 2220–2232. doi: 10.1056/NEJMoa0801869. pmid:19020324

49. Shepherd J, Blauw GJ, Murphy MB, Bollen EL, Buckley BM, Cobbe SM, et al. Pravastatin in elderly individuals at risk of vascular disease (PROSPER): a randomised controlled trial. Lancet 2002; 360: 1623–1630. pmid:12457784 doi: 10.1016/s0140-6736(02)11600-x

50. Djousse L, Driver JA, Gaziano JM. Relation between modifiable lifestyle factors and lifetime risk of heart failure. JAMA 2009; 302: 394–400. doi: 10.1001/jama.2009.1062. pmid:19622818

51. Cochran WG. The combination of estimates from different experiments. Biometrics 1954; 10: 101–129. doi: 10.2307/3001666

52. Higgins JP, Thompson SG. Quantifying heterogeneity in a meta-analysis. Statistics in medicine 2002; 21: 1539–1558. pmid:12111919 doi: 10.1002/sim.1186

53. Johnson AD, Handsaker RE, Pulit SL, Nizzari MM, O'Donnell CJ, de Bakker PI. SNAP: a web-based tool for identification and annotation of proxy SNPs using HapMap. Bioinformatics 2008; 24: 2938–2939. doi: 10.1093/bioinformatics/btn564. pmid:18974171

54. Ernst J, Kellis M. ChromHMM: automating chromatin-state discovery and characterization. Nat Methods 2012; 9: 215–216. doi: 10.1038/nmeth.1906. pmid:22373907

55. Ward LD, Kellis M. HaploReg: a resource for exploring chromatin states, conservation, and regulatory motif alterations within sets of genetically linked variants. Nucleic Acids Res 2012; 40: D930–934. doi: 10.1093/nar/gkr917. pmid:22064851

56. Vazquez AI, Bates DM, Rosa GJM, Gianol D, Weigel KA. Technical note: an R package for fitting generalized linear mixed models in animal breeding. J Anim Sci 2010; 88: 497–504. doi: 10.2527/jas.2009-1952. pmid:19820058

57. Houseman EA, Accomando WP, Koestler DC, Christensen BC, Marsit CJ, Nelson HH, et al. DNA

methylation arrays as surrogate measures of cell mixture distribution. BMC Bioinformatics 2012; 13: 86. doi: 10.1186/1471-2105-13-86. pmid:22568884

58. Leek JT, Storey JD. Capturing heterogeneity in gene expression studies by surrogate variable analysis. PLoS genetics 2007; 3: e161. doi: 10.1371/journal.pgen.0030161

59. Price AL, Patterson NJ, Plenge RM, Weinblatt ME, Shadick NA, Reich D. Principal components analysis corrects for stratification in genome-wide association studies. Nat Genet 2006; 38: 904–9. pmid:16862161 doi: 10.1038/ng1847

60. Stegle O, Parts L, Piipari M, Winn J, Durbin R. Using probabilistic estimation of expression residuals (PEER) to obtain increased power and interpretability of gene expression analyses. Nat Protoc 2012; 7: 500–7. doi: 10.1038/nprot.2011.457. pmid:22343431

61. Pfaffl MW. A new mathematical model for relative quantification in real-time RT-PCR. Nucleic Acids Res 2001; 29: e45. pmid:11328886 doi: 10.1093/nar/29.9.e45

62. International HapMap Consortium. A haplotype map of the human genome. Nature 2005; 437: 1299–1320. pmid:16255080 doi: 10.1038/nature04226

63. Li JZ, Absher DM, Tang H, Southwick AM, Casto AM, Ramachandran S, et al. Worldwide human relationships inferred from genome-wide patterns of variation. Science 2008; 319: 1100–1104. doi: 10.1126/science.1153717. pmid:18292342

64. Weir BS, Cockerham CC. Estimating F-statistics for the analysis of population structure. Evolution 1984; 38: 1358–70. doi: 10.2307/2408641

65. Kosova G, Pickrell JK, Kelley JL, McArdle PF, Shuldiner AR, Abney M, et al. The CFTR Met 470 allele is associated with lower birth rates in fertile men from a population isolate. PLoS genetics 2010; 6: e1000974. doi: 10.1371/journal.pgen.1000974. pmid:20532200

66. Pickrell JK, Coop G, Novembre J, Kudaravalli S, Li JZ, Absher D, et al. Signals of recent positive selection in a worldwide sample of human populations. Genome research 2009; 19: 826–837. doi: 10.1101/gr.087577.108. pmid:19307593

CHAPTER 2

APPLICATION OF TARGETED MULTI-GENE PANEL TESTING FOR THE DIAGNOSIS OF INHERITED PERIPHERAL NEUROPATHY PROVIDES A HIGH DIAGNOSTIC YIELD WITH UNEXPECTED PHENOTYPE-GENOTYPE VARIABILITY

Thalia Antoniadi, Chris Buxton, Gemma Dennis, Natalie Forrester, Debbie Smith, Peter Lunt and Sarah Burton-Jones

BACKGROUND

Inherited peripheral neuropathy (IPN) is the most common group of inherited neurological disorders with an estimated prevalence of 1 in 2500 individuals [1]. It is clinically and genetically heterogeneous; with over 90 genes and loci implicated in the normal function of the myelinated axons of the peripheral nervous system. Onset is typically in the first or second decade, but there are congenital and infantile onset forms of the disease, as well as late onset adult forms. The classical clinical phenotype may manifest as distal limb muscle wasting and weakness, mild to moderate sensory loss, abnormalities of deep tendon reflexes and foot deformities (*pes cavus* and hammer toes). Hearing loss, or respiratory impairment resulting from phrenic nerve involvement, may also be characteristic in some forms.

IPN classification is based on clinical phenotype, mode of inheritance, age of onset, electrophysiological studies and causal mutation. The main subtypes include hereditary motor and sensory neuropathy (HMSN), typically known as Charcot-Marie-Tooth disease (CMT); hereditary sensory and autonomic neuropathy (HSAN) also known as hereditary sensory neuropathy; hereditary motor neuropathy (HMN), also known as distal hereditary motor neuropathy, and hereditary neuropathy with liability to pressure palsy (HNPP).

CMT is the phenotype with the widest genetic heterogeneity. Nerve conduction velocity studies (NCV) subdivide CMT into type 1 (CMT1), a demyelinating form with median or ulnar motor NCV <38 m/s; type 2 (CMT2), an axonal form with NCV >38 m/s, and an intermediate form with both demyelinating and axonal features. Inheritance modes include autosomal dominant (AD), autosomal recessive (AR) and X-linked (XL). A single gene may be implicated in different phenotypes and present with different modes of inheritance, presenting a challenge to diagnose patients with specific types of IPN [2, 3].

Following the exclusion of a 1.5 Mb duplication at 17p11.2 including the *PMP22* gene as the most common cause of CMT1, the traditional strategy for genetic testing consisted of sequential sequencing of individual genes, selected according to the patient's clinical presentation and family history. This strategy, alongside the cost of serial testing and the limited breadth of genes available for testing, resulted in a low diagnostic yield.

Since the cost of next generation sequencing (NGS) has been decreasing dramatically over the last few years, this technology has found numerous applications in the diagnosis of heterogeneous disorders, including IPN. Whole genome sequencing (WGS) and whole exome sequencing (WES) have identified new genetic causes for many conditions, as demonstrated for IPN by the identification of *SH3TC2* as a cause of autosomal recessive CMT1 [4] and *DYNC1H1* as a new genetic cause for autosomal dominant CMT2 [5].

Citation: Thalia Antoniadi, Chris Buxton, Gemma Dennis, Natalie Forrester, Debbie Smith, Peter Lunt and Sarah Burton-Jones, "Application of targeted multi-gene panel testing for the diagnosis of inherited peripheral neuropathy provides a high diagnostic yield with unexpected phenotype-genotype variability," BMC Medical Genetics201516:84. DOI: 10.1186/s12881-015-0224-8.

The targeted panel approach, which restricts analysis to genes known to be implicated in a particular phenotype has also been also successfully applied to IPN. Choi et al. applied WES to a series of unrelated individuals with CMT and restricted analysis to genes already known to be causes of IPN [6]. WGS and WES, however, generate a huge amount of data. Management and storage of this data can present a challenge in a clinical diagnostic environment.

We designed and validated a targeted NGS panel assay including 56 genes associated with known causes of inherited neuropathy and evaluated this approach for the diagnosis of IPN. The results of the pilot project were submitted as a gene dossier to the UK Genetic Testing Network (UKGTN) [7]. This received approval in January 2013 and the diagnostic service was launched in July 2013. We present and summarise the results of 448 patients reported in the first 18 months of this diagnostic service.

METHODS

Patients

Blood or DNA samples from patients referred by neurologists and clinical geneticists were accepted for testing when the UKGTN approval criteria were met. These criteria are listed below:

- 'Idiopathic' peripheral neuropathy diagnosed by clinical presentation with progressive weakness in hands/wrists and/or feet/ankles and/or associated *pes cavus* or finger flexion contractures and/or peripheral sensory loss
- Supportive nerve conduction test result (defining type I or II according to NCV)
- Absence of other non-genetic causes (alcohol, B12 deficiency, diabetes, trauma)
- No associated CNS involvement

The referring clinicians were also asked to indicated the suspected mode of inheritance and provide a recent clinical letter with further details of the clinical phenotype. A small number of samples were accepted for testing from patients not strictly meeting these criteria, after discussion with the individual requesting clinician.

From July 2013 to December 2014, DNA samples from 448 unrelated probands with suspected IPN were tested and reported; a significant proportion of these patients had previously tested negative for the common causes of IPN.

Two hundred ninety nine patients were referred by neurologists (67 %) and 149 patients (33 %) by clinical geneticists. Approximately one third of the patients were under 18 years of age at referral (135 patients, 30 %).

Informed consent for IPN multi-gene panel testing was obtained from patients or their parents/legal guardians by the requesting clinician. The decision to request this test was made by each clinician according to their local ethical guidelines.

This diagnostic test has been assessed for validity, utility and socio-legal/ethical implications in the process of ratification by the UKGTN and UK NHS commissioners, and was undertaken in an accredited UK NHS Laboratory. Data presented pertains only to results of routine diagnostic testing; therefore this study was not subject to ethical approval.

Targeted Capture

Genes were selected following extensive searches of the literature and locus specific databases [8, 9] to ensure their clinical validity and utility. Two additional genes flanking *PMP22* that are commonly involved in the 1.5 Mb reciprocal deletion/duplication event occurring at 17p11.2 [10] were included, to assist in the copy number assessment of this region (*COX10, TEKT3*). All the genes had a disease OMIM (Online Mendelian Inheritance in Man) entry related to a subtype of peripheral neuropathy. Table 1 details the 56 genes included in the assay.

Table 1: Genes included in IPN NGS panel, with associated phenotype and inheritance pattern

GENE	OMIM	Locus	CMT1	CMT2	HMN	HS(A)N	OMIM	Inheritance
AARS	601065	16q22		CMT 2N			613287	AD
ARHGEF10	608136	8p23	Slowed NCV; hypomyelination				608236	AD
ATL1	606439	14q11-q21				HSN 1D	613708	AD
ATP7A	300011	Xq12-q13			dSMAX3		300489	XL
BAG3	603883	10q26.11	Myopathy; myofibrillar, BAG-3 related				612954	AD
BSCL2	606158	11q12.3			HMN 5		600794	AD
CCT5	610150	5p15.2				HSN with spastic paraplegia	256840	AR
CTDP1	604927	18q23	CCFDN: Congenital cataracts, facial dysmorphism, neuropathy				604168	AR
DCTN1	601143	2p13.1			HMN 7B		607641	AD

DNM2	602378	19p13.2		CMT DI B			606482	AD
				CMT 2 M				
DYNC1H1	600112	14q32.31		CMT 2O	SMA-LED		614228/158600	AD
EGR2	129010	10q21.1-q22.1	CMT 1D				607678	AD
			CMT 4E CHN				605253	AD
			DSS				145900	AR
FAM134B	613114	5p15.1				HSAN 2B	613115	AR
FGD4	611104	12p11.21	CMT 4H				609311	AR
FIG4	609390	6q21	CMT 4J				611228	AR
GAN	605379	16q23.2	Giant Axonal Neuropathy 1				256850	AR
GARS	600287	21q22.11		CMT 2D	HMN 5		601472/600794	AD
GDAP1	606598	8q21	CMTA RI				608340	AR
			CMT 4A				214400	AR
				CMT 2H			607706	AR
				CMT 2K			607831	AD
GJB1	304040	Xq13.1	CMT X1				302800	XL
HOXD10	142984	2q31.1	HMSN with Congenital vertical talus				192950	AD
HSPB1	602195	7q11		CMT 2F	HMN 2B		606595/608634	AD
HSPB3	604624	5q11.2			HMN 2C		613376	AD
HSPB8	608014	12q24		CMT 2L	HMN 2A		608673/158590	AD
IGHMBP2	600502	11q13.3			HMN 6		604320	AR
IKBKAP	603722	9q31.3				HSAN 3	223900	AR
KARS	601421	16q23.1	CMT RI B				613641	AR
KIF1B	605995	1p36.22		CMT 2A1			118210	AD
LITAF	603795	16p13.3-p12	CMT 1C				601098	AD
LMNA	150330	1q22		CMT 2B1			605588	AR
LRSAM1	610933	9q33.3		CMT 2P			614436	AD/AR
MED25	610197	19q33.13		CMT 2B2			605589	AR
MFN2	608507	1p35-36		CMT 2A2			609260	AD
				HMSN6			601152	
MPZ	159440	1q22	CMT 1B				118220	AD
			CHN				605253	
			CMT DI D				607791	
				CMT 2I			607677	
				CMT 2J			607736	
MTMR2	603557	11q21	CMT 4B1				601382	AR

NDRG1	605262	8q24.22	CMT 4D				601455	AR
NEFL	162280	8p21	CMT 1F	CMT 2E			607734/607684	AD
NGFB	162030	1q13.2				HSAN 5, absence of pain	608654	AR
NTRK1	191315	1q23.1				HSAN 4; anhidrosis, insensitivity to pain	256800	AR
PLEKHG5	611101	1p36.31	CMT RIC		dSMA 4		615376/611067	AR
PMP22	601097	17p11.2	CMT 1A				118220	AD
			HNPP				162500	AD
			CMT 1E				118300	AD
			DSS				145900	AR
PRPS1	311850	Xq22.3	CMT X5				311070	XL D/R
PRX	605725	19q13.1-q13.2	CMT 4F/DSS				145900	AR
RAB7A	602298	3q21.3		CMT 2B		HSN	600882	AD
REEP1	609139	2p11.2			HMN5B		614751	AD
SBF2	607697	11p15.4	CMT 4B2				604563	AR
SCN9A	603415	2q24.3	Absence of pain				243000	AR
			Small fibre neuropathy				133020	AD
SEPT9	604061	17q25.2-q25.3	Hereditary neuralgic amyotrophy, HNS, HNA & dysmorphic features				162100	AD
SH3TC2	608206	5q32	CMT 4C				601596	AR
SLC12A6	604878	15q14	PN with agenesis of the corpus callosum				218000	AR
SOX10	602229	22q13.1	PCWH syndrome				609136	AD
SPTLC1	605712	9q22.1-q22.3				HSAN 1	162400	AD
SPTLC2	605713	14q24.3				HSAN 1C	613640	AD
TDP1	607198	14q32.11	Spinocerebellar ataxia, with axonal neuropathy				607250	AR
TRPV4	605427	12q24.1		CMT 2C			606071	AD
WNK1	605232	12p13.33				HSAN 2A	201300	AR
YARS	603623	1p13.1	CMT DI C				608323	AD

A custom SureSelect (Agilent Technologies) solution-based oligonucleotide target capture assay was designed using the web-based tool eArray (version 7.7). Regions of interest (ROI) were designed to encompass coding regions of all alternate transcripts for each gene. 5' and 3' untranslated regions and non-coding exons were also included. Promoter sites were included for *GJB1* and *PMP22,* and also part of *MPZ* intron 1 [11]. Each ROI included 20 base pairs (bp) upstream and 10 bp downstream of the coding exon to capture canonical splicing donor and acceptor sites.

Library Preparation and Sequencing

Genomic DNA was extracted from whole blood using the Puregene protocol (Gentra Systems Incorporated), EZ1 DNA Blood kit (Qiagen) or a standard phenol-chloroform extraction. We also received DNA samples extracted in other laboratories. A Qubit 2.0 Fluorometer (Life Technologies) was used to quantify double stranded DNA concentration in genomic DNA samples. Sequencing libraries were prepared according to the manufacturer's standard protocol; SureSelectXT Target Enrichment System for Illumina Paired-End Sequencing Library Illumina HiSeq and MiSeq Multiplexed Sequencing Platforms Version 1.5, November 2012. Genomic DNA was sheared to a median size of 200 bp using the Bioruptor NGS sonicator (Diagenode). Fragment size was assessed using the Tapestation 2100 Bioanalyzer (Agilent Technologies). Sequencing was performed on a MiSeq instrument (Illumina) using Version 2 reagents, 2x150 paired-end reads in batches of 16 patients' samples.

Data analysis was performed using an open source in-house pipeline (alignment: BWA; alignment modification and variant calling: GATKv2; variant annotation: Annovar) with hg19 human genome as a reference, and followed the Association of Clinical Genetics Science (ACGS) Practice Guidelines [12]. Viewing of variants and recording of classification evidence was facilitated using Geneticist Assistant software (Soft Genetics). Copy number enumeration was performed for the 17p11.2 region using the CONTRA tool as a component of the analysis pipeline [13]. This was necessary to ensure that the most frequent cause of CMT1/HNPP (and therefore a positive genetic diagnosis) would not be missed if a patient had not been pre-screened for *PMP22*dosage, for reasons of clinical oversight, or an atypical clinical presentation.

The assay was validated using genomic DNA samples from nine patients previously tested in our laboratory; six of these had single nucleotide variants (SNVs) in six different genes (a total of 26 SNVs),previously identified by Sanger sequencing. A further three had the classical deletion or duplication of *PMP22,* identified previously by MLPA dosage analysis. All 26 SNV occurrences and the *PMP22* copy number variants (CNVs) were confirmed using this assay. Using 95 % confidence intervals for the binomial distribution, the sensitivity of this assay was determined to be between 87 and 100 % [14]. To date, all of the (410) variants detected by NGS and followed up by subsequent Sanger sequencing have been confirmed as true positives. Due to lack of CNV positive controls for genes other than *PMP22*, the analysis pipeline has not been validated as capable of CNV detection automatically. Visual checking of CONTRA data is undertaken when one pathogenic variant is detected in a recessive gene. Small insertions and deletions have been detected and confirmed, ranging from 2 bp to whole gene deletions; however this does not exclude the possibility that there are other CNVs present that were not detected. The report of results states clearly that the test has not excluded copy number variation in the genes examined.

Variant Filtering and Classification

Variants were managed using Genetic Assistant (SoftGenetics). Classification followed the Association of Clinical Genetics Science Practice Guidelines [15], and all variants were classified into five pathogenicity groups. Table 2 details the criteria applied to classify variants (Class 1: clearly not pathogenic; Class 2: unlikely to be pathogenic; Class 3: unknown significance; Class 4: likely to be pathogenic; Class 5; clearly pathogenic). Variants were filtered according to their frequency; assessment included comparison of frequency data from the database dbSNP (version 142) [16] and the Exome Variant Server (version 6500) [17]. All variants with frequency above 3 % were considered as clearly not pathogenic (Class 1). The remaining variants were further investigated for their clinical significance. Literature searches, the IPNMDB database [8] and our local laboratory database were interrogated. *In silico* analysis was assisted by AlamutVisual (Interactive Biosoftware), which incorporates multiple amino acid substitution and splice-prediction tools.

Table 2: Criteria applied in the classification of variants

Class	Pathogenicity	Criteria
5	Clearly pathogenic	1. Reported in the literature as pathogenic with supporting evidence; multiple independent cases, pedigree segregation studies and/or functional analysis *AND*
		2. Phenotype and inheritance pattern in patient correlates with the gene
4	Likely to be pathogenic	1. Not described in the literature, or weak evidence for pathogenicity in published literature; no segregation or functional analysis available *AND*
		2. Phenotype and inheritance pattern in patient correlates with the gene *AND*
		3. Location of variant in gene and pathogenic mechanism are compatible with previously described pathogenic variants in the gene *AND*
		4. Minor allele frequency (MAF) is <1 % in dbSNP (v142) *and* <1 % in Exome Variant Server (EVS, v6500) *AND*
		5. a) Missense variant; conserved amino acid, Polyphen 2 (HumVar) and SIFT concur in predicting deleterious effect *or* three or more of five splice prediction tools[a] return >10 % difference in splice site prediction value between wild type and variant *OR*
		b) Frameshift or nonsense variant; where expected mechanism is loss of function *OR*
		c) Synonymous or intronic change; nucleotide highly conserved across multiple species and three or more splice prediction tools return >10 % difference in splice site prediction value between wild type and variant

3	Unknown significance (VoUS)	1. Minor allele frequency <1 % in dbSNP (v142) and EVS (v6500) *AND*
		2. Phenotype and inheritance pattern in patient correlates with the gene *AND*
		3. Not described in the literature, or described in literature with inconclusive or no evidence *AND*
		a) In silico predictions score variants as Class 4 but phenotype *does not* correlate with the gene *OR*
		b) In silico predictions are conflicting (for example; conserved amino acid, very low MAF but Polyphen 2 and SIFT predict benign)
2	Unlikely to be pathogenic	1. a) Minor allele frequency is between 1 % and 3 % in dbSNP (v142) and/or between 1 % and 3 % in EVS (v6500) but phenotype and/or inheritance pattern in patient correlates with the gene *OR*
		b) Minor allele frequency is <1 % in dbSNP and EVS but phenotype and/or inheritance pattern in patient does not correlate with gene *AND*
		2. a) Missense variant; amino acid is weakly conserved across multiple species and/or Polyphen 2 (HumVar) and SIFT concur in predicting benign *OR*
		b) Synonymous or intronic change; nucleotide weakly conserved across multiple species and splice prediction tools[a] show no significant difference between wild type and variant, even if MAF is <1 % on dbSNP and EVS *AND*
		3. a) Some evidence for benign status in literature but weak or inconclusive *AND/OR* b) Some evidence that variant does not segregate with disease in pedigree of patient(s) tested in this cohort
1	Clearly not pathogenic	Frequency >3 % in dbSNP (v142) or >3 % in Exome Variant Server (v6500) *OR*
		Frequency <3 % but described and proven as not pathogenic in published literature

[a]Five splice prediction tools queried via Alamut software interface: SpliceSite Finder Like, MaxEntScan, NNSplice, GeneSplicer, Human Splicing Finder

Variants classified as pathogenic, likely pathogenic or of uncertain clinical significance were confirmed by Sanger sequencing and were detailed within the report of results. Candidate pathogenic CNVs were confirmed by MLPA analysis either using a commercially available probe mix (MRC Holland), or alternatively by designing bespoke MLPA probes to target the gene of interest, combining these with the MRC Holland P300-A2 reference probe mix. Bespoke MLPA probes were designed using the online tool MAPD [18].

For patients without candidate pathogenic variants, one unique variant was selected and confirmed by Sanger sequencing to ensure the correct identification of all samples in the batch.

RESULTS AND DISCUSSION

Analysis of the data demonstrated high read depth and target coverage. On average, 99.81 % of the targeted region was covered to a minimum of 30x reads, and 99.86 % to a minimum of 15x. The mean depth of coverage was 537x reads.

A total of 56,000 variants were detected in the 448 patients. Of those, 1830 variants had prevalence less than 3 % in dbSNP (version 142) or in Exome Variant Server (version 6500). These variants were individually assessed and classified according to the ACGS guidelines [14].

Gene Spectrum in the Diagnosis

A total of 195 variants in 31 genes provided a genetic diagnosis for 137 patients (diagnostic yield 31 %). Of these, 107 variants were previously reported in the literature as pathogenic with supporting evidence. The remaining 88 variants were novel and were classified as likely pathogenic (class 4) based on conservation, *in silico* predictions, phenotype compatibility and in several cases family studies. 215 variants were classified as of uncertain clinical significance and the remaining 1420 variants were assessed as unlikely pathogenic (class 2) or clearly not pathogenic (class 1).

Fifty patients had pathogenic variants in genes not previously available for genetic testing in a diagnostic setting in the UK, including six with variants in regions of the *DYNC1H1* gene not previously screened due to its large size; another 14 had pathogenic variants in genes where testing was previously available but did not feature in the regular IPN diagnostic strategy (Table 3).

Table 3: Diagnostic yield for each phenotypic group, mode of inheritance and gene

Phenotype	Inheritance	Gene	Patients
CMT1	**AD**	*MPZ*	5
		NEFL	3
		LITAF	3
		GDAP1	3
		MFN2	2
		AARS	**1**
		PMP22	1
	X-L	*GJB1*	3
		ATP7A	**1**
	AR	***SH3TC2***	**10**
		IGHMBP2 [a]	4
		SBF2	**2**
		FIG4	**2**
		FGD4	**1**
Tested	**101**	**Diagnosed**	**41**
CMT2	**AD**	*MFN2*	8
		BSCL2	5
		DYNC1H1 [b]	4
		GDAP1	3
		HSPB1	3
		PMP22	3
		SPTLC2	**2**
		NEFL	2
		SCN9A [a]	2
		GARS	**2**
		RAB7A	2
		TRPV4	1
		SH3TC2	**1**
		DNM2	**1**
		MPZ	1
	X-L	*GJB1*	6
	AR	*IGHMBP2* [a]	4
		GDAP1	2
		FIG4	**1**
		SLC12A6	**1**
Tested	**152**	**Diagnosed**	**54**

HMN	**AD**	*DYNC1H1* [b]	5
		GARS	**4**
		HSPB1	3
		REEP1	**2**
		MFN2	2
		SPTLC1	1
		DNM2	**1**
		PMP22	1
		DCTN1	**1**
		NEFL	1
		LMNA [a]	1
	AR	*IGHMBP2* [a]	2
tested	**91**	**Diagnosed**	**24**
HSN	**AD**	***SPTLC2***	**1**
		AARS	**1**
		PMP22	1
		RAB7A	1
	AR	***MED25***	**1**
Tested	**25**	**Diagnosed**	**5**
Mixed	**AD**	***AARS***	**5**
		MPZ	1
	X-L	*GJB1*	2
Tested	**22**	**Diagnosed**	**8**
Complex	**AD**	*DYNC1H1* [b]	1
		SEPT9	**1**
		SH3TC2	**1**
	AR	*SCN9A* [a]	1
		GAN	**1**
Tested	**57**	**Diagnosed**	**5**
Overall total	**448**	**Diagnosed**	**137**

Highlighted bold: gene not previously available for routine diagnostic testing in UK

[a]Gene not previously requested for IPN

[b]Only part of the gene was previously available for testing

Diagnostic Yield in the Different IPN Subtypes

The patients were grouped into a phenotypic subtype according to the information on the clinical proforma provided. Table 3 presents the positive diagnostic yield achieved in detail. The diagnostic yield was highest for patients with demyelinating neuropathy (41/101, 41 %). This cohort included CMT1 patients in whom a 17 p11.2 *PMP22* copy number variant had previously been excluded. The group of the patients with axonal neuropathy had a diagnostic yield of 35.5 % (54/152), similar to the group of patients with mixed neuropathy (36 %, 8/22). The diagnostic yield in patients with HMN was 26 % (24/91) and for HSN 20 % (5/25). Only 5/57 patients received a genetic diagnosis (9 %) in the group with the complex phenotypes. This group included patients with multisystem disorders, where neuropathy was not the primary cause of disease, and UKGTN criteria were not strictly met; however testing was performed as the clinical teams felt it would be beneficial.

The majority (93/137, 68 %) of the genetically diagnosed patients had an autosomal dominant form of neuropathy, while 12 patients presented with X-linked disease (9 %). It has been previously estimated that in Northern Europe and North American populations, approximately 90 % of cases of CMT are either autosomal dominant or X-linked [3, 19]. Autosomal recessive disease is estimated to account for significantly less, although in populations with a high rate of consanguineous marriages, autosomal recessive forms can account for up to 40 % [20]. We identified 32 patients with recessive aetiology, representing almost one quarter of our positive cases (23 %). The age of the patients with recessive neuropathy ranged from 3 to 68 years at the time of diagnosis. Fifteen patients were under 18 years of age (47 %) while 17 were adults (53 %). Evidently autosomal recessive peripheral neuropathy is not exclusively associated with very early onset, severe progressive disease. Two genes, *SH3TC2* and *IGHMBP2*, accounted for 62 % of all recessive diagnoses (20/32 patients).

There was no significant difference in the overall diagnostic yield achieved between patients under 18 years of age and those aged 18 and over (36/135 vs 102/313; Fisher's exact test value 0.222, $p<0.05$).

The clinical and genetic heterogeneity of IPN has always presented a challenge for the clinical classification. Specialist clinics have played a significant role in guiding the genetic testing. A positive diagnostic yield of 62.6 % in CMT patients attending specialist clinics was reported by Murphy et al. [21] and 67 % by Saporta et al. [22]. This proportion is reported to be significantly lower at 37.7 % in patients that have not been assessed in specialist clinics [21]. These figures include patients positive for the *PMP22* duplication. Our overall diagnostic yield of 31 % does not include *PMP22* duplication positive patients, as the majority of our patients are referred to us for gene panel testing following a normal result for *PMP22* copy number in their local laboratories. For our local patients, the pick-up rate was estimated to be approximately 28 % (125/443) for the*PMP22* duplication and 27 % (90/332) for the *PMP22* deletion (patients tested in a three year period 2007–2010, unpublished data). This figure is slightly higher than the 20.9 % *PMP22* CNVs identified by Murphy et al. [21] in those not attending a specialist inherited neuropathy clinic, and almost twice the level seen in the cohort of DiVincenzo et al., at 14.5 % [23].

DiVincenzo et al. have reported the positive rate of mutations in 14 genes (*PMP22, GJB1, MPZ, MFN2, SH3TC2,GDAP1, NEFL, LITAF, GARS, HSPB1, FIG4, EGR2, PRX and RAB7A*) in a very large cohort of 17,789 individuals [23]. Among these patients, 4 genes accounted for 94.9 % of positive results: *PMP22, GJB1, MFN2, MPZ*. Murphy et al. also reported a 94 % pick up rate of those four genes, including point mutations and rearrangements [21]. The equivalent positive pick up rate in our patients, excluding the rearrangements is 43.4 % (we have 36 positive patients in the four genes versus 83 positive in the 14 genes). This possibly reflects the fact that in this first year that the gene panel was available, a significant proportion of the patients referred for testing had already undergone testing for these common genes, and only those without a mutation were referred to us for further testing on the NGS panel.

Copy Number Variation

Copy number variation (CNV) is considered rare except for the common 17p11.2 *PMP22* copy number variants, and accounts for about 1 % of diagnoses [24]. We detected two patients with whole *PMP22* gene duplication, and three patients with a whole gene deletion (*PMP22, GJB1, SLC12A6*). We also detected one patient homozygous for *GAN* exon 1 deletion and one compound heterozygous for a partial deletion of the*SBF2* gene encompassing exons 14 to 27.

Our existing pipeline is set up to detect whole gene deletions and duplications; for smaller CNVs we currently manually check the data. However, it has been proven particularly useful to have this ability to check for CNVs in the cases where one pathogenic variant was detected in a gene associated with recessive inheritance.

The *PMP22* c.353C>T, p.(Thr118Met) Variant

We detected the *PMP22* c.353C>T, p.(Thr118Met) variant in five patients (patients 1–5, Table 4). This variant has been widely documented; however its pathogenicity has been controversial in the literature [25, 26, 27]. The latest evidence suggested that it is associated with neuropathy, albeit with reduced penetrance [28]. This variant is present in dbSNP (rs104894619) with a minor allele frequency (MAF) of 0.08 %, in the Exome Variant Server with a MAF of 0.53 % (European-American cohort) and in the ExAC browser with MAF 0.73 % (European non-Finnish, including one homozygote). Four out of five of our patients had another class 4 or 5 variant and only one patient had no other variants.

Table 4: Interesting patients; examples to highlight the added value of gene panel testing

Patient	Phenotype	Gene and RefSeq Transcript		Variant		Pathogenicity class	Reference
1	CMT2	*PMP22*	NM_000304.2	c.353C>T	p.(Thr118Met)	C5	[25, 26,27, 28]
		GJB1	NM_000166.5	c.-17G>A	p.?	C5	[35]
2	CMT2	*PMP22*	NM_000304.2	c.353C>T	p.(Thr118Met)	C5	[25, 26,27, 28]
		PMP22	NM_000304.2	c.281delG	p.(Gly94Alafs*17)	C5	[36, 37,38]
3	HMN	*PMP22*	NM_000304.2	c.353C>T	p.(Thr118Met)	C5	[25, 26,27, 28]
		REEP1	NM_001164730.1	c.*50G>A	p.?	C5	[39]
		DYNC1H1	NM_001376.4	c.3500 T>A	p.(Val1167Glu)	C4	This study
4	HMN	*PMP22*	NM_000304.2	c.353C>T	p.(Thr118Met)	C5	[25, 26,27, 28]
		GARS	NM_002047.2	c.485A>G	p.(His162Arg)	C4	This study
5	HSN	*PMP22*	NM_000304.2	c.353C>T	p.(Thr118Met)	C5	[25, 26,27, 28]
6	CMT2	*MFN2*	NM_014874.3	c.1403G>A	p.(Arg468His)	C5	[29, 30,31]
		MFN2	NM_014874.3	c.809 T>C	p.(Met270Thr)	C3	This study
		MFN2	NM_014874.3	c.1029_1032delGAG	p.(Arg344del)	C3	This study
7	HMN	*MFN2*	NM_014874.3	c.1403G>A	p.(Arg468His)	C5	[29, 30,31]
8	CMT1	*MFN2*	NM_014874.3	c.1403G>A	p.(Arg468His)	C5	[29, 30,31]
9	CMT1	*MFN2*	NM_014874.3	c.2119A>G	p.(Arg707Trp)	C5	[31]
10	CMT2	*PMP22*	NM_000304.2	c.448G>C	p.(Gly150Arg)	C4	This study
11	CMT2	*PMP22*	NM_000304.2	c.(?_-1)_(*1_?) del	p.0	C5	[40]
12	CMT1	*AARS*	NM_001605.2	c.986G>A	p.(Arg329His)	C5	[41]
13	HMN	*PMP22*	NM_000304.2	c.185 T>G	p.(Leu62Arg)	C4	This study
		MED25	NM_030973.3	c.1004C>T	p.(Ala335Val)	C5	[42]
14	CMT2	*PMP22*	NM_000304.2	c.(?_-1)_(*1_?) dup	p.(=) dup	C5	[43]
		SCN9A	NM_002977.3	c.3369G>T	p.(Leu1123Phe)	C4	This study
15	CMT complex	*PMP22*	NM_000304.2	c.(?_-1)_(*1_?) dup	p.(=) dup	C5	[43]
		SH3TC2	NM_024577.3	c.505 T>C	p.(Tyr169His)	C5	[4]
		MFN2	NM_014874.3	c.1936G>A	p.(Val646Ile)	C3	This study
16	CMT2	*SH3TC2*	NM_024577.3	c.505 T>C	p.(Tyr169His)	C5	[4]
		GDAP1	NM_018972.2	c.501dupA	p.(Glu168Argfs*3)	C4	This study
17	HSN	*RAB7*	NM_004637.5	c.484G>A	p.(Val162Met)	C5	[32]
		SPTLC2	NM_004863.3	c.1142 T>C	p.(Phe381Ser)	C4	This study

The diversity of the phenotypes in our patients with the c.353C>T, p.(Thr118Met) *PMP22* variant, the variant's co-existence with pathogenic mutations in other genes and its high MAF in the general population, challenge it being a causative variant, although its contribution to a phenotype cannot be excluded.

The *MFN2* c.1403G>A, p.(Arg468His) Variant

This variant was detected in three patients (patients 6–8, Table 4). The c.1403G>A, p.(Arg468His) variant is recorded on dbSNP (rs138382758) with MAF 0.20 %, on the Exome Variant Server with MAF 0.24 % (European-American cohort) and in the ExAC with MAF 0.32 % (European, non-Finnish cohort) including two homozygotes. It was originally reported by Engelfried K et al. in a patient with distal weakness and atrophy of the legs and also her symptomatic father, but as it was also detected on one allele in the population study (260 chromosomes), it was considered to be a benign polymorphism [29]. In a later study by Casasnovas et al. the variant c.1403G>A, p.(Arg468His) was identified in six of 14 unrelated Spanish families presenting mild or moderate CMT2, with dominant inheritance and onset of disease in the third to fifth decade [30]. Functional studies were conducted on fibroblasts from a skin biopsy and it was demonstrated that this variant decreased efficiency of ATP synthesis leading to decreased ATP production; this supported the pathogenicity of this variant. The authors suggested that the anonymous control subject identified by Engelfried et al. could be a CMT2 patient who had not yet reached the age of onset of symptoms. Braathen et al. described a patient with CMT1 (reduced NCVs) presenting at age 2, who had the *MFN2* c.1403G>A variant, suggesting that it may also associate with demyelinating CMT [31]. The presence of this variant was consistent with axonal phenotype in two of our patients while the third presented with early onset CMT1, matching the patient described by Braathen et al.

Broadening the Phenotypic Spectrum Associated with Specific Genes

A total of six patients were found to have pathogenic variants in genes that would not have been traditionally tested for their phenotype (patients 8–14, Table 4). Two patients referred as having CMT1 were found to have pathogenic variants in *MFN2*. Two patients referred with axonal neuropathy were found to have a *PMP22*variant; one of them the classical *PMP22* deletion and the other the *PMP22* duplication. A patient referred with HMN was also found to have a likely pathogenic variant in *PMP22*, and was carrier of a recessive pathogenic variant in *MED25*. A patient with CMT1 was found to have the recurrent pathogenic variant in *AARS*. Our IPN multigene testing is an unbiased approach; these patients would have not received a diagnosis based on the phenotypically - led genetic testing, either by the traditional sequential testing or by CMT phenotype-specific panels. A few of these patients might have been misclassified in the local clinic, without the benefit of specialist expertise at a regional or national centre; however the NGS panel approach for genetic diagnosis can help to compensate for limited access to specialist diagnostic expertise.

Identification of Multiple Genetic Causes

Four adult patients were found to have potentially more than one causative variant in different genes (patients 15–17, Table 4). One patient with atypical CMT and known to have the *PMP22* duplication was found to have the known pathogenic variant in *SH3TC2*, c.505 T>C p.(Tyr169His). This patient had also one variant of unknown clinical significance in *MFN2*, c.1936G>A, p.(Val646Ile). Another adult patient with axonal neuropathy was found to have the *SH3TC2* c.505 T>C, p.(Tyr169His) variant alongside a likely pathogenic (predicted truncating) variant in *GDAP1*, c.501dupA, p.(Glu168Argfs*3). While homozygous or compound heterozygous pathogenic variants in *SH3TC2* are associated with autosomal recessive CMT4C, there is growing evidence that heterozygosity for some *SH3TC2* variants can cause axonal neuropathy. Lupski et al. identified the p.(Tyr169His) variant in the heterozygous state in a parent and a grandparent of the proband [4]. The p.(Tyr169His) heterozygotes did not have CMT1, but were found to have patchy axonal polyneuropathy with definite median-nerve mononeuropathy at the wrist on neurophysiology. A patient referred with CMT2 was found to have the *PMP22* duplication and a *SCN9A* likely pathogenic variant c.3369G>T, p.(Leu1123Phe). Another patient presenting with HSN was found to have the known *RAB7A* pathogenic variant c.484G>A, p.(Val162Met) [32], and a likely pathogenic variant in *SPTLC2*, c.1142 T>C, p.(Phe381Ser). These results support the effectiveness of multi-gene testing, since traditional or phenotype-led testing would not have picked up these variants, and have implications for recurrence risk and genetic counselling.

Focused genetic testing has been supported as an approach to provide diagnosis either in the form of small panels or as a tiered approach, by exclusion of variants in common genes before proceeding to NGS testing. The review and recommendations set out by Murphy et al. and Saporta et al. provided essential guidance for clinicians navigating a sea of individual genes in pursuit of a genetic diagnosis [21, 22]. Elsewhere the approach of CMT phenotype-specific panel testing has been proposed for CMT diagnosis as a means to avoid high cost and difficulty in result interpretation [33]. In our experience NGS is efficient and removes the need for serial gene sequencing in most cases. Testing specific genes according to phenotype association may be beneficial if the patients have a very definite neuropathy subtype *and* are referred from expert neurology clinics; for example *GJB1* in a clearly X-linked pedigree, or *PMP22* in HNPP. The cost of design and validation for one large panel is significantly lower than the sum costs of multiple smaller panels. The challenge of variant interpretation should not be underestimated; however prioritization of variants according to phenotypic compatibility, and the use of *in silico* tools and public databases has proven to be an efficient approach. By incorporating all 56 genes in one assay, we revealed mutations in genes that would have not emerged had we been limited to a phenotype-derived subset of genes. Other authors have also recently been advocating the benefits of expanded multi-gene testing. Hoyer et al. have reported findings similar to ours, including detection of dual pathology and CMT1 patients with *MFN2* mutations [34].

CONCLUSIONS

We developed a 56-gene IPN NGS targeted panel assay as a specialist UK Genetic Testing Network service. This is a frontline diagnostic tool and has largely replaced single-gene Sanger sequencing. Testing was completed for 448 patients in the first 18 months post launch. Genetic diagnosis was achieved in 137 patients (31 %). Testing revealed high heterogeneity, dual pathology and less-tight phenotype-genotype associations.

Assessment and classification of variants is currently a time consuming process, however this task becomes easier as the tools improve and the databases expand. Detailed clinical information is advantageous for variant interpretation, nonetheless we have highlighted cases that would not have achieved a diagnosis had the phenotype been used to guide gene selection.

Clinical-exome or whole-exome sequencing may be appropriate for patients in whom no pathogenic variant is detected on the NGS panel. However, the diagnostic yield achieved at this stage does not support sequencing the exome immediately, despite comparable cost, due to the complexity of analysis of larger data sets. This targeted panel approach has the advantage of producing smaller data sets than exome or genome sequencing, but simultaneously it overcomes the limitation imposed by basing testing decisions on the limited genotype-phenotype data available for the rare genes. It facilitates the testing of category-equivocal cases where the patients do not fit in a particular phenotypic subgroup. It is an efficient approach from the perspective of an accredited diagnostic laboratory, as only one assay needs to be validated. Redesign of the panel is relatively straightforward, allowing inclusion of genes newly identified in IPN pedigrees. Genes with very recently established clinical associations tend not to feature on commercial clinical exome panels which are updated infrequently.

The clinical and genetic heterogeneity of IPN makes both diagnosis and genetic counselling quite challenging [21]. The benefits of obtaining a genetic diagnosis include provision of a definite clinical classification (including clarification of equivocal cases) and guidance on prognosis. Furthermore, accurate genetic risk assessment and cascade testing is beneficial not only for the patient but also for their family. As clinical trials progress for some types of CMT, genotyping will be essential. Advances in sequencing technology have made it time and cost effective to screen large numbers of causative genes simultaneously.

ACKNOWLEDGEMENTS

The authors would like to thank the North Bristol NHS Trust Showering Fund for the grant awarded to TA which made this study possible. The authors would like to thank all the referring clinicians and their patients who have supported this study.

AUTHORS' CONTRIBUTIONS

TA led the project and prepared the manuscript; CB designed the assay and the bioinformatics pipeline; GD performed the technical validation and supervised the technical work; NF and DS carried out analysis, interpretation and reporting; PWL provided the clinical support during the pilot study and reviewed the manuscript; SBJ performed the scientific validation, carried out analysis, interpretation and reporting. All authors read and approved the final manuscript.

REFERENCES

1. Barisic N, Claeys KG, Sirotkovic-Skerlev M, Löfgren A, Nelis E, De Jonghe P, et al. Charcot-Marie-Tooth disease: a clinico-genetic confrontation. Ann Hum Genet. 2008;72(Pt 3):416–41.
2. Reilly MM. Sorting out the inherited neuropathies. PractNeurol. 2007;7(2):93–105.
3. Saporta MA, Shy ME. Inherited Peripheral Neuropathies. NeurolClin. 2013;31(2):597–619.
4. Lupski JR, Reid JG, Gonzaga-Jauregui C, Rio Deiros D, Chen DC, Nazareth L, et al. Whole-genome sequencing in a patient with Charcot-Marie-Tooth neuropathy. N Engl J Med. 2010;362(13):1181–91.
5. Weedon MN, Caswell R, Hastings R, Xie W, Paszkiewicz K, Antoniadi T, et al. Exome sequencing identifies a *DYNC1H1*mutation in a large pedigree with dominant Charcot-Marie-Tooth disease. Am J Hum Genet. 2011;89:1–5.
6. Byung-Ok Choi BO, Kyung Koo S, Park MH, Rhee H, Yang SJ, Choi KJ, et al. Exome sequencing is an efficient tool for genetic screening of Charcot–Marie–tooth disease. Hum Mutat. 2012;11:1610–5.
7. UK Genetic Testing Network. http://ukgtn.nhs.uk. Accessed 02 Apr 2015.
8. Inherited Peripheral Neuropathies Database. http://www.molgen.ua.ac.be/CMTmutations/. Accessed 02 Apr 2015.
9. Hamosh A, Scott AF, Amberger JS, Bocchini CA, McKusick V. Online Mendelian Inheritance in Man (OMIM), a knowledgebase of human genes and genetic disorders. Nucleic Acids Res. 2005;33:D514–7.
10. Zhang F, Seeman P, Liu P, Weterman MAJ, Gonzaga-Jauregui C, Towne CF, et al. Mechanisms for nonrecurrent genomic rearrangements associated with CMT1A or HNPP: rare CNVs as a cause for missing heritability. Am J Hum Genet. 2010;86:892–903.
11. Antonellis A, Dennis MY, Burzynski G, Huynh J, Maduro V, Hodonsky CJ, et al. A Rare Myelin Protein Zero (MPZ) Variant Alters Enhancer Activity In Vitro and In Vivo. PLoS One. 2011;5(12):e14346.
12. Association of Clinical Genetics Science: Practice Guidelines for Targeted Next Generation Sequencing Analysis and Interpretation.2014;http://www.acgs.uk.com/media/774807/bpg_for_targeted_next_generation_sequencing_may_2014_final.pdf. Accessed 2 Apr 2015.
13. Li J, Lupat R, Amarasinghe KC, Thompson ER, Doyle MA, Ryland G. et al. CONTRA: copy number analysis for targeted resequencing. Bioinformatics. 2012;28(10):1307–13.
14. Online tool for Diagnostic Test Evaluation. http://www.medcalc.net. Accessed 02 Apr 2015
15. Association of Clinical Genetics Science: Practice Guidelines for the Evaluation of Pathogenicity and the Reporting of Sequence Variants in Clinical Molecular Genetics. 2013.http://www.acgs.uk.com/media/774853/evaluation_and_reporting_of_sequence_variants_bpgs_june_2013_-_finalpdf.pdf. Accessed 2 Apr 2015.
16. Database of Single Nucleotide Polymorphisms (dbSNP). Bethesda (MD): National Center for Biotechnology Information, National Library of Medicine. http://www.ncbi.nlm.nih.gov/SNP/. Accessed 02 Apr 2015.
17. Exome Variant Server, NHLBI Exome Sequencing Project (ESP), Seattle, WA http://evs.gs.washington.edu/EVS/. Accessed 02 Apr 2015.
18. Zhi J. MAPD: a probe design suite for multiplex ligation-dependent probe amplification assays. BMC Res Notes. 2010;3:137.
19. Szigeti K, Lupski JR. Charcot-Marie-Tooth disease. Eur J Hum Genet. 2009;17(6):703–10.
20. Dubourg O, Azzedine H, Verny C, Durosier G, Birouk N, Gouider R, et al. Autosomal-recessive forms of demyelinating Charcot-Marie-Tooth disease. Neuromolecular Med. 2006;8(1-2):75–86.
21. Murphy SM, Laura M, Fawcett K, Pandraud A, Liu YT, Davidson GL, et al. J Charcot-Marie-Tooth disease: frequency of genetic subtypes and guidelines for genetic testing. NeurolNeurosurg Psychiatry. 2012;83:706–10.

22. Saporta ASD, Sottile SL, Miller LJ, Feely SME, Siskind CE, Shy ME. Charcot Marie Tooth (CMT) subtypes and genetic testing strategies. Ann Neurol. 2011;69(1):22–33.

23. DiVincenzo C, Elzinga CD, Medeiros AC, Karbassi I, Jones JR, Evans MC, et al. Mol Genet Genomic Med. 2014;2(6):522–9.

24. HoyerH, Braathern GJ, EekAK, Nordang GBN, Skjelbred CF, Russell MB. Copy Number Variation in a Population-Based Study of Charcot-Marie-Tooth Disease. Biomed Res Int. 2015; doi:10.1155/2015/960404.

25. Nelis E, Holmberg B, Adolfsson R, Holmgren G, Van Broeckhoven C. PMP22 Thr118Met: recessive CMT1 mutation or polymorphism? Nat Genet. 1997;15:13–4.

26. Young P, Stögbauer F, Eller B, de Jonghe P, Lofgren A, Timmerman V, et al. PMP22 Thr118Met is not a clinically relevant CMT1 marker. J Neurol. 2000;247:696–700.

27. Shy ME, Scavina MT, Clark A, Krajewski KM, Li J, Kamholz J, et al. T118M PMP22 mutation causes partial loss of function and HNPP-like neuropathy. Ann Neurol. 2006;59:358–64.

28. Russo M, Laura M, Polke JM, Davis MB, Blake J, Bradner S, et al. Variable phenotypes are associated with PMP22 missense mutations. NeuromusculDisord. 2011;21(2):106–14.

29. Engelfried K, Vorgerd M, Hagedorn M, Haas G, Gilles J, Epplen JT, et al. Charcot-Marie-Tooth neuropathy type 2A: novel mutations in the mitofusin 2 gene (MFN2). BMC Med Genet. 2006;7:53.

30. Casasnovas C, Banchs I, Cassereau J, Gueguen N, Chevrollier A, Martinez-Matos JA, et al. Phenotypic spectrum of MFN2 mutations in the Spanish population. J Med Genet. 2009;47:249–56.

31. Braathen G, Sand JC, Lobato A, Hoyer H, Russell MB. *MFN2* point mutations occur in 3.4% of Charcot-Marie-Tooth families. An investigation of 232 Norwegian CMT families. BMC Med Genet. 2010;11:48.

32. Verhoeven K, De Jonghe P, Coen K, Verpoorten N, Auer-Grumbach M, Kwon JM, et al. Mutations in the small GTP-ase late endosomal protein RAB7 cause Charcot-Marie-tooth type 2B neuropathy. Am J Hum Genet. 2003;72:722–7.

33. Verma A. Next-generation sequencing and genetic diagnosis of Charcot-Marie-Tooth disease. Ann Indian AcadNeurol. 2014;17(4):383–6.

34. Hoyer H, Braathern GJ, Busk OI, Oystein HL, Svedsen M, Hilmarsen HT et al. Genetic diagnosis of Charcot-Marie-Tooth disease in a population by next-generation sequencing. Biomed Res Int. 2014; doi:10.1155/2014/210401.

35. Murphy SM, Polke J, Manji H, Blake J, LillaReiniger L, Sweeney M, et al. A novel mutation in the nerve-specific 5_UTRof the*GJB1* gene causes X-linked Charcot-Marie-Tooth disease. J Peripher Nerv Syst. 2011;16:65–70.

36. Lonasescu VV, Searby C, Ionanescu R, Reisin R, Ruggieri V, Arberas C. Severe Charcot-Marie-Tooth neuropathy type 1A with 1-base pair deletion and frameshift mutation in the peripheral myelin protein 22 gene. MuscleNerve. 1997;20:1308–10.

37. Boerkoel CF, Takashima H, Garcia CA, Olney RK, Johnson J, Berry K, et al. Charcot-Marie-Tooth disease and related neuropathies: mutation distribution and genotype-phenotype correlation. Ann Neurol. 2002;51:190–201.

38. Niedrist D, Joncourt F, Matyas G, Muller A. Severe phenotype with cis-acting heterozygous *PMP22* mutations. Clin Genet. 2009;75:286–9.

39. Zuchner S, Wang G, Tran-Viet KN, Nance MA, Gaskell PC, Vance JM, et al. Mutations in the Novel Mitochondrial Protein REEP1 Cause Hereditary Spastic Paraplegia Type 31. Am J Hum Genet. 2006;79:365–9.

40. Chance PF, Alderson MK, Leppig KA, Lensch MW, Matsunami N, Smith B, et al. DNA deletion associated with hereditary neuropathy with liability to pressure palsies. Cell. 1993;72:143–51.

41. Latour P, Thauvin-Robinet C, Baudelet-Mery C, Soichot P, Cusin V, Faivre L, et al. A major determinant for binding and aminoacylation of tRNA-Ala in cytoplasmic alanyl-tRNA synthetase is mutated in dominant axonal Charcot-Marie-Tooth Disease. Am J Hum Genet. 2010;86:77–82.

42. Leal A, Huehne K, Bauer F, Sticht H, Berger P, Suter U, et al. Identification of the variant Ala335Val of *MED25* as responsible for CMT2B2: molecular data, functional studies of the SH3 recognition motif and correlation between wild-type *MED25* and PMP22 RNA levels in CMT1A animal models. Neurogenetics. 2009;10:275–87.

43. Lupski JR, Montes de Oca-Luna R, Slaugenhaupt S, Pentao L, Guzzetta V, Trask BJ, et al. DNA duplication associated with Charcot-Marie-Tooth disease type 1A. Cell. 1991;66:219–32.

CHAPTER 3

ABCA7 P.G215S AS POTENTIAL PROTECTIVE FACTOR FOR ALZHEIMER'S DISEASE

Celeste Sassi, Michael A. Nalls, Perry G. Ridge, Jesse R. Gibbs, Jinhui Ding, Michelle K. Lupton, Claire Troakes, Katie Lunnon, Safa Al-Sarraj, Kristelle S. Brown[i], Christopher Medway, Naomi Clement, Jenny Lord, James Turton, Jose Bras, Maria R. Almeida, Henne Holstege, Eva Louwersheimer, Wiesje M. van der Flier, Philip Scheltens, John C. Van Swieten, Isabel Santana, Catarina Oliveira, Kevin Morgan, John F. Powell, John S. Kauwe, Carlos Cruchaga, Alison M. Goate, Andrew B. Singleton, Rita Guerreiro, John Hardy

Keywords: Alzheimer's disease (AD); Genome-wide association studies (GWASs); *ABCA7*; Whole exome sequencing (WES); Whole genome sequencing (WGS); Protective variant

INTRODUCTION

Alzheimer's disease (AD) is the most common cause of progressive dementia in the elderly. Aging and genetic factors play a critical role for the disease development. Rare coding and fully penetrant mutations in *APP*, *PSEN1*, and *PSEN2* explain part of the AD autosomal-dominant cases. On the other hand, *APOE* ε4 allele and rare coding variants in *TREM2* represent the main risk factors for late-onset and apparently sporadic AD (Chartier-Harlin et al., 1994 and Guerreiro et al., 2013). In the last 5 years, genome-wide association studies (GWASs) identified over 20 main risk loci influencing AD susceptibility (Harold et al., 2009, Hollingworth et al., 2011, Lambert et al., 2009, Lambert et al., 2013,Naj et al., 2011 and Seshadri et al., 2010). Among these, 9 have been replicated by at least 2 independent GWASs and present the strongest effect sizes after *APOE* (*BIN1*,*CLU*, *CR1*, *PICALM*, *MS4A6A*, *ABCA7*, *EPHA1*, *CD33*, and *CD2AP*).

GWASs have been a successful strategy to identify loci associated to a common trait, shedding light on disease pathways and for AD these include the following:

- immune response (*CR1*, *MSA4A/MSA7A*, *CD2AP*, *CD33*, *EPHA1*, and *ABCA7*);
- vesicles trafficking (*PICALM* and *BIN1*);
- lipid metabolism (*CLU* and *ABCA7*); and
- amyloid beta peripheral clearance (*PICALM*, *BIN1*, *CD33*, and *ABCA7*; http://www.alzgene.org/).

Nevertheless, the functional variant(s) within these risk loci have not yet been fully defined.

GWAS arrays tag common, low penetrant, and generally noncoding variants that likely exert a subtle regulatory effect (0.8 < odds ratio [OR] < 1.5) on a trait, affecting gene expression, CpG islands methylation and splicing, in *cis* or *trans* (Ramasamy et al., 2014 and Visscher et al., 2012). Whereas, low frequency (1% < minor allele frequency [MAF] < 5%) and rare variants (MAF < 1%) with a modest penetrance remain mostly undetected either because they are not in the array or because, even with the implementation of imputation, the detection of variants with MAF < 2% is not sufficiently accurate. As an illustrative example, *APOE* GWAS hit maps to an intronic region and it is likely driven by the *APOE* ε4 allele, which is a common coding haplotype (rs429358, p.C130R and rs7412, p.R176R, MAF = 15%) in exon 4, that is not tagged by the custom genotyping arrays mostly used.

Recently, resequencing studies have been powerful strategies to bridge the gap between susceptibility loci identified and actual disease-modifying variant(s) (Beaudoin et al., 2013, Lohmueller et al., 2013, Rivas et al., 2011 and Service et al., 2014).

Therefore, we have used exome and genome sequencing data (1) to identify rare and low-frequency coding variants in

Citation: Celeste Sassi, Michael A. Nalls, Perry G. Ridge, Jesse R. Gibbs, Jinhui Ding, Michelle K. Lupton, Claire Troakes, Katie Lunnon, Safa Al-Sarraj, Kristelle S. Brown, Christopher Medway, Naomi Clement, Jenny Lord, James Turton, Jose Bras, Maria R. Almeida, Peter Passmore, David Craig, Janet Johnston, Bernadette McGuinness, Stephen Todd, Reinhard Heun, Heike Kölsch, Patrick G. Kehoe, Emma R.L.C. Vardy, Nigel M. Hooper, David M. Mann, Stuart Pickering-Brown, Kristelle Brown, James Lowe, Kevin Morgan, A. David Smith, Gordon Wilcock, Donald Warden, Clive Holmes, Henne Holstege, Eva Louwersheimer, Wiesje M. van der Flier, Philip Scheltens, John C. Van Swieten, Isabel Santana, Catarina Oliveira, Kevin Morgan, John F. Powell, John S. Kauwe, Carlos Cruchaga, Alison M. Goate, Andrew B. Singleton, Rita Guerreiro, John Hardy, ABCA7 p.G215S as potential protective factor for Alzheimer's disease, Neurobiology of Aging, Available online 20 April 2016, ISSN 0197-4580, http://dx.doi.org/10.1016/j.neurobiolaging.2016.04.004.

BIN1, *CLU*, *CR1*, *PICALM*, *MS4A6A*, *ABCA7*, *EPHA1*, *CD33*, and *CD2AP* and (2) to investigate their single independent and combined effect on AD susceptibility. Both the single-variant and the gene-based association tests confirmed*ABCA7* as susceptibility locus associated with AD. Importantly, although *ABCA7* loss of function (LoF) mutations (indels, nonsense, and splice-site mutations) have been recently reported as main mechanism increasing AD risk at this locus (Steinberg et al., 2015), our results are not driven by such variants. Whereas, we report an enrichment for*ABCA7* common and low-frequency coding variants with a potential protective effect, that is mainly responsible for our gene-based signal. Among these, *ABCA7* p.G215S is the main low-frequency missense hit in the single-variant analysis in the discovery cohort. The potential protective role of this variant has been further confirmed in an independent European and North-American cohort. Our results show that *ABCA7* p.G215S exerts a mild but statistically significant influence, lowering the risk for AD. Thus, confirming*ABCA7* to be a good potential target to address functional studies.

MATERIALS AND METHODS

The discovery cohort was composed of 332 apparently sporadic AD cases and 676 elderly controls, neuropathologically and clinically confirmed, originating from the UK and North America. The mean age at disease onset was 71.66 years (range 41–94 years) for cases and the mean age of ascertainment was 78.15 years (range 60–102 years) for controls (Table 1). Most of the AD cases (77%) were late onset (>65 years at onset) (LOAD).

Table 1. Description of the different cohorts used in this study

Cohorts	N	Type	Sequencing strategy	Origin	Age (y), Mean ± SD (range)	Male (%)	*APOE*, E4+ (%)
Discovery set							
NIH-UCL							
Cases	127	Neuropath	Exome sequencing	Caucasian (British)	65.5 (41–94)	46.4	58
Controls	204	Neuropath	Exome sequencing	Caucasian (British, North American)	79.8 (61–102)	58.3	45
WashU							
Cases	23	Clinical	Exome sequencing	Caucasian (North American)	57 (46–75)	52.17	NA
Controls	16	Clinical	Exome sequencing	Caucasian (North American)	79.5 (75–92)	43.7	NA
ADNI							
Cases	182	Clinical	Exome sequencing	Caucasian (North American)	74.65 (55–90)	67	56.6
Controls	257	Clinical	Exome sequencing	Caucasian (North American)	74.68 (60–90)	50.1	27.6
BYU							
Controls	199	Clinical	Genome sequencing	Caucasian (North American)	80.8 (75–94.5)	37.7	100
Follow-up genotyping set *ABCA7* rs72973581							
NIH-NIA							
Cases	307	Clinical	Sanger sequencing	Caucasian (North American, British, Dutch, Italian, Portuguese)	Average >65 y		
Controls	501	Clinical	Sanger sequencing	Caucasian, (North American, British, Greek, German, Polish, Australian, Canadian)	>60 y		

Key: *ABCA7*, ATP-binding cassette subfamily A member 7; N, number; NA, not applicable; UCL, University College London; WashU, Washington University; ADNI, Alzheimer's Disease Neuroimaging Initiative; BYU, Brigham Young University; NIA, National Institute on Aging; NIH, National Institutes of Health; SD, standard deviation.

Among the cases and controls, 42% and 51% were female, respectively; 58% and 47% of the cases and controls carried the *APOE* ε4 allele, respectively. The *APOE* ε4 allele was significantly associated to the disease status in the National Institutes of Health (NIH) and Alzheimer›s Disease Neuroimaging Initiative (ADNI) series (p-value = 0.02 and 1.19×10^{-9}, respectively). The threshold call rate for inclusion of the subject in analysis was 95%. On this cohort, we performed (1) gene-based analysis (SNP-set Sequence Kernel Association Test [SKAT] and c-alpha tests) and (2) single-variant association analysis, targeting 23.5 kilobase pairs (Kbs) of coding sequence. Finally, we followed up, in an independent Caucasian data set, *ABCA7* p. G215S, the only nominal significant low-frequency missense variant in the single-marker association test in our discovery set (Fig. 1).

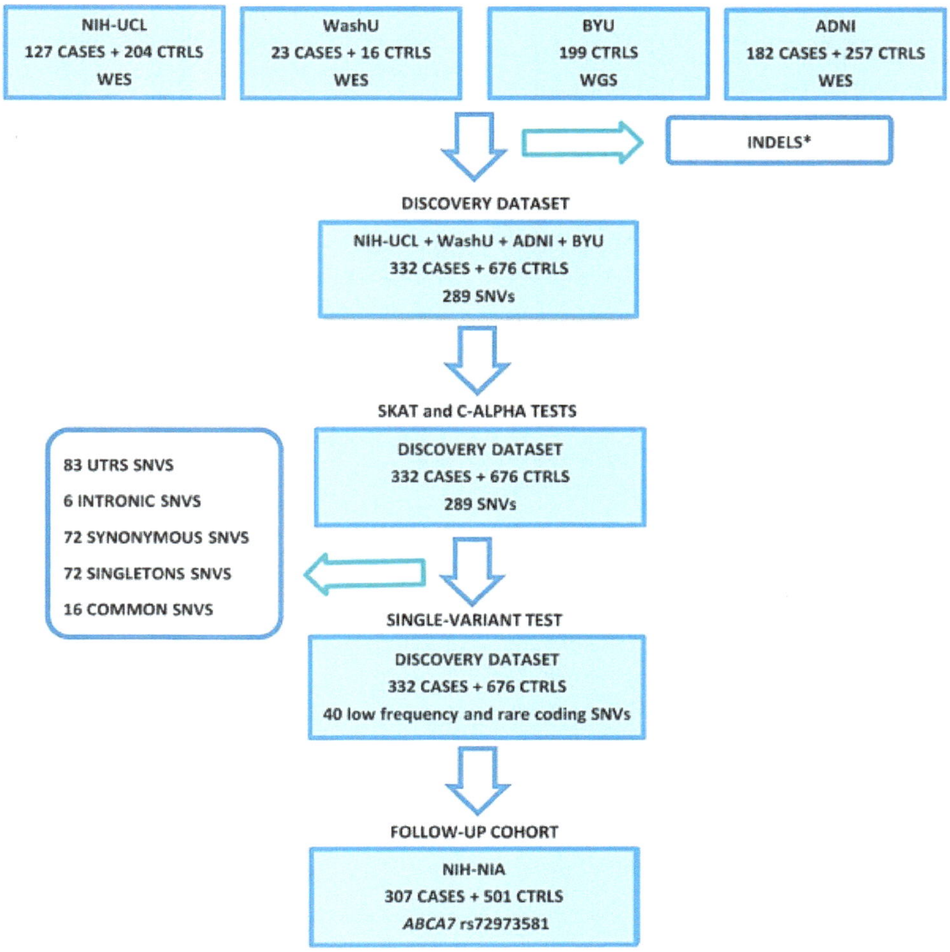

Figure 1. Pipeline of our study design. *INDELS have been excluded from the discovery cohort because not targeted in the ADNI data set. Abbreviations: ADNI, Alzheimer's Disease Neuroimaging Initiative; CTRLS, controls; INDELS, in-frame insertions and deletions; SNVs, single-nucleotide variants; UTRs, untranslated regions; WES, whole exome sequencing; WGS, whole genome sequencing.

The follow-up data set was composed of 307 late-onset apparently sporadic AD cases from North America and Europe and 501 elderly Caucasian controls from North America (Coriell repositories), Europe, Australia, and Canada (Table 1). Written informed consent was obtained for each clinically assessed individual, and the study was approved by the appropriate institutional review boards. All samples had fully informed consent for retrieval and were authorized for ethically approved scientific investigation (UCLH Research Ethics Committee number 10/H0716/3, BYU IRB, Cardiff REC for Wales 08/MRE09/38+5, REC Reference 04/Q2404/130, National Research Ethics Service).

Exome Sequencing

DNA was extracted from blood or brain for cases and brain only for controls using standard protocols. Library preparation for next generation sequencing used DNA (between 1 μg and 3 μg) fragmented in a Covaris E210 (Covaris Inc). DNA was end-repaired by 5›phosphorylation, using the Klenow polymerase. A polyadenine tail was added to the 3›end of the phosphorylated fragment and ligated to Illumina adapters. After purification using an AMPure DNA Purification kit (Beckman Coulter, Inc), adapter-ligated products were amplified. The DNA library was then hybridized to an exome capture library (NimbleGen SeqCap EZ Exome v2.0, Roche Nimblegen Inc or TruSeq, Illumina Inc) and precipitated using streptavidin-coated magnetic beads (Dynal Magnetic Beads, Invitrogen). These exome libraries were polymerase chain reaction amplified and then DNA hybridized to paired-end flow cells using a cBot (Illumina, Inc) cluster generation system. Samples were sequenced on the Illumina HiSeq 2000 using 2 × 100 paired-end reads cycles.

Whole Genome Sequencing

Genome sequencing was performed in 199 controls, from the Cache County Study on Memory in Aging. All samples were sequenced with the use of Illumina HiSeq technology.

Sanger Sequencing

ABCA7 p.G215S (rs72973581) was screened in an additional follow-up cohort composed of 307 late-onset AD cases and 501 elderly controls. Primers were designed in Primer3 (http://bioinfo.ut.ee/primer3-0.4.0/) using the University California Santa Cruz (UCSC) (http://genome.ucsc.edu/) reference sequences NM_019112 (*ABCA7*).

Purified sequences were analyzed on an ABI 3730 DNA Analyzer (Applied Biosystems, CA, USA) and chromatograms were visualized in Sequencher software (version 4.2 Gene Codes Corporation, MI, USA).

Bioinformatics

Sequence alignment and variant calling were performed against the reference human genome (UCSC hg19). Alignment was performed with the use of CASAVA software, and variant calling was performed with the use of SAMtools (Li et al., 2009) and the Genome Analysis Toolkit (GATK) (McKenna et al., 2010). Paired-end sequence reads (2 × 100 bp paired-end read cycles) were aligned using the Burrows-Wheeler aligner (Li and Durbin, 2009). Format conversion and indexing were performed with Picard (www.picard.sourceforge.net/index.shtml). GATK was used to recalibrate base quality scores, perform local realignments around indels, and to call and filter the variants (McKenna et al., 2010). VCFtools was used to annotate gene information for the remaining novel variants. We used ANNOVAR software to annotate the variants (Wang et al., 2010). Variants were checked against established databases (1000 Genomes Project and dbSNP v.134). Calling algorithms, pipelines, and reference panels were the same as the pooled data sets. The protein coding effects of variants were predicted using SIFT, Polyphen2, and SeattleSeq Annotation (gvs.gs.washington.edu/SeattleSeqAnnotation). All variants within the coding regions of the 9 risk loci (*ABCA7* [NM_019112]; *CD2AP* [NM_012120]; *MS4A6A* [NM_152851]; *CR1* [NM_000573]; *BIN1* [NM_139343]; *PICALM* [NM_001206946]; *EPHA1* [NM_005232]; *CLU* [NM_001831]; and *CD33* [NM_001772]) have been collected and analyzed. Indels were excluded from the merged data set because they were not targeted in the ADNI subcohort (Fig. 1).

Statistical Analysis

In the single-variant analysis, allele frequencies were calculated for each low frequency and rare coding variant in cases and controls, and Fisher's exact test on allelic association was performed. To study the joint effect of the variants detected, we performed a gene-based analysis with SKAT and c-alpha test, and we analyzed together for each gene the whole spectrum of allelic variability (common, low frequency, rare, coding, and noncoding).

C-alpha test and SKAT are closely related, being both nonburden tests, analyzing and collapsing the effect of genetic variants of different frequency (common and rare), effect (protective, damaging, and neutral), and effect size (modest, moderate, and strong). SKAT can be considered an expansion of the c-alpha test because it overcomes some of its limits. Indeed, SKAT (1) can be applied also to the study of continuous traits and (2) does not need any permutation.

Low frequency and rare variants were defined as having a 1% < MAF < 5% and MAF <1%, respectively, either in cases or controls.

All computations, c-alpha, and SKAT tests were performed in R (version ×64 3.0.2, http://www.r-project.org/) and PLINK/SEQ.

A *p*-value of 0.05 was set as a nominal significance threshold. Based on multiple testing correction, the thresholds for single-variant and gene-based association tests are defined by *p*-value = 1.25×10^{-3} (0.05/40 [total number of coding low frequency and rare variants detected in our study]) and 5.5×10^{-3} (0.05/9 genes), respectively. Furthermore, we excluded singletons from the single-variant analysis because a variant observed only once is not largely informative about the overall distribution (Neale et al., 2011). However, we pooled the singletons together and analyzed their collective effect in the gene-based analysis (SKAT and c-alpha test).

RESULTS

The discovery set consisted of a total of 332 sporadic and mainly late-onset AD cases and 676 elderly controls of British and North-American ancestry (Table 1).

A total of 289 single-nucleotide variants were identified. Among these, 128 (44.3%) were nonsynonymous, 72 (24.9%) synonymous, 83 (28.7%) were untranslated region (UTR) and 6 (2%) intronic variants. Among the missense variants, 99 (77.34%) were rare (MAF < 1%) and 72 of these (72.72%) were singletons (a variant observed only once either in cases or controls). Fifteen nonsynonymous variants (11.7%) were low frequency (1% < MAF < 5%) and 16 (12.5%) were common (MAF ≥ 5%). In addition, we report 14 novel coding variants (not reported in ExAC, released 13 January 2015, or dbSNP 137). None of the detected low frequency and rare coding variants clusters within common haplotypes (MAF ≥ 5%) and, therefore, could have been missed by GWASs and chip based fine-mapping approaches. We report the presence of 2 or 3 low frequency and/or rare variants in the studied genes in the same individual, both for cases and controls.

Overall, the total variant frequency of the 9 GWAS loci in our study was in line with the one reported for the European-American cohort in the Exome Variant Server (http://evs.gs.washington.edu/EVS/). The only exception was represented by *CR1*, that showed a 2.7-fold higher relative frequency of total variants, compared to the Exome Variant Server database.

PICALM harbors the lowest burden of low frequency and rare coding variants (3.27 coding variants per kb of coding sequence). By contrast, *CD33*, presents the highest relative frequency of coding variants and the lowest relative frequency of damaging variants (9.14 and 0.91 coding and damaging variants per kb of coding sequence, respectively), suggesting that most coding variability in *CD33* is likely nonfunctional.

BIN and *ABCA7* display the highest relative proportion of damaging variants (3.92 [87.3% of its coding variability] and 3.72 [60% of its coding variability] damaging coding variants per kb of coding sequence, respectively), thus arguing for a potential functional impact of missense mutations at these loci. Moreover, *ABCA7*was the only gene harboring nonsense mutations.

Most of the low frequency and rare coding variability identified within these loci exerts generally a relatively modest effect (mean OR = 1.1) that is comparable to those observed for common and generally noncoding variants identified by GWASs.

Single Coding Variant Association Test

The main hits of the single variants association test map mainly to *ABCA7* (Table 2). Particularly, we report *ABCA7* p.G215S (rs72973581), that was the only low-frequency (MAF = 4.3%) missense variant showing a trend toward significance in the single-marker association test (*p*-value = 0.02 and corrected *p*-value = 0.8) in the discovery set and was statistically significant after Bonferroni correction (*p*-value = 6×10^{-4} and corrected *p*-value = 0.024) in the combined data sets (discovery set and follow-up data set).

Table 2. Most significant variants detected in our discovery set

Gene	Position	MA	cDNA change	Aa change	Rs	MAF cases-controls (%)	MAF ExAC (%)	SIFT	Polyphen	Mutation assessor	aa/Aa/AA cases	aa/Aa/AA controls	*p*-value	Corr.*p*-value	OR (95% CI)
ABCA7	19:1043103	A	c.G643A	p.G215S	rs72973581	4.66–7.24[a]	4.31[b]	Tolerated	Benign	Low	0/31/301	1/96/579	0.02	0.8	0.61 (0.38–0.95)
											0/55/584[c]	1/164/1012[c]	0.0006[c]	0.024	0.57[c](0.41–0.80)
ABCA7	19:1050996	A	c.G2629A	p.A877T[d]	rs74176364	0.3–1.18	1.69	Deleterious	Benign	Low	0/2/330	0/16/660	0.07	2.8	0.25 (0.02–1.07)
EPHA1	7:143095153	A	c.G1475A	p.R492Q[e]	rs11768549	2.56–1.47	1.21	Tolerated	Benign		0/17/315	1/18/657	0.07	2.8	1.86 (0.89–3.84)
ABCA7	19:1059056	A	c.G5435A	p.R1812H[d]	rs114782266	1.5–0.81	1.05	Tolerated	Benign	Neutral	0/10/322	0/11/665	0.16	6.4	1.87 (0.70–4.92)
ABCA7	19:1057343	A	c.G4795A	p.V1599M[d]	rs117187003	0.6–0.22	0.3	Deleterious	Possibly damaging	Medium	0/4/328	0/3/673	0.22	8.8	2.73 (0.45–18.7)
CD2AP	6:47573971	A	c.G1488A	p.M496I	rs143297472	0.3–0.07	NA	Tolerated	Benign		0/2/330	0/1/675	0.25	10	4.08 (0.21–241.3)
ABCA7	19:1047537	C	c.A2153C	p.N718T	rs3752239	1.65–2.44	7.02	Deleterious	Benign	Low	0/11/321	0/33/641	0.32	12.8	0.66 (0.29–1.37)

Position is in hg19/GRCh37.

Key: cDNA, complementary DNA; CI, confidence interval; Corr, corrected *p*-value, *p*-value after Bonferroni correction (*p*-value* 40 [number of variants considered in the single-variant association test]); ExAC, Exome Aggregation Consortium; MA, minor allele; MAF, minor allele frequency; ExAC, Exome Aggregation Consortium; OR, odds ratio.

[a]MAF cases-controls reported a Belgian cohort = 4.66%–6.27% (Cuyvers et al., 2015).

[b]MAF in ExAC (European non-Finnish) = 6.14% and MAF in EVS (European American) = 6.24%.

[c]Combined results discovery and follow-up data set.

[d]Variants reported associated also with autism spectrum disorders (ASD) (He et al., 2014).

[e]Variant reported associated to a more rapid disease progression.

Rs72973581 [A] results in a glycine to serine amino acid change at the position 215 of ATP-binding cassette subfamily A member 7 (ABCA7; G215S) and its frequency was 1.56-fold higher in controls compared to cases (MAF = 4.66% and 7.24% for cases and controls, respectively), arguing for a protective effect (OR = 0.6, 95% confidence interval [CI] = 0.38–0.95). This variant was present in homozygosity in one control. The study possessed relatively low power to detect a significant association between cases and controls for low frequency and rare variants. Therefore, we have followed up *ABCA7*p. G215S, carrying out Sanger sequencing in an independent data set composed of 307 Caucasian late-onset AD cases and 501 elderly Caucasian controls (*p*-value = 0.012; OR = 0.54, 95% CI 0.31–0.89). In this follow-up data set, we confirmed a higher frequency of the *ABCA7* p.G215S variant in controls compared to cases (carrier frequency = 13.5% vs. 7.8% [1.7-fold] and MAF = 7% vs. 4.3% [1.6-fold], respectively).

Finally, we also report a common coding polymorphism in *ABCA7* (p.R1349Q, rs3745842), that maps 1.3 kb from a reported GWAS hit, rs3752246, but clusters within a different common haplotype block (MAF > 5%) (Naj et al., 2011). Rs3745842 major allele (G) was more frequent in cases compared to controls, although the association was nominally significant after multiple testing correction (*p*-value = 1.4×10^{-3}, corrected *p*-value = 0.081).

LoF Mutations in *ABCA7*

LoF mutations in *ABCA7* have been recently reported as main mechanism explaining the GWAS signal and the increased susceptibility to AD.

In our cohort, we detected 5 LoF mutations in *ABCA7*: 2 stopgain mutations (p.Y1579X and p.E1974X) and 3 splice-site or near splice-site mutations (c.7-2A>G, c.7-7T>C and c.231-12C>A). *ABCA7* p.E1974X and c.231-12C>A are novel variants and, together with p.Y1579X, are singletons, detected only in controls.

Importantly, the enrichment for *ABCA7* LoF mutations and novel variants in controls did not rely on the sequencing strategy (exome sequencing vs. genome sequencing). These variants have not been indeed mainly detected in the 199 BYU controls that underwent genome sequencing.

Moreover, 3 very rare indels have been identified in controls in the NIH-UCL cohort (p.1402delT, p.1638delCTT, and p.1749delCTACTG). *ABCA7* p.1749delCTACTG is a novel mutation, and *ABCA7* p.1402delT was also present in one case. These indels have been excluded from the pooled data set because they are not targeted in the ADNI subcohort (Fig. 1).

Finally, 3 nonsense mutations (p.W749X, p.W903X, and p.R1754X) and one splice-site mutation, (c.4416+2T>G) did not pass either the sample or variant quality control (QC) criteria and, therefore, have not been included in the study.

Gene-Based Association Test

In addition to single-marker analysis, we carried out gene-wide analysis to combine the joint signal from multiple variants (coding variants and flanking UTRs) within a gene and to provide greater statistical power than that for single-marker tests. All the variants (nonsynonymous, synonymous, UTRs, and singletons) located within the studied genes and their exon-intron flanking regions were collapsed together and their combined effect was studied. *ABCA7* was the main hit both in the SKAT and c-alpha test, nominally and statistically significant, respectively, after multiple testing correction (corrected *p*-value = 0.6 and 5.3 × 10^{-3}, respectively) (Table 3 and Table 4). Importantly, given the exclusion of indels in the merged discovery data set, the presence of only 2 nonsense singleton mutations in *ABCA7* (p.Y1579X and p.E1974X, both detected in controls) and 4 putative splice-site mutations (rs3752229, rs2242437, c.231–12C>A, and rs182233998, the latter one nominally significant in controls), our findings are not influenced by a burden of LoF mutations in *ABCA7*. Considering the very rare frequency of these LoF variants, their detection with sufficient power would have required a very large sample size. By contrast, the top signals are represented mainly by common and low-frequency coding variants with a higher frequency in controls compared to cases and with a modest to intermediate protective effect (0.329 < OR < 0.755).

Table 3. Results from the c-alpha test performed

Transcript ID	Position	Gene	N.variants	Test	*p*-value	Corrected *p*-value
NM_019112	chr19:1040131...1065563	*ABCA7*	72	c-alpha	0.0006	0.0053
NM_012120	chr6:47445789...47594915	*CD2AP*	20	c-alpha	0.0353	0.31
NM_152851	chr11:59939123...59950523	*MS4A6A*	11	c-alpha	0.0548	0.49
NM_000573	chr1:207669709...207814864	*CR1*	72	c-alpha	0.0677	0.6
NM_139343	chr2:127805799...127864546	*BIN1*	27	c-alpha	0.0730	0.65
NM_001206946	chr11:85668697...85779900	*PICALM*	19	c-alpha	0.0742	0.66
NM_005232	chr7:143088365...143105830	*EPHA1*	30	c-alpha	0.1065	0.95
NM_001831	chr8:27454493...27472251	*CLU*	29	c-alpha	0.4444	3.99
NM_001772	chr19:51728380...51743144	*CD33*	13	c-alpha	0.7142	6.42

Position is in hg19/GRCh37. Corrected *p*-value, *p*-value after Bonferroni correction (*p*-value* 9 [number of genes considered in the single gene-based analysis]).

Table 4. Results from the SKAT test performed

Transcript ID	Position	Gene	N.variants	Test	*p*-value	Corrected *p*-value
NM_019112	chr19:1040131...1065563	*ABCA7*	72	SKAT	0.0737	0.66
NM_005232	chr7:143088365...143105830	*EPHA1*	30	SKAT	0.2981	2.68
NM_139343	chr2:127805799...127864546	*BIN1*	27	SKAT	0.4472	4.02
NM_012120	chr6:47445789...47594915	*CD2AP*	20	SKAT	0.4489	4.04
NM_000573	chr1:207669709...207814864	*CR1*	72	SKAT	0.5105	4.59
NM_001831	chr8:27454493...27472251	*CLU*	29	SKAT	0.5902	5.31
NM_152851	chr11:59939123...59950523	*MS4A6A*	11	SKAT	0.9377	8.43
NM_001772	chr19:51728380...51743144	*CD33*	13	SKAT	0.9389	8.45
NM_001206946	chr11:85668697...85779900	*PICALM*	19	SKAT	0.9437	8.49

Position is in hg19/GRCh37. Corrected *p*-value, *p*-value after Bonferroni correction (*p*-value* 9 [number of genes considered in the single gene-based analysis]).

DISCUSSION

We report the results of single-variant and gene-based association tests performed in*BIN1*, *CLU*, *CR1*, *PICALM*, *MS4A6A*, *ABCA7*, *EPHA1*, *CD33*, and *CD2AP* in a cohort composed of 332 apparently sporadic and mainly late-onset AD cases and 676 elderly Caucasian controls from North America and the UK. In the single-variant association test, we have analyzed the effect of low frequency and rare coding variants (MAF < 5%), aiming to identify potential functional variant(s) underlying the GWAS hit(s). In the gene-based analysis (SKAT and c-alpha test), we collapsed the full spectrum of variants identified in these loci to study their collective effect.

We do not report any pathogenic mutation in *APP*, *PSEN1*, and *PSEN2* in our cohort. However, one of the controls was a heterozygous carrier of the protective variant *APP*p.A673T (MAF 7 × 10^{-4} in our cohort and MAF 5 × 10^{-4} among the European non-Finnish, ExAC database, released 13 January, 2015) (Jonsson et al., 2012).

TREM2 p.R47H, the second most common risk factor for sporadic AD, has been detected in 6 cases (1.8%) and 4 controls (0.59%) and, likely given our small sample size, with a MAF = 0.2%, was not significantly associated to AD (*p*-value = 0.09).

ABCA7 was the only significant hit in the c-alpha test and harbors a low-frequency coding variant (p.G215S, rs72973581), whose minor allele confers a modest (OR = 0.57, 95% CI = 0.41–0.80) but statistically significant protection (corrected *p*-value =0.024) against AD. Importantly, this single-nucleotide polymorphism (SNP) was not present in several GWAS or exome SNP arrays and does not cluster within common haplotypes identified by tagging SNPs, whereas it has been detected through *ABCA7* direct sequencing in the present study. Therefore, rs72973581 would have stayed likely undetected using common fine-mapping genotyping arrays. In addition, it does not cluster in the risk haplotypes identified by GWAS main hits (rs3764650, rs115550680, rs3752246, and rs4147929), suggesting an independent signal and a likely different pathogenic mechanism of the major allele (Hollingworth et al., 2011, Liu et al., 2014, Naj et al., 2011 and Reitz et al., 2013). Importantly, the *ABCA7* p.G215S significant protective role against AD is supported by a targeted resequencing study of *ABCA7* in a Belgian cohort, where rs72973581 (A) frequency was 1.34-fold higher in controls compared to cases (*p*-value = 0.055) (Cuyvers et al., 2015). Notably, the main variant associated to LOAD in this Belgian cohort was a low-frequency intronic variant (rs78117248) that did not pass our QC filter. However, in line with our findings, Cuyvers et al. report an enrichment for common and low-frequency polymorphism with a modest protective role in *ABCA7*. Importantly, among the top 10 genetic variants identified in our study, 3 missense mutations (rs74176364, rs114782266, and rs117187003) have been described associated also to autism spectrum disorder, strongly pointing toward a functional role of these amino acid changes and suggesting a possible shared pathogenic mechanisms underpinning neurodegenerative and neurodevelopmental diseases (He et al., 2014).

Interestingly, several lines of evidence reported that a significant decrease in ABCA7 levels is associated to AD. At this regard, different and likely not mutually exclusive mechanisms have been described to influence the protein level: (1) common and generally noncoding variants in regulatory regions; (2) alternative splicing; (3) increased CpG island methylation (Humphries et al., 2015 and Vasquez et al., 2013). Recently, also LoF mutations in *ABCA7* have been shown to significantly increase the susceptibility to AD in the Islandic population (Steinberg et al., 2015). This has been replicated in 2 different populations (Caucasian North American and Belgian) by 2 independent studies (Cuyvers et al., 2015 and Vardarajan et al., 2015). Therefore, we report another potential mechanism, through which low-frequency protein coding variability in *ABCA7* may influence AD risk.

Notably, *ABCA7* p.G215S provides critical insights into the genetic architecture of diseases, reinforcing the view that GWAS loci, likewise Mendelian genes, harbor low frequency and rare protective coding variants that can counteract with a similar effect size the damaging alleles (OR ≈0.6 vs. ≈1.1 and ≈0.2 vs. ≈5, for GWAS loci and Mendelian genes, respectively; Asante et al., 2015, Jonsson et al., 2012, Nejentsev et al., 2009 and Rivas et al., 2011).

ABCA7 is mainly expressed in leukocytes and in myelolymphatic tissues (thymus, spleen, and bone marrow) and microglia in the brain (http://web.stanford.edu/group/barres_lab/brain_rnaseq.html; http://www.uniprot.org/; Kim et al., 2008). *ABCA7* encodes for ABCA7, a multi-pass protein, present on the cell, Golgi, and endosome membranes (http://www.uniprot.org/). In vitro and in vivo experiments have shown ABCA7 pivotal role in phagocytosis and a likely modest role in high-density lipoprotein biogenesis. In *Abca7*$^{-/-}$mice, macrophages and microglia display impaired phagocytosis and clearance of amyloid from the brain, which leads to cognitive impairment (Iwamoto et al., 2006 and Tanaka et al., 2011).

Therefore, *ABCA7*, likewise *TREM2* and *CD33*, may play an important role in regulating microglial uptake and clearance of amyloid-beta debris.

ABCA7 p.G215S clusters within the extracellular topological domain of ABCA7. Remarkably, at the homologous residue, the serine is the reference amino acid in ABCA7 in different mammals and in the homologous protein ABCA1 in humans. Thus, suggesting that this amino acid may confer some biologic advantage and may have been positively selected during the evolution.

Likewise other low-frequency and rare protective variants at the GWAS loci, *ABCA7* p.G215 is a relatively conserved residue among different species and this amino acid change (glycine to serine) may only slightly modify the protein activity (−5.86, 56, Gerd and Grantham score, respectively). Moreover, it has been reported as a tolerated change and benign, arguing against any possible LoF or significant impairment of ABCA7, that has been indeed associated to increased risk for AD (Steinberg et al., 2015). The biological effect of this substitution may therefore lead to a mild *ABCA7* gain of function, possibly strengthening the interaction with a binding protein or regulating its expression. Although *ABCA7*p.G215 has not been predicted to be a coding target for miRNA (https://www.umm.uni-heidelberg.de), a possible posttranscriptional or posttranslational regulation should not be excluded. Importantly, the substitution of a glycine with a serine may imply an additional substrate for serine-kinases or proteases. Moreover, in a similar way, *ABCA1*, whose LoF variants have been associated to AD (Kim et al., 2012 and Nordestgaard et al., 2015), has been reported to be particularly enriched for low frequency and rare coding variants with an average 1.5-fold higher frequency in controls compared to LOAD cases and a modest protective effect in a Greek cohort (OR = 0.96–0.38; Lupton et al., 2014).

Thus, understanding the effect of *ABCA7* p.G215S has the potential of unraveling new pathogenic mechanisms underpinning AD and may provide a promising therapeutic target that would not significantly alter ABCA7 overall physiological function, which is critical for AD development.

Finally, we support the resequencing study of the GWAS loci by Vardarajan et al. (2015), confirming a burden of damaging variants in *ABCA7* and *BIN1* and to a lesser extent in *CD2AP*, *EPHA1*, and *MS4A6A* (main hits in the gene-based analysis; Table 3 and Table 4), highlighting their potential role as susceptibility loci for LOAD.

However, we could not replicate the main hits detected by Vardarajan et al. in the single-variant analysis, either because such variants have been targeted but not detected in our cohort

(*ABCA7* p.E1679X, *EPHA1* p.P460L, and *BIN1* p.K358R) or because the variants have been targeted but eliminated by the QC filter (*CD2AP* p.K633R). Thus, suggesting a possible lack of replication compared to the previous studies attributable to the different population but also different sequencing strategies, capture, and coverage. Nevertheless, *EPHA1* and *CD2AP* harbor 2 of the main hits detected in the single-variant analysis in our cohort (rs11768549 and rs143297472, respectively) (Table 2), with rs11768549 already associated with the rapid progression of the disease in a cohort of Caucasian North American LOAD cases (Wang et al., 2015).

In summary, we support previous studies, suggesting that (1) *ABCA7* significantly influences AD risk; (2) *ABCA7* p.G215S is likely to reduce the susceptibility to AD; (3) GWAS hits are pleomorphic loci harboring a complex spectrum of variants synergistically contributing to the disease phenotype with different mechanisms, effects (damaging, protective, and neutral), and effect sizes (0 < OR < 4); and (4) gene-based approaches are effective methods to mine genetic data and to accurately filter potential candidate genes.

ACKNOWLEDGEMENTS

The authors thank contributors, including the Alzheimer's Disease Centers who collected samples used in this study, as well as patients and their families, whose help and participation made this work possible.

This study was supported by the Alzheimer's Research UK, the Medical Research Council (MRC), the Wellcome Trust/ MRC Joint Call in Neurodegeneration award (WT089698) to the UK Parkinson's Disease Consortium (whose members are from the University College London Institute of Neurology, the University of Sheffield, and the MRC Protein Phosphorylation Unit at the University of Dundee), grants (P50 AG016574, U01 AG006786, and R01 AG18023), the National Institute for Health Research Biomedical Research Unit in Dementia at University College London Hospitals, University College London; an anonymous donor, the Big Lottery (to Dr. Morgan); a fellowship from Alzheimer's Research UK (to Dr. Guerreiro); and the Intramural Research Programs of the National Institute on Aging and the National Institute of Neurological Disease and Stroke, National Institutes of Health (Department of Health and Human Services Project number, ZO1 AG000950-10). The MRC London Neurodegenerative Diseases Brain Bank and the Manchester Brain Bank from Brains for Dementia Research are jointly funded from ARUK and AS. This work was supported in part by the Intramural Research Program of the National Institute on Aging, National Institutes of Health, Department of Health and Human Services, project number ZO1 AG000950-10. Samples from the National Cell Repository for Alzheimer's Disease (NCRAD), which receives government support under a cooperative agreement grant (U24 AG21886) awarded by the National Institute on Aging (NIA), were used in this study. NIH grant R01 AG042611 to Kauwe J.

REFERENCES

1. Asante, E.A., Smidak, M., Grimshaw, A., Houghton, R., Tomlinson, A., Jeelani, A., Jakubcova, T., Hamdan, S., Richard-Londt, A., Linehan, J.M., Brandner, S., Alpers, M., Whitfield, J., Mead, S., Wadsworth, J.D.F., Collinge, J., 2015. A naturally occurring variant of the human prion protein completely prevents prion dis-ease. Nature 522, 478e481.
2. Beaudoin, M., Goyette, P., Boucher, G., Lo, K.S., Rivas, M.A., Stevens, C., Alikashani, A., Ladouceur, M., Ellinghaus, D., Törkvist, L., Goel, G., Lagacé, C., Annese, V., Bitton, A., Begun, J., Brant, S.R., Bresso, F., Cho, J.H., Duerr, R.H., Halfvarson, J., McGovern, D.P.B., Radford-Smith, G., Schreiber, S., Schumm, P.L., Sharma, Y., Silverberg, M.S., Weersma, R.K., Quebec IBD Genetics Consortium NIDDK IBD Genetics Consortium International IBD Genetics Consortium, D'Amato, M., Vermeire, S., Franke, A., Lettre, G., Xavier, R.J., Daly, M.J., Rioux, J.D., 2013. Deep resequencing of GWAS loci identifies rare variants in CARD9, IL23R and RNF186 that are associated with ulcerative colitis. PLoS Genet. 9, e1003723.
3. Chartier-Harlin, M.C., Parfitt, M., Legrain, S., Pérez-Tur, J., Brousseau, T., Evans, A., Berr, C., Vidal, O., Roques, P., Gourlet, V., 1994. Apolipoprotein E, epsilon 4 allele as a major risk factor for sporadic early and late-onset forms of Alzheimer's disease: analysis of the 19q13.2 chromosomal region. Hum. Mol. Genet. 3, 569e574.
4. Cuyvers, E., De Roeck, A., Van den Bossche, T., Van Cauwenberghe, C., Bettens, K., Vermeulen, S., Mattheijssens, M., Peeters, K., Engelborghs, S., Vandenbulcke, M., Vandenberghe, R., De Deyn, P.P., Van Broeckhoven, C., Sleegers, K., 2015. Mu-tations in ABCA7 in a Belgian cohort of Alzheimer's disease patients: a targeted resequencing study. Lancet Neurol. 14, 814e822.
5. Guerreiro, R., Wojtas, A., Bras, J., Carrasquillo, M., Rogaeva, E., Majounie, E., Cruchaga, C., Sassi, C., Kauwe, J.S.K., Younkin, S., Hazrati, L., Collinge, J., Pocock, J., Lashley, T., Williams, J., Lambert, J.-C., Amouyel, P., Goate, A., Rademakers, R., Morgan, K., Powell, J., St George-Hyslop, P., Singleton, A., Hardy, J. Alzheimer Genetic Analysis Group, 2013. TREM2 variants in Alzheimer's disease. N. Engl. J. Med. 368, 117e127.
6. Harold, D., Abraham, R., Hollingworth, P., Sims, R., Gerrish, A., Hamshere, M.L., Pahwa, J.S., Moskvina, V., Dowzell, K., Williams, A., Jones, N., Thomas, C., Stretton, A., Morgan, A.R., Lovestone, S., Powell, J., Proitsi, P., Lupton, M.K., Brayne, C., Rubinsztein, D.C., Gill, M., Lawlor, B., Lynch, A., Morgan, K., Brown, K.S., Passmore, P.A., Craig, D., McGuinness, B., Todd, S., Holmes, C., Mann, D., Smith, A.D., Love, S., Kehoe, P.G., Hardy, J., Mead, S., Fox, N., Rossor, M., Collinge, J., Maier, W., Jessen, F., Schürmann, B., Heun, R., van den Bussche, H., Heuser, I., Kornhuber, J., Wiltfang, J., Dichgans, M., Frölich, L., Hampel, H., Hüll, M., Rujescu, D., Goate, A.M., Kauwe, J.S.K., Cruchaga, C., Nowotny, P., Morris, J.C., Mayo, K., Sleegers, K., Bettens, K., Engelborghs, S., De Deyn, P.P., Van Broeckhoven, C., Livingston, G., Bass, N.J., Gurling, H., McQuillin, A., Gwilliam, R., Deloukas, P., Al-Chalabi, A., Shaw, C.E., Tsolaki, M., Singleton, A.B., Guerreiro, R., Mühleisen, T.W., Nöthen, M.M., Moebus, S., Jöckel, K.-H., Klopp, N., Wichmann, H.-E., Carrasquillo, M.M., Pankratz, V.S., Younkin, S.G., Holmans, P.A., O'Donovan, M., Owen,

M.J., Williams, J., 2009. Genome-wide association study identifies variants at CLU and PICALM associated with Alzheimer's disease. Nat. Genet. 41, 1088e1093.

7. He, Z., O'Roak, B.J., Smith, J.D., Wang, G., Hooker, S., Santos-Cortez, R.L.P., Li, B., Kan, M., Krumm, N., Nickerson, D.A., Shendure, J., Eichler, E.E., Leal, S.M., 2014. Rare-variant extensions of the transmission disequilibrium test: application to autism exome sequence data. Am. J. Hum. Genet. 94, 33e46.

8. Hollingworth, P., Harold, D., Sims, R., Gerrish, A., Lambert, J.-C., Carrasquillo, M.M., Abraham, R., Hamshere, M.L., Pahwa, J.S., Moskvina, V., Dowzell, K., Jones, N., Stretton, A., Thomas, C., Richards, A., Ivanov, D., Widdowson, C., Chapman, J., Lovestone, S., Powell, J., Proitsi, P., Lupton, M.K., Brayne, C., Rubinsztein, D.C., Gill, M., Lawlor, B., Lynch, A., Brown, K.S., Passmore, P.A., Craig, D., McGuinness, B., Todd, S., Holmes, C., Mann, D., Smith, A.D., Beaumont, H., Warden, D., Wilcock, G., Love, S., Kehoe, P.G., Hooper, N.M., Vardy, E.R.L.C., Hardy, J., Mead, S., Fox, N.C., Rossor, M., Collinge, J., Maier, W., Jessen, F., Rüther, E., Schürmann, B., Heun, R., Kölsch, H., van den Bussche, H., Heuser, I., Kornhuber, J., Wiltfang, J., Dichgans, M., Frölich, L., Hampel, H., Gallacher, J., Hüll, M., Rujescu, D., Giegling, I., Goate, A.M., Kauwe, J.S.K., Cruchaga, C., Nowotny, P., Morris, J.C., Mayo, K., Sleegers, K., Bettens, K., Engelborghs, S., De Deyn, P.P., Van Broeckhoven, C., Livingston, G., Bass, N.J., Gurling, H., McQuillin, A., Gwilliam, R., Deloukas, P., Al-Chalabi, A., Shaw, C.E., Tsolaki, M., Singleton, A.B., Guerreiro, R., Mühleisen, T.W., Nöthen, M.M., Moebus, S., Jöckel, K.-H., Klopp, N., Wichmann, H.-E., Pankratz, V.S., Sando, S.B., Aasly, J.O., Barcikowska, M., Wszolek, Z.K., Dickson, D.W., Graff-Radford, N.R., Petersen, R.C., Alzheimer's Disease Neuroimaging Initiative, van Duijn, C.M., Breteler, M.M.B., Ikram, M.A., DeStefano, A.L., Fitzpatrick, A.L., Lopez, O., Launer, L.J., Seshadri, S., CHARGE consortium, Berr, C., Campion, D., Epelbaum, J., Dartigues, J.-F., Tzourio, C., Alpérovitch, A., Lathrop, M., EADI1 consortium, Feulner, T.M., Friedrich, P., Riehle, C., Krawczak, M., Schreiber, S., Mayhaus, M., Nicolhaus, S., Wagenpfeil, S., Steinberg, S., Stefansson, H., Stefansson, K., Snaedal, J., Björnsson, S., Jonsson, P.V., Chouraki, V., Genier-Boley, B., Hiltunen, M., Soininen, H., Combarros, O., Zelenika, D., Delepine, M., Bullido, M.J., Pasquier, F., Mateo, I., Frank-Garcia, A., Porcellini, E., Hanon, O., Coto, E., Alvarez, V., Bosco, P., Siciliano, G., Mancuso, M., Panza, F., Solfrizzi, V., Nacmias, B., Sorbi, S., Bossù, P., Piccardi, P., Arosio, B., Annoni, G., Seripa, D., Pilotto, A., Scarpini, E., Galimberti, D., Brice, A., Hannequin, D., Licastro, F., Jones, L., Holmans, P.A., Jonsson, T., Riemenschneider, M., Morgan, K., Younkin, S.G., Owen, M.J., O'Donovan, M., Amouyel, P., Williams, J., 2011. Com-mon variants at ABCA7, MS4A6A/MS4A4E, EPHA1, CD33 and CD2AP are asso-ciated with Alzheimer's disease. Nat. Genet. 43, 429e435.

9. Humphries, C., Kohli, M.A., Whitehead, P., Mash, D.C., Pericak-Vance, M.A., Gilbert, J., 2015. Alzheimer disease (AD) specific transcription, DNA methylation and splicing in twenty AD associated loci. Mol. Cell. Neurosci 67, 37e45.

10. Iwamoto, N., Abe-Dohmae, S., Sato, R., Yokoyama, S., 2006. ABCA7 expression is regulated by cellular cholesterol through the SREBP2 pathway and associated with phagocytosis. J. Lipid Res. 47, 1915e1927.

11. Jonsson, T., Atwal, J.K., Steinberg, S., Snaedal, J., Jonsson, P.V., Bjornsson, S., Stefansson, H., Sulem, P., Gudbjartsson, D., Maloney, J., Hoyte, K., Gustafson, A., Liu, Y., Lu, Y., Bhangale, T., Graham, R.R., Huttenlocher, J., Bjornsdottir, G., Andreassen, O.A., Jönsson, E.G., Palotie, A., Behrens, T.W., Magnusson, O.T., Kong, A., Thorsteinsdottir, U., Watts, R.J., Stefansson, K., 2012. A mutation in APP protects against Alzheimer's disease and age-related cognitive decline. Nature 488, 96e99.

12. Kim, J., Yoon, H., Ramírez, C.M., Lee, S.-M., Hoe, H.-S., Fernández-Hernando, C., Kim, J., 2012. MiR-106b impairs cholesterol efflux and increases Ab levels by repressing ABCA1 expression. Exp. Neurol. 235, 476e483.

13. Kim, W.S., Weickert, C.S., Garner, B., 2008. Role of ATP-binding cassette transporters in brain lipid transport and neurological disease. J. Neurochem. 104, 1145e1166.

14. Lambert, J.-C., Heath, S., Even, G., Campion, D., Sleegers, K., Hiltunen, M., Combarros, O., Zelenika, D., Bullido, M.J., Tavernier, B., Letenneur, L., Bettens, K., Berr, C., Pasquier, F., Fiévet, N., Barberger-Gateau, P., Engelborghs, S., De Deyn, P., Mateo, I., Franck, A., Helisalmi, S., Porcellini, E., Hanon, O., European Alzheimer's Disease Initiative Investigators, de Pancorbo, M.M., Lendon, C., Dufouil, C., Jaillard, C., Leveillard, T., Alvarez, V., Bosco, P., Mancuso, M., Panza, F., Nacmias, B., Bossù, P., Piccardi, P., Annoni, G., Seripa, D., Galimberti, D., Hannequin, D., Licastro, F., Soininen, H., Ritchie, K., Blanché, H., Dartigues, J.-F., Tzourio, C., Gut, I., Van Broeckhoven, C., Alpérovitch, A., Lathrop, M., Amouyel, P., 2009. Genome-wide association study identifies variants at CLU and CR1 associated with Alzheimer's disease. Nat. Genet. 41, 1094e1099.

15. Lambert, J.C., Ibrahim-Verbaas, C.A., Harold, D., Naj, A.C., Sims, R., Bellenguez, C., DeStafano, A.L., Bis, J.C., Beecham, G.W., Grenier-Boley, B., Russo, G., Thorton-Wells, T.A., Jones, N., Smith, A.V., Chouraki, V., Thomas, C., Ikram, M.A., Zelenika, D., Vardarajan, B.N., Kamatani, Y., Lin, C.F., Gerrish, A., Schmidt, H., Kunkle, B., Dunstan, M.L., Ruiz, A., Bihoreau, M.T., Choi, S.H., Reitz, C., Pasquier, F., Cruchaga, C., Craig, D., Amin, N., Berr, C., Lopez, O.L., De Jager, P.L., Deramecourt, V., Johnston, J.A., Evans, D., Lovestone, S., Letenneur, L., Morón, F.J., Rubinsztein, D.C., Eiriksdottir, G., Sleegers, K., Goate, A.M., Fiévet, N., Huentelman, M.W., Gill, M., Brown, K., Kamboh, M.I., Keller, L., Barberger-Gateau, P., McGuiness, B., Larson, E.B., Green, R., Myers, A.J., Dufouil, C., Todd, S., Wallon, D., Love, S., Rogaeva, E., Gallacher, J., St George-Hyslop, P., Clarimon, J., Lleo, A., Bayer, A., Tsuang, D.W., Yu, L., Tsolaki, M., Bossù, P., Spalletta, G., Proitsi, P., Collinge, J., Sorbi, S., Sanchez-Garcia, F., Fox, N.C., Hardy, J., Deniz Naranjo, M.C., Bosco,

P., Clarke, R., Brayne, C., Galimberti, D., Mancuso, M., Matthews, F., European Alzheimer's Disease Initiative (EADI) Genetic and Environmental Risk in Alzheimer's Disease Alzheimer's Disease Genetic Con-sortium Cohorts for Heart and Aging Research in Genomic Epidemiology, Moebus, S., Mecocci, P., Del Zompo, M., Maier, W., Hampel, H., Pilotto, A., Bullido, M., Panza, F., Caffarra, P., Nacmias, B., Gilbert, J.R., Mayhaus, M., Lannefelt, L., Hakonarson, H., Pichler, S., Carrasquillo, M.M., Ingelsson, M., Beekly, D., Alvarez, V., Zou, F., Valladares, O., Younkin, S.G., Coto, E., Hamilton-Nelson, K.L., Gu, W., Razquin, C., Pastor, P., Mateo, I., Owen, M.J., Faber, K.M., Jonsson, P.V., Combarros, O., O'Donovan, M.C., Cantwell, L.B., Soininen, H., Blacker, D., Mead, S., Mosley, T.H., Bennett, D.A., Harris, T.B., Fratiglioni, L., Holmes, C., de Bruijn, R.F., Passmore, P., Montine, T.J., Bettens, K., Rotter, J.I., Brice, A., Morgan, K., Foroud, T.M., Kukull, W.A., Hannequin, D., Powell, J.F., Nalls, M.A., Ritchie, K., Lunetta, K.L., Kauwe, J.S., Boerwinkle, E., Riemenschneider, M., Boada, M., Hiltuenen, M., Martin, E.R., Schmidt, R.,

16. Rujescu, D., Wang, L.S., Dartigues, J.F., Mayeux, R., Tzourio, C., Hofman, A., Nöthen, M.M., Graff, C., Psaty, B.M., Jones, L., Haines, J.L., Holmans, P.A., Lathrop, M., Pericak-Vance, M.A., Launer, L.J., Farrer, L.A., van Duijn, C.M., Van Broeckhoven, C., Moskvina, V., Seshadri, S., Williams, J., Schellenberg, G.D., Amouyel, P., 2013. Meta-analysis of 74,046 individuals identifies 11 new sus-ceptibility loci for Alzheimer's disease. Nat. Genet. 45, 1452e1458.

17. Li, H., Durbin, R., 2009. Fast and accurate short read alignment with Burrows-Wheeler transform. Bioinforma. Oxf. Engl. 25, 1754e1760.

18. Li, H., Handsaker, B., Wysoker, A., Fennell, T., Ruan, J., Homer, N., Marth, G., Abecasis, G., Durbin, R. 1000 Genome Project Data Processing Subgroup, 2009. The sequence alignment/map format and SAMtools. Bioinforma. Oxf. Engl. 25, 2078e2079.

19. Liu, G., Li, F., Zhang, S., Jiang, Y., Ma, G., Shang, H., Liu, J., Feng, R., Zhang, L., Liao, M., Zhao, B., Li, K., 2014. Analyzing large-scale samples confirms the association between the ABCA7 rs3764650 polymorphism and Alzheimer's disease sus-ceptibility. Mol. Neurobiol. 50, 757e764.

20. Lohmueller, K.E., Sparsø, T., Li, Q., Andersson, E., Korneliussen, T., Albrechtsen, A., Banasik, K., Grarup, N., Hallgrimsdottir, I., Kiil, K., Kilpeläinen, T.O., Krarup, N.T., Pers, T.H., Sanchez, G., Hu, Y., Degiorgio, M., Jørgensen, T., Sandbæk, A., Lauritzen, T., Brunak, S., Kristiansen, K., Li, Y., Hansen, T., Wang, J., Nielsen, R., Pedersen, O., 2013. Whole-exome sequencing of 2,000 Danish individuals and the role of rare coding variants in type 2 diabetes. Am. J. Hum. Genet. 93, 1072e1086.

21. Lupton, M.K., Proitsi, P., Lin, K., Hamilton, G., Daniilidou, M., Tsolaki, M., Powell, J.F., 2014. The role of ABCA1 gene sequence variants on risk of Alzheimer's disease. J. Alzheimers Dis. 38, 897e906.

22. McKenna, A., Hanna, M., Banks, E., Sivachenko, A., Cibulskis, K., Kernytsky, A., Garimella, K., Altshuler, D., Gabriel, S., Daly, M., DePristo, M.A., 2010. The genome analysis toolkit: a MapReduce framework for analyzing next-generation DNA sequencing data. Genome Res. 20, 1297e1303.

23. Naj, A.C., Jun, G., Beecham, G.W., Wang, L.-S., Vardarajan, B.N., Buros, J., Gallins, P.J., Buxbaum, J.D., Jarvik, G.P., Crane, P.K., Larson, E.B., Bird, T.D., Boeve, B.F., Graff-Radford, N.R., De Jager, P.L., Evans, D., Schneider, J.A., Carrasquillo, M.M., Ertekin-Taner, N., Younkin, S.G., Cruchaga, C., Kauwe, J.S.K., Nowotny, P., Kramer, P., Hardy, J., Huentelman, M.J., Myers, A.J., Barmada, M.M., Demirci, F.Y., Baldwin, C.T., Green, R.C., Rogaeva, E., St George-Hyslop, P., Arnold, S.E., Barber, R., Beach, T., Bigio, E.H., Bowen, J.D., Boxer, A., Burke, J.R., Cairns, N.J., Carlson, C.S., Carney, R.M., Carroll, S.L., Chui, H.C., Clark, D.G., Corneveaux, J., Cotman, C.W., Cummings, J.L., DeCarli, C., DeKosky, S.T., Diaz-Arrastia, R., Dick, M., Dickson, D.W., Ellis, W.G., Faber, K.M., Fallon, K.B., Farlow, M.R., Ferris, S., Frosch, M.P., Galasko, D.R., Ganguli, M., Gearing, M., Geschwind, D.H., Ghetti, B., Gilbert, J.R., Gilman, S., Giordani, B., Glass, J.D., Growdon, J.H., Hamilton, R.L., Harrell, L.E., Head, E., Honig, L.S., Hulette, C.M., Hyman, B.T., Jicha, G.A., Jin, L.-W., Johnson, N., Karlawish, J., Karydas, A., Kaye, J.A., Kim, R., Koo, E.H., Kowall, N.W., Lah, J.J., Levey, A.I., Lieberman, A.P., Lopez, O.L., Mack, W.J., Marson, D.C., Martiniuk, F., Mash, D.C., Masliah, E., McCormick, W.C., McCurry, S.M., McDavid, A.N., McKee, A.C., Mesulam, M., Miller, B.L., Miller, C.A., Miller, J.W., Parisi, J.E., Perl, D.P., Peskind, E., Petersen, R.C., Poon, W.W., Quinn, J.F., Rajbhandary, R.A., Raskind, M., Reisberg, B., Ringman, J.M., Roberson, E.D., Rosenberg, R.N., Sano, M., Schneider, L.S., Seeley, W., Shelanski, M.L., Slifer, M.A., Smith, C.D., Sonnen, J.A., Spina, S., Stern, R.A., Tanzi, R.E., Trojanowski, J.Q., Troncoso, J.C., Van Deerlin, V.M., Vinters, H.V., Vonsattel, J.P., Weintraub, S., Welsh-Bohmer, K.A., Williamson, J., Woltjer, R.L., Cantwell, L.B., Dombroski, B.A., Beekly, D., Lunetta, K.L., Martin, E.R., Kamboh, M.I., Saykin, A.J., Reiman, E.M., Bennett, D.A., Morris, J.C., Montine, T.J., Goate, A.M., Blacker, D., Tsuang, D.W., Hakonarson, H., Kukull, W.A., Foroud, T.M., Haines, J.L., Mayeux, R., Pericak-Vance, M.A., Farrer, L.A., Schellenberg, G.D., 2011. Common variants at MS4A4/MS4A6E, CD2AP, CD33 and EPHA1 are associated with late-onset Alzheimer's disease. Nat. Genet. 43, 436e441.

24. Neale, B.M., Rivas, M.A., Voight, B.F., Altshuler, D., Devlin, B., Orho-Melander, M., Kathiresan, S., Purcell, S.M., Roeder, K., Daly, M.J., 2011. Testing for an unusual distribution of rare variants. PLoS Genet. 7, e1001322.

25. Nejentsev, S., Walker, N., Riches, D., Egholm, M., Todd, J.A., 2009. Rare variants of IFIH1, a gene implicated in antiviral responses, protect against type 1 diabetes. Science 324, 387e389.

26. Nordestgaard, L.T., Tybjærg-Hansen, A., Nordestgaard, B.G., Frikke-Schmidt, R., 2015. Loss-of-function mutation in ABCA1 and risk of Alzheimer's disease and cere-brovascular disease. Alzheimers Dement. 11, 1430e1438.

27. Ramasamy, A., Trabzuni, D., Guelfi, S., Varghese, V., Smith, C., Walker, R., De, T., UK Brain Expression Consortium North American Brain Expression Consortium, Coin, L., de Silva, R., Cookson, M.R., Singleton, A.B., Hardy, J., Ryten, M., Weale, M.E., 2014. Genetic variability in the regulation of gene expression in ten regions of the human brain. Nat. Neurosci. 17, 1418e1428.

28. Reitz, C., Jun, G., Naj, A., Rajbhandary, R., Vardarajan, B.N., Wang, L.-S., Valladares, O., Lin, C.-F., Larson, E.B., Graff-Radford, N.R., Evans, D., De Jager, P.L., Crane, P.K., Buxbaum, J.D., Murrell, J.R., Raj, T., Ertekin-Taner, N., Logue, M., Baldwin, C.T., Green, R.C., Barnes, L.L., Cantwell, L.B., Fallin, M.D., Go, R.C.P., Griffith, P., Obisesan, T.O., Manly, J.J., Lunetta, K.L., Kamboh, M.I., Lopez, O.L., Bennett, D.A., Hendrie, H., Hall, K.S., Goate, A.M., Byrd, G.S., Kukull, W.A., Foroud, T.M., Haines, J.L., Farrer, L.A., Pericak-Vance, M.A., Schellenberg, G.D., Mayeux, R. Alzheimer Disease Genetics Consortium, 2013. Variants in the ATP-binding cassette transporter (ABCA7), apolipoprotein E ε4, and the risk of late-onset Alzheimer disease in African Americans. JAMA 309, 1483e1492.

29. Rivas, M.A., Beaudoin, M., Gardet, A., Stevens, C., Sharma, Y., Zhang, C.K., Boucher, G., Ripke, S., Ellinghaus, D., Burtt, N., Fennell, T., Kirby, A., Latiano, A., Goyette, P., Green, T., Halfvarson, J., Haritunians, T., Korn, J.M., Kuruvilla, F., Lagacé, C., Neale, B., Lo, K.S., Schumm, P., Törkvist, L., National Institute of Diabetes and Digestive Kidney Diseases Inflammatory Bowel Disease Genetics Consortium (NIDDK IBDGC) United Kingdom Inflammatory Bowel Disease Genetics Con-sortium International Inflammatory Bowel Disease Genetics Consortium, Dubinsky, M.C., Brant, S.R., Silverberg, M.S., Duerr, R.H., Altshuler, D., Gabriel, S., Lettre, G., Franke, A., D'Amato, M., McGovern, D.P.B., Cho, J.H., Rioux, J.D., Xavier, R.J., Daly, M.J., 2011. Deep resequencing of GWAS loci identifies independent rare variants associated with inflammatory bowel disease. Nat. Genet. 43, 1066e1073.

30. Service, S.K., Teslovich, T.M., Fuchsberger, C., Ramensky, V., Yajnik, P., Koboldt, D.C., Larson, D.E., Zhang, Q., Lin, L., Welch, R., Ding, L., McLellan, M.D., O'Laughlin, M., Fronick, C., Fulton, L.L., Magrini, V., Swift, A., Elliott, P., Jarvelin, M.-R., Kaakinen, M., McCarthy, M.I., Peltonen, L., Pouta, A., Bonnycastle, L.L., Collins, F.S., Narisu, N., Stringham, H.M., Tuomilehto, J., Ripatti, S., Fulton, R.S., Sabatti, C., Wilson, R.K., Boehnke, M., Freimer, N.B., 2014. Re-sequencing ex-pands our understanding of the phenotypic impact of variants at GWAS loci. Plos Genet. 10, e1004147.

31. Seshadri, S., Fitzpatrick, A.L., Ikram, M.A., DeStefano, A.L., Gudnason, V., Boada, M., Bis, J.C., Smith, A.V., Carassquillo, M.M., Lambert, J.C., Harold, D., Schrijvers, E.M.C., Ramirez-Lorca, R., Debette, S., Longstreth, W.T., Janssens, A.C.J.W., Pankratz, V.S., Dartigues, J.F., Hollingworth, P., Aspelund, T., Hernandez, I., Beiser, A., Kuller, L.H., Koudstaal, P.J., Dickson, D.W., Tzourio, C., Abraham, R., Antunez, C., Du, Y., Rotter, J.I., Aulchenko, Y.S., Harris, T.B., Petersen, R.C., Berr, C., Owen, M.J., Lopez-Arrieta, J., Varadarajan, B.N., Becker, J.T., Rivadeneira, F., Nalls, M.A., Graff-Radford, N.R., Campion, D., Auerbach, S., Rice, K., Hofman, A., Jonsson, P.V., Schmidt, H., Lathrop, M., Mosley, T.H., Au, R., Psaty, B.M., Uitterlinden, A.G., Farrer, L.A., Lumley, T., Ruiz, A., Williams, J., Amouyel, P., Younkin, S.G., Wolf, P.A., Launer, L.J., Lopez, O.L., van Duijn, C.M., Breteler, M.M.B. CHARGE Consortium, GERAD1 Consortium EADI1 Consortium, 2010. Genome-wide analysis of genetic loci associated with Alzheimer disease. JAMA 303, 1832e1840.

32. Steinberg, S., Stefansson, H., Jonsson, T., Johannsdottir, H., Ingason, A., Helgason, H., Sulem, P., Magnusson, O.T., Gudjonsson, S.A., Unnsteinsdottir, U., Kong, A., Helisalmi, S., Soininen, H., Lah, J.J., DemGene, Aarsland, D., Fladby, T., Ulstein, I.D., Djurovic, S., Sando, S.B., White, L.R., Knudsen, G.-P., Westlye, L.T., Selbæk, G., Giegling, I., Hampel, H., Hiltunen, M., Levey, A.I., Andreassen, O.A., Rujescu, D., Jonsson, P.V., Bjornsson, S., Snaedal, J., Stefansson, K., 2015. Loss-of-function variants in ABCA7 confer risk of Alzheimer's disease. Nat. Genet. 47, 445e447.

33. Tanaka, N., Abe-Dohmae, S., Iwamoto, N., Fitzgerald, M.L., Yokoyama, S., 2011. HMG-CoA reductase inhibitors enhance phagocytosis by upregulating ATP-binding cassette transporter A7. Atherosclerosis 217, 407e414.

34. Vardarajan, B.N., Ghani, M., Kahn, A., Sheikh, S., Sato, C., Barral, S., Lee, J.H., Cheng, R., Reitz, C., Lantigua, R., Reyes-Dumeyer, D., Medrano, M., Jimenez-Velazquez, I.Z., Rogaeva, E., St George-Hyslop, P., Mayeux, R., 2015. Rare coding mutations identified by sequencing of Alzheimer disease genome-wide association studies loci. Ann. Neurol. 78, 487e498.

35. Vasquez, J.B., Fardo, D.W., Estus, S., 2013. ABCA7 expression is associated with Alzheimer's disease polymorphism and disease status. Neurosci. Lett. 556, 58e62.

36. Visscher, P.M., Brown, M.A., McCarthy, M.I., Yang, J., 2012. Five years of GWAS dis-covery. Am. J. Hum. Genet. 90, 7e24.

37. Wang, K., Li, M., Hakonarson, H., 2010. ANNOVAR: functional annotation of genetic variants from high-throughput sequencing data. Nucleic Acids Res. 38, e164.

38. Wang, X., Lopez, O.L., Sweet, R.A., Becker, J.T., DeKosky, S.T., Barmada, M.M., Demirci, F.Y., Kamboh, M.I., 2015. Genetic determinants of disease progression in Alzheimer's disease. J. Alzheimers Dis. 43, 649e655.

CHAPTER 4

WHOLE EXOME SEQUENCING AND SEGREGATION ANALYSIS CONFIRMS THAT A MUTATION IN COL17A1 IS THE CAUSE OF EPITHELIAL RECURRENT EROSION DYSTROPHY IN A LARGE DOMINANT PEDIGREE PREVIOUSLY MAPPED TO CHROMOSOME 10Q23-Q24

Benjamin R. Lin, Derek J. Le, Yabin Chen, Qiwei Wang, D. Doug Chung, Ricardo F. Frausto, Christopher Croasdale, Richard W. Yee, Fielding J. Hejtmancik, Anthony J. Aldave

INTRODUCTION

The corneal dystrophies are a group of inherited disorders associated with bilateral, symmetric, and progressive loss of visual acuity due to the loss of corneal clarity. Reis-Bücklers corneal dystrophy (RBCD; MIM 608470) and Thiel-Behnke corneal dystrophy (TBCD; MIM 602082) are two forms of epithelial-stromal transforming growth factor beta induced (TGFBI) corneal dystrophies. RBCD and TBCD are associated with similar phenotypes, presenting with recurrent corneal erosions and the development of a "geographic" (RBCD) or "honeycomb" (TBCD) pattern of the anterior corneal layer.[1–3] These two dystrophies share a common genetic origin, with RBCD and TBCD associated with the p.(Arg124Leu) and p.(Arg555Gln) mutations in the transforming growth factor beta induced (*TGFBI*) gene on chromosome 5q31, respectively.[4–7]

A previous report by Yee et al. described a large pedigree affected with a dystrophy of the anterior corneal layer in which a *TGFBI* mutation was not identified.[8, 9] The majority of affected family members were described as demonstrating bilateral "honeycomb" opacities, and curly fibers were noted on electron microscopic examination of a corneal specimen from an affected individual, leading to the diagnosis of TBCD.[9] However, the resultant classification of this family as having TBCD has been questioned, primarily due to disagreement regarding whether the published photographs demonstrate a honeycomb-like pattern.[10, 11] In addition, the electron microscopic images of a corneal specimen from the proband's sister were never published, and thus could not be independently evaluated. Linkage analysis mapped the dystrophy to an approximately 25 Mbp region on chromosome 10q23-q24 with a maximum multipoint LOD score of 5.5.[8] Examination of the genes mapped to the linked interval on chromosome 10q23-q24 led to the identification of *COL17A1* as both a positional and functional candidate, but screening of the gene did not reveal any presumed pathogenic variants.

Citation: Lin BR, Le DJ, Chen Y, Wang Q, Chung DD, Frausto RF, et al. (2016) Whole Exome Sequencing and Segregation Analysis Confirms That a Mutation in COL17A1 Is the Cause of Epithelial Recurrent Erosion Dystrophy in a Large Dominant Pedigree Previously Mapped to Chromosome 10q23-q24. PLoS ONE 11(6): e0157418. doi:10.1371/journal.pone.0157418.

[12] Thus, both the clinical and genetic characterization of chromosome 10q23-q24 linked corneal dystrophy remained ambiguous.

Given the failure to identify the genetic basis of chromosome 10q23-q24 linked corneal dystrophy using a positional candidate gene approach, we decided to use whole exome sequencing (WES), focusing initially on variants identified in the previously linked interval. This led to the identification of three variants in two different genes that met all of the filtering criteria, with only one of the variants, c.3156C>T (p.(Gly1052 =)) in *COL17A1*, segregating with the affected phenotype in additional family members. Jonsson and colleagues recently reported the results of *in silico* modeling and an *in vitro* splice assay demonstrating that the c.3156C>T (p.(Gly1052 =)) synonymous variant in *COL17A1* leads to the introduction of a splice donor site, resulting in an aberrantly spliced transcript.[11] Even more recently, Oliver and colleagues independently identified the c.3156C>T variant in *COL17A1* segregating with the affected phenotype in three families with corneal dystrophies characterized by the early onset recurrent corneal erosions.[13, 14] Here, we report the use of whole exome sequencing (WES) to identify the genetic basis of what should be classified as epithelial recurrent erosion dystrophy in the original large pedigree that mapped the condition to chromosome 10q23-q24.

METHODS

This study followed the Declaration of Helsinki, adhered to the ARVO statement on human subjects, and was approved by the Institutional Review Board at the University of California at Los Angeles (UCLA IRB#11–000020). Informed written consent was obtained from all subjects in this study.

Determination of Affected Status

The most recently published pedigree of the family with chromosome 10q23-q24 linked corneal dystrophy was used as a basis for determining the affected status of family members. [9] For previously examined family members who were subsequently reexamined, and for previously unexamined family members, determination of the affected status was based on a history of recurrent corneal erosions and the presence of subepithelial opacification, likely representing scarring in the location of a prior corneal erosion.

DNA Collection

DNA samples were collected for WES from five affected family members (Figure 1: III-3, IV-1, IV-6, IV-8, V-3). Previously collected DNA samples were used for Sanger sequencing of potentially pathogenic variants identified with WES.

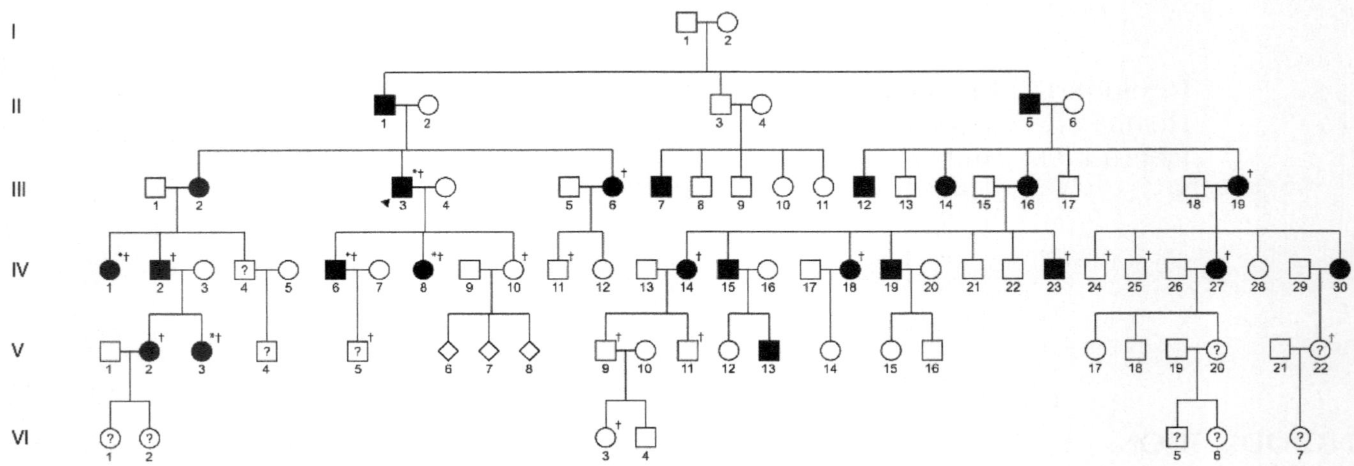

Figure 1. Pedigree of a family with chromosome 10q23-q24 linked dystrophy. Females are represented by circles, males by squares, and individuals of unknown gender by diamonds. Affected individuals are shown with filled symbols and unaffected individuals are shown with open symbols. Individuals of unknown affected status are shown with a question mark. The arrow head indicates the proband. An asterisk (*) indicates individuals that underwent WES. A dagger (†) indicates individuals that underwent Sanger sequencing for identified candidate variants in *COL17A1* and *DNMBP*.

WES and Variant Calling

WES was performed using genomic DNA samples from the aforementioned five affected individuals and two unrelated, unaffected controls at the UCLA Clinical Microarray Core. DNA libraries were prepared using the TruSeq DNA Sample Preparation Kit v2 (Illumina, Inc., San Diego, CA) and exome capture was performed with the SeqCap EZ Exome Library v3.0 (Roche NimbleGen, Inc., Madison, WI). Paired-end sequencing (2x100 bp) was performed with Illumina's HiSeq 2500. Using the Partek Flow platform (Partek Inc., St. Louis, MO), the resulting reads were aligned to the human hg19 reference genome with Bowtie2, and variant identification was performed with SAMtools mpileup.

WES Data Analysis and Filtering

Using Partek Genomics Suite, multiple criteria were used to filter detected single nucleotide variants (SNVs) to identify those that were: located in a region of ≥ 5X read depth and ≥ 20 base quality score; located in the linked interval on chromosome 10q23-q24; novel or rare (minor allele frequency ≤ 0.01 per The Database of Single Nucleotide Polymorphisms [dbSNP Build ID 138], National Center for Biotechnology Information, National Library of Medicine, Bethesda, MD); heterozygous, identified by allele frequency of 0.5; present in all five affected individuals; absent in both unrelated, unaffected controls; and expressed in *ex vivo* human corneal epithelial cells (reads per kilobase per million ≥1) (Frausto and Aldave, unpublished data recorded in 2014). Quality score values were calculated by Partek Genomics Suite using default settings. Filtered variants were then annotated with RefSeq database compiled on May 7, 2015. While dbSNP was used to determine the minor allele frequency of previously identified variants and perform the filtering for novel or rare variants, the NHLBI Exome Sequencing Project (ESP) (Seattle, WA), EVS data release ESP6500SI-V2, was used to validate the minor allele frequency of filtered variants.

Polymerase Chain Reaction (PCR) Amplification and Automated Sequencing

DNA samples from 13 affected family members, seven unaffected family members and two family members of undetermined affected status were screened for sequence variants that met each of the aforementioned filtering criteria using custom designed PCR primers (Table 1). PCR was performed in 10μL reaction volumes containing 40 ng of genomic DNA, 2.5 picomoles of each primer, 1 μL GeneAmp 10×PCR Gold buffer (Applied Biosystems, Grand Island, NY), 0.8 μL 10 mM dNTP mix (Applied Biosystems, Grand Island, NY), 0.6 μl 25 mM $MgCl_2$ and 1 μL Taq DNA polymerase. Touchdown PCR amplification consisted of a denaturizing step at 95°C for 5 min, followed by decreasing the annealing temperature from an initial 64°C in a stepwise fashion by 0.5°C every second cycle for 15 cycles, then an annealing temperature of 57°C for 20 cycles and finally a prolonged elongation step at 72°C for 10 min. The PCR primers were used for bidirectional sequencing using the BigDye Terminator Ready reaction mix according to the manufacturer's instructions (Applied Biosystems, Grand Island, NY). Sequencing was performed on an ABI PRISM 3130 Automated Sequencer (Applied Biosystems, Grand Island, NY).

Table 1. Filtered coding region variants in positional candidate genes for the corneal dystrophy linked to chromosome 10q23-q24

Gene	Genomic Position	Protein Change	Transcript	Reference	Alternative	RefSeq number	Minor allele frequency
COL17A1[a]	g.105,797,446	p.(Gly1052 =)	NM_000494	G	A	None	N/A
DNMBP[b]	g.101,667,814	p.(Met831Thr)	NM_015221	A	G	rs17854134	0.0088[c] 0.0235[d]
DNMBP[b]	g.101,667,792	p.(Ser2606 =)	NM_015221	C	T	rs17854135	0.0096[c] 0.0258[d]

[a]Sequenced with (5' to 3'): forward primer CTTTGTTCCTTGGTCGGCAG; reverse primer CAGCAAACGAGGAGATGAGG
[b]Sequenced with (5' to 3'): forward primer ATCTTCCCGGAGGCATAGTT; reverse primer GCTCATTTCCGTGCATTTTT
[c]Minor allele frequency from The Database of Single Nucleotide Polymorphisms [dbSNP Build ID 138]
[d]Minor allele frequency from NHLBI Exome Sequencing Project (Seattle, WA), ESV data release ESP6500SI-V2

doi:10.1371/journal.pone.0157418.t001

In silico Protein Analysis

Polyphen-2, SIFT and PANTHER were used to predict the functional impact of segregating missense variants that met the filtering criteria.[15–17] In addition, splice site prediction programs, NetGene2 and MutPred Splice, were used to predict if synonymous substitutions created a cryptic splice donor site. [18, 19]

RESULTS

Determination of Affected Status

Examination of younger affected individuals revealed clear corneas, with the diagnosis made based on the development of spontaneous corneal erosions (Figure 2A–2C). Several decades after the onset of the recurrent corneal erosions, scattered or diffuse subepithelial opacification secondary to subepithelial scarring was observed (Figure 2D–2F). While such scarring did impair vision, it was also typically associated with a decrease in the frequency and severity of episodes of corneal erosion.

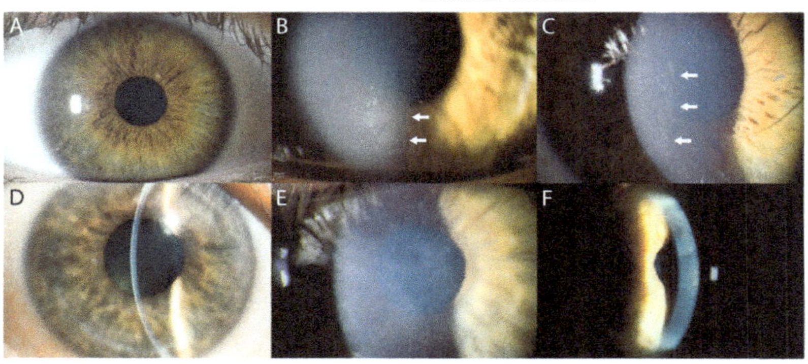

Figure 2. Clinical features of chromosome 10q23-q24 linked dystrophy. A. Affected individual V-2 (28 years old) demonstrates a clear cornea in the right eye. B. Three months later, she developed a corneal erosion (arrows). C. Affected individual V-3 (24 years old) also demonstrated clear corneas prior to the development of a corneal erosion, the location of which can be identified by the opacification of the corneal epithelium (arrows). D. Affected individual IV-1 (72 years old) demonstrates diffuse subepithelial opacification secondary to chronic recurrent corneal erosions seen with slit and diffuse illumination in the right (D and E) and left (F) eyes.

Two individuals (Figure 1: V-5 and V-22) were diagnosed as unaffected when initially examined at ages 20 (V-5) and 22 (V-22). Thus, both appear as unaffected in previously published pedigrees.[8, 12] However, when reexamined nine years later, both were subsequently reclassified as affected (Dr. Yee, unpublished communication). In the most recently published pedigree, individual V-22 is denoted as affected, although individual V-5 is still denoted as unaffected.[9] For individual V-22, the change in classification was made based on the presence of 3 focal corneal opacities, although no explanation was given for the change in affected status of individual V-5. Of note, this individual's status remained unaffected in the most recently published pedigree (individual III-4).[9] Based on the uncertain affected status of both individuals V-5 and V-22, we have changed the status of both to undetermined in the revised pedigree (Figure 1).

Identification of Candidate Genes by WES Analysis

Each sample that underwent WES resulted in an average of 44 million paired-end (2x100 bp) sequence reads with an average quality score of 35 and exome coverage of 43X. An average of 124,730 variants were identified per individual. A total of 1,466 heterozygous SNVs in the chromosome 10q23-q24 linked interval were present in all five affected individuals and absent in the two unaffected controls. While 1,192 of the identified variants were identified in untranslated and other non-coding regions, 153 were predicted to result in synonymous substitutions and 121 were predicted to produce missense or nonsense changes. Three of these SNVs located in two genes survived all of the filtering criteria: one rare variant, c.3156C>T (p.(Gly1052 =)) in exon 46 of *COL17A1* with an undefined minor allele frequency; and two rare variants in *DNMBP*, c.2492T>C (p.(Met831Thr)) and c.2514G>A (p.(Ser2606 =)), with minor allele frequencies of 0.0088 and 0.0096 (dbSNP) and 0.0235 and 0.0258 (ESP), respectively (Table 1). These three remaining SNVs were screened for via Sanger sequencing in affected and unaffected family members (Table 2).

Table 2. Sanger sequencing results of filtered candidate variants for the corneal dystrophy linked to chromosome 10q23-q24

Gene		*COL17A1*	*DNMBP*	*DNMBP*
Strand		-	-	-
Genomic Position		g.105,797,446	g.101,667,814	g.101,667,792
Affected Individuals	III-3	G/A	A/G	C/T
	III-6	G/A	A/G	C/T
	III-19	G/A	A/A	C/C
	IV-1	G/A	A/G	C/T
	IV-2	G/A	A/G	C/T
	IV-6	G/A	A/G	C/T
	IV-8	G/A	A/G	C/T
	IV-14	G/A	A/A	C/C
	IV-18	G/A	A/A	C/C
	IV-23	G/A	A/A	C/C
	IV-27	G/A	A/A	C/C
	V-2	G/A	A/G	C/T
	V-3	G/A	A/G	C/T
Unaffected Individuals	IV-10	G/G	A/A	C/C
	IV-11	G/G	A/A	C/C
	IV-24	G/G	A/A	C/C
	IV-25	G/G	A/A	C/C
	V-9	G/G	A/A	C/C
	V-11	G/G	A/A	C/C
	VI-3	G/G	A/A	C/C
Unknown Affected Status	V-5	G/G	A/A	C/C
	V-22	G/G	A/A	C/C

Screening of Candidate Genes in Additional Affected and Unaffected Individuals

DNA samples from 13 affected family members (including the five in which WES was performed), seven unaffected family members and two family members of undetermined affected status were screened for the c.3156C>T variant in *COL17A1* and the c.2492T>C and c.2514G>A variants in *DNMBP* using Sanger sequencing (Table 2). The *COL17A1* variant was identified in all 13 of the affected family members, none of the seven unaffected family members, and in neither of the two individuals of undetermined affected status. While both *DNMBP* variants were identified in eight of the 13 affected family members, neither was present in the other five affected family members, in the seven unaffected family members or the two individuals of undetermined affected status.

In silico Protein Analysis

SIFT and Panther predict the missense variant in *DNMBP* coding for a p.(Met831Thr) substitution to be tolerated. Polyphen-2 predicts this variant to be possibly damaging, but only with a score of 0.485 out of 1. Both NetGene2 and MutPred Splice predict that the synonymous c.2514G>A (p.(Ser2606 =)) variant in *DNMBP* does not give rise to a splice site variant.

DISCUSSION

Given that COL17A1 protein is a component of hemidesmosomes that are responsible for anchoring corneal epithelial cells to the epithelial basement membrane, Sullivan and colleagues identified *COL17A1* as a positional and functional candidate gene for the cause of chromosome 10q23-q24 linked corneal dystrophy.[12] While screening of the 56 exons and exon/intron junctions of *COL17A1* did not identify any presumed pathogenic mutations, one novel coding region variant was identified in the affected individual and absent in the unaffected individual that was screened. However, as this c.3156C>T variant was predicted to encode a synonymous amino acid substitution, the investigators concluded that it "cannot be considered disease-specific."[12]

Instead of utilizing a candidate gene approach to identifying the genetic basis of chromosome 10q23-q24 linked corneal dystrophy, we used a WES approach to identify the genetic basis of this dominantly inherited disorder. As Sanger sequencing of the positional candidate genes had failed to reveal a potentially pathogenic variant, the accuracy of the boundaries determined by the previous linkage analysis were questioned.[12] We have previously encountered such an error in our search for the genetic basis of another corneal dystrophy, Schnyder corneal dystrophy. After we were unable to identify a pathogenic mutation in any of the genes in the linked interval for Schnyder corneal dystrophy, Dr. Weiss and colleagues reexamined the linkage data and identified an error in the assignment of the candidate locus due to misclassification of the affected status in one pedigree. Exclusion of this family and reanalysis resulted in the expansion of the candidate interval and the identification of pathogenic mutations in the *UBIAD1* gene, located outside of the original interval.[20, 21] Therefore, WES was employed to identify the genetic basis of chromosome 10q23-q24 linked corneal dystrophy, with the initial filtering performed for variants identified in the previously linked interval. We identified three sequence variants that met all of the filtering criteria that we employed: a missense and a synonymous variant in *DNMBP*, and a synonymous variant in *COL17A1*. However, screening of an additional 15 affected and unaffected family members demonstrated that only c.3156C>T variant in *COL17A1* segregated with the affected phenotype. This, in addition to *in silico* analysis predicting that both *DNMBP* variants are most likely tolerated and the fact that both *DNMBP* variants are identified in 2–3% of the population based on the minor allele frequency from ESP (significantly higher that the population prevalence of recurrent corneal erosions), indicates that neither is related to the pathogenesis of chromosome 10q23-q24 linked corneal dystrophy.

The synonymous c.3156C>T (p.(Gly1052 =)) variant in *COL17A1* has been determined to be pathogenic by Jonsson and colleagues.[11] After using WES to identify a missense mutation in*COL17A1* that segregated with the affected phenotype in a large family with epithelial recurrent erosion dystrophy (ERED), they reexamined the potential role of *COL17A1* in chromosome 10q23-q24 linked corneal dystrophy. Specifically, they examined whether the c.3156C>T variant could be disease-causing by altering splicing. The authors noted that three different splice site prediction programs predicted that this variant created a cryptic splice donor site. Additionally, they demonstrated using an *in vitro* splice assay that the c.3156C>T variant produced an aberrantly spliced transcript that corresponded to the introduction of a splice donor site 54 nucleotides 5' of the wild-type exon 46 donor site, as predicted by the splice site prediction programs.[11]

Additional evidence of the pathogenic role of the c.3156C>T variant in *COL17A1* in ERED has been provided by Oliver and colleagues.[13, 14] After a genome-wide association study in a New Zealand family with a corneal dystrophy associated with early onset of recurrent corneal erosions demonstrated a signal on chromosome 10, exome sequencing identified candidate variants in three genes, including *COL17A1*. Screening of *COL17A1* in this family and in two other families with a similar phenotype confirmed segregation of the c.3156C>T variant with the affected phenotype in each family.

The vast majority of previously identified mutations associated with corneal dystrophies have been non-synonymous and within the coding regions of their associated genes.[22] To the best of our knowledge, pathogenic splice site mutations have been previously reported in only two other dystrophies: posterior polymorphous corneal dystrophy 3 and congenital hereditary endothelial dystrophy.[23, 24] The c.3156C>T variant in exon 46 of *COL17A1* described herein is only the third instance in which a mutation affecting a splice site has been associated with a corneal dystrophy and the first in which a cryptic splice site is directly introduced. While the cell-based splice site assays performed by Jonsson et al. likely represent what is occurring *in vivo*, analysis of the COL17A1 protein in tissue or cells obtained from the affected family members would validate the cell based experimental results.[11]

CONCLUSIONS

Given the similar phenotype and clinical course in the family that we report and those recently reported by Jonsson and colleagues and Oliver and colleagues, and the fact that each has been associated with the same variant in *COL17A1*, we propose that chromosome 10q23-q24 linked corneal dystrophy be classified as ERED.[11, 14] The current International Committee for Classification of Corneal Dystrophies (IC3D) classification of the corneal dystrophies has grouped three dystrophies with erosions developing in the first decade of life followed by the development of subepithelial fibrosis under the heading of ERED: Franceschetti corneal dystrophy (FRCD), Dystrophia Smolandiensis (DS) and Dystrophia Helsinglandica (DH).[22] We join Jonsson and colleagues and Oliver and colleagues in encouraging the screening of*COL17A1* in families affected with FRCD, DS and DH to determine if a common genetic basis exists for each.

ACKNOWLEDGMENTS

We thank Dr. Marian Macsai for facilitating patient recruitment.

AUTHOR CONTRIBUTIONS

Conceived and designed the experiments: BL DL YC QW DC RF CC RY FH AA. Performed the experiments: BL DL YC QW DC RF. Analyzed the data: BL DL YC QW DC RF. Contributed reagents/materials/analysis tools: CC RY FH AA. Wrote the paper: BL DL YC QW DC RF CC RY FH AA.

REFERENCES

1. Reis W. Familiäre, fleckige Hornhautentartung. Dtsch Med Wochenschr. 1917;43:575.
2. Bücklers M. Über eine weitere familiäre Hornhaut dystrophie (Reis). Klin Monatsbl Augenkeilkd. 1949;114:386–97.
3. Thiel HJ, Behnke H. Eine bisher unbekannte subepitheliale hereditare Hornhautdystrophie. Klin Monatsbl Augenheilkd. 1967;150(6):862–74. Epub 1967/01/01.
4. Munier FL, Korvatska E, Djemai A, Le Paslier D, Zografos L, Pescia G, et al. Kerato-epithelin mutations in four 5q31-linked corneal dystrophies. Nature genetics. 1997;15(3):247–51. Epub 1997/03/01. doi: 10.1038/ng0397-247 pmid:9054935.
5. Okada M, Yamamoto S, Tsujikawa M, Watanabe H, Inoue Y, Maeda N, et al. Two distinct kerato-epithelin mutations in Reis-Bucklers corneal dystrophy. American journal of ophthalmology. 1998;126(4):535–42. Epub 1998/10/21. pmid:9780098. doi: 10.1016/s0002-9394(98)00135-4
6. Dighiero P, Valleix S, D›Hermies F, Drunat S, Ellies P, Savoldelli M, et al. Clinical, histologic, and ultrastructural features of the corneal dystrophy caused by the R124L mutation of the BIGH3 gene. Ophthalmology. 2000;107(7):1353–7. pmid:10889112. doi: 10.1016/s0161-6420(00)00149-4
7. Ridgway AE, Akhtar S, Munier FL, Schorderet DF, Stewart H, Perveen R, et al. Ultrastructural and molecular analysis of Bowman›s layer corneal dystrophies: an epithelial origin? Investigative ophthalmology & visual science. 2000;41(11):3286–92. Epub 2000/09/28. pmid:11006215.
8. Yee RW, Sullivan LS, Lai HT, Stock EL, Lu Y, Khan MN, et al. Linkage mapping of Thiel-Behnke corneal dystrophy (CDB2) to chromosome 10q23-q24. Genomics. 1997;46(1):152–4. Epub 1997/12/24. doi: 10.1006/geno.1997.5028 pmid:9403072.
9. Nakamura H, Li FT, Foltermann MO, Macsai M, Ma X, Zhao XC, et al. Individual phenotypic variances in a family with Thiel-Behnke corneal dystrophy. Cornea. 2012;31(11):1217–22. Epub 2012/09/12. doi: 10.1097/ICO.0b013e3181c32efd pmid:22965308.
10. Lisch W, Kivela T. Individual phenotypic variances in a family with Thiel-Behnke corneal dystrophy. Cornea. 2013;32(12):e192–3. doi: 10.1097/ICO.0b013e3182a73865 pmid:24113370.
11. Jonsson F, Bystrom B, Davidson AE, Backman LJ, Kellgren TG, Tuft SJ, et al. Mutations in Collagen, Type XVII, Alpha 1 (COL17A1) Cause Epithelial Recurrent Erosion Dystrophy (ERED). Human mutation. 2015;36(4):463–73. doi: 10.1002/humu.22764 pmid:25676728.
12. Sullivan LS, Zhao X, Bowne SJ, Xu X, Daiger SP, Yee SB, et al. Exclusion of the human collagen type XVII (COL17A1) gene as the cause of Thiel-Behnke corneal dystrophy (CDB2) on chromosome 10q23-q25. Current eye research. 2003;27(4):223–6. pmid:14562173. doi: 10.1076/ceyr.27.4.223.16595
13. Vincent AL, Markie DM, De Karolyi B, Wheeldon CE, Patel DV, Grupcheva CN, et al. Exclusion of known corneal dystrophy genes in an autosomal dominant pedigree of a unique anterior membrane corneal dystrophy. Molecular vision. 2009;15:1700–8. pmid:19710953; PubMed Central PMCID: PMC2731832.
14. Oliver VF, van Bysterveldt KA, Cadzow M, Steger B, Romano V, Markie D, et al. A COL17A1 Splice-Altering Mutation Is Prevalent in Inherited Recurrent Corneal Erosions. Ophthalmology. 2016;123(4):709–22. doi: 10.1016/j.ophtha.2015.12.008 pmid:26786512.
15. Adzhubei IA, Schmidt S, Peshkin L, Ramensky VE, Gerasimova A, Bork P, et al. A method and server for predicting damaging missense mutations. Nature methods. 2010;7(4):248–9. doi: 10.1038/nmeth0410-248 pmid:20354512; PubMed Central PMCID: PMC2855889.
16. Ng PC, Henikoff S. Predicting deleterious amino acid substitutions. Genome research. 2001;11(5):863–74. doi: 10.1101/gr.176601 pmid:11337480; PubMed Central PMCID: PMC311071.
17. Mi H, Muruganujan A, Thomas PD. PANTHER in 2013: modeling the evolution of gene function, and other gene attributes, in the context of phylogenetic trees. Nucleic acids research. 2013;41(Database issue):D377–86. doi: 10.1093/nar/gks1118 pmid:23193289; PubMed Central PMCID: PMC3531194.
18. Hebsgaard SM, Korning PG, Tolstrup N, Engelbrecht J, Rouze P, Brunak S. Splice site prediction in Arabidopsis thaliana pre-mRNA by combining local and global sequence information. Nucleic acids research. 1996;24(17):3439–52. pmid:8811101; PubMed Central PMCID: PMC146109. doi: 10.1093/nar/24.17.3439
19. Mort M, Sterne-Weiler T, Li B, Ball EV, Cooper DN, Radivojac P, et al. MutPred Splice: machine learning-based prediction of exonic variants that disrupt splicing. Genome biology. 2014;15(1):R19. doi: 10.1186/gb-2014-15-1-r19 pmid:24451234; PubMed Central PMCID: PMC4054890.
20. Aldave AJ, Rayner SA, Principe AH, Affeldt JA, Katsev D, Yellore VS. Analysis of fifteen positional candidate genes for Schnyder crystalline corneal dystrophy. Molecular vision. 2005;11:713–6. pmid:16163269.
21. Weiss JS, Kruth HS, Kuivaniemi H, Tromp G, White PS, Winters RS, et al. Mutations in the UBIAD1 gene on chromosome short arm 1, region 36, cause Schnyder crystalline corneal dystrophy. Investigative ophthalmology & visual science. 2007;48(11):5007–12. doi: 10.1167/iovs.07-0845 pmid:17962451.
22. Weiss JS, Moller HU, Aldave AJ, Seitz B, Bredrup C, Kivela T, et al. IC3D classification of corneal

dystrophies—edition 2. Cornea. 2015;34(2):117–59. doi: 10.1097/ICO.0000000000000307 pmid:25564336.

23. Evans CJ, Liskova P, Dudakova L, Hrabcikova P, Horinek A, Jirsova K, et al. Identification of six novel mutations in ZEB1 and description of the associated phenotypes in patients with posterior polymorphous corneal dystrophy 3. Annals of human genetics. 2015;79(1):1–9. doi: 10.1111/ahg.12090 pmid:25441224.

24. Kodaganur SG, Kapoor S, Veerappa AM, Tontanahal SJ, Sarda A, Yathish S, et al. Mutation analysis of the SLC4A11 gene in Indian families with congenital hereditary endothelial dystrophy 2 and a review of the literature. Molecular vision. 2013;19:1694–706. pmid:23922488; PubMed Central PMCID: PMC3733908.

CHAPTER 5

WHOLE-EXOME SEQUENCING OF PANCREATIC CANCER DEFINES GENETIC DIVERSITY AND THERAPEUTIC TARGETS

Agnieszka K. Witkiewicz, Elizabeth A. McMillan, Uthra Balaji, GuemHee Baek, Wan-Chi Lin, John Mansour, Mehri Mollaee, Kay-Uwe Wagner, Prasad Koduru, Adam Yopp, Michael A. Choti, Charles J. Yeo, Peter McCue, Michael A. White & Erik S. Knudsen

Keywords: Biological sciences, Cancer, Genetics

INTRODUCTION

Pancreatic ductal adenocarcinoma (PDA) portends a poor prognosis, with a 5-year survival rate of ~6% (refs 1, 2, 3). This poor outcome is related to multiple factors, including the relatively late stage of diagnosis, many patients presenting with unresectable disease, and therapy recalcitrance resulting in disease recurrence in spite of operable disease and systemic therapy1. Genetic analysis of PDA yielded insights related to altered signalling pathways4·5; however, unlike other cancers the number of sequenced PDA genomes has been relatively modest. This reflects the difficulty of sequencing a tumour that is dominated by presence of stromal and inflammatory cells2·6. Tumour cellularity in samples subjected to whole-exome or -genome sequencing represents a known barrier in obtaining high quality data; therefore, many studies recommended >60% tumour cell content in the samples7. To facilitate mutation detection and remove contaminating non-neoplastic tissue the initial PDA sequencing effort employed patient-derived xenografts and cell lines5. This approach resulted in detecting an average of 48 somatic mutations per model sequenced. Subsequent work employed exome sequencing on 99 clinical cases of PDA, and applied the q-pure algorithm to overcome contamination with non-neoplastic tissue, yielding an average of 26 mutations per case4.

Here 109 micro-dissected PDA cases annotated with outcome data were subjected to whole-exome sequencing. The data from this analyses demonstrates that there is substantial genetic diversity within PDA with a mutation frequency comparable to other solid tumours. Alterations in mutation spectra and burden are associated with hallmark pancreatic cancer risk factors. Increased copy number variation indicative of chromosomal instability is associated with poor outcome. Amplification of the locus involving the *MYC* oncogene are specifically associated with poor outcome and association with adenosquamous subtype of PDA. In contrast, select mutated genes (for example, *RBM10*) are associated with longer patient survival. Diversity within the canonical KRAS signalling pathway defined prognostic features of the disease and the potential for targeted therapy of BRAF-mutated cases. The mutation and copy number alterations frequently involved important signalling pathways that contribute to tumour development and represent therapeutic targets.

RESULTS

Tissue Cell Enrichment and Mutation Burden

To enhance tumour cellularity in PDA patient samples, needle dissection was employed to significantly enrich PDA tumour cells from the surrounding microenvironment (Fig. 1a). This approach was applied to a total of 109 surgically resected

Citation: Witkiewicz, A. K. et al. Whole-exome sequencing of pancreatic cancer defines genetic diversity and therapeutic targets. Nat. Commun. 6:6744 doi: 10.1038/ncomms7744 (2015).

PDA cases annotated with clinical outcome and etiological features (Table 1). The cohort had a median survival of 21 months that is consistent with historical survival of resected PDA8·9. Of standard pathological features, grade 3 was associated with poor outcome, while nodal status and adenosquamous subtype trended towards poor outcome. Enriched tumour cells were utilized as the substrate for whole-exome sequencing (Methods). Average tumour purity exceeded 50% and the average sequencing depth was ~50 × . In prior studies sequencing PDA4, 26 mutations were detected per case representing a relatively low mutational burden when compared with the majority of solid tumours. The average mutation burden observed in our cohort was 67 non-synonymous events per case, equivalent to multiple other solid-tumour types (Fig. 1b). Sanger sequencing was used to validate 248 genetic events with validation occurring in 92% of cases. In addition, p53 immunohistochemistry performed across the cohort demonstrated the expected strong correlation between *TP53* mutation and stabilization of the protein. To determine whether deeper sequencing would further increase the number of mutations detected, 21 cases were sequenced to ~120 × depth. The majority of events identified with deeper sequencing had low allelic frequency or coverage depth, and only a small fraction of statistically significant mutations were identified.

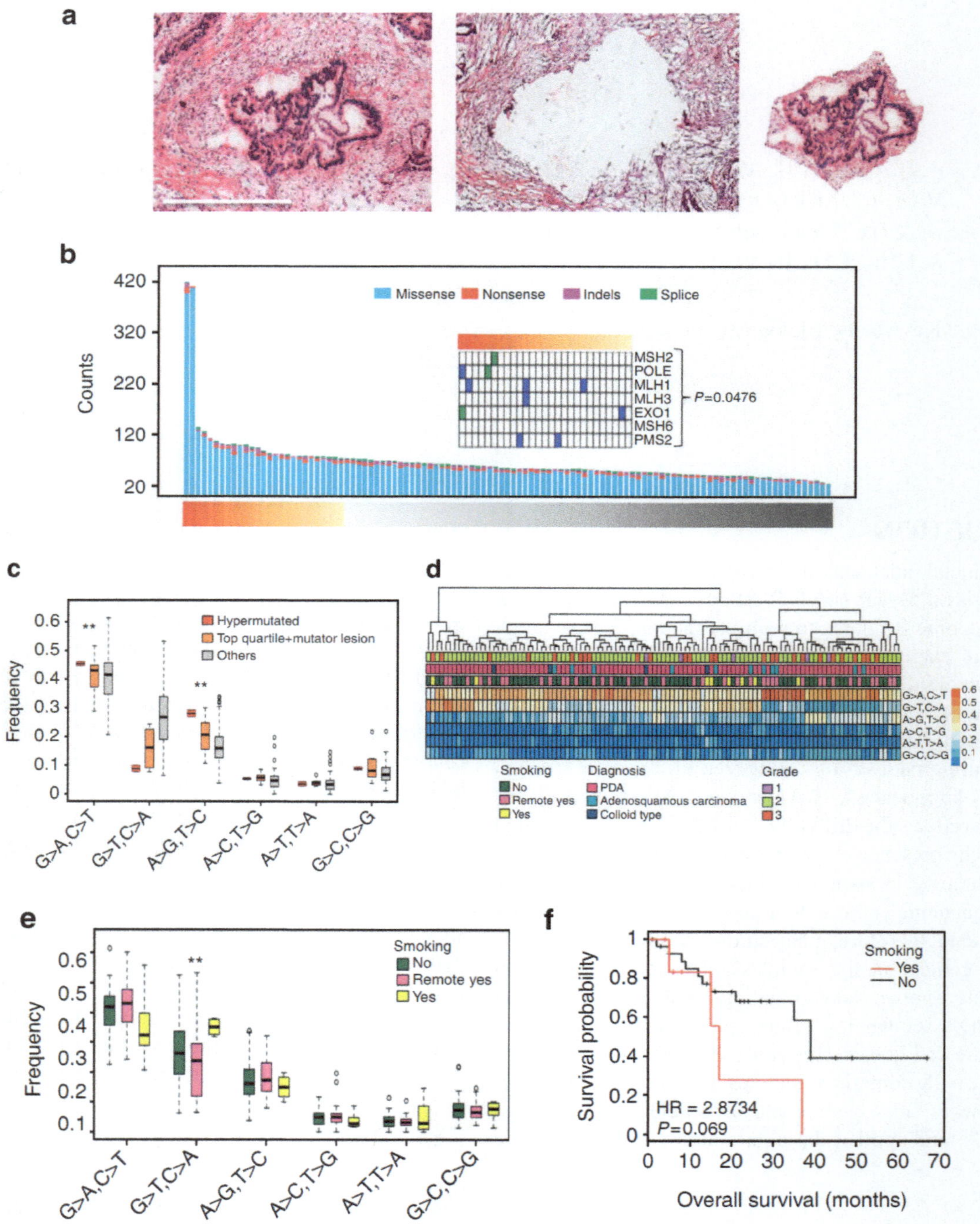

Figure 1: Overall mutational landscape of PDA and association with etiological features of disease. (**a**) Representative image of the needle dissection approach used to enrich tumour cells for whole-exome sequencing (scale bar, 600 μM) (**b**) Mutational burden across the sequenced cohort. The presence of genetic alterations in genes associated with mutator phenotypes are shown for the cases with the top 25% of muta-

tional burden. Association with mutational burden was determined by a hypergeometric test. (**c**) Mutation spectra of the hypermutated cases, top quartile cases with mutator mutations and others (** denotes $P<0.05$, Student's t-test). The boxes show the distance between the first and third quartile with the whiskers extending up to 1.5 times the interquartile range. (**d**) Unsupervised hierarchical clustering of cases based on mutation spectra. (**e**) Mutation spectra based on smoking status, the increase in G>T transversion is significant ($P<0.05$, Student's t-test). The boxes show the distance between the first and third quartile with the whiskers extending up to 1.5 times the interquartile range. The unfilled circles represent possible outliers. (**f**) Association of PDA smoking status with overall survival. Hazard ratio and P value were obtained from Cox proportional hazard test.

Table 1: Demographic and clinicopathologic features of the PDA 109 patient cohort analyzed

Variable	N (109)
Age (years)	
Average (range)	66 (29-86)
Gender	
Male	55 (50%)
Female	54 (50%)
Diagnosis	
PDA	94 (86%)
Adenosquamous	11 (10%)
Colloid	4 (4%)
Grade	
1	5
2	77
3	27
Survival (months)	
Median	21

PDA, pancreatic ductal adenocarcinoma.

Determinants of Mutation Burden and Spectrum

Consistent with observations in other solid malignancies[10,11,12], cases with the highest relative mutational burden harboured genetic lesions in mismatch repair genes, known to be associated with a mutator phenotype (Fig. 1c). These cases also displayed a mutation spectrum consistent with mismatch repair deficiency, T>C transition at CTG tri-nucleotides[11,13] (Fig. 1c). The overarching mutational spectra of the PDA cohort were dominated by C>T mutations associated with age, and had minimal evidence for enriched APOBEC-associated mutagenesis at TCW tri-nucleotides. By hierarchical clustering it was apparent that the variable mutational spectra of individual PDA cases specified discrete subtypes (Fig. 1d). Smoking status was associated with an enrichment of C>A transversions that resembled the mutational 'smoking signature' observed in other cancers for which smoking is a significant risk factor (Fig. 1e)[12,14]. Smokers had generally poor outcome, with a median survival half that of non-smokers (Fig. 1f). Thus, etiological events associated with the development of PDA were identifiably associated with the mutational spectrum of the tumours.

Copy Number Variation and the Role of MYC

Affinity propagation clustering (APC) was used to delineate deterministic patterns of commonality associated with copy number alterations (Fig. 2a, left panel) and partitioned cases in discrete subtypes with increasing genetic complexity (Fig. 2a, right panel). Cases with high levels of amplifications/deletions (clusters 5 and 6) were significantly over-represented for alterations of genes involved in DNA break repair, but not *TP53* (Fig. 2b). Similar results were observed with standard hierarchical clustering methods based on Euclidean distance. Both cluster 5 and 6 harboured poor outcome relative to other clusters with fewer copy number changes (Fig. 2c). GISTIC analysis defined significant common regions of amplification and deletion that harbour multiple oncogenes (for example, *MYC* and *CCND1*) and tumour suppressors (for example, *SMAD4* and *CDKN2A*) (Fig. 2d). Interrogation of amplified and deleted regions for association with survival revealed that the amplification of the 8q24 locus, harboring the *MYC* oncogene, was uniquely associated with poor outcome (Fig. 2e). While *MYC* overexpression has been shown to facilitate the development of pancreatic cancer in mouse models[15], little analysis has been performed in patient specimens. The amplification of *MYC* was confirmed by fluorescent *in situ* hybridization (Fig. 2f). The *MYC* amplified cases did not have a higher mutation burden or association with other hallmark mutations of PDA; however, amplification was significantly over-represented in the adenosquamous subtype of pancreatic carcinoma. Evaluation of precursor lesions pancreatic intraepithelial neoplasia (PanIN) associated with invasive disease also revealed *MYC* amplification. These data suggest a role of *MYC* in initiation and progression of this exceedingly aggressive form of PDA. Consistent with this observation, in a *MYC*-driven mouse model of pancreatic cancer, the tumours exhibited adenosquamous histology and stained positively for p63, an established marker of squamous differentiation.

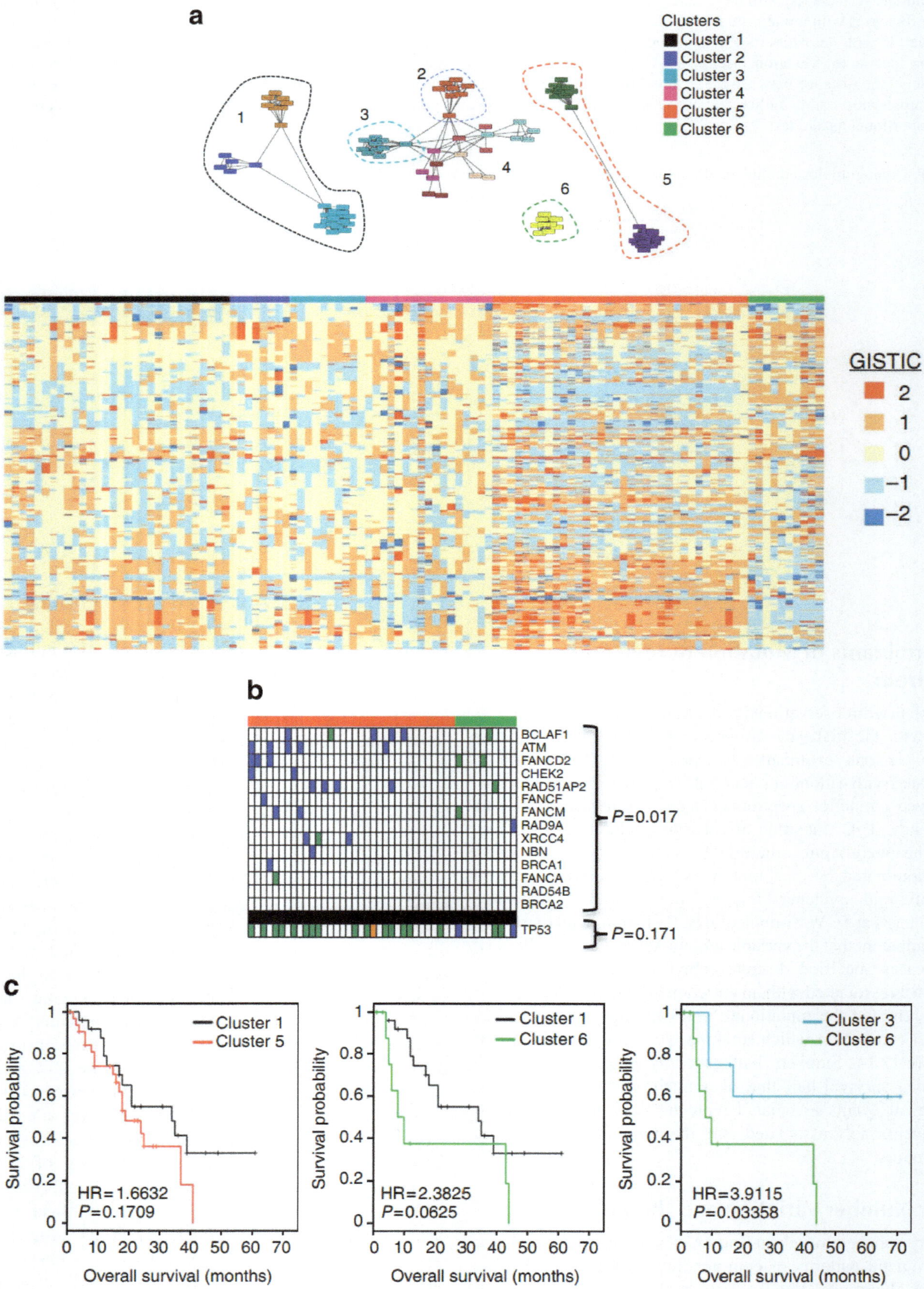
a
Clusters
Cluster 1
Cluster 2
Cluster 3
Cluster 4
Cluster 5
Cluster 6
1
2
3
4
5
6
GISTIC
2
1
0
−1
−2
b
BCLAF1
ATM
FANCD2
CHEK2
RAD51AP2
FANCF
FANCM
RAD9A
XRCC4
NBN
BRCA1
FANCA
RAD54B
BRCA2
P=0.017
TP53
P=0.171
c
Cluster 1
Cluster 5
HR=1.6632
P=0.1709
Cluster 1
Cluster 6
HR=2.3825
P=0.0625
Cluster 3
Cluster 6
HR=3.9115
P=0.03358
Survival probability
Overall survival (months)

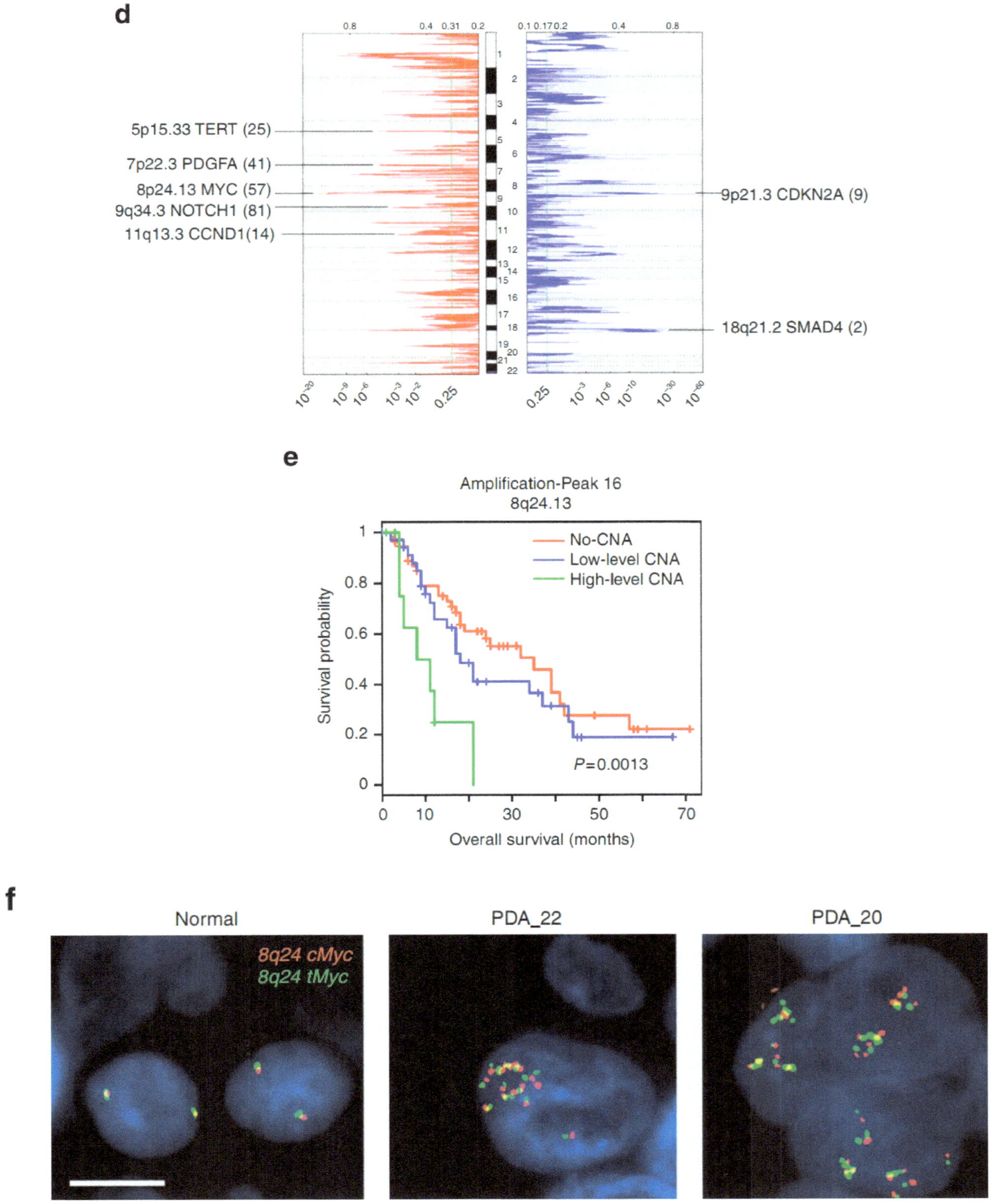

Figure 2: Copy number alterations in PDA. (**a**) APC clustering of PDA cases by CNV revealed several distinct clades and associated clusters. Clusters 5 and 6 exhibit higher overall CNV relative to the other clusters. (**b**) Clusters 5 and 6 are enriched for mutations or homozygous deletion of genes involved in double-strand break repair, but not in TP53. Significance was determined by hypergeometric test. (**c**) Cluster 5 and 6 trend towards poor survival relative to clusters with less CNV. Hazard ratio and *P* values were obtained from Cox proportional hazard test. (**d**) GISTIC analysis of cases reveals chromosomal regions that are significantly deleted/amplified. (**e**) Kaplan–Meier analysis of the association of the 8q24.13 locus amplification with overall survival. *P* value was obtained from Cox proportional hazard test. (**f**) Fluorescence *in situ* hybridization with MYC break apart probes demonstrates amplification in the absence of translocation at the MYC locus (Scale bar, 5 μM).

Significantly Mutated Genes in Pancreatic Cancer

MutSigCV analysis16 of 109 PDA/normal tissue pairs revealed 24 significantly mutated genes occurring in >3.5% of cases (Fig. 3a). For comparison. similar analysis performed using data from prior sequencing effort4 revealed only four non-synonymous mutations. The identification of*KRAS, TP53, CDKN2A* and *SMAD4* confirmed that our approach detected known alterations promoting PDA tumorigenesis. *GNAS* mutations were present in all four colloid carcinomas and two conventional PDAs. Most*GNAS* mutations were in hotspot codon 201 (R201C and R201H). Consistent with recent studies17, all*GNAS*-mutated invasive carcinoma cases were derived from a common precursor, intraductal pancreatic neoplasm (IPMN). Four of the six PDAs harbouring *GNAS* alteration had concomitant mutations in *KRAS*. Two other genes reported as mutated in IPMN,*RNF43* and *RBM10*, were detected in six and four conventional PDA cases, respectively. Notably, PDAs harbouring *RNF43* and *RBM10*alterations did not arise from IPMN and there was no association between these genes and *GNAS* mutations in this cohort. We also identified a number of 'cancer genes'18 as significantly mutated in PDA, including *BCLAF1* (5% of cases), *IRF6* (4% of cases), *FLG* (10% of cases),*AXIN1* (5% of cases), *GLI3* (6% of cases) and *PIK3CA* (4% of cases).

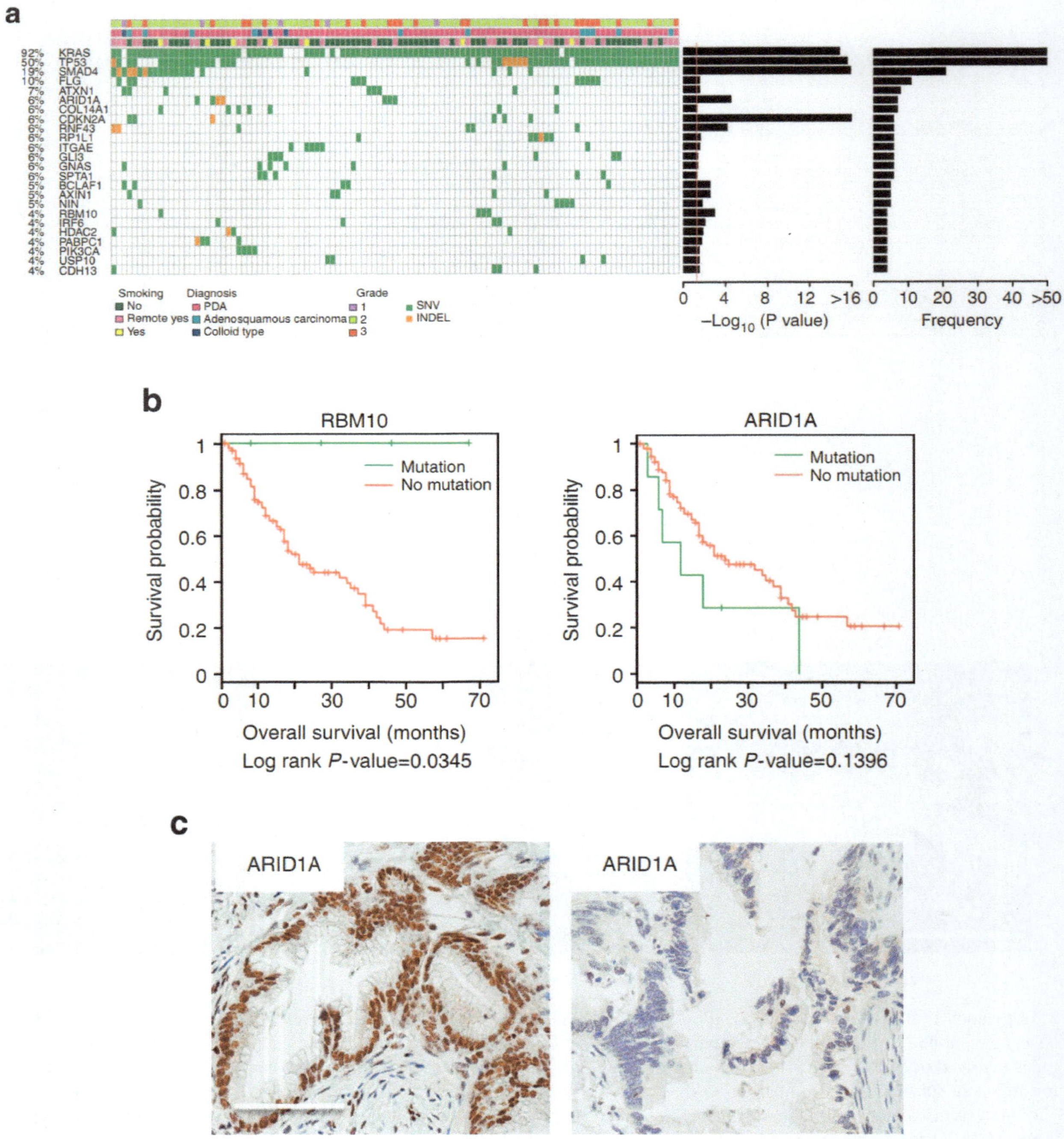

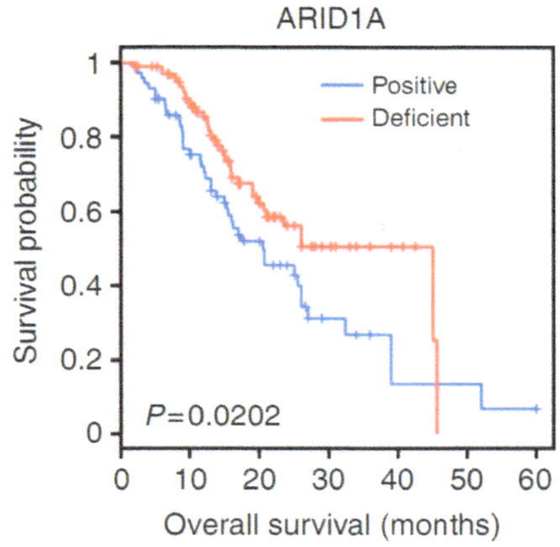

Figure 3: Significantly mutated genes in PDA. (**a**) Mutational significance was determined for SNV and INDELS from the 109 sequenced cases using MutsigCV. Genes were subjected to unsupervised clustering, and the frequency and *P* value as determined by MutsigCV are shown. (**b**) Kaplan–Meier analysis of select significantly mutated genes. *P* value was obtained from Cox proportional hazard test. (**c**) IHC was used to confirm loss of ARID1A protein in a large cohort of PDA cases (scale bar, 150 μM). Diminished ARID1A protein level was associated with overall survival. *P* value was obtained from Cox proportional hazard test.

Mutated Genes Associated with Pancreatic Cancer Prognosis

Selected mutations were associated with pathological features of disease or survival (Fig. 3b). *RBM10* is critical regulator of alternative splicing that was recently identified as significantly mutated in lung cancer19,20. Remarkably, *RBM10* mutations were associated with enhanced survival (Fig. 3b), although all *RBM10*-mutated cases were of high grade, pT3 stage and harboured lymph node metastasis in three of four cases. Mutations in the chromatin remodelling gene ARID1A trended towards poor survival (Fig. 3b). Immunohistochemical analyses of additional 296 PDA cases demonstrated that ARID1A protein deficiency was significantly associated with poor outcome in this expanded cohort (Fig. 3c). The loss of ARID1A protein is similarly observed in multiple PDA cell lines, although whether this event is functionally significant is unknown. Loss of ARID1A is known to induce vulnerability to ARID1B depletion in other cancer models21and we could confirm this finding in ARID1A-deficient PDA cell lines, indicating that loss of protein has a functional consequence. To further explore gene interactions, deterministic clustering was performed followed by analysis of exemplar cases. This work further reinforced the significance of genetic alterations beyond *KRAS* and *TP53*, as coordinate mutation of *ARID1A* delineated a cluster with worse prognosis. Correlation of mutations with clinicopathologic data revealed specific somatic aberrations associated with PDA grade, presence of lymph node metastasis and adenosquamous histology.

To comprehensively interrogate significant mutations in PDA clinical cases, MutSigCV analysis was performed using combined data from current and 99 previously reported PDA cases4. The combined 208 cases yielded no additional high-frequency mutated genes, however, multiple lower frequency (<2.5% recurrence) cancer-related genes were identified including *ATM, ARID2, TGFBR2* and *ACVR1B*. This additional discovery is consistent with the added power of combined data analysis being required for a full accounting of cancer genomes18.

Genetic Diversity in the KRAS Pathway

PDA is dominated by mutations in *KRAS* that were detected in 92% of the cases in our cohort (Fig. 4a). The majority of KRAS mutations occurred in codon 12, although mutations also occurred in codons 13 and 61 (Fig. 4b and c), consistent with other sequencing studies4,5 (Fig. 4c). It has been suggested that different *KRAS* mutations harbor diverse biological activity22. Notably, while all mutations affecting codon 12 exhibited similar association with survival, cases mutated at codon 61 had a remarkably favorable prognosis in this cohort (Fig. 4d). Interestingly, as we observed in the *RBM10* mutant tumours, cases harboring mutations at codon 61 paradoxically exhibited histological and clinical features indicative of poor survival. When assessed for pERK staining, cases with *KRAS* mutations at codon 61 exhibited less ERK activation as compared with cases with other *KRAS* alleles (Fig. 4e). This finding suggests that in PDA, *KRAS* mutant allele status potentially determines the quality of RAS pathway activation and the prognosis of disease.

a

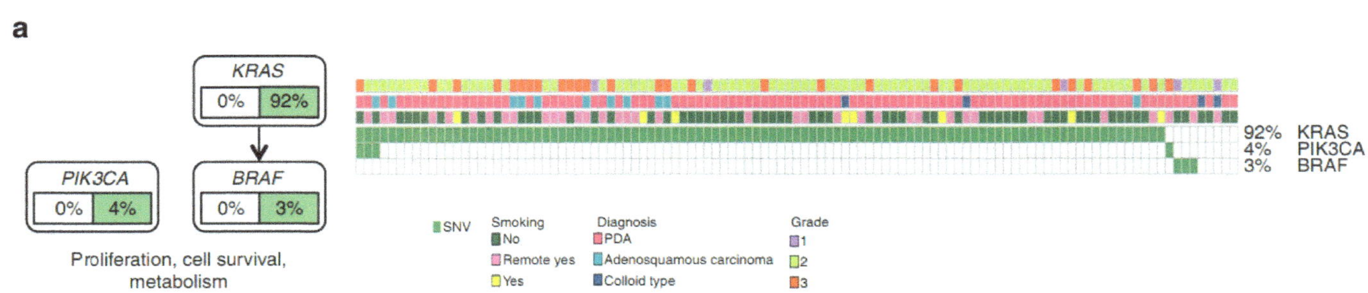

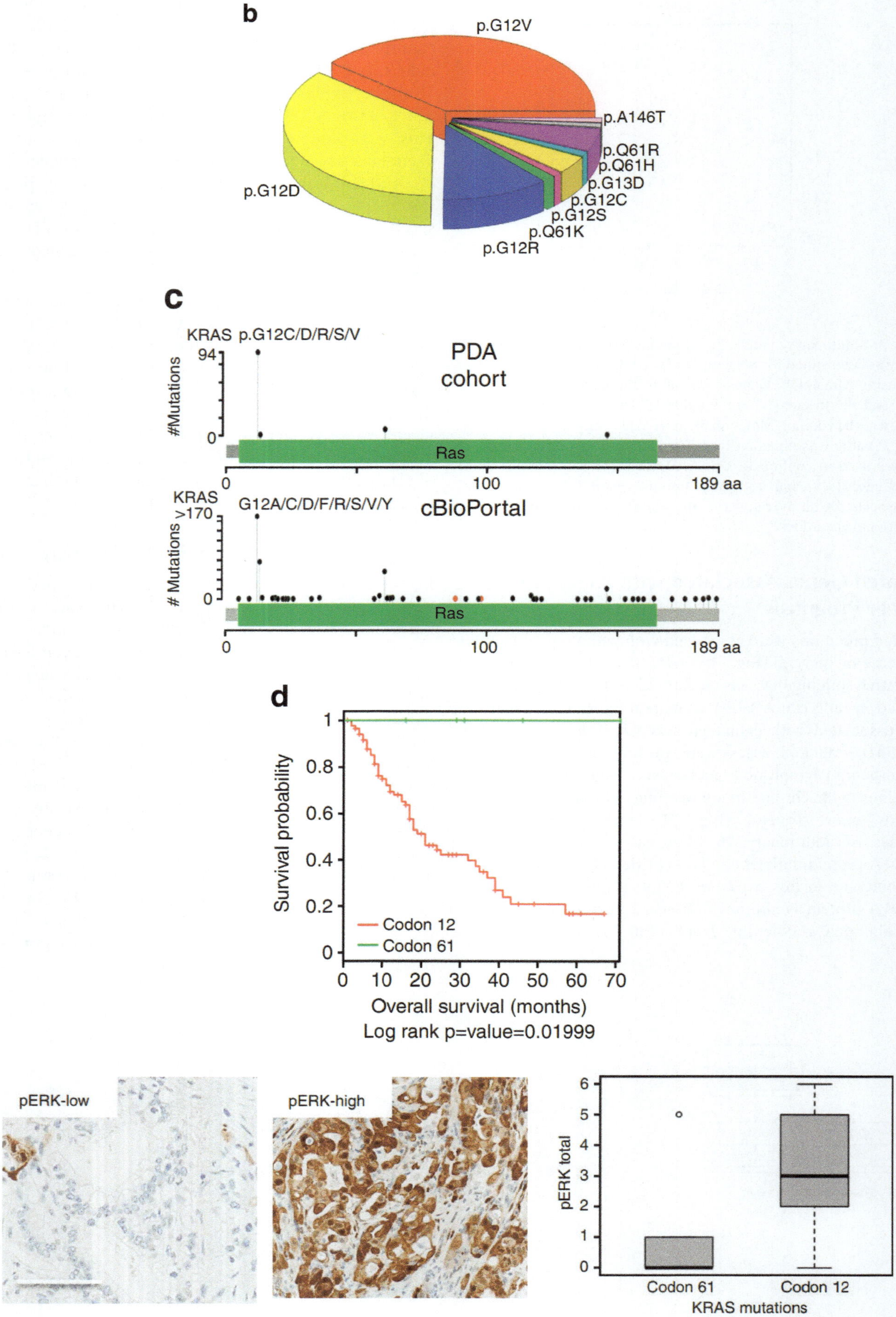
b
p.G12V
p.A146T
p.Q61R
p.Q61H
p.G13D
p.G12C
p.G12S
p.Q61K
p.G12R
p.G12D
c
KRAS p.G12C/D/R/S/V
PDA
cohort
#Mutations
94
0
Ras
0
100
189 aa
KRAS
G12A/C/D/F/R/S/V/Y
cBioPortal
Mutations
>170
0
Ras
0
100
189 aa
d
Survival probability
1
0.8
0.6
0.4
0.2
0
Codon 12
Codon 61
0 10 20 30 40 50 60 70
Overall survival (months)
Log rank p=value=0.01999
pERK-low
pERK-high
pERK total
6
5
4
3
2
1
0
Codon 61
Codon 12
KRAS mutations

(e)

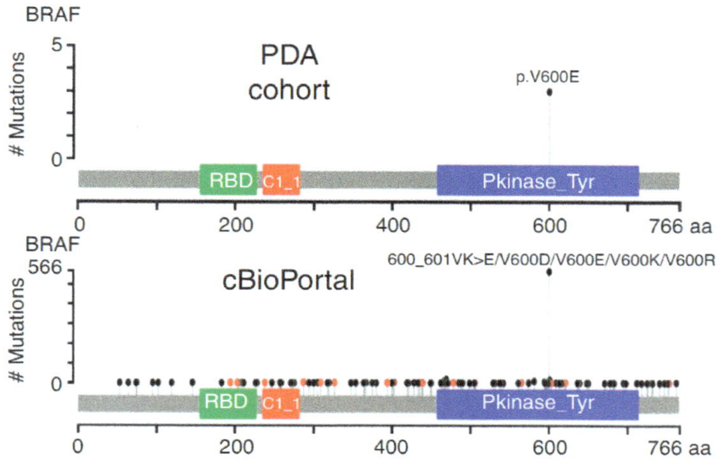

(f)

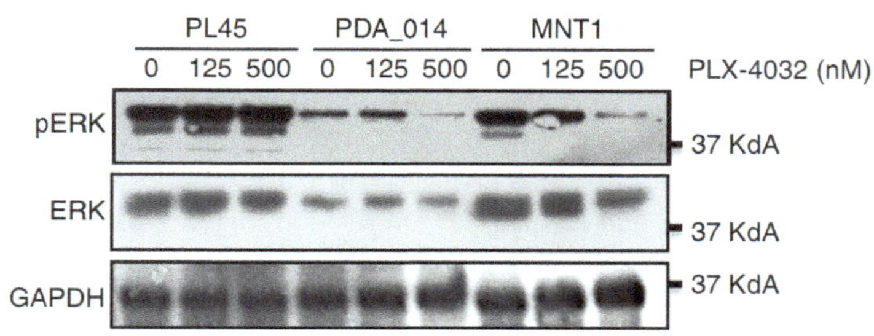

(g)

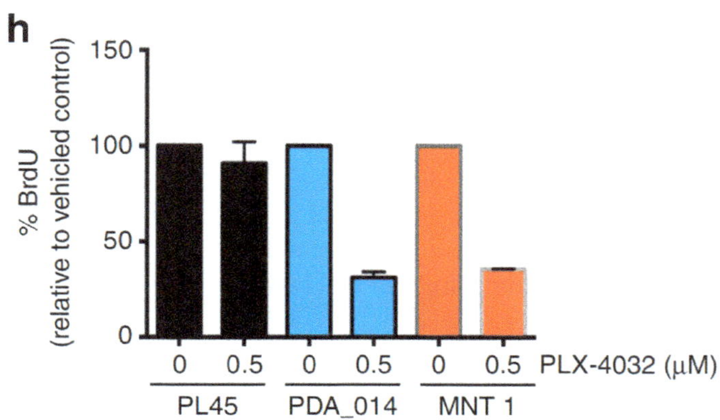

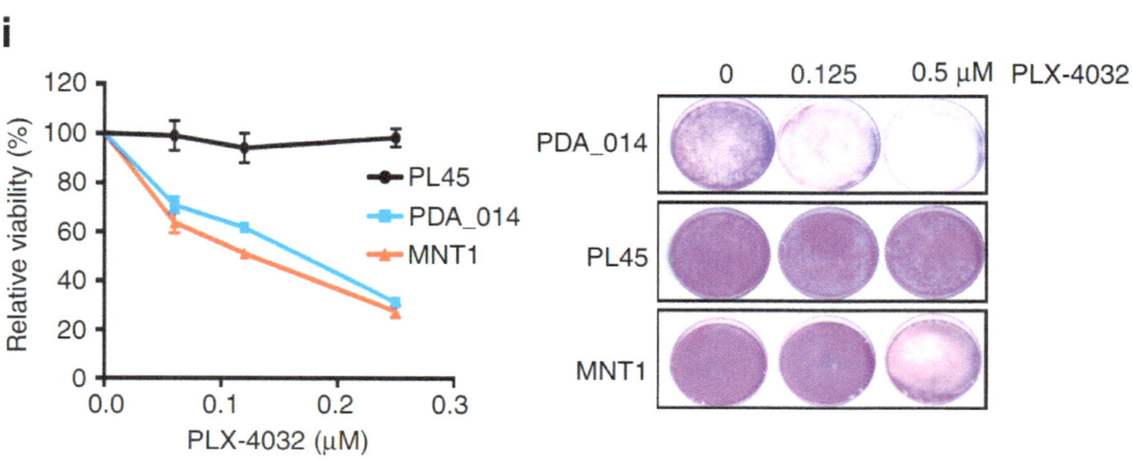

Figure 4: KRAS/BRAF-Pathway in PDA. (**a**) Oncomap showing the distribution of KRAS, BRAF and PIK3CA mutation in the cohort. (**b**) KRAS mutations in the cohort exhibit a diverse spectrum concentrated at known oncogenic codons 12, 13 and 61 (**c**) Analysis of KRAS mutations in the PDA cohort in comparison with all cancer cases in cbioportal (**d**) Kaplan–Meier analysis of codon 12 versus codon 61

mutated cases. Kaplan–Meier analysis of codon 12 versus codon 61 mutated cases. *P* value was obtained from Cox proportional hazard test. (**e**) Representative staining of pERK in pancreatic cancer cases (scale bar, 150 μM), and the association of pERK staining in PDA cases with KRAS codon 12 versus codon 61 mutations. The boxes show the distance between the first and third quartile with the whiskers extending up to 1.5 times the interquartile range. (**f**) Analysis of BRAF mutations identified in the PDA cohort in comparison with the analysis of all cancer cases in bioportal. (**g**) Impact of PLX-4032 on the phosphorylation of ERK in the indicated cell lines. PL45 is KRAS mutated, PDA_014 is a patient-derived cell line harbouring BRAF V600E and MNT1 is a BRAF V600E melanoma cell line (**h**) Impact of PLX-4032 on cell cycle progression in the indicated cell lines (data analysis was performed in triplicate and error bars denote s.d.). (**i**) Impact of PLX-4032 on viability in the indicated cell lines (data points are from at least six values and error bars denote s.d.).

While prior sequencing studies have shown that a small subset of PDA cases harbour wild-type *KRAS*, the genomic driver(s) of these tumours has remained obscure. Here we found that *KRAS* wild-type PDA exhibit somatic lesions in known oncogenes that are also RAS effector proteins (Fig. 4a). Importantly, the *PIK3CA*mutations and *BRAF* mutations, occurring in cases with wild-type KRAS, are known to be mechanistically activating and are enriched in other tumour types (Fig. 4f). In cases with wild-type *KRAS, BRAF* and *PIK3CA* a number of cancer-associated genes representing potential drivers were identified (for example, *STK11, GNAS, CHEK2* and *RB1*). *BRAF* V600E mutations, occurring at a frequency of 3%, were mutually exclusive with *KRAS*mutations. While the expression of *BRAF* V600E would suggest that it is a driver event based on mouse models23, no endogenous model of this form of pancreatic cancer had been developed. One of the *BRAF* V600E cases (PDA_014) was utilized to develop a cell line that maintained the *BRAF* mutation. *BRAF* V600E PDA cell line was equally sensitive as MNT1 melanoma cells to the FDA-approved BRAF inhibitor PLX-4032, while KRAS driven cells (PL45) are resistant (Fig. 4g-i). Together, these data suggest that genetic analysis of PDA could identify a subset of patients, *albeit* small, who may benefit from targeted therapy along the KRAS/BRAF axis.

Frequently Mutated Pathways in Pancreatic Cancer

In addition to heterogeneity within the core KRAS pathway, we detected multiple additional pathways that were genetically altered at high frequency (>20%) in PDA (Fig. 5a). Specifically, we found that the TGF-ß pathway is disrupted via a combination of largely mutually exclusive events in addition to the frequent disruption of *SMAD4*. This includes loss of *TGFBR2/ TGFBR1*, as well as mutations in *ACVR1B*, which is a newly identified cancer-associated gene in the pathway18. Similarly, in concert with frequent deletion of *CDKN2A* and *CDKN2B, CDK4* and *CCND1* amplification and*RB1* loss was observed in the RB pathway. Among signalling pathways, beta-catenin and NOTCH pathways exhibited frequent alterations indicative of oncogenic activation. Multiple Fanconi Anaemia genes as well as *ATM, CHEK2, BCLAF1, BRCA1* and *BRCA2* were observed to be mutated or deleted at relatively high frequency with alterations targeting some facet of DNA repair occurring in >35% of cases. Genetic lesions in chromatin remodelling SWI/ SNF pathway occurred in 42% of the cases. In addition to these well-described pathways, we observed frequent aberrations that impinge on histone modification and the FAT/HIPPO pathway. Each pathway was interrogated for association with outcome. Only DNA repair-associated pathways had a trend towards poor outcome, while beta-catenin signalling trended towards improved outcome. In general, the associations between pathways were limited as determined by Pearson's correlation (Fig. 5b). Random Forest and APC clustering were used to define subtypes of PDA based on altered pathways, and demonstrated marked diversity of combinations of deregulated pathways (Fig. 5c). While tumours with isolated KRAS-pathway alterations alone or in combination with TP53 have a poor prognosis, tumours with more complex pathway deregulations trended towards even poorer outcome (Fig. 5d). Many of the highlighted pathways represent therapeutic targets that are actionable in preclinical models and, in some cases, in the clinic (Table 2). For example, BRAF V600E has been shown to be a potential target in melanoma and other cancers. Loss of *RNF43* or *AXIN1*are associated with sensitivity to porcupine and tankyrase inhibitors targeting the beta-catenin pathway24, while deletion of *CDKN2A* or amplification of *CDK4/CCND1* confer sensitivity to CDK4/6 inhibitors25. Deficits in Fanconi Anaemia genes can be targeted by cross-linking agents26, a finding readily observed in cell lines harbouring homozygous deletion of *FANCF*. Similarly, loss of BRCA function is associated with response to PARP inhibitors27. A number of these genetic alterations remain actionable in the presence of an activating KRAS mutation and could provide the opportunity for genetically targeted therapy in PDA.

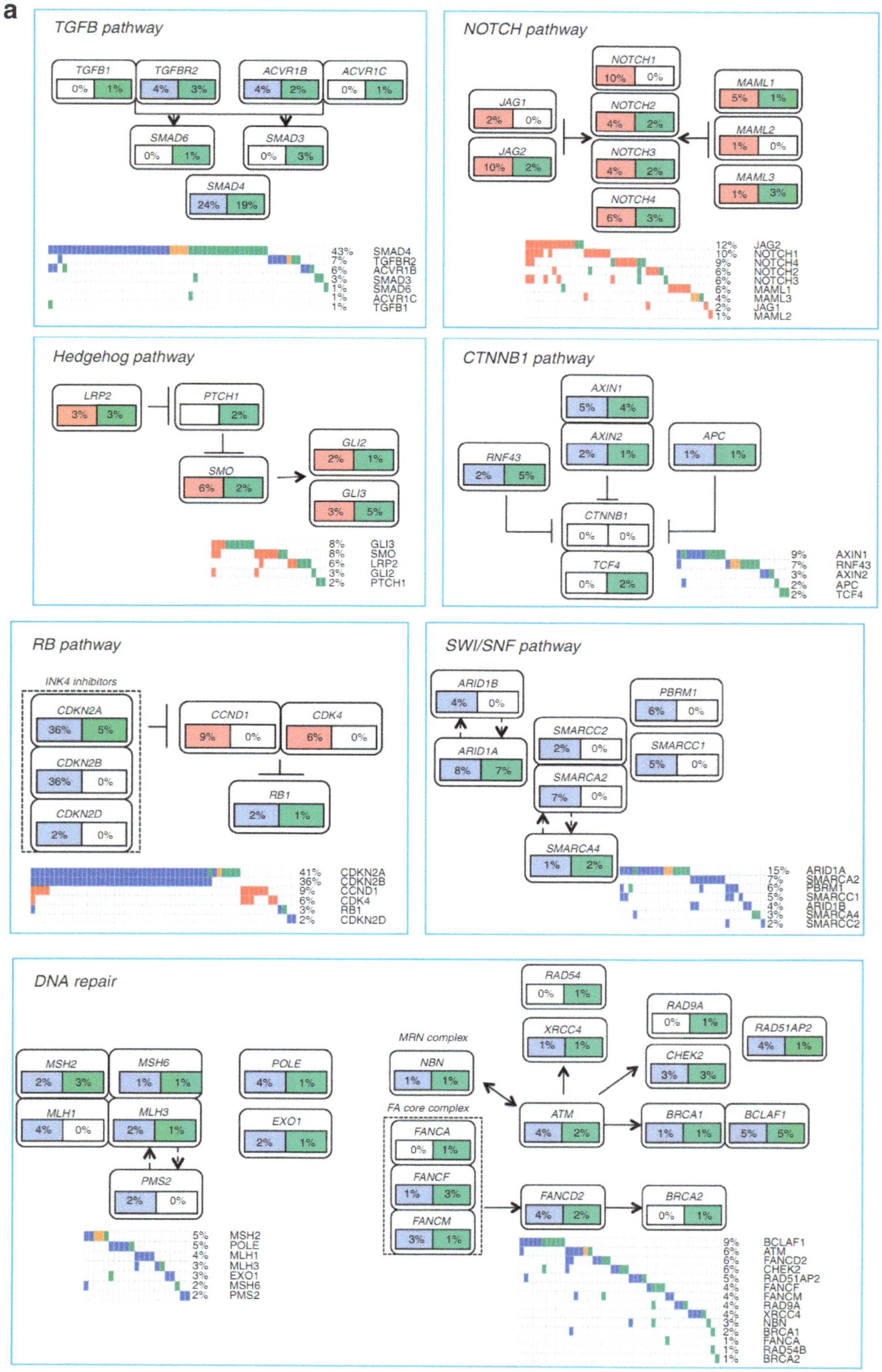
a
TGFB pathway
TGFB1 0% 1%
TGFBR2 4% 3%
ACVR1B 4% 2%
ACVR1C 0% 1%
SMAD6 0% 1%
SMAD3 0% 3%
SMAD4 24% 19%
43% SMAD4
7% TGFBR2
6% ACVR1B
3% SMAD3
1% SMAD6
1% ACVR1C
1% TGFB1
NOTCH pathway
NOTCH1 10% 0%
JAG1 2% 0%
NOTCH2 4% 2%
MAML1 5% 1%
MAML2 1% 0%
JAG2 10% 2%
NOTCH3 4% 2%
MAML3 1% 3%
NOTCH4 6% 3%
12% JAG2
10% NOTCH1
9% NOTCH4
6% NOTCH2
6% NOTCH3
6% MAML1
4% MAML3
2% JAG1
1% MAML2
Hedgehog pathway
LRP2 3% 3%
PTCH1 2%
GLI2 2% 1%
SMO 6% 2%
GLI3 3% 5%
8% GLI3
8% SMO
6% LRP2
3% GLI2
2% PTCH1
CTNNB1 pathway
AXIN1 5% 4%
AXIN2 2% 1%
APC 1% 1%
RNF43 2% 5%
CTNNB1 0% 0%
TCF4 0% 2%
9% AXIN1
7% RNF43
3% AXIN2
2% APC
2% TCF4
RB pathway
INK4 inhibitors
CDKN2A 36% 5%
CCND1 9% 0%
CDK4 6% 0%
CDKN2B 36% 0%
RB1 2% 1%
CDKN2D 2% 0%
41% CDKN2A
36% CDKN2B
9% CCND1
6% CDK4
3% RB1
2% CDKN2D
SWI/SNF pathway
ARID1B 4% 0%
PBRM1 6% 0%
SMARCC2 2% 0%
SMARCC1 5% 0%
ARID1A 8% 7%
SMARCA2 7% 0%
SMARCA4 1% 2%
15% ARID1A
7% SMARCA2
6% PBRM1
5% SMARCC1
4% ARID1B
3% SMARCA4
2% SMARCC2
DNA repair
RAD54 0% 1%
RAD9A 0% 1%
XRCC4 1% 1%
RAD51AP2 4% 1%
MRN complex
MSH2 2% 3%
MSH6 1% 1%
POLE 4% 1%
NBN 1% 1%
CHEK2 3% 3%
FA core complex
MLH1 4% 0%
MLH3 2% 1%
EXO1 2% 1%
ATM 4% 2%
BRCA1 1% 1%
BCLAF1 5% 5%
FANCA 0% 1%
PMS2 2% 0%
FANCF 1% 3%
FANCD2 4% 2%
BRCA2 0% 1%
FANCM 3% 1%
5% MSH2
5% POLE
4% MLH1
3% MLH3
3% EXO1
3% MSH6
2% PMS2
9% BCLAF1
6% ATM
6% FANCD2
6% CHEK2
5% RAD51AP2
4% FANCF
4% FANCM
4% RAD9A
3% XRCC4
2% NBN
1% BRCA1
1% FANCA
1% RAD54B
1% BRCA2

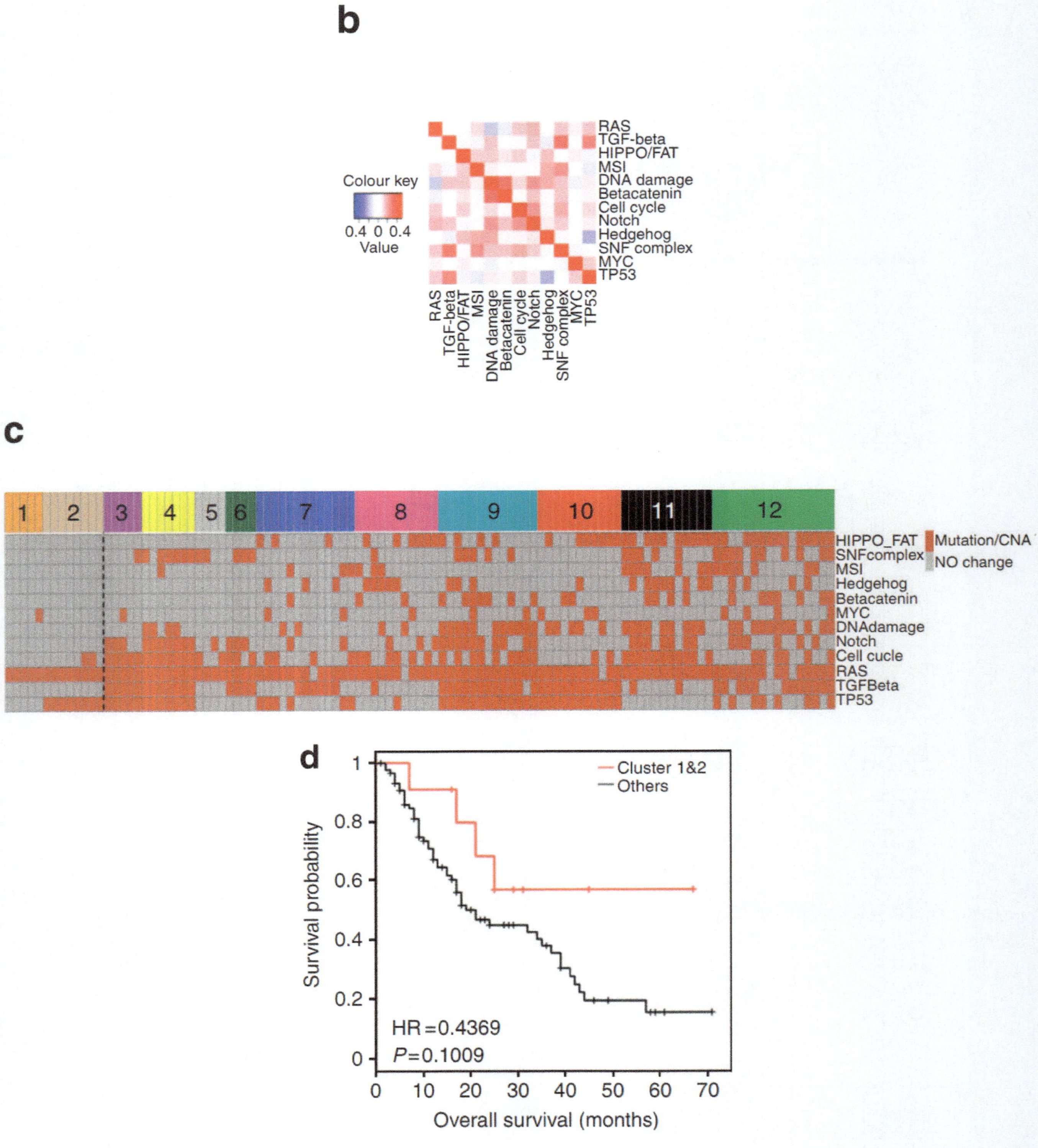

Figure 5: Pathways in PDA. (**a**) Oncomaps and schematics of pathways with combined genetic lesions occurring at a frequency >15% in the PDA cohort. (**b**) Correlation between pathways in the PDA cohort reveals relatively weak interactions between pathways. (**c**) Random forest-based clustering of pathways. (**d**) Overall survival related to genetic alterations targeting KRAS and TP53 in clusters 1 and 2 versus all others. *P* value was obtained from Cox proportional hazard test.

Table 2: Tabular summary of potential pharmacological strategies for specific genetic pathway deregulation

	Genetic event (frequency)	Target: drugs
BRAF	BRAF V600E (3%)	BRAF: Vemurafinib
PIK3CA	PIK3CA activating mutation (1%)	PIK3CA: BKM120, GDC0941
RB pathway	CDKN2A loss (41%) CDK4 amp, CCND1 amp. (9%)	CDK4/6: PD-0332991, LEE-11 CDK4/6: PD-0332991, LEE-11
Myc	Myc amp (14%)	CDK9: PHA767491 BET-Bromodomain: JQ1
Double-strand break repair	FANC gene family (14%) BRCA gene family (3%)	DNA: Mitomycin C PARP: Olaparib
Beta-Catenin	RNF43 loss (7%) AXIN1/2 or APC loss (11%)	Porcupine: LGK974 Tankyrase: XAV939
Notch	Notch family amp (31%)	Gamma secretase: MK0752

DISCUSSION

In total, the data herein demonstrates that PDA is a genetically diverse disease harbouring mutational burden similar to other solid malignancies. Due to the enrichment approach utilized, we identified multiple significantly mutated genes that have not been previously described in PDA and provide genetic insights into the etiology, prognostic features and potential therapeutic targets.

Risk factors for PDA, include smoking and deficits in DNA mismatch repair as exemplified by Lynch Syndrome28·29·30. We observed that smoking status in PDA was associated with a mutation spectra similar to other smoking-associated tumours13·14. We also found that select cases exhibit a 'mutator' phenotype, suggesting a contribution of mismatch repair deficiency in tumorigenesis12.

In the analysis of copy number, we identified many regions that have been previously identified by CGH and other methods31. In general, tumors with more copy number alterations (indicative of chromosomal instability) exhibited mutations in DNA break repair genes and trended toward poor prognosis. Importantly, we discovered that *MYC* amplification was associated with poor prognosis and adenosquamous histology. We did not identify mutations in *UPF1,* which has been recently reported to be selectively mutated with high frequency in pancreatic adenosquamous carcinoma32. In spite of a comprehensive unbiased survey of amplicons/deleted regions only one locus (MYC) was associated with outcome; however, clearly there is variation in the overall magnitude of copy number alterations, suggesting that there are perhaps distinct drivers for the chromosomal instability observed in PDA.

Using MutSigCV and pathway analysis, multiple genes and pathways altered at high frequency were identified, a subset of which associated with survival. *ARID1A* was confirmed as a marker of poorer outcome using an independent cohort, and *RBM10* mutation was associated with longer survival. We also uncovered a number of novel *albeit* rare alterations in signalling pathways important in PDA biology. While*CDKN2A* loss is common in PDA and contributes to the aberrant cell cycle progression, loss of *RB1* or *CCND1* amplification has not been previously reported. In PDA tumours harbouring wild-type KRAS, we identified *BRAF* and *PIK3CA* mutations at oncogenic hotspots expanding the spectrum of oncogenic drivers in PDA. Importantly, a cell line developed from BRAF V600E-mutated PDA was equally sensitive as melanoma cells to FDA-approved BRAF inhibitor PLX-4032. In addition to these rare, potentially, clinically actionable events, several pathways that were genetically altered at high frequency (>20%) in our study could yield selective therapeutic sensitivities. Genetic lesions targeting some facet of DNA repair or chromatin remodelling SWI/SNF pathway occurred in >35% and 42% of PDA cases, respectively. Alteration in WNT/beta-catenin pathway were common and included loss of RNF43 or AXIN1 conferring sensitivity to porcupine and tankyrase inhibitors targeting. A number of these genetic alterations remain actionable in the presence of an activating KRAS mutation and could provide the opportunity for genetically targeted therapy in PDA.

Together this expanded genetic framework of PDA reveals disease complexity that will contribute to future efforts of disease modelling and patient stratification for treatment.

METHODS

Tissue Collection and Microdissection

All tissues were collected under approved protocols of the Thomas Jefferson University Institutional Review Board and UT Southwestern Medical Center Institutional Review Board that require the informed consent of all patients. Cases represented diverse histological subtypes including colloid (n=4), adenosquamous (n=11) and ductal carcinoma, NOS (n=94). The majority of the analysed tumours were stage I (n=5) or II (n= 97); the remainder were stage III (n=6) or IV (n=1). For stage IV case tissue form the primary pancreatic tumour was analysed. About 7-μM thick slides were cut from frozen tissue and lightly stained with haematoxylin. Tumour epithelial cells were dissected manually using a 14 gauge needle under a dissecting microscope (Leica Microsystems). DNA was extracted using the QIAamp DNA Mini Kit. Between 500 ng and 1.5 μg of enriched tumour DNA was recovered from each case. Analysis of KRAS mutation and other tumour-associated mutations demonstrated that average tumour cellularity >50%. Germline DNA was obtained from normal tissue resected at surgery (n=105) or peripheral blood (n=4).

DNA Quality Control and Sequencing

Recovered DNA was quantified by Nanodrop and Qubit using standard procedures. Integrity of genomic DNA was confirmed by gel electrophoresis. A total of 109 tumour/normal pairs passed the quality control to move forward for whole-exome sequencing. TruSeq Exome Enrichment (FC-121-1048) and Nextera Exome Enrichment (FC-140-1003) kits were utilized to build the sequencing libraries as per manufacturer's protocol. The DNA was then subjected to paired-end whole-exome sequencing using Illumina HiSeq2500 instruments. The sequence reads were aligned to the human genome (Build-UCSC hg19) using the BWA alignment algorithm33. A general sequencing depth of 51.28 × (95% confidence interval (CI): 49.5 and 53.1) was observed for sequenced cases, and represented 99.26% (95% CI: 99.20–

99.31%) of the target sequences. Across the target sequences, 91.51% (95% CI: 90.97–92.05%) had >15 read coverage, and on average 2.178 × 10^9 bases were sequenced per case. For 21 cases, the sequencing depth was doubled with an average number of 5.523 × 10^9 bases sequenced per case. The average fold coverage was 123.01 (95% CI: 115.4–130.6). Across target sequences, 94.74% (95% CI: 93.41–96.07) had >15 read coverage. With the deeper sequencing relatively few additional mutation were discovered. The increased depth largely served to detect mutations with low sequence coverage and allele frequency.

Informatics Pipeline

Somatic point mutations and INDELS in the tumour tissues compared with their corresponding normal tissues were identified using the MuTect34 and VarScan2 algorithms35, respectively, with default parameter settings. The mutations were annotated using Broad Institute's Oncotator programme. Minimum of 14 reads covering a site in the tumour and 8 in the normal were required for mutation calling. Only INDELS with reference allele counts greater than eight and tumour variant allele counts greater than three were considered.

Genome Analysis Toolkit (v2.4, DepthOfCoverage) was used to calculate the coverage for all sequenced cases36. Log ratio of coverage between tumour and normal were normalized by circular binary segmentation using ExomeCNV and DNAcopy packages in R/Bioconductor. To obtain copy number alterations in tumours compared with normal tissue, GISTIC 2.0 module version 6.2 with join segment size=4 and confidence level=0.95 was used37. SNV and INDEL analyses were summarized into tables using R (R Core Team-2014; R Foundation for Statistical Computing, Vienna, Austria, http://www.R-project.org).

Computational and Statistical Analysis

Hierarchical clustering and Kaplan–Meier analyses were performed using the stats and survival packages in R. Associations of genes with features of disease was determined using hypergeometric testing, with$P<0.05$ considered significant. Survival between two groups was determined using log-rank statistic, with $P<0.05$ considered significant. Heatmaps for mutation spectrum and pathway oncoprints were generated using the R package pheatmap. All other plots were generated using the graphics package in R.

The MutSigCV algorithm16 was run locally with default settings to define significantly mutated genes with input MAF files developed from the merge of MuTect and VarScan outputs. To remove any gene paralogue or alignment driven artifacts, the mutations at all positions for high-frequency genes were interrogated by BLAST against the human genome. This approach removed a number of high-frequency false positives, including CDC27 and FRG1. Mutations in significantly mutated genes and all INDELs were manually interrogated by examining BAM files in Integrative Genomics Viewer38. Subsequent filtering was employed were genes with at least four nonsilent mutations (>3.5%) were retained. A P value cut-off of 0.05 was used to obtain a list of top significantly mutated genes. Subsequent analysis of sequenced paralogues or analogous sequences was used to further restrict the results to 24 genes. Unsupervised Random Forest clustering approach was used to cluster the pathways39. Random Forest package in R was used to obtain the Random forest dissimilarity measure and the data was classified into clusters using the partitioning around medoids implemented in the cluster package in R.

Clustering analysis was performed with the APC algorithm using the 'apcluster' package in R. APC is a deterministic clustering method, which identifies the number of clusters, and cluster 'exemplars' (that is, the cluster centroid or the data point that is the best representative of all the other data points within that cluster) from the data40. APC performs clustering by passing messages between the data points. It takes as input a square matrix representing pairwise similarity measures between all data points. The algorithm views each data point as a node in a network and is initialized by connecting all the nodes together where edges between nodes are proportional to Pearson correlations. The algorithm then iteratively transmits messages along the edges, pruning edges with each iteration until a set of clusters and exemplars emerges.

Two real valued messages are passed between nodes. The 'responsibility' message computes how well-suited it is for point i to choose point k as an exemplar, given all the other candidate exemplars, k', and is updated by:

$$r(i,k) \leftarrow s(i,k) - \max_{k' \, st \, k' \neq k}\{a(i,k') + s(i,k')\} \quad (1)$$

The availability message, $a(i, k)$, computes how appropriate it is for point i to select point k as an exemplar, taking into account all the other points for which k is an exemplar, i', and is given by:

$$a(i,k) \leftarrow \min\left\{0, r(k,k) + \sum_{i' st \; i' \notin \{i,k\}} \max(0, r(i',k)\right\} \quad (2)$$

In the above equation, $a(i, k)$ is set to the self responsibility, $r(k,k)$, plus the sum of the positive responsibilities candidate k receives from other points. The entire sum is thresholded at 0, and a negative availability indicates that it is inappropriate for point i to chose point k as an exemplar and the tie is severed. The self-availability, $a(k,k)$, reflects the accumulated evidence that point k is an exemplar and is updated with the following rule, which reflects the evidence that k is an exemplar based on the positive responsibilities sent to k from all points, and is updated by:

$$a(k,k) \leftarrow \sum_{i' st \; i' \notin \{i,k\}} \max(0, r(i',k) \quad (3)$$

In the first iteration, all points are considered equally likely to be candidate exemplars, and $a(i, k)$ is set to 0 and $s(i, k)$ is set to the input similarity measure between points i and k. The above rules are then iteratively updated until a clear, stable set of clusters and exemplars emerges.

In our implementation of the algorithm, we first ran the algorithm to identify an initial set of exemplars and clusters from the data. The exemplars were then clustered together and this procedure was repeated until no more clusters emerged to identify a hierarchical structure of clusters. Networks were drawn with cytoscape41.

To generate mutation peg plots, protein domain information was downloaded from the Pfam database (http://

pfam.xfam.org/) and isoforms were matched between data sets with Uniprot IDs. Mutation information for various TCGA tumour data sets were gathered using the cBioPortal R package, 'cgdsr' (cbioportal.org).

To compare the mutational spectra of the PDA_COHORT data set to the TCGA tumour data sets, individual MAF files from the most current TCGA data were downloaded from the TCGA website (http://cancergenome.nih.gov/) and nucleotide substitution information was compared.

Validation

A total of 248 nonsilent mutations from 84 of the 109 tumour/normal tissue pairs were selected to validate by Sanger sequencing. This validation encompassed 132 different genes, including most that were referenced in the text (for example *BRAF*, *BRCA2*, *RB1*, *ARID1A*). We aimed to validate mutations with allelic frequency of >0.15, recognizing that Sanger sequencing sensitivity is impaired by low mutant allele frequency. Mutations confirmation was 92% (95% CI: 87.64–94.58%). This range is consistent with other studies using MuTect to define mutations42, validated mutations are denoted.

Immunohistochemistry and FISH

ARID1A immunohistochemistry was performed in a clinical laboratory using a BenchMark Ultra autostainer (Ventana Medical Systems). The ARID1A antibody (cat# sc-32761, Santa Cruz Biotechnology—dilution 1:150) specificity was confirmed by negative staining in cases with known mutation in ARID1A. All scores were read blind to outcome, and data were dichotomized based on loss of the homogeneous ARID1A staining observed in normal tissue or known cases with intact ARID1A gene. Association with survival was performed by Kaplan–Meier analysis. Phospho-ERK immunohistochemistry (cat#4370, Cell Signaling Technology-dilution 1:150) was performed on a DAKO autostainer. All scores were read blind to outcome or association with KRAS mutation.

Tumours from the Pdx1-Cre/CAG-βgeo-tTA/TetO-Myc transgenic mice were formalin fixed15, embedded and stained for haematoxylin and eosin. Immunohistochemical staining for p63 (cat#619002, Biolegend-1dilution :100) was performed on a DAKO autostainer.

Fluorecence *in situ* hybridization (FISH) at the MYC locus was performed in a CLIA clinical laboratory using Vysis LSI MYC dual colour 'break apart' rearrangement probe. The probes detect regions directly centromeric and telomeric of the MYC genes, facilitating the detection of amplifications (both colours amplified in proximity) and/or translations (resulting in the separation of the two colours spatially).

Functional Studies

The established PDA cell lines PL45, CAPAN2, PL5, ASPC1 and HS7662 were obtained from ATCC and cultured in DMEM supplemented with 10% FBS. The cell line derived from PDA_014 was established from primary tumour fresh tissue that was digested with trypsin and plated on collagen coated plates in Keratinocyte Serum-Free Media supplemented with Bovine Pituitary Extract, EGF and 2% FBS. Tumour cells were expanded and subjected to sequencing analysis, which confirmed the retention of select tumour-associated mutations including BRAF V600E. The melanoma cell line MNT1 was cultured in DMEM supplemented with 10% FBS. Sensitivity to PLX-4032 was determined by flow-cytometry, crystal violet and CTG analysis as we have previously published25. For immunoblotting, 30 μg of total protein were resolved by SDS–PAGE and transferred to polyvinylidene fluoride membranes (Immobilon-P, Millipore). Membranes were immunoblotted with anti-ARID1A (SC-32761, Santa Cruz, dilution 1:500), β-actin (SC47778, Santa Cruz, dilution 1:500), anti-phospho-ERK (SC-7383, Cell Signaling dilution 1:500), anti-ERK (SC-94, Cell Signaling, dilution 1:500) and anti-horse radish peroxidase-conjugated antibodies (Jackson ImmunoResearch-dilution 1:2,500).).

For knockdown experiments, 5,000 cells were plated in each well of 96-well plate followed by transfection of siARID1B (L-013970-01-0005, GE HealthCare/Dharmacon: pool of 3′-UGAUCAACAUGGCGGACAA-5′, 3′-CCGAAUUACAAACGCCAUA-5′, 3′-UCUCAAAGCAGACGGCAAA-5′ and 3′-ACGAGCAUCCAGAGAGAAA-5′) with transfection reagent (Lipofectamine RNAiMAX, Lif Technologies) and incubated for 4 days at 37°C. Non-targeting siRNA pools (D001810-10-20, GE Healthcare/Dharmacon) were used as negative control. Cell Titer-Glo reagent (Promega) was used to see cell viability. About 15 μl of reagent was added to each well and mixed by orbital shaker. Luminescence of each well was measured in a microplate reader (BioTek).

REFERENCES

1. Paulson, A. S., Tran Cao, H. S., Tempero, M. A. & Lowy, A. M. Therapeutic advances in pancreatic cancer. Gastroenterology 144, 1316–1326 (2013).
2. Cowan, R. W. & Maitra, A. Genetic progression of pancreatic cancer. Cancer J. 20, 80–84 (2014).
3. Vincent, A., Herman, J., Schulick, R., Hruban, R. H. & Goggins, M. Pancreatic cancer. Lancet 378, 607–620 (2011).
4. Biankin, A. V. et al. Pancreatic cancer genomes reveal aberrations in axon guidance pathway genes. Nature 491, 399–405 (2012).
5. Jones, S. et al. Core signalling pathways in human pancreatic cancers revealed by global genomic analyses. Science 321, 1801–1806 (2008).
6. Erkan, M., Reiser-Erkan, C., Michalski, C. W. & Kleeff, J. Tumour microenvironment and progression of pancreatic cancer. Exp. Oncol. 32, 128–131 (2010).
7. Cancer Genome Atlas Research, N. Comprehensive genomic characterization of squamous cell lung cancers. Nature 489, 519–525 (2012).
8. Yeo, C. J. et al. Six hundred fifty consecutive pancreaticoduodenectomies in the 1990s: pathology, complications, and outcomes. Ann. Surg. 226, 248–257 (1997).
9. Weber, C. E. et al. Clinical and pathologic features influencing survival in patients undergoing pancreaticoduodenectomy for pancreatic adenocarcinoma. J. Gastrointest. Surg. 18, 340–347 (2014).

10. Barbieri, C. E. et al. Exome sequencing identifies recurrent SPOP, FOXA1 and MED12 mutations in prostate cancer. Nat. Genet. 44, 685–689 (2012).
11. Cancer Genome Atlas Research Network et al. Integrated genomic characterization of endometrial carcinoma. Nature 497, 67–73 (2013).
12. Alexandrov, L. B. et al. Signatures of mutational processes in human cancer. Nature 500, 415–421 (2013).
13. Helleday, T., Eshtad, S. & Nik-Zainal, S. Mechanisms underlying mutational signatures in human cancers. Nat. Rev. Genet. 15, 585–598 (2014).
14. Henderson, S., Chakravarthy, A., Su, X., Boshoff, C. & Fenton, T. R. APOBEC-mediated cytosine deamination links PIK3CA helical domain mutations to human papillomavirus-driven tumour development. Cell Rep. 7, 1833–1841 (2014).
15. Lin, W. C. et al. Dormant cancer cells contribute to residual disease in a model of reversible pancreatic cancer. Cancer Res. 73, 1821–1830 (2013).
16. Lawrence, M. S. et al. Mutational heterogeneity in cancer and the search for new cancer-associated genes. Nature 499, 214–218 (2013).
17. Furukawa, T. et al. Whole-exome sequencing uncovers frequent GNAS mutations in intraductal papillary mucinous neoplasms of the pancreas. Sci. Rep. 1, 161 (2011).
18. Lawrence, M. S. et al. Discovery and saturation analysis of cancer genes across 21 tumour types. Nature 505, 495–501 (2014).
19. Imielinski, M. et al. Mapping the hallmarks of lung adenocarcinoma with massively parallel sequencing. Cell 150, 1107–1120 (2012).
20. Inoue, A. et al. RBM10 regulates alternative splicing. FEBS Lett. 588, 942–947 (2014).
21. Helming, K. C. et al. ARID1B is a specific vulnerability in ARID1A-mutant cancers. Nat. Med. 20, 251–254 (2014).
22. Ihle, N. T. et al. Effect of KRAS oncogene substitutions on protein behavior: implications for signalling and clinical outcome. J. Natl. Cancer Inst. 104, 228–239 (2012).
23. Collisson, E. A. et al. A central role for RAF--4MEK--4ERK signalling in the genesis of pancreatic ductal adenocarcinoma. Cancer Discov. 2, 685–693 (2012).
24. Jiang, X. et al. Inactivating mutations of RNF43 confer Wnt dependency in pancreatic ductal adenocarcinoma. Proc. Natl Acad. Sci. USA 110, 12649–12654 (2013).
25. Franco, J., Witkiewicz, A. K. & Knudsen, E. S. CDK4/6 inhibitors have potent activity in combination with pathway selective therapeutic agents in models of pancreatic cancer. Oncotarget 5, 6512–6525 (2014).
26. van der Heijden, M. S. et al. In vivo therapeutic responses contingent on Fanconi anemia/BRCA2 status of the tumour. Clin. Cancer Res. 11, 7508–7515 (2005).
27. Farmer, H. et al. Targeting the DNA repair defect in BRCA mutant cells as a therapeutic strategy. Nature 434, 917–921 (2005).
28. Yadav, D. & Lowenfels, A. B. The epidemiology of pancreatitis and pancreatic cancer. Gastroenterology 144, 1252–1261 (2013).
29. Klein, A. P., Hruban, R. H., Brune, K. A., Petersen, G. M. & Goggins, M. Familial pancreatic cancer. Cancer J. 7, 266–273 (2001).
30. Jones, S. et al. Exomic sequencing identifies PALB2 as a pancreatic cancer susceptibility gene. Science 324, 217 (2009).
31. Holzmann, K. et al. Genomic DNA-chip hybridization reveals a higher incidence of genomic amplifications in pancreatic cancer than conventional comparative genomic hybridization and leads to the identification of novel candidate genes. Cancer Res. 64, 4428–4433 (2004).
32. Liu, C. et al. The UPF1 R. N. A. surveillance gene is commonly mutated in pancreatic adenosquamous carcinoma. Nat. Med. 20, 596–598 (2014).
33. Li, H. & Durbin, R. Fast and accurate short read alignment with Burrows-Wheeler transform. Bioinformatics 25, 1754–1760 (2009).
34. Cibulskis, K. et al. Sensitive detection of somatic point mutations in impure and heterogeneous cancer samples. Nat. Biotechnol. 31, 213–219 (2013).
35. Koboldt, D. C. et al. VarScan 2: somatic mutation and copy number alteration discovery in cancer by exome sequencing. Genome Res. 22, 568–576 (2012).
36. McKenna, A. et al. The Genome Analysis Toolkit: a MapReduce framework for analyzing next-generation DNA sequencing data. Genome Res. 20, 1297–1303 (2010).
37. Mermel, C. H. et al. GISTIC2.0 facilitates sensitive and confident localization of the targets of focal somatic copy-number alteration in human cancers. Genome Biol. 12, R41 (2011).
38. Robinson, J. T. et al. Integrative genomics viewer. Nat. Biotechnol. 29, 24–26 (2011).
39. Shi, T., Seligson, D., Belldegrun, A. S., Palotie, A. & Horvath, S. Tumour classification by tissue microarray profiling: random forest clustering applied to renal cell carcinoma. Mod. Pathol. 18, 547–557 (2005).
40. Frey, B. J. & Dueck, D. Clustering by passing messages between data points. Science 315, 972–976 (2007).
41. Shannon, P. et al. Cytoscape: a software environment for integrated models of biomolecular interaction networks. Genome Res. 13, 2498–2504 (2003).
42. Goode, D. L. et al. A simple consensus approach improves somatic mutation prediction accuracy. Genome Med. 5, 90 (2013).

CHAPTER 6

EXPLORING THE RELATIONSHIP BETWEEN LIFESTYLES, DIETS AND GENETIC ADAPTATIONS IN HUMANS

Cristina Valente, Luis Alvarez, Sarah J. Marks, Ana M. Lopez-Parra, Walther Parson, Ockie Oosthuizen, Erica Oosthuizen, António Amorim, Cristian Capelli, Eduardo Arroyo-Pardo, Leonor Gusmão and Maria J. Prata

Keywords: Diet adaptations, Signals of natural selection, Africa Sub-Saharan

BACKGROUND

The most remarkable dietary change over the recent history of human populations was that associated with the change from food collection to food production [1], which occurred independently and in different times in separate parts of the world marking the beginning of the Neolithic, a transition that in some regions dates back to 12,000 years ago. The domestication of plants and animals prompted the conditions that would brought about new modes of subsistence as well as new food habits as a consequence of the shift in the availability and exploitation of dietary resources [1, 2]. Genetic adaptations to dietary specializations are thought to have represented advantageous evolutionary solutions in humans, however it is still unclear the extent to which dietary factors have created selective pressures acting on genes that play roles in food-related metabolic pathways. Recent studies have revealed genomic signatures of adaptations likely driven by diet-related pressures [1, 3, 4]. In addition, candidate genes approaches had already provided tight evidence for genetic adaptations to differences in nutrient consumption such as at the lactase and amylase genes [5-10].

Other metabolic-related genes have been hypothesized to constitute dietary adaptations, among which are included: *AGXT,* coding for alanine:glyoxylate aminotransferase, the enzyme responsible for the transamination of glyoxylate into glycine [11-13]; *PLRP2,* coding for pancreatic lipase-related protein 2, involved in galactolipids hydrolysis, [14-17]; *MTRR,* encoding for methionine synthase reductase, an enzyme acting in the complex folate pathway [15, 18]; *NAT2* coding for *N*-acetyltransferase 2, a phase-II enzyme involved in the detoxification of a wide number of xenobiotics [15, 19-23]; and *CYP3A5,* coding for cytochrome P-450 3A5, a member of the CYP3A enzymes that are involved in the oxidative metabolism of many endogenous substrates and xenobiotics, which is implied in sodium homeostasis [24-27].

Genetic variation in *AGXT* was tentatively linked with meat content in diets, *PLRP2* with richness in cereals [15], both *MTRR* and *NAT2* with availability of folate in foods and *CYP3A5* with health conditions that are influenced by dietary salt intake [24, 27]. However, for these 5 genes results so far obtained were either contradictory (e.g. *AGXT*), or not yet replicated (e.g. *MTRR* and *PLRP2*), or not clear enough to ascertain whether they can indeed represent genetic adaptations to any dietary variable. This prompted us to address the issue applying of genetic adaptation within those genes.

Thus, assuming that current modes of subsistence are still good surrogates of main diets in which populations have traditionally relied, the aim of this study was to gain further insights into the relationship between diet-related variables

Citation: Cristina Valente, Luis Alvarez, Sarah J. Marks, Ana M. Lopez-Parra, Walther Parson, Ockie Oosthuizen, Erica Oosthuizen, António Amorim, Cristian Capelli, Eduardo Arroyo-Pardo, Leonor Gusmão and Maria J. Prata, "Exploring the relationship between lifestyles, diets and genetic adaptations in humans," BMC Genetics201516:55. DOI: 10.1186/s12863-015-0212-1.

in populations and patterns of diversity at variations in above mentioned five genes.

Functional variants within *AGXT*, *PLRP2*, *MTRR*, *NAT2* and *CYP3A5* were examined in six sub-Saharan populations with distinct modes of subsistence and also in one European population that was also screened to generate a non-African reference group. Results were then combined with previously published information for other African and Eurasian populations to evaluate the contribution of geography and mode of subsistence or other diet-related variables to explain the patterns of genetic diversity observed for the five genes.

RESULTS

Locus by Locus Analysis

The observed genotypic distributions did not revealed significant departures from Hardy-Weinberg expectations after applied the Bonferroni's correction for multiple tests. Estimates of allele frequencies for the five *loci* in the seven studied populations are shown in Table 1 and for each locus results here and previously obtained will be dissected in the following sections.

Table 1: Derived allele frequencies

POPULATION	c.32C > T (*AGXT*)	c.1074G > A (*PLRP2*)	c.1130A > G (*MTRR*)	c.191G > A (*NAT2*14*)	c.341 T > C (*NAT2*5*)	c.590G > A (*NAT2*6*)	c.857G > A (*NAT2*7*)	c.219-237G > A (*CYP3A5*)
ANG	0.0000 ± 0.0000	0.3261 ± 0.0691	0.5294 ± 0.0856	0.1522 ± 0.0530	0.2046 ± 0.0748	0.3636 ± 0.0725	0.0000 ± 0.0000	0.2400 ± 0.0604
EQG	0.0482 ± 0.0166	0.3214 ± 0.0360	0.3563 ± 0.0363	0.0977 ± 0.0225	0.3588 ± 0.0536	0.1786 ± 0.0296	0.0233 ± 0.0115	0.1429 ± 0.0270
MOZ	0.0370 ± 0.0257	0.2333 ± 0.0546	0.5500 ± 0.0642	0.1429 ± 0.0540	0.2500 ± 0.0884	0.2857 ± 0.0697	0.0000 ± 0.0000	0.1167 ± 0.0414
UGN	0.0727 ± 0.01751	0.3945 ± 0.0331	0.3835 ± 0.0339	0.0699 ± 0.0187	0.3902 ± 0.0575	0.3085 ± 0.0337	0.0055 ± 0.0055	0.2336 ± 0.0289
BPY	0.0147 ± 0.0146	0.18912 ± 0.0455	0.3846 ± 0.0551	0.0263 ± 0.0184	0.1842 ± 0.0536	0.2568 ± 0.0508	0.0000 ± 0.0000	0.1447 ± 0.0404
KNA	0.0000 ± 0.0000	0.0242 ± 0.0138	0.1371 ± 0.0309	0.0000 ± 0.0000	0.0656 ± 0.0239	0.0484 ± 0.0193	0.0968 ± 0.0266	0.2097 ± 0.0366
PTG	0.1915 ± 0.0406	0.5106 ± 0.0516	0.1383 ± 0.0356	0.0000 ± 0.0000	0.5000 ± 0.0903	0.2021 ± 0.0414	0.0532 ± 0.0232	0.9022 ± 0.03010

Populations' abbreviations as referred in material and methods section

AGXT

In the *AGXT* gene, we studied the variant c.32C > T, concerning which the derived allele T had been previously suggested to play an adaptive role in populations traditionally relying in meat-rich diets [11, 28]. The hypothesis was specifically investigated by Caldwell et al. [11] who reported on frequency data sustaining the model, a conclusion for which much accounted the observation of the highest frequency of the derived allele in the Sweden Saami, who have a long history of consuming high amounts of animal products [11, 28]. Though, later, revisiting the question with a better coverage of Central Asian populations Ségurel et al. [13] failed to find increased allele frequencies across populations with diets richer in meat comparatively to those less meat rich, challenging this way the adaptive model proposed for the variation.

In this study, in terms of meat content in diets of African populations, we have assumed that in general farmers rely less in meat than pastoralists or hunter-gatherers, in accordance with a recent review from ethnographic compilations of hunter-gatherer diets indicating that animal food comprises their dominant energy source [29]. Among the 6 sub-Saharan populations examined, the frequency of the derived allele at c.32C > T ranged from 0 to 7.27 % without showing any pattern of variation that could be connected with mode of subsistence or meat content in diets of populations. For instance, it was absent both from the farmers from Angola and from the hunter-gatherers Khoisan, although the first are representative of less meat consumers groups while the second are from more meat consumers ones. In the sample from Portugal, considered to be a farming population with a mixed diet reasonably balanced regarding animal and plant food resources, the derived allele reached 19.15 %, a frequency higher than registered in any of the African populations regardless of its mode of subsistence or reliance upon meat.

To integrate our results in a more comprehensive distribution, data for c.32C > T was retrieved from the literature on populations for which information on the relative predominance of meat in their diets was available. There were results only for populations from Africa and Eurasia, among which the average frequency of the derived allele was 0.081 across the set of populations assigned to have high meat consumption, while it was, 0.133, across the populations with low-meat consumption. Actually neither the overall differences in allele frequencies within the "low-meat" and "high-meat" groups were statistically significant (P = 0.0710, One-Way ANOVA), nor the trend in the frequency distribution sustained the hypothesis that the allele could be positively selected in meat-rich diet populations.

Furthermore, if the broad geographical distribution of c.32C > T in Africa and Eurasia conformed well the major population clusters commonly identified by random neutral genetic markers, intriguingly in Asia, where there is a high dispersion of gene frequencies, the extreme values were reported for two populations in rather close geographical proximity but with distinct traditional lifestyles: in the Tajiks, a group of sedentary agriculturalists from Western Tajikistan the derived allele was very well represented (26.9 %), whereas in the Kazaks from Western Uzbekistan, who are traditionally nomadic herders whose diet mainly consists of meat, milk and dairy products, the allele only occurred marginally (1.7 %).

From these analyses, no connection emerged between the frequency distribution of c.32C > T in *AGXT* and lifestyle of populations.

PLRP2

In this gene we focused on c.1074G > A, a variant that causes a premature truncation of the pancreatic lipase-related protein 2 resulting in a more active version of the enzyme. In a very recent genome-wide scan for selection in human populations, Hancock et al. [15] identified in this variant a convincing signal of adaptation to a dietary specialization, since the derived allele was found to be significantly more common in populations relying in diets with high content in cereals (farmers) than in other populations.

As long as we know, the association was not further investigated except in the present study, where among the screened African groups, the derived allele was detected to be quite common in the three farmers' groups (23.3 % - 32.6 %) as well as in the herders from Uganda (39.5 %). Comparatively, the two hunter-gatherers groups showed lower frequencies, specially the Ju/hoansi (2.42 %). The sample from Portugal showed the highest frequency in this study with, 51.1 % (Table 1).

As a whole, our results do not conflict with the hypothesis that the distribution of c.1074G > A might be related to the weight of cereals in diets, in the sense that at least within Africa, farmers populations tended to have higher frequencies of the derived allele compared to hunters-gatherers who rely less in cereals. These results were then put in a wide-ranging context, recruiting information on c.1074G > A for African and Eurasian populations from several sources, and maintaining the classification in populations that specialize and that do not specialize on cereals when originally presented. As shown in Figure 1A and B, the frequency of the truncated allele was found to be more common across populations with cereal-rich diets (average frequency 35.8 % in Africa; 49.1 % in Eurasia + Africa) than across those less dependent on cereals (average frequency 22.7 % in Africa; 22.9 % in Eurasia + Africa), differences that were statistically significant either in Africa (P = 0.0050, One-Way ANOVA) or in Eurasia + Africa (P = <0.0001, One-Way ANOVA). Comparing herders and hunter-gatherers, both integrated in the group of cereal less rich populations (Figure 1), mean frequency was respectively 41.3 % and 18.6 % in Africa, and 46 % and 19.5 % in Eurasia + Africa, with both differences being again statistically significant (P = 0.0060 for Africa; P = <0.0001, for Eurasia + Africa, One-Way ANOVA). Considering Africa and Eurasia together, the trend that can be extracted from the whole data points to a decreasing frequency gradient of the derived allele at c.1074G > A from populations more specialized on cereals towards those less relying on them, as was also captured by the MDS plot shown in Figure 1C, where it is visible some structure between hunter-gatherer, herder and farmer populations.

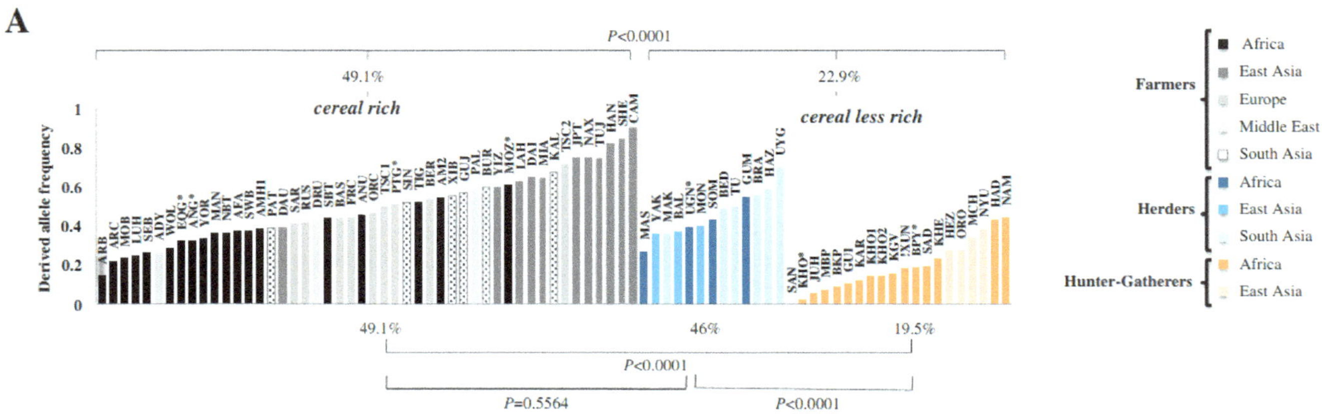

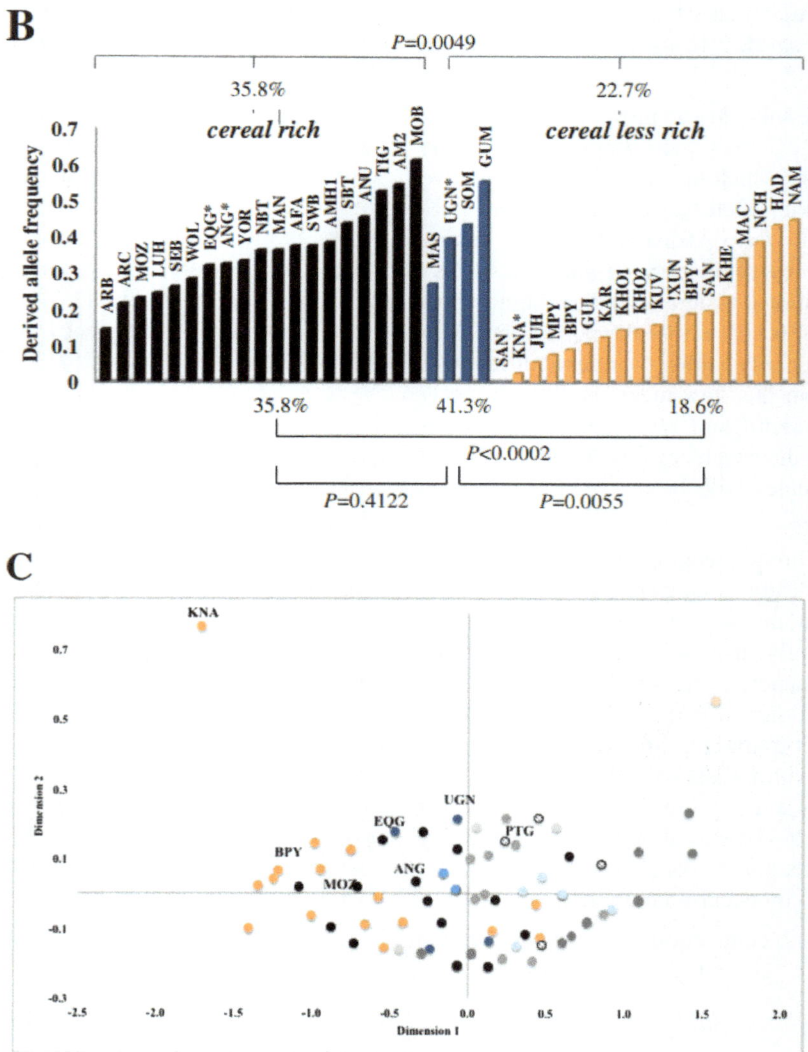

Figure 1: Allele frequencies and MDS plot for *PLRP2*. *P* values of ANOVA One-Way test in **(A)** African + Eurasian and **(B)** African populations' group; MDS plot of pairwise genetic distances between populations **(C)**. In the MDS plot different colors represent distinct lifestyles: hunter-gatherers (*orange*), herders (*blue*) and farmers populations (*black*). *populations addressed in this study. Populations' abbreviations are referred in Material and Methods section.

As a whole, these results suggest that diversity at *PLRP2* was shaped by selective pressures that differed according to populations' lifestyle.

MTRR

Within *MTRR* we examined the common variation affecting levels of enzymatic activity c.1130A > G, since it was another candidate adaptive genetic variation identified in the before mentioned genome-wide study [15]. Before, *MTRR* had received high attention in association studies, having been implicated, for instance, with risk for *spina bifida* [18]. However, its adaptive role to dietary specializations was addressed in only one work where c.1130A > G was found to be strongly correlated with diets containing mainly the folate-poor foods roots and tubers [15]. The results obtained in this work revealed that the derived allele was quite common in most African groups, peaking in the agriculturists from Angola and Mozambique with values of 0.529 and 0.550, respectively (Table 1). Both estimates are similar to that described in the Yoruba (0.548) the only African group with a diet principally relying on roots and tubers addressed in a previous study [15]. So, at least in Africa high frequencies of this allele can be found in populations without having such a dietary specialization. Furthermore, no indication arose that the distribution of c.1130A > G could be correlated to the dietary availability in folates, which is generally thought to be lower in non-forager populations (agricultural and pastoral) than in hunter-gatherers [22]. In fact, in the hunter-gatherers Baka, in the herders from Uganda and in the farmers from Equatorial Guinea, the derived allele occurred at similar frequencies (0.385, 0.384, 0.356, respectively) despite the differences in mode of food production. In the hunter-gatherers Ju/honasi from Namibia, the allele occurred at the lowest frequency in Africa (0.137) but with a magnitude similar to that found in the European sample (0.138), considered as a representative of an agriculturalist society (Table 1). To interpret our results under a wide framework of African and Eurasian populations, frequency data were recruited once more from the literature, and the combined information allowed to realize that the distribution of c.1130A > G fitted well the pattern generally provided by neutral markers, not appearing to be influenced by the mode of subsistence or the relative folate content in diets of populations from Eurasia and Africa. In East Asia, for instance, the two highest values of the derive allele were present in the Tu (0.4), nomadic herders, and in the Hezhen (0.333), mainly hunters and fishers, but nonetheless in the foragers Orogen and Yakut, who also live in East Asia, the allele was absent or very rare.

So, for the variation c.1130A > G in *MTRR*, the current patterns of diversity do not indicates that it could represent an adaptation to the mode of subsistence of human populations.

NAT2

The dietary availability in folates had also been previously hypothesized to be a modulator of genetic diversity at the gene that encodes for *NAT2* (*N*-acetyltransferase 2) [22]. Individuals can be classified in fast, intermediate or slow acetylator phenotypes, which are determined by the haplotypic composition defined by genetic variations at the *NAT2* locus. Evidence for the diet-related hypothesis provided by Luca et al. [22] was reinforced with the recent findings by the same people [1], based on a more comprehensive analysis of *NAT2*worldwide genetic diversity, that were also compatible with a model holding that the slow acetylator phenotypes were selectively favored in populations relying in dietary regimens with reduced folate supply, whereas the fast acetylators were neutral or even advantageous in the presence of folate-rich diets, as those thought to be fulfilled by hunter-gatherers. To extent the population coverage of previous works, frequencies of *NAT2* haplotypes and acetylator phenotypes were also estimated in this study. The distribution of haplotypes was very heterogeneous across African populations, but in line with previous observations the prevalence of the slow acetylator phenotype in the two hunters-gatherers groups (Khoisan, 1.6 %; Baka Pygmies, 13.5 %) was significantly much lower than in the three agriculturalists groups or in the Ugandan pastoralists, all displaying values up to 37.4 % (P = 0.0139, One-Way ANOVA). In the Portuguese the slow acetylator phenotype accounted for the high proportion of 52.2 %, which falls within the range typical from other European populations [21].

Next, we contrasted our data with other results before published for Eurasian and African populations, confining the analysis to c.590G > A, which defines allele *NAT2*6*, because it was the variation with more information accumulated for populations representatives of the three modes of subsistence.

From Figure 2A, which shows the allelic distribution of c.590G > A across Africa and Eurasia, it becomes clear that its prevalence is scarcely influenced by the continent where populations are located. However, some connection arises with systems of food production and acquisition given that in the whole set of African and Eurasian populations foraging groups tended to exhibit statistically significant lower frequencies of the derived allele compared to populations dependent on agricultural and pastoral resources (see in Figure 2B the P-values of One-Way ANOVA). Between pastoralists and agriculturalists, no significant differences were detected, which means that the clustering of c.590G > A frequencies only showed correspondence with populations that are food producers or food collectors, an observation that otherwise fully meets that reported by Sabbagh et al. [21], and the results even more recent published by the same team [30].

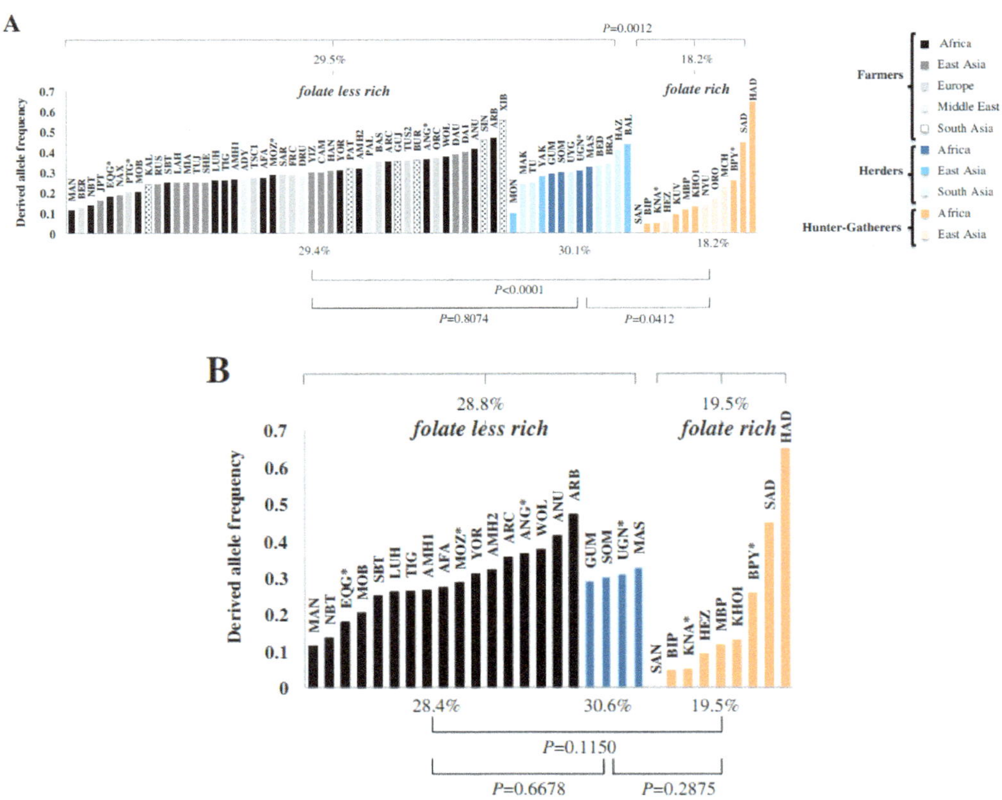

Figure 2: Allele frequencies for *NAT2*. *P* values of ANOVA One-Way test in **(A)** worldwide and **(B)** African populations' group hunter-gatherers (*orange*), herders (*blue*) and farmers populations (*black*), *populations addressed in this study. Populations' abbreviations are referred in material and methods section.

In brief, our analyses reinforce previous indications that *NAT2* has evolved under a selective factor influenced by human diet.

CYP3A5

With regard to *CYP3A5,* we screened the intronic variation c.219-237G > A, commonly referred to *CYP3A5*1/*3* polymorphism, in which the derived allele A results in a premature stop codon that reduces protein expression. It has been firmly demonstrated that the variation possesses a very unusual worldwide distribution whereby the frequency of *CYP3A5*3* is significantly correlated with latitude [24].

*CYP3A5*1/*3* likely influences salt and water retention and risk for salt-sensitive hypertension [24], exerting an effect on blood pressure that is determined by interactions with dietary salt intake [27,31]. Since anthropological evidence indicates that diet of hunting and gathering people is usually characterized by low level of salt intake, being often considered as a surrogate of the preagricultural humans' diet, lately praised as a model of well balanced food consumption [32], we asked whether diversity at *CYP3A5*1/*3* could be related with diet of populations.

Thus, we screened the variation in the six African populations, among whom the derived allele was only moderately represented, but suggestively it was in two farmer groups that the lowest and the highest frequencies were found (11.7 % and 24.0 % in the groups from Mozambique and Angola, respectively), disfavoring thus any link between lifestyle and differences in allele frequency across populations. In the Portuguese, the allele reached the very elevated value of 90.2 %, which it is usual in populations from Europe where *CYP3A5*3* varies quiet narrowly being near-fixation in most populations [24]. Again, our data were combined with those retrieved from the literature, and with an enlarged coverage of African and Eurasian populations, we confirmed in fact that the frequency of the low expressor allele significantly increased with distance from the equator (Figure 3A) (SRCSC = 0.7540; $P < 0.0001$). When the relationship was assessed separately in each of the three continents, no significant rank correlation was observed in Africa (SRCSC = 0.1058; P = 0.2438) or in Europe (SRCSC = 0.4183; P = 0.1310), but in Asia the correlation coefficient was again statistically significant (SRCSC = 0.5724; $P < 0.0002$). Interestingly, in Asia, where the average allele frequency was 0.793, the significant correlation can be explained since the lowest values are consistently present in populations from the South of the continent, located very near or already inside the intertropical zone. In Africa, the frequency of the allele drastically declines to an average value of 0.286 when inferred from a panel of populations' majority located inside the tropical zone. In Europe, which is fully situated in a temperate climatic region, the average frequency reaches 0.903. Therefore, being or not located in the tropical zone seems to be a factor that strongly influences the distribution of *CYP3A5*1/*3* alleles (see Figure 3B).

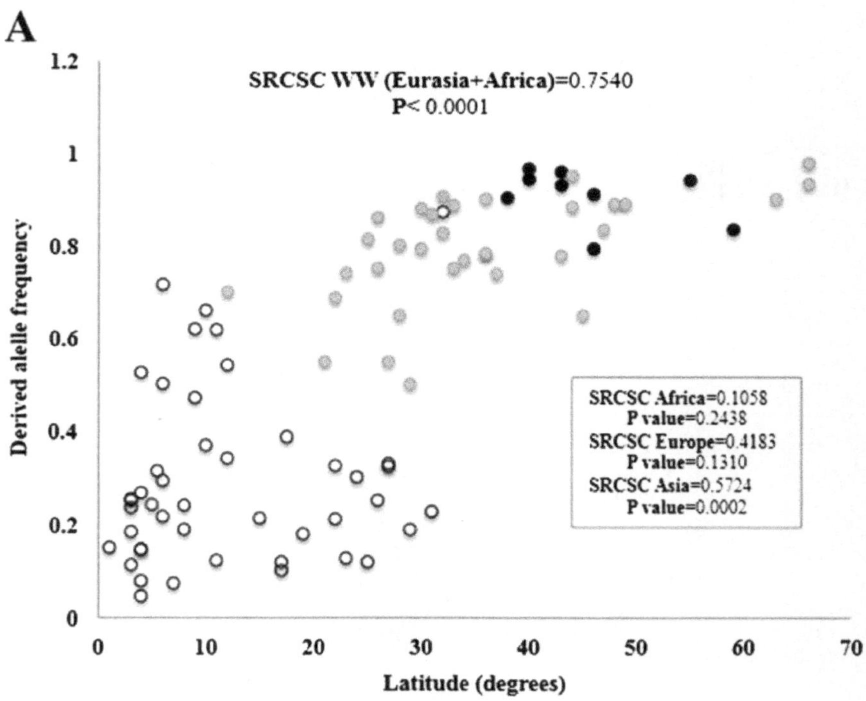

B

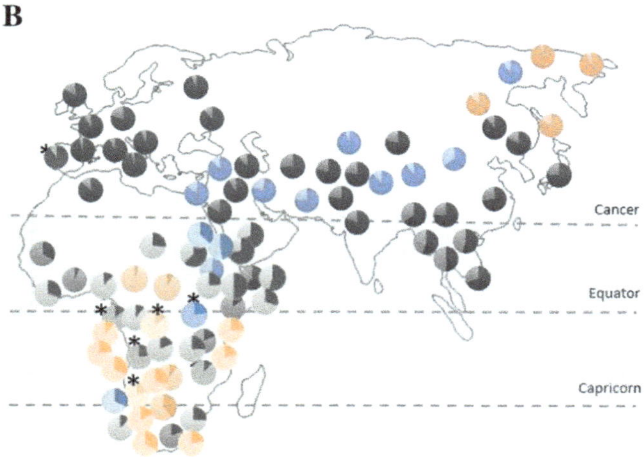

Figure 3: Distribution of *CYP3A5*3* in Africa and Eurasia and correlation with latitude. Correlation plot between latitude and allele frequencies in African (*open dots*), European (*black dots*) and Asian populations (*grey dots*) **(A)**. Map representing the distribution of CYP3A5*3 across Africa and Eurasia **(B)**. ancestral allele frequency (*light pie*) and the derived allele (*dark pie*); hunter-gatherers (*orange pie*), herders (*blue pie*) and farmers populations (*black pie*), *populations addressed in this study.

These analyses led to conclude that *CYP3A5* was the target of a selective factor determined by the geographic location of human populations.

Hierarchical AMOVA

Hierarchical AMOVA was performed to determine the relative contribution of geography, mode of subsistence and different diet-related variables to the genetic structure observed in the SNPs at *AGXT, PLRP2, MTRR, NAT2*(only for that defining *NAT2*6*) and *CYP3A5*, hereinafter referred for simplicity as uniquely by their gene symbols (Table 2).

Table 2: AMOVA analysis under different criteria

	c.32C > T (*AGXT*)	***P*-value**	**c.1074G > A (*PLRP2*)**	***P*-value**	**c.1130A > G (*MTRR*)**	***P*-value**	**c.590G > A (*NAT2*)**	***P*-value**	**c.219-237G > A (*CYP3A5*)**	***P*-value**
Mode of subsistence	**1.5**	**0.0355**	**8.8**	**0.0000**	−0.6	0.8793	**1.6**	**0.0046**	4.9	0.0542
Main diet component	0.4	0.1478	**6.5**	**0.0001**	−0.2	0.4712	**3.0**	**0.0019**	2.5	0.1004
Geography	**3.9**	**0.0001**	**8.5**	**0.0000**	**3.4**	**0.0003**	0.2	0.2149	**40.4**	0.0000
Latitude	-	-	-	-	-	-	-	-	**37.8**	**0.0000**
Above/below Tropic of Cancer	-	-	-	-	-	-	-	-	**44.9**	**0.0000**

Significant differences are highlighted in **bold**

Geography was found to significantly account to explain the total genetic variance across Africa and Eurasia at*AGXT, PLPR2, MTRR,* and *CYP3A5*, but not at *NAT2*. The contribution of geography was especially high in*CYP3A5* in which it amounted to a very high proportion, 40.4 % of total diversity. For this variation it was further assessed the effect of i) latitude and ii) the location North and South the Tropic of Cancer, leading to realize that for *CYP3A5* the highest value of F_{CT} (which measures the proportion of variance among groups) was achieved when populations North of the Tropic of Cancer were grouped against the southern ones, attaining then 44.9 % of total diversity.

Concerning mode of subsistence, it was found to be a considerable modulator of diversity at *PLRP2*, explaining 8.8 % of the total diversity at the locus, while also accounting to residual proportions of diversity at *NAT2*(1.6 %) and *AGXT* (1.5 %). When the criterion to group populations was the content in diets of cereals (for*PLPR2*), meat (for *AGXT*), folates (for *MTRR* and *NAT2*) or salt (for *CYP3A5*), significant F_{CT} values were only observed at *PLRP2,* in which the more or less reliance in cereals contributed to 6.5 % of the total variance, and at *NAT2,* where differences in the dietary richness in folates explained 3 % of the locus diversity.

Signals of Selection

To dissect better whether from the levels of genetic differentiation across Africa and Eurasia signs of selection could be captured, we used a conventional F_{ST}-based approach that assumes that genetic differentiation among populations is expectedly higher or lower for *loci* under directional or balanced selection, respectively, expected under neutrality.

Viewing that, we have firstly generated null sampling distribution of the empirical F_{ST} employing two different models, the finite Island Model (IM), which assumes the classical island model at migration-drift equilibrium [33]; and the Hierarchical Island Model (HIM), in which populations samples are assigned to different groups, allowing for increased migration rates between populations within groups than between groups [34]. Besides portraying more realistically the demographic history of human populations, HIM was shown to produce a low rate of false positive signs comparatively to IM, when used to test *loci* for selection [34].

The simulated null-distributions are presented in Figure 4 where are also shown the F_{ST} values plotted against scaled heterozygosity estimated for the SNPs at *AGXT, PLRP2, MTRR, CYP3A5* and *NAT2* [35].

A

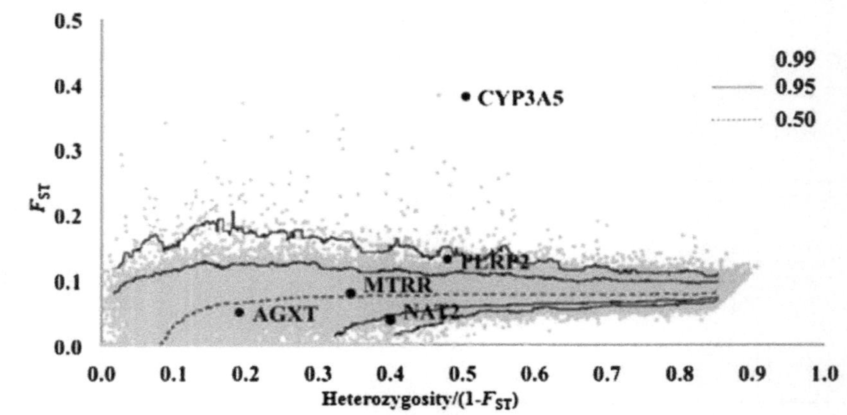

	F_{ST}	*Het*	*P*-value
AGXT	0.0500	0.1803	0.3382
CYP3A5	0.3813	0.3113	**0.0000**
MTRR	0.0794	0.3174	0.4052
PLRP2	0.1314	0.4156	**0.0170**
NAT2	0.0382	0.3845	**0.0365**

B

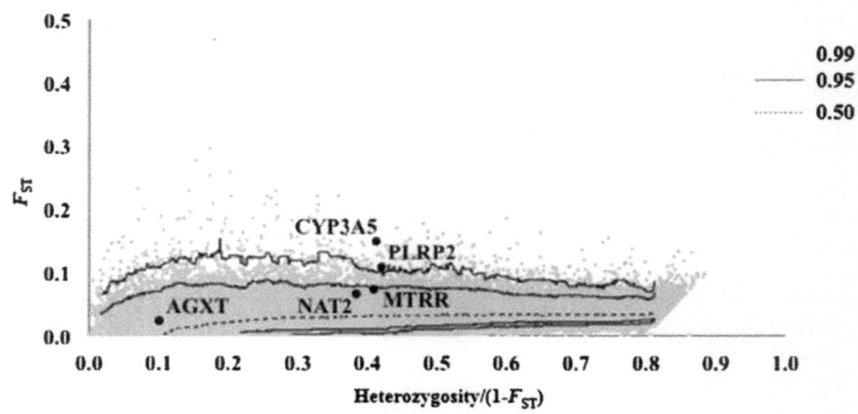

	F_{ST}	*Het*	*P*-value
AGXT	0.0224	0.0977	0.2766
CYP3A5	0.1477	0.3509	**0.0005**
MTRR	0.0717	0.3796	0.0555
PLRP2	0.1000	0.3744	**0.0055**
NAT2	0.0654	0.3581	0.0720

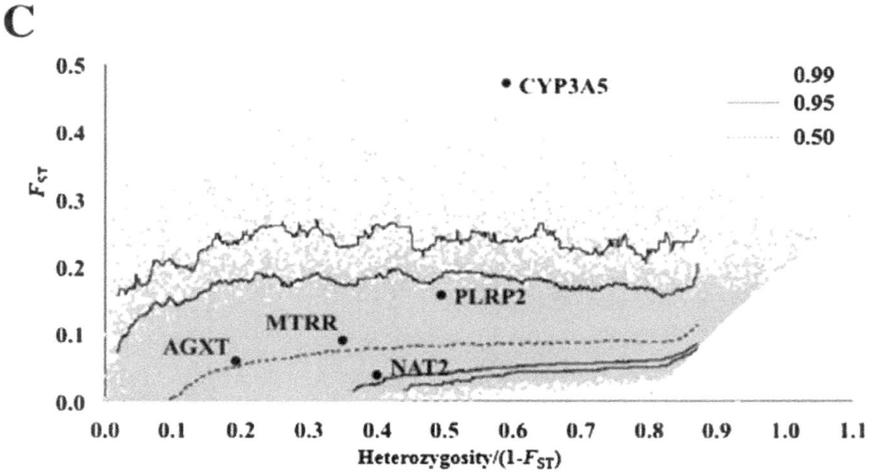

	F_{ST}	*Het*	*P*-value
AGXT	0.0579	0.1803	0.4372
CYP3A5	0.4724	0.3113	**0.0000**
MTRR	0.0898	0.3174	0.3602
PLRP2	0.1577	0.4156	0.0710
NAT2	0.0387	0.3845	0.1351

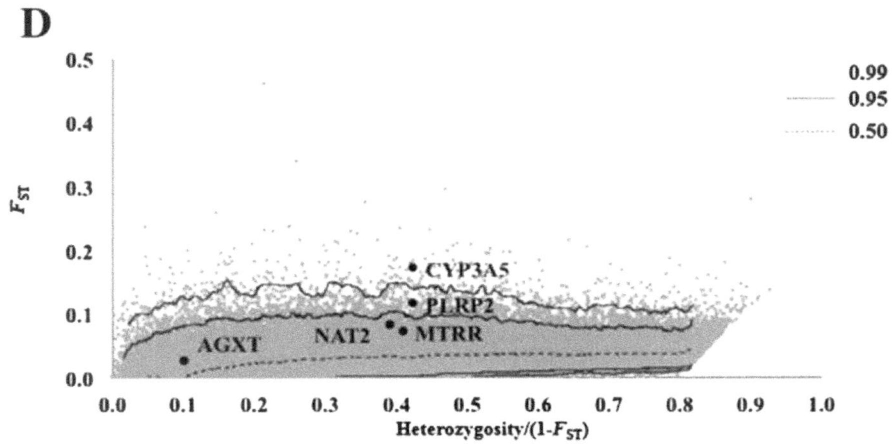

	F_{ST}	*Het*	*P*-value
AGXT	0.0261	0.0977	0.2651
CYP3A5	0.1717	0.3509	**0.0010**
MTRR	0.0718	0.3796	0.0966
PLRP2	0.1159	0.3744	**0.0150**
NAT2	0.0822	0.3581	0.0725

Figure 4: Joint distribution of F_{ST} vs. Scaled Heterozygosity expected under two neutral models. Joint distributions in African + Eurasian (**A**) and African populations (**B**) under Island Model (IM); and joint distributions in African + Eurasian (**C**) and African populations (**D**) under Hierarquical Island Model (HIM). It is represented the 99 % confidence regions of the null distribution. Black dots represent the observed measures in the studied genes, referred for simplicity as uniquely by their gene symbols; significant differences after Bonferroni's correction for multiple tests are highlighted in bold.

Considering simultaneously Africa and Eurasia and using as reference the IM distribution, the F_{ST}s for *MTRR* and *AGXT* did not differed significantly from the null expectations (Figure 4A). By contrast, the global differentiations at *PLRP2, NAT2* and *CYP3A5*, all lied outside the 95 % confidence region of the neutral distribution, though showing departures with opposite directions: whereas the F_{ST} coefficient for *NAT2* was significantly smaller than expected, the coefficients for *CYP3A5* and *PLRP2* were both significantly larger (*P*-values in Figure 4A). The outlier position is especially remarkable in the case of *CYP3A5* that presented the exceedingly high F_{ST} coefficient of 0.3813, almost five times greater compared to the average empirical neutral level of 0.079 between African and Eurasian populations. These results suggest that *NAT2* could have been under balanced or negative selection whist both *PLRP2* and *CYP3A5* might well have been modeled by positive selection. Taken into account the F_{ST} null distribution simulated under the HIM (Figure 4C), the F_{ST}s for*NAT2* and *PLRP2* lost the condition of significant outliers and the unique differentiation that remained significantly higher than the neutral expectations was at *CYP3A5*. Simulations were also carried out considering separately Africa and Eurasia. While in Eurasia none of the five assessed SNPs revealed to be outsiders in the distributions simulated under the simple or the hierarchical island models (results not shown), noteworthy in Africa the differentiations at *PLRP* and *CYP3A5* were significantly higher than expected under the neutral expectations derived from the two demographical models (Figure 4B and D).

LD Patterns

In order to assess whether the examined genetic variants were in fact those responsible for the selective signals detected *PLRP2, NAT2* and *CYP3A5,* we explored the patterns of *linkage disequilibrium* (LD) surrounding each of the three genes, viewing which a genomic window was considered that encompassed the adjacent genes. In Table 3 are presented the non-synonymous variants showing significant *D'* and r^2 values with our target SNPs, identified in African populations, which were the unique with genome data available. The correspondent LD plots for each gene across different African populations are present. For *CYP3A5* and*NAT2* no significant LD was detected with neighbor genes. Within each of the two genes, high LD was only found between our target SNP at *NAT2* and the linked variants rs1801280 and rs1208, both associated with decreased enzyme activity like rs1799930. Although this makes it difficult to discriminate the effects of the three variants, we can conclude that the selective signal detected at *NAT2* is related with variations that affect enzyme activity in a similar direction. As for the gene *PLRP2,* it was found to be located in a region of considerable LD with *PLRP1*, a downstream gene that codes for pancreatic lipase-related protein 1. Within*PLRP1* two non-synonymous (rs2305204 and rs1049125), whose functional consequences are unknown, are in strong LD with our target SNP at *PLRP2*, which in addition was at high LD with rs475199, a non-synonymous substitution of unknown functional effect, also located in *PLRP2*.

Table 3: Linkage Disequilibrium including *D'* and r^2 parameters

Gene (target SNP)	Non-synonymous SNP	*D'*	r^2	Gene location	DNA changes	Functional consequences	Reference
PLRP2(rs4451995)	rs2305204	1.0	0.036	*PLRP1*	c.1242G > C	-	[58]
		n.d.	n.d.				[57]
		n.d.	n.d.				[59]
	rs1049125	0.884	0.052	*PLRP1*	c.1382 T > C	-	[58]
		n.d.	n.d.				[57]
		n.d.	n.d.				[59]
	rs4751996	0.981	0.962	*PLRP2*	c.1084G > A	-	[58]
		n.d.	n.d.				[57]
		n.d.	n.d.				[59]
NAT2(rs1799930)	rs1801280	n.d.	n.d.	*NAT2*	c.341 T > C	Associated with slow acetylator due to N-acetyltransferase enzyme variant (acetylation slow phenotype)	[58]
		1.0	0.142				[57]
		n.d.	n.d.				[59]
	rs1208	n.d.	n.d.	*NAT2*	c.803G > A		[58]
		1.0	0.360				[57]
		n.d.	n.d.				[59]

n.d. no data available

DISCUSSION

The analysis of patterns of human genetic diversity at wide geographical scales can disclose remarkable features difficultly explained by demographic events or pure neutral processes, that rather might represent the first symptoms of environmental adaptations.

In this study, we draw attention to variations in *AGXT*, *PLRP2*, *MTRR*, *NAT2* and *CYP3A5*, five genes assumedly involved in the metabolism of substances (including xenobiotics) that gain entry into the organism through dietary food stuffs, for which it has been previously posited that they could represent instances of gene-culture coevolution in humans [13, 15, 20].

Out of those genes, *PLRP2*, *NAT2* and *CYP3A5* were found to present signs in their distribution patterns evoking the action of environmental selective pressures, though of diverse nature and strength.

The most unequivocal signature of selection was associated with *CYP3A5* that displayed a level of inter-population differentiation dramatically surmounting even the most conservative neutral expectations. Contrarily to our starting hypothesis, however, the amount of salt presumed to be ingested across main dietary habits did not accounted for the distribution of *CYP3A5*, which instead was highly determined by the geographical location of populations in the North or in the South of the Tropic of Cancer. So, the analyses here undertaken fully support previous findings indicating that *CYP3A5*3* evolved under a selective pressure determined by an environmental factor correlated with latitude [24], but also add accuracy to the interpretation pointing toward a factor shared by regions located above or below the Northern Tropic. *CYP3A5*has been intensively explored in the context of the genetic factors contributing to hypertension susceptibility, known to vary widely across different human populations. Nearly 40 years ago Gleibermann [36] proposed the "sodium retention" hypothesis, according to which the high rate of hypertension in certain populations could partially be due to a genetic background that was environmental adaptive, presuming that efficient salt retaining mechanisms might had been advantageous in the hot savanna climate where humans first emerged. More recently, it was argued that hypertension susceptibility was ancestral in humans, and that differential susceptibility arose due to distinct selective pressures after the Out-of-Africa expansion of modern humans [27]. *CYP3A5* is being often quoted to address the evolutionary perspective of hypertension susceptibility, due to the demonstrated role of *CYP3A5* enzymes in sodium homeostasis, even though the many studies that analyzed the relationship between *CYP3A5* genotypes and blood pressure/hypertension have provided quite inconsistent results (reviewed in Lamba et al. [37]). So, together with the clarification of the link between*CYP3A5* and blood pressure, future lines of research should pay more attention to the role of *CYP3A5* enzymes in the physiological processes related with thermoregulation and/or with neutralization of effects of sunlight exposure. In the highly heat stressful intertropical region, there is a regular need to deal with the threat of dehydration, which may raise complicated physiological responses in wet or dry climates under which the efficient control of heat loss likely differs. Interestingly, the involvement of *CYP3A5* in such responses seems to obtain support from the recent discovery of an osmosensitive transcriptional control of human *CYP3A4*, *CYP3A7*, and *CYP3A5* that revealed increased mRNA expressions under ambient hypertonicity [38].

Concerning *PLRP2*, the explorations here undertaken led in essence to corroborate the findings of Hancock et al. [15], indicating that diversity at the locus is somehow connected with mode of subsistence in populations. In fact, the assessed truncated allele showed to be significantly more frequent in farmers comparatively to groups not relying in farming, with the general trend, inferred from the whole set of African and Eurasian populations, pointing to a clinal decrease in frequency from farmers, next pastoralists towards agriculturalists. In addition, the global differentiation at this variant fell outside the neutral expectations, except when the HIM model was used in the tests for selection in Africa plus Eurasia. Hancock et al. [15] have associated the worldwide distribution of *PLRP2* to the content in cereals in diets of populations, on the grounds of the important role of the protein encoded by *PLRP2* in plant-based diets once, unlike other pancreatic lipases, this enzyme hydrolyzes galactolipids, which are the main triglyceride component in plants [15]. However, the recent demonstration that the truncated allele addressed in their (and our) study exhibits near absence of secretion makes it unlikely that the encoded product may contribute to plant lipid digestion in humans [39], which seemingly undermines the biological basis originally proposed. In the meanwhile, new insights arose on the physiological role of *PLRP2*, suggesting for instance a likely major influence in fat digestion in newborns [40]. This refreshed information opens new perspectives that deserve future investigation to clarify whether cereal content or other *PLRP2* substrate, or amount of substrate, differing in hunter-gatherers, herders and farmers is the factor that exerted the selective pressure contributing to shape the current pattern of *PLRP2*diversity in human populations.

Before, however, it is necessary to overcome the uncertainty raised by the presence of significant LD between*PLRP2* and *PLRP1*. The two genes encode lipases that show high sequence homology and that assumedly participate in dietary fat metabolism, although not being yet clarified their differences in substrate specificity. Consequently, for the moment is not possible to discriminate between which variations at *PLRP2* or *PLRP1* are the best candidates to be causative of the selective signals detected.

With respect to *NAT2*, in agreement with earlier studies [21, 22] we also detected that the average frequency of a slow acetylator allele was statistically lower in hunter-gatherers than in food-producer populations, when Eurasian and African populations were taken as a set. Furthermore, the same slow acetylator variant, which is the widespread *NAT2*6* allele, revealed an unusual low level of geographical structure across Africa and Eurasia, indicating that it was subjected to drifting constrains that likely could arise under the action of a mode of selection resulting in such a homogeneous allele distribution. In the tests for selection involving *NAT2*, significant departure from the neutral expectation was captured when the simple IM demographic model was assumed, which can raise uncertainty on whether signs of more subtle selective pressures might mistakenly escape the stringency of the HIM. The adaptive evolution of *NAT2* has been supported by a number of different studies [19-23, 41, 42], including the examination of *NAT2* sequence data whose patterns of diversity made it plausible that slow-acetylating variants have been subject to weak selective pressures [23]. However, it is yet to be clarified the nature of such pressures. Luca et al. [22] tentatively claimed that it could be related with the diminished availability of folates in diet brought with the shift from economies relying in hunting and gathering to those based on farming and herding of domesticates. In line with the hypothesis, very recently a significant correlation was reported between *NAT2* acetylator phenotypes and historical dietary habitudes in India with the slow acetylator prevalence being higher in regions where is higher the proportion of vegetarians populations [43]. A major problem with this folate-related model is the still non-demonstrated role of *NAT2*human enzymes in folate metabolism [44]. Endogenous substrates for human *NAT2* are not known, although being well established that *NAT2* catalyzes the acetylation of many xenobiotics [44]. Since the exposure to xenobiotic substances or to concentration of xenobiotics likewise must had altered along the change in diets experienced by producers of food resources, the role of this kind of substances in shaping diversity at *NAT2*deserves further attention.

In summary, we provided added evidence that diversity at *PLRP2* and *NAT2* harbor signatures of genetic adaptations that might have been triggered by the diversification of modes of subsistence and diets that human populations began to experience after the rise of the Neolithic. Before that, modern humans had started to leave their original homeland in Africa, traveling out of the continent to colonize all regions in the globe. They progressively reached a wide range of new environments, facing new selective pressures that may have contributed to shape human genetic diversity. In *CYP3A5*, the compelling correlation with regions North and South the Tropic of Cancer, makes it likely that it belongs to the yet not fully understood catalog of genetic adaptations triggered by environmental stresses. Furthermore, the genetic signatures that *CYP3A5* harbors seem strength enough to have been driven by very long-lasting selection.

Finally, we were unable to confirm the hypotheses at stake that *AGXT* and *MTRR* could also be diet-selected genes, since their diversity patterns could be well reconciled with demographic history at least of African and Eurasian populations.

CONCLUSIONS

In this study, we found signs that *PLRP2*, *NAT2* and *CYP3A5*, three genes assumedly involved in the metabolism of substances (including xenobiotics) that gain entry into the organism through dietary food stuffs, can represent instances of gene-culture coevolution in humans. Concerning *PLRP2*, it is still needed to clarify whether the signal detected is not a hitchhiking effect of its neighbor *PLRP1*. In addition, it is also necessary to demonstrate which were the biological mechanisms, and the environmental factors involved as well as their interactions, to understand the nature of the selective pressures that contributed to shape current patterns of genetic diversity at those *loci*. Furthermore, functional studies are needed to demonstrate the putative biological impact of the variations assessed, which ultimately would also exclude that the detected signs of selection could be due to other variations in *linkage disequilibrium* with those that were here examined.

METHODS

Ethics Statement

The current study was approved by the Institute of Molecular Pathology and Immunology of the University of Porto institutional review board. All samples involved were anonymised DNA extracts previously obtained from healthy unrelated individuals. The samples were collected under written informed consent. The current study complies with the ethical principles of the 2000 Helsinki Declaration of the 206 World Medical Association (http://jama.jamanetwork.com).

Samples and DNA Extraction

A total amount of 361 individuals from six sub-Saharan African populations with different modes of subsistence were analyzed, i) 144 were farmers/agriculturalists from three populations: 32 from Angola (ANG), 82 from Equatorial Guinea (EQG) and 30 from Mozambique (MOZ); ii) 116 were herders/pastoralists from the Karamoja region in Northern Uganda (UGN); and iii) 101 belonged to two hunter-gatherer/forager populations: 39 Baka Pygmies from Gabon (BPY) and 62 Ju/hoansi from Tsumkwe, a small settlement in North Eastern Namibia (KNA). Forty-eight individuals from the Portuguese population (PTG) were additionally studied. Each population was assigned to a specific mode of subsistence and main diet component according to Murdock Ethnographic Atlas (http://lucy.ukc.ac.uk/cgi-bin/uncgi/Ethnoatlas/atlas.vopts), Encyclopedia of World Cultures, Africa: An Encyclopedia for Students [45].

From blood samples stored in FTA™ cards (Whatman), total DNA was extracted using a standard phenol-chlorophorm protocol [46] in the sample from Equatorial Guinea, while in the remaining samples it was done as described in previous works: Angola [47], Mozambique [48], Uganda [49], Baka Pygmies [50], Khoisan (Marks et al., submitted) and Portugal.

Genotyping

The screened SNPs were c.32C > T (rs34116584; NM_000030.2 *AGXT* gene), c.1074G > A (rs4751995; NM_005396.3 *PLRP2* gene), c.1130A > G (rs162036; NM_024010.2 *MTRR* gene), c.219-237G > A (rs776746; NM_000777.3 *CYP3A5* gene) and the following 4 variants in *NAT2* gene (NM_000015.2): c.191G > A (rs1801279), c.341 T > C (rs1801280), c.590G > A (rs1799930) and c.857G > A (rs1799931). The last 4 SNPS, defining the alleles*NAT2*14*, *NAT2*5*, *NAT2*6* and *NAT2*7*, respectively, were selected in this study because reportedly they allow inferring the acetylator phenotypes with high accuracy [41].

A pair of primers was designed for each SNP using Primer3 software ver. 4.0. Possible secondary structures or interactions between primers were checked with AutoDimer software ver. 1.0 [51]. A multiplex Polymerase Chain Reaction (PCR) system was developed to co-amplify the regions containing the eight SNPs. The genotyping methodology was based in a multiplex minisequencing reaction, using the Single Base Extension (SBE) [31] reaction kit (Applied Biosystems). SBE primers were designed and tested identically as described above. Poly tails of varying lengths were attached to the 5′ end of each primer in order to avoid identical fragment sizes, allowing simultaneous typing of multiple variants in the same reaction. SBE products were run on ABI 3130 Genetic Analyser (AB Applied Biosystems) and the electropherograms were analyzed using GeneMapper software ver. 4.0. (Applied Biosystems, Foster City, USA), based on fragment size inferred with GeneScan-120 size standard.

Chromosomal locations and genomic segments from the 5 genes were obtained using the latest version of the human genome assembly GRCh37 (http://www.ensembl.org/).

Statistical Analysis

The Arlequin software ver. 3.5.1 [52] was used to estimate allele and haplotype frequencies, to test for Hardy-Weinberg Equilibrium (HWE) and to calculate genetic distances (F_{ST}). Regarding the *NAT2* gene, to account for *linkage disequilibrium* (LD) between SNPs, from the unphased multilocus genotypic data, haplotypic frequencies defined by the 4 SNPs were computationally estimated also with Arlequin software ver. 3.5.1. [52] and the acetylation phenotypes deduced from the pair of haplotypes carried by each subject [41].

In order to determine whether means of allele frequencies were statistically different between groups of populations defined according to various criteria, One-Way ANOVA tests were performed. The Spearman Rank Correlation Score Corrected (SRCSC) was used to evaluate the correlation between allele frequencies and latitude. Both statistical analyses were performed on the website for statistical computation VassarStats (http://vassarstats.net).

The graphical representation of a F_{ST} distance matrix was constructed by means of the Multidimensional Scaling (MDS) procedure implemented in StatSoft, Inc. (2007), Statistica version 8.0 (www.statsoft.com).

To investigate possible signals of selection we obtained the neutral distribution of F_{ST}s conditional on heterozygozity based on genotypic data from a validated panel of 52 assumedly neutral SNPs for human identification [53], using the 61 African and Eurasian populations contained in the SNP*for* ID browser, to draw a neutral dispersion cloud 50,000 coalescent simulations of 100 demes were carried on Arlequin software. For variants at *AGXT*, *PLRP2*, *MTRR*, *NAT2* and *CYP3A5*, F_{ST}, average heterozygosity within populations (*Het*) and the scaled heterozygosity *Het*/(1-F_{ST}) were also computed. The null sampling distribution of the empirical F_{ST}values was calculated using two distinct models, both implemented in Arlequin software ver. 3.5.1 [52]: i) the classical Island Model at migration-drift equilibrium (IM), conventionally referred to as Fdist approach, proposed by [33], considering together all populations; and ii) the Hierarchical Island Model (HIM) more recently recommended by Excoffier and Hofer [35], for which populations were clustered in 5 groups (Europe, Africa, Middle East, South and East Asia) when a worldwide scale was assumed, or in 4 groups (Northern, Southern, Western and Eastern Africa) when only Africa was considered. Since F_{ST} strongly correlates with heterozygosity [33], the empirical *P*-value for each locus was calculated within bins of 2,0000 SNPs grouped according to Minimum Allele Frequency (MAF), as the proportion of the bivariate probability distribution which was less probable than the estimated values, in the same way as calculated in the DFdist software package [33].

Hierarchical Analysis of Molecular Variance (AMOVA) was performed in the Arlequin software ver. 3.5.1. defining population groups according to: i) mode of subsistence, ii) main diet component, and iii) geography in the context of Eurasia and Africa. Additionally for *CYP3A5,* populations groups were also defined according to latitude and location North or South the Tropic of Cancer.

Finally for LD patterns analyses, the .ped files containing the data sets were first manipulated with gPLINK software (http://pngu.mgh.harvard.edu/purcell/plink/) [54] and then exported to Haploview software ver. 4.1 [55] to calculate *D'* (normalized *D*, where *D'* = *D*/*D*max measures the LD strength) and r^2 (squared correlation coefficient measure of LD between the two *loci*) parameters and visualize the LD plots, considering a window of 200, 300 and 500 Kb for *PLRP2*, *CYP3A5* and *NAT2* genes, respectively.

Comparative Data

Viewing comparative analyses, data for other populations were retrieved from the database dbCLINE (http://genapps.uchicago.edu/software.html) from the Di Rienzo laboratory, which includes information from the International HaPMap Project Phase III (http://hapmap.ncbi.nlm.nih.gov/), the Human Genome Diversity Project Panel-Centre d'Etude du Polymorphisme Humain (HGDP-CEPH); and also we included data already published in other works [56-59]. The populations' abbreviations used in Figs. 1 and 2 are MAN (Mandenka), NBT (North Bantu), MOB (Mozabite), SBT (South Bantu), LUH (Luhya), TIG (Tigray), AMH1 (Amhara1), AFA (Afar), YOR (Yoruba), AMH2 (Amhara2), ARC (Ari Cultivator), WOL (Wolayta), ANU (Anuak), ARB (Ari Blacksmith), JPT (Japanese), NAX (Naxi), LAH (Lahu), MIA (Miaozu), TUJ (Tujia), SHE (She), YIZ (Yizu), CAM (Cambodian), HAN (Han), DAU (Daur), DAI (Dai), BER (Bergamo), RUS (Russian), TSC1 (Tuscan1), SAR (Sardinian), FRC (French), BAS (Basque), TUSC2 (Tuscan2), ORC (Orcadian), ADY (Adygei), DRU (Druze), PAL (Palestinian) KAL (Kalash), PAT (Pathan), GUJ (Gujarati), BUR (Burusho), SIN (Sindhi), XIB (Xibo), GUM (Gumuz), SOM (Somali), MAS (Maasai), MON (Mongola), YAK (Yakut), BAL (Balochi), MAK (Makrani), TU (Tu), UYG (Uygur), BED (Bedouin), BRA (Brahui), HAZ (Hazara), SAN (San), BIP (Biaka Pygmies), KUV (Kung Vasekela), MBP (Mbuti Pygmies) KHO1 (Khomani2), SAD (Sadawe), HAD (Hadza), HEZ (Hezhen), NYU (Naukan Yup'ik), ORO (Orogen) and MCH (Maritime Chukchee).

ACKNOWLEDGMENTS

We thank David Comas (Universitat Pompeu Fabra, Barcelona), Jean-Marie Hombert and Lolke van der Veen (Dynamique du Langage, Institut des Sciences De l'Homme, Lyon France) as well as Patrick Mouguiama Daouda (University Omar Bongo, Libreville, Gabon) and the Centre International des Recherches Médicales de Franceville (CIRMF, Gabon) for sharing the Baka Pygmies data from Gabon. We also would like to thank Renaud Vitallis from Centre de Biologie pour la Gestion des Populations, Université Montpellier for sharing his modified version for codominant data of the Dfdist software. Finally we are really grateful to all volunteers who contributed with their DNA to this study.

AUTHORS' CONTRIBUTIONS

CV and MJP conceived, designed and directed the experiment; CV, LA and MJP analyzed the data; CV and MJP wrote the paper. SJM, CC, OO, EO, AMLP and EAP contributed with African samples. CV, LA, CC, WP, AA, LG and MJP revised the paper. All authors read and approved the final manuscript.

REFERENCES

1. Luca F, Perry GH, Di Rienzo A. Evolutionary adaptations to dietary changes. Annu Rev Nutr. 2010;30:291-314.
2. Patin E, Quintana-Murci L. Demeter's legacy: rapid changes to our genome imposed by diet. Trends Ecol Evol. 2008;23(2):56-9.
3. Lachance J, Vernot B, Elbers CC, Ferwerda B, Froment A, Bodo J-M, et al. Evolutionary history and adaptation from high-coverage whole-genome sequences of diverse African hunter-gatherers. Cell. 2012;150(3):457-69.
4. Balaresque PL, Ballereau SJ, Jobling MA. Challenges in human genetic diversity: demographic history and adaptation. Hum Mol Genet. 2007;16 Spec No. 2:R134-9.
5. Bersaglieri T, Sabeti PC, Patterson N, Vanderploeg T, Schaffner SF, Drake JA, et al. Genetic signatures of strong recent positive selection at the lactase gene. Am J Hum Genet. 2004;74(6):1111-20.
6. Jones BL, Raga TO, Liebert A, Zmarz P, Bekele E, Danielsen ET, et al. Diversity of lactase persistence alleles in Ethiopia: signature of a soft selective sweep. Am J Hum Genet. 2013;93(3):538-44.
7. Liu X, Ong RT-H, Pillai EN, Elzein AM, Small KS, Clark TG, et al. Detecting and characterizing genomic signatures of positive selection in global populations. Am J Hum Genet. 2013;92(6):866-81.
8. Vitalis R, Gautier M, Dawson KJ, Beaumont MA. Detecting and measuring selection from gene frequency data. Genetics. 2014;196(3):799-817.
9. Tishkoff SA, Reed FA, Ranciaro A, Voight BF, Babbitt CC, Silverman JS, et al. Convergent adaptation of human lactase persistence in Africa and Europe. Nat Genet. 2007;39(1):31-40.
10. Perry GH, Dominy NJ, Claw KG, Lee AS, Fiegler H, Redon R, et al. Diet and the evolution of human amylase gene copy number variation. Nat Genet. 2007;39(10):1256-60.
11. Caldwell EF, Mayor LR, Thomas MG, Danpure CJ. Diet and the frequency of the alanine:glyoxylate aminotransferase Pro11Leu polymorphism in different human populations. Hum Genet. 2004;115(6):504-9.
12. Danpure CJ. Variable peroxisomal and mitochondrial targeting of alanine: glyoxylate aminotransferase in mammalian evolution and disease. Bioessays. 1997;19(4):317-26.
13. Segurel L, Lafosse S, Heyer E, Vitalis R. Frequency of the AGT Pro11Leu polymorphism in humans: does diet matter? Ann Hum Genet. 2010;74(1):57-64.
14. De Caro J, Eydoux C, Cherif S, Lebrun R, Gargouri Y, Carriere F, et al. Occurrence of pancreatic lipase-related protein-2 in various species and its relationship with herbivore diet. Comp Biochem Physiol B Biochem Mol Biol. 2008;150(1):1-9.
15. Hancock AM, Witonsky DB, Ehler E, Alkorta-Aranburu G, Beall C, Gebremedhin A, et al. Colloquium paper: human adaptations to diet, subsistence, and ecoregion are due to subtle shifts in allele frequency. Proc Natl Acad Sci U S A. 2010;107 Suppl 2:8924-30.
16. Lowe ME. The triglyceride lipases of the pancreas. J Lipid Res. 2002;43(12):2007-16.
17. Sias B. Human pancreatic lipase-related protein 2 is a galactolipase. Biochemistry. 2004;43(31):10138-48.
18. Shaw GM, Lu W, Zhu H, Yang W, Briggs FB, Carmichael SL, et al. 118 SNPs of folate-related genes and risks of spina bifida and conotruncal heart defects. BMC Med Genet. 2009;10:49.
19. Magalon H, Patin E, Austerlitz F, Hegay T, Aldashev A, Quintana-Murci L, et al. Population genetic diversity of the NAT2 gene supports a role of acetylation in human adaptation to farming in Central Asia. Eur J Hum Genet. 2008;16(2):243-51.
20. Sabbagh A, Langaney A, Darlu P, Gerard N, Krishnamoorthy R, Poloni ES. Worldwide distribution of NAT2 diversity: implications for NAT2 evolutionary history. BMC Genet. 2008;9:21.
21. Sabbagh A, Darlu P, Crouau-Roy B, Poloni ES. Arylamine N-acetyltransferase 2 (NAT2) genetic diversity and traditional subsistence: a worldwide population survey. PLoS One. 2011;6(4):e18507.
22. Luca F, Bubba G, Basile M, Brdicka R, Michalodimitrakis E, Richards O, Vershubsky G, Quintana-Murci L, Kozlov AI, Novelleto A: Multiple Advantageous Amino Acid Variants in the NAT2 Gene in Human Populations. *PloS one* 2008, 3(9).
23. Patin E, Barreiro LB, Sabeti PC, Austerlitz F, Luca F, Sajantila A, Behar DM, Semino O, Sakuntabhai A, Guiso N *et al*: Deciphering the Ancient and Complex Evolutionary History of Human Arylamine N-Acetyltransferase Genes. American Journal of Human Genetics 2006, 78.
24. Thompson EE, Kuttab-Boulos H, Witonsky D, Yang L, Roe BA, Di Rienzo A. CYP3A variation and the evolution of salt-sensitivity variants. Am J Hum Genet. 2004;75(6):1059-69.
25. Thompson EE, Kuttab-Boulos H, Yang L, Roe BA, Di Rienzo A. Sequence diversity and haplotype structure at the human CYP3A cluster. Pharmacogenomics J. 2006;6(2):105-14.
26. Young JH, Chang YP, Kim JD, Chretien JP, Klag MJ, Levine MA, et al. Differential susceptibility to hypertension is due to selection during the out-of-Africa expansion. PLoS Genet. 2005;1(6):e82.
27. Zhang L, Miyaki K, Wang W, Muramatsu M. CYP3A5 polymorphism and sensitivity of blood pressure to dietary salt in Japanese men. J Hum Hypertens. 2010;24(5):345-

50.

28. Danpure CJ. Primary hyperoxaluria type 1: AGT mistargeting highlights the fundamental differences between the peroxisomal and mitochondrial protein import pathways. Biochim Biophys Acta. 2006;1763(12):1776-84.
29. Cordain L, Eaton SB, Miller JB, Mann N, Hill K. The paradoxical nature of hunter-gatherer diets: meat-based, yet non-atherogenic. Eur J Clin Nutr. 2002;56 Suppl 1:S42-52.
30. Patillon B, Luisi P, Poloni ES, Boukouvala S, Darlu P, Genin E, et al. A homogenizing process of selection Has maintained an "ultra-slow" acetylation NAT2 variant in humans. Hum Biol. 2014;86(3):185-214.
31. Eap CB, Bochud M, Elston RC, Bovet P, Maillard MP, Nussberger J, et al. CYP3A5 and ABCB1 genes influence blood pressure and response to treatment, and their effect is modified by salt. Hypertension. 2007;49(5):1007-14.
32. Eaton SB, Eaton 3rd SB. Paleolithic vs. modern diets-selected pathophysiological implications. Eur J Nutr. 2000;39(2):67-70.
33. Beaumont MA, Nichols RA. Evaluating loci for use in the genetic analysis of population structure. P Roy Soc B-Biol Sci. 1996;263(1377):1619-26.
34. Excoffier LH, T.; Foll, M.: Detecting loci under selection in a hierarchically structured population. Heredity 2009, 103(4).
35. Excoffier L, Hofer T, Foll M. Detecting loci under selection in a hierarchically structured population. Heredity (Edinb). 2009;103(4):285-98.
36. Gleibermann L. Blood pressure and dietary salt in human populations. Ecol Food Nutr. 1973;2(2):143-56.
37. Lamba JK, Lin YS, Schuetz EG, Thummel KE: Genetic contribution to variable human CYP3A-mediated metabolism. Advanced drug delivery reviews 2012.
38. Kosuge K, Chuang AI, Uematsu S, Tan KP, Ohashi K, Ko BC, Ito S: Discovery of Osmo-sensitive Transcriptional Regulation of Human Cytochrome P450 3As (CYP3As) by the Tonicity-Responsive Enhancer Binding Protein (TonEBP/ NFAT5). Molecular Pharmacology 2007.
39. Xiao X, Mukherjee A, Ross LE, Lowe ME. Pancreatic lipase-related protein-2 (PLRP2) Can contribute to dietary Fat digestion in human newborns. J Biol Chem. 2011;286(30):26353-63.
40. Berton A, Sebban-Kreuzer C, Crenon I. Role of the structural domains in the functional properties of pancreatic lipase-related protein 2. FEBS J. 2007;274(22):6011-23.
41. Hein DW, Doll MA. Accuracy of various human NAT2 SNP genotyping panels to infer rapid, intermediate and slow acetylator phenotypes. Pharmacogenomics. 2012;13(1):31-41.
42. Talbot J, Magno LA, Santana CV, Sousa SM, Melo PR, Correa RX, et al. Interethnic diversity of NAT2 polymorphisms in Brazilian admixed populations. BMC Genet. 2010;11:87.
43. Khan N, Pande V, Das A. NAT2 sequence polymorphisms and acetylation profiles in Indians. Pharmacogenomics. 2013;14(3):289-303.
44. Sim E, Lack N, Wang CJ, Long H, Westwood I, Fullam E, et al. Arylamine N-acetyltransferases: structural and functional implications of polymorphisms. Toxicology. 2008;254(3):170-83.
45. John Haley CR, Rebecca Stefoff, Joseph Ziegler: Africa: An Encyclopedia for Students. In: Africa: An Encyclopedia for Students. Edited by Middleton J: Charles Scribner's Sons; 2002.
46. Sambrook J. Molecular cloning : a laboratory manual, vol. 3. 2nd ed. NY: Cold Spring Harbor, N.Y: Cold Spring Harbor Laboratory Press; 1989.
47. Beleza S, Alves C, Reis F, Amorim A, Carracedo A, Gusmao L. 17 STR data (AmpF/STR identifiler and powerplex 16 system) from Cabinda (Angola). Forensic Sci Int. 2004;141(2-3):193-6.
48. Alves C, Gusmao L, Amorim A. STR data (AmpFlSTR profiler plus and GenePrint CTTv) from Mozambique. Forensic Sci Int. 2001;119(1):131-3.
49. Gomes V, Sanchez-Diz P, Alves C, Gomes I, Amorim A, Carracedo A, et al. Population data defined by 15 autosomal STR loci in Karamoja population (Uganda) using AmpF/STR Identifiler kit. Forensic Sci Int Genet. 2009;3(2):e55-8.
50. Batini C, Lopes J, Behar DM, Calafell F, Jorde LB, van der Veen L, et al. Insights into the demographic history of African Pygmies from complete mitochondrial genomes. Mol Biol Evol. 2011;28(2):1099-110.
51. Vallone PM, Butler JM. AutoDimer: a screening tool for primer-dimer and hairpin structures. Biotechniques. 2004;37(2):226-31.
52. Excoffier L, Lischer HE. Arlequin suite ver 3.5: a new series of programs to perform population genetics analyses under Linux and Windows. Mol Ecol Resour. 2010;10(3):564-7.
53. Sanchez JJ, Phillips C, Børsting C, Balogh K, Bogus M, Fondevila M, et al. A multiplex assay with 52 single nucleotide polymorphisms for human identification. Electrophor. 2006;27(9):1713-24.
54. Purcell S, Neale B, Todd-Brown K, Thomas L, Ferreira MA, Bender D, et al. PLINK: a tool set for whole-genome association and population-based linkage analyses. Am J Hum Genet. 2007;81(3):559-75.
55. Barrett JC, Fry B, Maller J, Daly MJ. Haploview: analysis and visualization of LD and haplotype maps. Bioinformatics. 2005;21(2):263-5.
56. Bryc K, Auton A, Nelson MR, Oksenberg JR, Hauser SL, Williams S, et al. Genome-wide patterns of population structure and admixture in West Africans and African Americans. Proc Natl Acad Sci U S A. 2010;107(2):786-91.
57. Henn BM, Gignoux CR, Jobin M, Granka JM, Macpherson JM, Kidd JM, et al. Hunter-gatherer genomic diversity suggests a southern African origin for modern humans. Proc Natl Acad Sci U S A. 2011;108(13):5154-62.
58. Pagani L, Kivisild T, Tarekegn A, Ekong R, Plaster C, Gallego Romero I, et al. Ethiopian genetic diversity

reveals linguistic stratification and complex influences on the Ethiopian gene pool. Am J Hum Genet. 2012;91(1):83-96.

59. Schlebusch CM, Skoglund P, Sjodin P, Gattepaille LM, Hernandez D, Jay F, et al. Genomic variation in seven Khoe-San groups reveals adaptation and complex African history. Science. 2012;338(6105):374-9.

SECTION III: GENETIC DIVERSITY IN PLANTS AND ANIMALS: CHARACTERIZATION, PRESERVATION AND UTILIZATION IN THE CONTEXTS OF AGRICULTURE AND CONSERVATION

CHAPTER 7

GENOME-WIDE DETECTION OF CNVS AND THEIR ASSOCIATION WITH MEAT TENDERNESS IN NELORE CATTLE

Vinicius Henrique da Silva, Luciana Correia de Almeida Regitano, Ludwig Geistlinger, Fábio Pértille, Poliana Fernanda Giachetto, Ricardo Augusto Brassaloti, Natália Silva Morosini, Ralf Zimmer, Luiz Lehmann Coutinho

Keywords: Genome-wide association analysis, Functional analysis, Holstein cattle, Female fertility and milk production

INTRODUCTION

Bos taurus is a well-studied model organism [1,2] and a species of great agricultural relevance, especially for Brazil, which is one of the world's largest beef exporter [3] with a herd of estimated 209,541 million head [4]. Nelore (*Bos taurus indicus*) is the main cattle breed in Brazil [5] and is, like most *Bos taurus indicus* subspecies, adapted to the tropical Brazilian climate. However, meat tenderness (MT) of Nelore is not comparable to taurine breeds (*Bos taurus taurus*) [6].

The MT phenotype was first investigated in the 1920s [7], and is already well-studied for several livestock species [8–13]. An established measure of MT is shear force, and *Bos taurus indicus* typically requires a higher shear force compared with *Bos taurus taurus* to disrupt the beef fibers [14]. Larger muscle fibers, cross-bridges between filaments and reduced myofibrillar proteolysis are features that influence MT of indicus breeds [15,16]. Stress induced by genetic or environmental sources is also known to negatively affect MT [17].

Structural genetic variation associated with traits of interest are promising targets for animal breeding [18]. Copy number variation (CNV) is one of the frequently observed structural genomic variations and is, thus, increasingly being studied in cattle [19–28]. CNVs are defined as large genomic regions (conventionally >1 kb) with deviation from the normal diploid state due to duplication or deletion events [29]. CNVs are associated with several important phenotypes in humans [30–32] and livestock animals [33–35]. In cattle, chronic interstitial nephritis [36] as well as osteopetrosis [37] and birth defects [38] have been previously associated with CNVs. A CNV in the gene encoding CASL-like protein 2 (*MICAL-L2*), which performs a critical role in the development of muscle fibers, is correlated with gene expression in cattle making it an important molecular marker [39].

Experimental detection of CNVs on a large scale can be done using comparative genomic hybridization (CGH, [40]) or next-generation sequencing (CNV-seq, [41]). Although these are the most accurate methods to detect CNVs, they are still expensive to apply on many samples [42]. Thus, several studies in cattle inferred CNVs from SNP-chip data with subsequent validation of selected regions by quantitative polymerase chain reaction (qPCR) [19–28].

In this article, we infer CNVs in the genome of Nelore cattle, concatenate them into CNV regions (CNVRs) and validate a subset with qPCR. In addition, we perform a genome-wide association study of CN state with a quantitative MT phenotype derived from meat shear force. We identify MT-related genes and quantitative trait loci (QTLs) overlapping new and previously reported cattle CNVRs to establish relationships between structural genomic variation and MT.

Citation: Silva VHd, Regitano LCdA, Geistlinger L, Pértille F, Giachetto PF, Brassaloti RA, et al. (2016) Genome-Wide Detection of CNVs and Their Association with Meat Tenderness in Nelore Cattle. PLoS ONE 11(6): e0157711. doi:10.1371/journal.pone.0157711.

RESULTS

A Genome-Wide Map of CNV Regions in Nelore

We used SNP genotype data from 723 male Nelore animals to infer 49,997 CNV calls and 2,649 CNVRs using PennCNV [43] and CNVRuler [44], respectively (see Materials and Methods for genotyping, CNV calling and CNVR concatenation). The CNVRs represent 6.5% of the *Bos taurus* genome. The chromosomal proportion covered by CNVRs varies between chromosomes (from 2.3% to 19.7% for BTA22 and BTA15, respectively). The number of regions with copy loss and gain were 1,454 and 891, respectively. Presence of both types occurred in 304 regions. Average CNVR size was 64.4 kb, ranging from 5 kb (minimum threshold for CNV calls, see Materials and Methods) to 4.3 Mb. For each CNVR, the relative frequency of animals with an overlapping CNV ranges from 0.1% (1 out of 723) to 99.8% (722 out of 723). CNVRs with size between 5 and 50 kb represent the majority of our findings (71.4%) whereas CNVRs larger than 1 Mb were rarely observed (0.5%, Fig 1). We found 521 CNVRs to occur in more than 1% of the population and denote them as 'polymorphic CNVRs' in the following. These regions represent 3.2% of the *Bos taurus* genome (86.4 Mb, Fig 2). To exclude the possibility that polymorphic CNVRs with high frequency (occurring in >75% of our population) are technical artifacts of the mapping of indicus data onto the taurus assembly, we checked whether these CNVRs contained exclusively events of a particular CNV state (indicating rather genomic differences between taurus and indicus than individual CNVs within indicus). However, we did not find cases for which >95% of the contained CNV calls displayed the same CNV state, arguing against false positive detections due to the mapping.

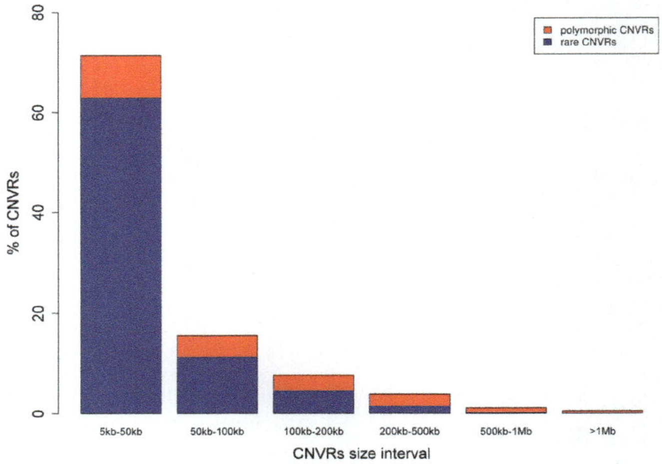

Figure 1. Distribution of CNVR length.

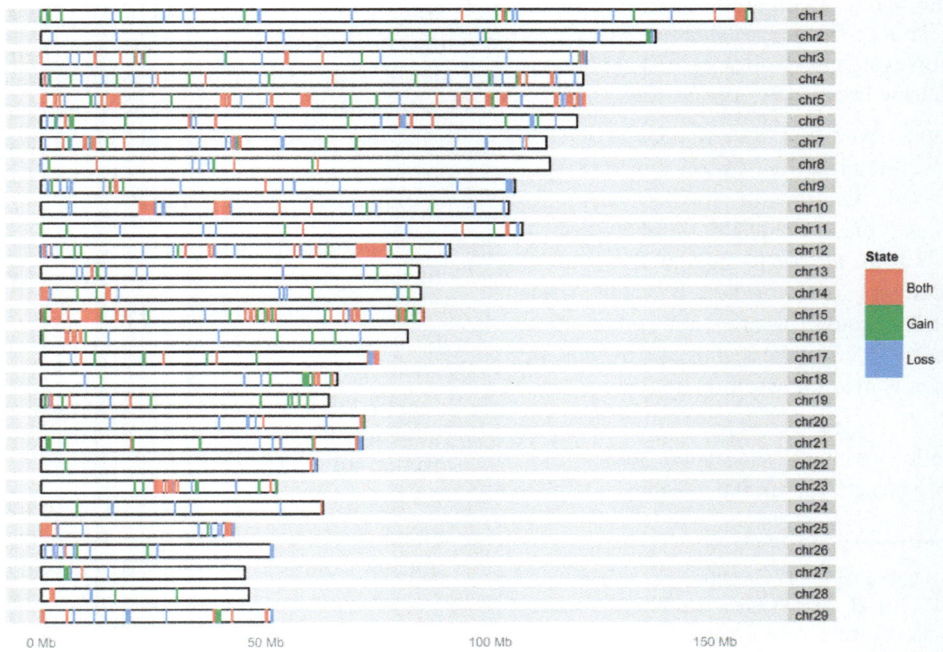

Figure 2. Chromosomal distribution of 521 polymorphic CNVRs (>1% of the population). The regions were categorized into loss or gain of copy number and states for which both events were observed.

When comparing our CNVRs to previously reported cattle CNVRs (denoted herein as 'known CNVRs), we found 1,387 (52.3%) overlapping regions. The total overlap corresponds to 79 Mb (46.3%) of the genomic area covered by the Nelore CNVRs (Fig 3). Repeated sampling of random genomic regions, matching our CNVRs in size and chromosomal distribution, showed that the overlap with known CNVRs is significantly larger than expected by chance.

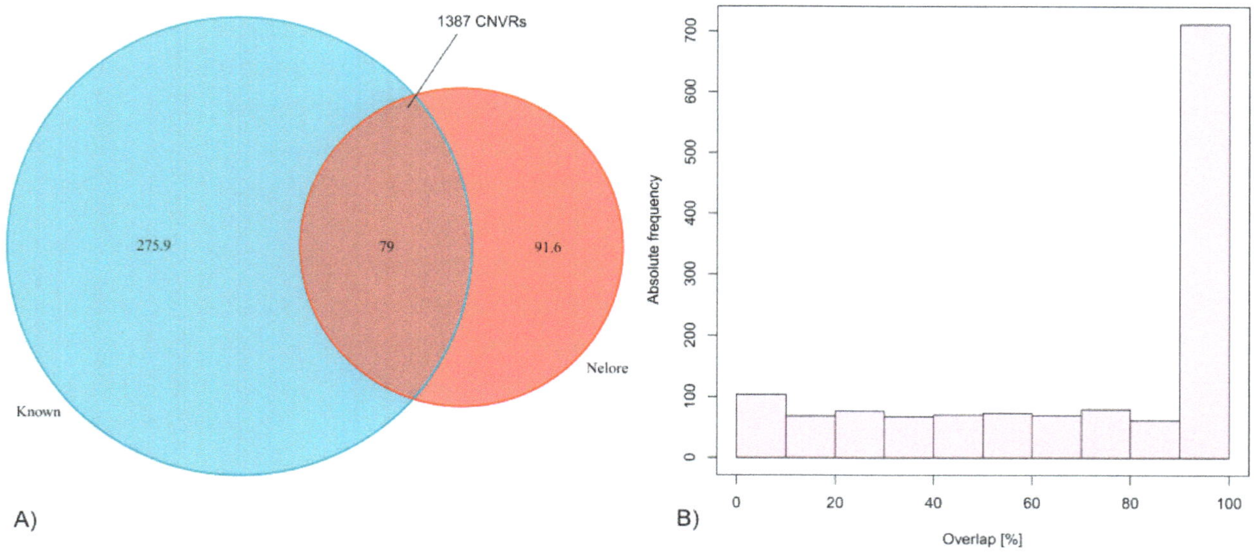

Figure 3. Overlap of the detected Nelore CNVRs with known cattle CNVRs. **A)** The Venn diagram shows the genomic size in Mb that is overlapped. Known CNVRs listed were reduced to unique and non-overlapping CNVRs. The overlapped fraction of 79 Mb corresponds to 1,387 Nelore CNVRs. **(B)** For these 1,387 CNVRs, the histogram shows the number of CNVRs (*y*-axis) overlapping with known CNVRs by the percentage shown on the *x*-axis. For example, genomic locations of >700 Nelore CNVRs overlap individually >90% with genomic locations of known CNVRs.

As different platforms can yield different results, we specifically compared the Nelore CNVRs with CNVRs previously obtained using the same High Density Illumina SNP chip [23]. We found 1,048 overlapping Nelore CNVRs, accounting for 53.9 Mb (31.6%) of the genomic area covered by the Nelore CNVRs.

Association of CNVs with biological functions and processes

We found 2,531 Nelore CNVRs (95.5%) overlapping at least one base with QTLs from Cattle Animal QTLdb [45]. However, most of the cases are CNVRs residing (completely overlapped) in a QTL region. A total of 482 polymorphic CNVRs overlapped with 2,310 QTLs, corresponding to 282 traits including several milk-related phenotypes (Fig 4). Regarding meat quality, marbling score (intramuscular fat) is frequently found among those overlapped QTL traits. Despite the presence of CNVRs in QTL regions, this is not as common as expected given that genome coverage of QTLs is large and not well defined for cattle. Compared to randomly sampled genomic regions the detected CNVRs overlap significantly less (permutation p-value < 0.001) with the QTLs than expected by chance (Material and Methods). The same tendency was observed for the individual traits depicted in Fig 4.

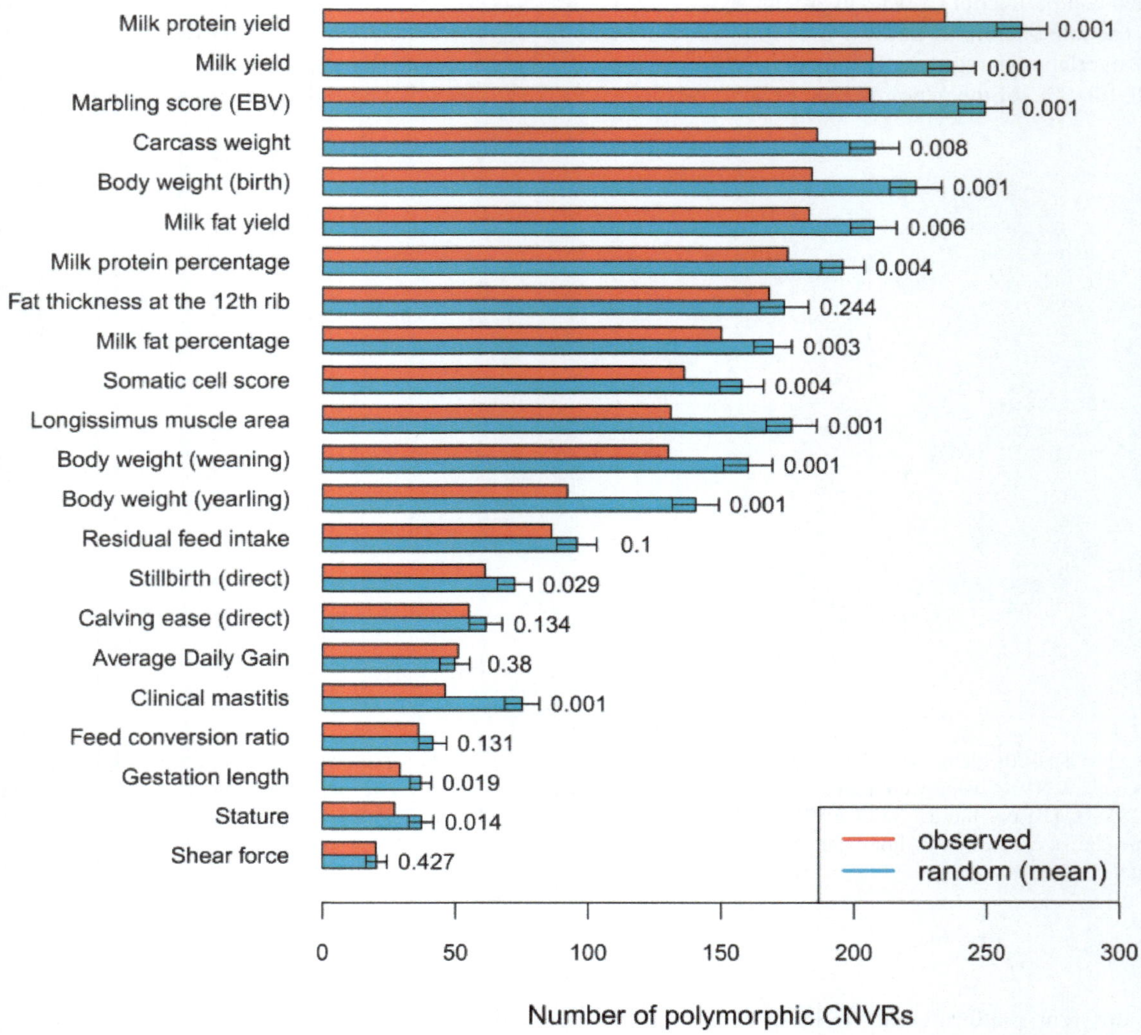

Figure 4. QTL-traits that overlap with polymorphic CNVRs in the Nelore population. Depicted are the most frequently overlapped traits. Red bars correspond to the observed number of polymorphic CNVRs overlapping with the respective trait, and cyan bars indicate the mean overlap when sampling 1000x random regions matching the polymorphic CNVRs in size and chromosomal location. The error bars indicate the standard deviation and permutation *p*-values are listed on the right.

Genomic positions of 44% of our CNVRs overlapped with 2,750 different genes, which were found most frequently inside CNVRs. On the other hand, 64 genes are overlapped by more than one CNVR. As observed for cattle QTLs, overlaps between CNVRs and genes were not as common as expected. We found that in comparison to randomly sampled regions the detected CNVRs overlap significantly less (permutation *p*-value < 0.001) with the annotated genes than expected by chance. As depicted in Table 1, we found genes overlapping with polymorphic CNVRs enriched for immunological functions (including the following gene sets: major histocompatibility complex, antigen processing and presentation, and immunoglobulin). A similar observation has been reported for other mammals, where immunological genes were enriched inside CNVRs [46–49] and associated with susceptibility to diseases [50,51].

Table 1. Functional groups associated with polymorphic CNVRs (>1% of the population).

The statistical significance level was set to 0.01. The enrichment *p*-values listed in the third column were computed using DAVID [62].

GO terms		
GO ID	**Description**	***P*-value**
GO:0004984	Olfactory receptor activity	1e-23
GO:0007186	G-protein coupled receptor protein signaling pathway	3.5e-15
GO:0007166	Cell surface receptor linked signal transduction	4.7e-12
GO:0016021	Integral component of membrane	2.5e-06
GO:0031224	Intrinsic component of membrane	2.7e-06
GO:0019882	Antigen processing and presentation	4.5e-06
GO:0042611	MHC protein complex	0.000011
GO:0032561	Guanyl ribonucleotide binding	0.00018
GO:0019001	Guanyl nucleotide binding	0.00022
GO:0042612	MHC class I protein complex	0.00059
GO:0005525	GTP-binding	0.0018
GO:0050909	Sensory perception of taste	0.0024
GO:0002474	Antigen processing and presentation of peptide antigen via MHC class I	0.0025
GO:0005833	Hemoglobin complex	0.0076
GO:0007606	Sensory perception of chemical stimulus	0.0087
Protein sets		
Protein set ID	**Description**	***P*-value**
IPR003597	Immunoglobulin C1-set	6.1e-06
IPR007110	Immunoglobulin-like domain	0.000012
IPR003006	Immunoglobulin/major histocompatibility complex, conserved site	0.000026
IPR013783	Immunoglobulin-like fold	0.000086
IPR010579	MHC class I, alpha chain, C-terminal	0.00018
IPR011161	MHC class I-like antigen recognition	0.001
IPR001400	Somatotropin hormone	0.0016
IPR007960	Mammalian taste receptor	0.0016
IPR018116	Somatotropin hormone, conserved site	0.0021
IPR013106	Immunoglobulin V-set domain	0.0036
IPR002338	Haemoglobin, alpha	0.0038
IPR001039	MHC class I, alpha chain, alpha1 and alpha2	0.004
IPR001461	Propeptide, peptidase A1	0.0099

doi:10.1371/journal.pone.0157711.t001

Olfactory receptors (OR) were among the significantly enriched GO terms (DAVID enrichment *p*-value = 1e-23). This is again in agreement with previous studies, in which the OR gene family was found enriched in mammalian CNVRs [52–55], and in cattle CNVRs in particular [21,24–28,56–58]. Biological functions involving GTP are also significantly enriched, which is consistent with previous findings in cattle [59,60] and goat [61].

Association of CNVs with MT

MT-related QTLs from the Cattle Animal QTLdb ("Tenderness score" and "Shear force") fell into 236 of the detected 2649 CNVRs (8.9%, corresponding to 18.2 Mb). The genes overlapping those CNVRs are enriched for peptidase and protease related pathways (*Aspartic peptidase*, *N-terminal*: IPR012848; *Aspartic peptidase*, *active site*: IPR001969,*aspartic-type peptidase and endopeptidase activity*: GO:0070001 and GO:0004190).

Additionally, we found that 75 out of the 132 (56.8%) Nelore-specific QTLs associated with shear force (measured at 0, 7 and 14 days after slaughter in a previous study conducted with the same population [63]) overlapped with 109 of our Nelore CNVRs (24 polymorphic). We also found the CNVRs to overlap unexpectedly less with the Nelore QTLs compared to randomly sampled regions (only 56 out of 1000 permutations resulted in an overlap as small or smaller as the observed one, corresponding to a permutation *p*-value of 0.057, Material and Methods). Similar to the findings for QTLs from the cattle QTLdb, we observed that overlapping CNVRs locate predominantly within the Nelore-specific QTLs.

We validated 10 MT-related CNVRs using qPCR (see Material and Methods). The concordance of copy number state (loss, normal and gain) between qPCR and SNP-array results was in average 66.4%, ranging from 46% to 86% of samples for each region tested. The inferred copy number state (from 0n to 4n) was calculated as a normalized ratio of PCR cycle threshold values.

To test the genome-wide association of the detected CNVs with MT, we used estimated breeding values derived from meat shear force (SF-EBV) as the quantitative MT phenotype (see Material and Methods for details). As the CNVRs are a result of concatenated individual CNVs and can thus contain different CN states, we constructed CNV segments from subsequent probes with highly similar genotype (constructed without parental information, see Material and Methods for details). This yielded 447 CNV segments constructed from 3,242 probes that deviate from the 2n state in ≥5% of the population. The number of probes in each segment ranged from 1 to 9, with 7.25 probes on average. The association level for each segment after multiple testing correction is depicted in Fig 5.

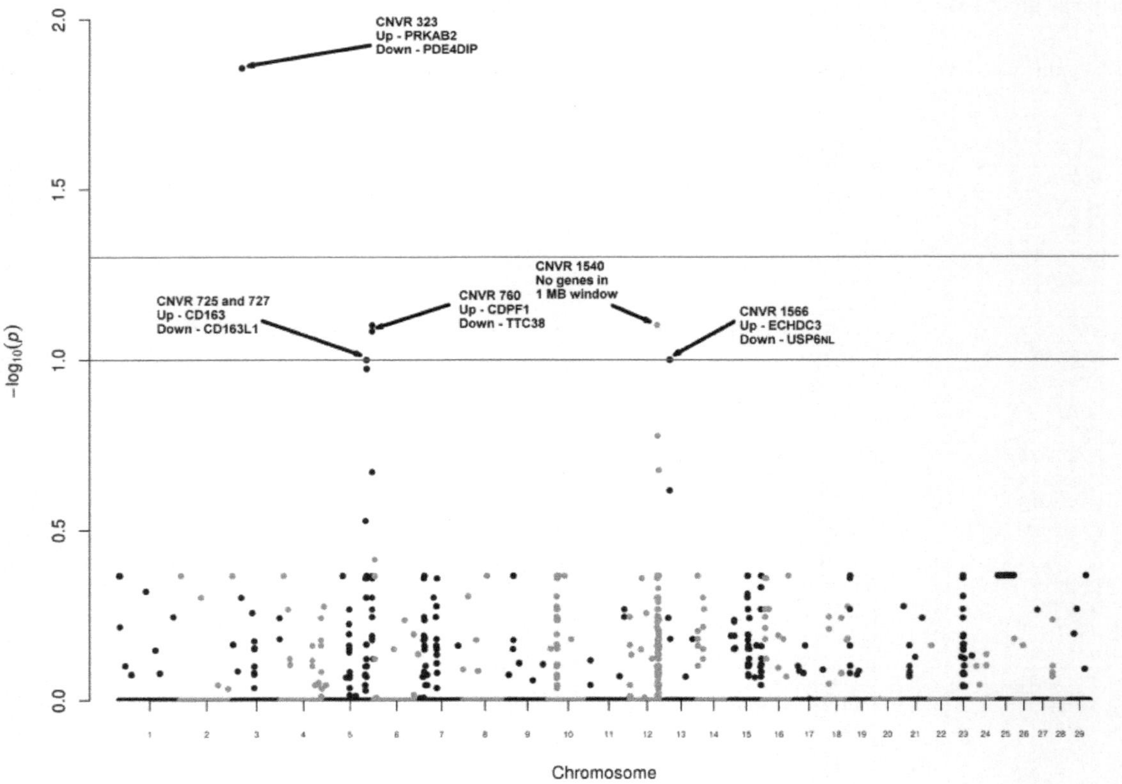

Figure 5. Genome wide association of CNV state with shear force estimated breeding value (SF-EBV). Manhattan plot for 447 CNV segments on somatic chromosomes (*x*-axis) and the corresponding -$\log_{10}$ *p*-value (*y*-axis) indicating the association strength with SF-EBV. Multiple testing correction was performed, controlling the false discovery rate (FDR, [64]). Red and blue lines indicate FDR-corrected *p*-values of 0.05 and 0.1, respectively. The black arrows indicate CNVRs with *p*-value <0.1 and list the nearest protein-coding genes up- and downstream of the segment (in a genomic window ≤1Mb).

Considering the FDR-corrected *p*-values depicted in Fig 5, we found a significantly associated CNV segment inside CNVR 323 (association *p*-value = 0.014). Gain state in CNVR 323 was validated by qPCR in 3 of 6 samples. Additional candidates were found when ranking all 447 CNV segments by FDR-corrected *p*-value and relaxing the significance level. This yielded 4 CNV segments located in CNVRs 725, 727, 1540, 1566, respectively, and 2 CNV segments located in CNVR 760 with association *p*-value <0.1. Although a significance level of 0.1 can be considered anti-conservative from a statistical point of view, we further investigated the additional candidates and checked whether they coincide with the aforementioned CNVRs validated by qPCR. CNVR 760 and 1566, containing 3 of the 6 candidate segments, were tested and successfully validated by qPCR in 23 of the 30 sires.

Fig 6 shows the SF-EBV distribution stratified by observed CN state for the significantly associated segment in CNVR 323 as well as the candidate segments in the corresponding 5 CNVRs (for CNVR 760, the higher ranked segment was selected). Four of the segments correspond to annotated Ensembl CNVRs. It has been previously reported that the expression of genes up to 450 kb distal from CNVR boundaries can be affected [65]. Based on that, we investigated a genomic window 450 kb up- and downstream, which corresponds to a 900 kb ≈ 1 Mb window, and found a total of 32 protein-coding genes to be located in the vicinity of the segments.

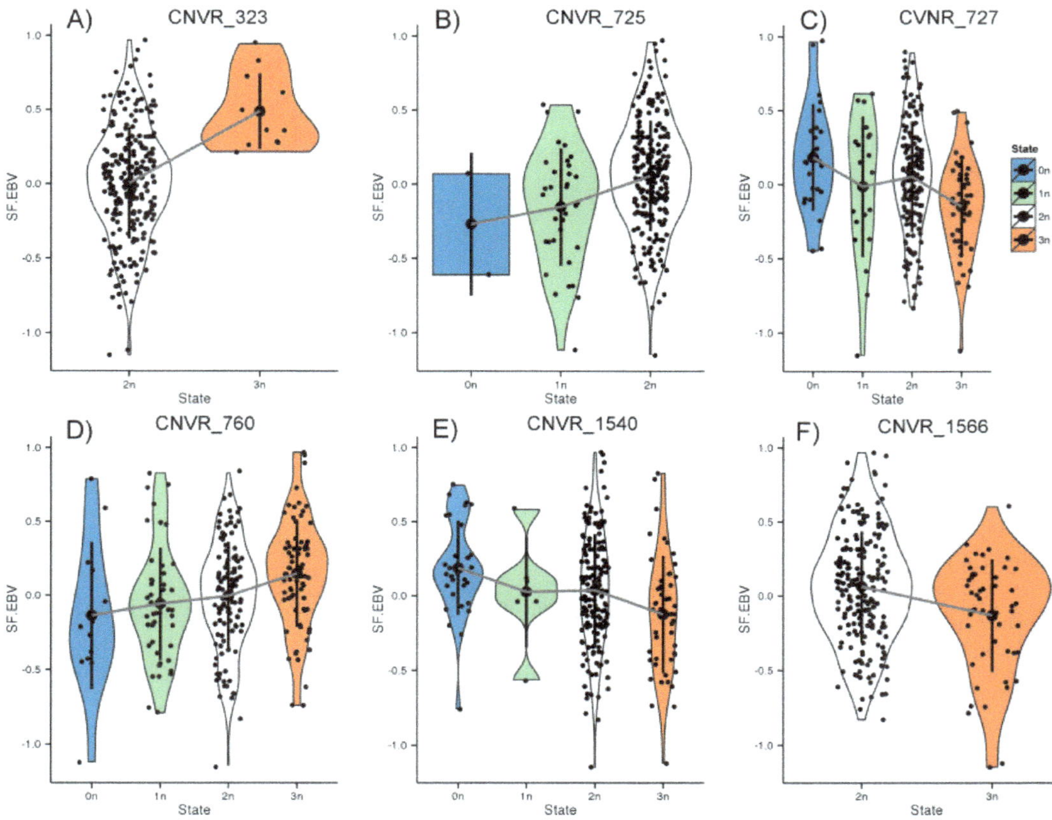

Figure 6. Shear force estimated breeding value (SF-EBV) distribution in each CN state for six significant CNV segments. Each dot represents an animal in the corresponding CN state (0-3n) on the *x*-axis and the assigned SF-EBV on the *y*-axis. The legend on the top right displays the color code for the CN state.

The CNV segment in CNVR 323 is located on BTA3 and CN is positively correlated with SF-EBV, i.e. increase in CN is associated with genetic potential for tougher meat. Interestingly, we found the region to be an Ensembl gain CNVR with several adjacent protein-coding genes. The same tendency is observed for the segments in CNVR 725 and CNVR 760 on BTA5, i.e. CN is positively correlated with SF-EBV. In addition, the segment in CNVR 725 matches an Ensembl CNVR and is highly similar to a previously described CNVR.

On the other hand, the segments in CNVR 727, 1540, and 1566 located on BTA5, BTA12, and BTA13, respectively, show the reverse tendency, i.e. CN is negatively associated with SF-EBV. This indicates that for these regions, increase in copy number is associated with genetic potential for tender meat. CNVR 1566 coincides with an Ensembl CNVR and is highly similar to a previously described CNVR.

DISCUSSION

In recent years, the main genetic variant investigated in genome-wide association studies is single nucleotide polymorphism (SNP); however, the extension to larger regions of variation such as CNVs is feasible and beneficial [66,67]. To investigate structural variation in the Nelore breed, we applied a computational approach and inferred a genome-wide CNV map from SNP-chip genotyping data. Our results revealed that variable regions are scattered across the genome and represent a large portion of it (6.5%), as reported for other cattle breeds [19,28].

Comparison of CNVs among studies, even within the same species, is not trivial as their mapping is typically based on different reference genomes, and the exact genomic start and end coordinates (CNV boundaries) are subject to biological and technical variation [57,68,69]. CNV studies in *Bos taurus* reported CNVRs that are predominantly based on the UMD_3.1 and Btau_4.0 reference assemblies. Conversion of genomic coordinates resulted in considerable data loss, as we could not convert 36.3% of the CNVRs from Btau_4.0 to UMD_3.1 using liftOver [70]. Nevertheless, we found significantly more CNVRs to overlap with known CNVRs than expected by chance (permutation p-value < 0.001), indicating that a considerable fraction of CNVRs is conserved between Nelore and other cattle breeds.

Genes and QTLs are important functional regions of the genome and are thus not expected to be subject to wide-range rearrangements, such as CNVs. This is in agreement with our finding that Nelore CNVRs overlap less frequently with genes and QTLs than random regions of the genome. Therefore, CNVRs located in genes and QTLs are of special interest. Several polymorphic CNVRs (found in more than 1% of the population) overlap with MT-QTLs from cattle QTLdb [45]. The enriched peptidase and protease gene families at these regions have a remarkable role in post-mortem tenderization as they are involved in MT-associated proteolysis and connective tissue turnover [71]. Investigation of genes at Nelore MT-QTLs for this population [63] overlapping our CNVRs showed an enrichment for immunological functions. A notable exception to immunological genes is the diaphanous-related formin *3* gene (*DIAPH3*), which is overlapped by two polymorphic CNVRs. Actin-based processes are regulated by diaphanous-related formins [72] and influence MT-associated proteolysis [73]. Moreover, diaphanous-related formins affect specific GTPases [74], which showed a different mRNA expression level between animals with tough and tender meat [60].

We also found enriched processes involving GTP (Table 1, *G-protein coupled receptor protein signaling pathway* and *GTP binding*). Previous MT and skeletal muscle studies described the influence of GTP-related genes on myotube morphology, skeletal muscle myogenesis, regeneration and calcium physiology [75–78]. The impact on calcium content is especially notable as it affects MT through postmortem protein degradation in skeletal muscle [79,80]. A representative example is the proteolysis-associated ATP/GTP binding protein-like 3 gene (*AGBL3*), present in fewer copies in Korean beef cattle (Hanwoo) when compared to Holstein dairy breed [59]. As beef and dairy breeds differ in tenderness [81], further investigation is desired on whether CNV in the AGBL3 locus affects MT. In Nelore, we found a region of polymorphic CNV (CNVR_507) 96 kb downstream of *AGBL3*.

The GTP-binding protein RAD-like gene is a validated meat quality biomarker [82], and GTP-related genes are also targets of differentially expressed miRNAs between MT phenotypes [83]. As the number of miRNA-binding sites is significantly higher inside CNVRs [84], this indicates a potential regulatory mechanism that could modulate transcripts synthesized at CNVRs.

To detect CNV events linked with MT, we performed a genome-wide association analysis between SF-EBV and CNV segments composed of highly similar probes, which are thus identical by state (IBS) and not necessarily by descent (IBD) as parental information was not available. The CNV segments associated with SF-EBV did not overlap QTLs previously found in the same population ([63]) indicating that GWAS studies using only SNP association may miss important QTL associated with CNV state. Exclusion of relevant CNV probes from the SNP-GWAS due to deviation from Hardy-Weinberg equilibrium and stringent multiple testing correction (with a typically conservatively chosen significance level) are likely causes for that.

Among the resulting significantly associated segments, we found segments located in the vicinity of cAMP and PKA-related genes. In addition, the small GTP-binding protein GTPase RAC1 (located in CNVR 2447) interacts directly with cAMP-dependent protein kinase A (PKA) [85]. The detected statistically significant CNV segment on chromosome 3 is located 87 kb upstream of PDE4DIP, a gene involved in cAMP and cGMP signaling [86]. It has been reported that cGMP signaling is involved in muscle contraction mechanisms [87], and cAMP signaling influences hypertrophy, metabolism and regeneration of skeletal muscle [88]. Located 150 kb upstream of the significantly associated CNV segment on chromosome 3 is*PRKAB2*, a subunit of the AMP-activated protein kinase. Presence of this kinase stimulates calcium transportation in sarcolemma [89] and affects calpastatin gene transcription and protein phosphorylation [90]. The Calpain-calpastatin system is known to be of key importance for postmortem proteolysis and meat tenderization [91].

The two CNV segments of CNVR 725 and 727, for which we found significant association between CN state and SF-EBV, are located in close proximity to *CD163* and *CD163L1*(located within CNVR 725). These genes influence the intra-cellular hemoglobin level [92] known to be associated with the density of cross-sectional muscle areas [93]. Interestingly, we also found the hemoglobin complex (GO:0005833) and hemoglobin alpha (IPR002338) enriched in genes overlapped by polymorphic CNVRs. Hemoglobin homeostasis is tightly linked to carbon monoxide (CO) oxidative stress signaling [94,95]. CO induces cGMP and antioxidant agents such as reduced gluthathione [95]. We also found two polymorphic CNVRs (CNVR 2061 and 2443) that overlapped two QTLs for glutathione redox balance [96], which is an interesting case for further targeted investigation as the glutathione dimer has been reported as a MT-predictor [97]. Moreover, those QTLs are overlapped by the gene encoding SMAD specific E3 ubiquitin protein ligase 1 (*SMURF1*), which inhibits the transduction pathway of myostatin [98]. The gene encoding chaperonin containing TCP1, subunit 6B (*CCT6B*) was also mapped to the glutathione redox balance QTLs [96] and overlapped with polymorphic CNVRs. Chaperonins play a crucial role in the protein folding process [99], which is in turn of key importance for meat tenderization [8].

The enriched *Somatotropin hormone* (Table 1) is significantly involved in growth and differentiation of skeletal muscle [100]. Interestingly, the gene encoding Insulin-like growth factor 2 hormone (*IGF2*) was found inside the polymorphic CNVR 2647. IGF2 is also a predicted target of the bta-let-7b miRNA (MI0005453), found inside another polymorphic CNVR (CNVR 760), which contains two CNV segments associated with SF-EBV. Notably,*IGF2* is a known stimulator of myoblast differentiation [101,102] and reduced *IGF2*expression is associated with improved tenderness [103].

The supplemental discussion contains further examples and also discusses influences on genes frequently reported to be associated with MT such as μ-calpain and calpastatin (*CAPN1* and *CAST*, [104–107]) and certain heat shock proteins (HSPs, [108,109]). However, they did not overlap with Nelore-specific CNVRs reported here.

In conclusion, we conducted the first comprehensive CNV study in *Bos taurus indicus* cattle and found that a considerable portion of the Nelore genome contains CNVRs. Concentrating on the MT phenotype, we systematically investigated the detected CNVRs with respect to the overlap with functional regions such as QTLs, the enrichment of gene functions, and the genome-wide association with MT. We identified several regions previously associated with MT as well as new regions potentially important for MT. We thus assume our results to serve as a good starting point for future studies on structural variations related with MT and skeletal muscle physiology.

MATERIALS AND METHODS

Ethical Statement

All experimental procedures involving steers in this study were approved by the Institutional Animal Care and Use Committee Guidelines (IACUC) from Brazilian Agricultural Research

Corporation (EMBRAPA) and sanctioned by the president Dr. Rui Machado.

Population and Genotyping

The population under study comprises 777 Nelore males including 30 founding sires used to produce the population. For SNP-array analysis, genomic DNA samples were extracted from semen straws (sires) and blood (offspring). All animals were genotyped using Illumina Bovine HD Beadchip arrays (>770,000 SNP markers). The call rate (percentage of successfully genotyped SNPs for a given animal) threshold was >95%. We excluded 54 animals from the analysis, as they did not satisfy the CNV calling filter criteria described in the next section, and applied all subsequent analysis to the remaining population of 723 animals. Raw genotyping data is available upon request (requires a signed declaration of exclusive research purpose).

CNV calling

Individual CNVs were called using PennCNV [43]. PennCNV incorporates Log R Ratio (LRR) and B Allele Frequency (BAF), which denote the log2 ratio of R (normalized total intensity of 2 SNP alleles) and the frequency of allele B defined as the normalized intensity ratio for each SNP allele, respectively. Application of PennCNV yielded 766 animals for which at least one CNV was inferred. LRR values were corrected for genomic wave bias of SNP-arrays [110] via "*genomic_wave.pl -adjust*", considering a genomic window of 1 kb (500 bp up- and downstream of each investigated SNP). Sexual chromosomes were excluded from the analysis. The Population Frequency of B Allele (PFB) file was generated using a subset of the same Nelore population (671 animals).

A total of 43 samples with >150 CNV calls ("-qcnumcnv 150") were excluded as they were expected to be of low quality (http://penncnv.openbioinformatics.org/en/latest/misc/faq/) and thus inflate CNVR length due to concatenation of false positively detected CNVs. Similarly, CNV calls <5kb ("-length 5k") were excluded as such regions seem to be less concordant with CGH [111] and confidential CNV inference from SNP arrays has been reported accordingly for >5kb [112]. Default settings were used for B allele frequency drift (BAF_drift) and quality control waviness factor (QC_WC). LRR standard deviation (LRR_SD) was set to <0.3 ("*-qclrrsd* 0.3") as previously suggested [19,23,113,114].

CNVR Compilation

Filtered individual calls from PennCNV were concatenated into CNV regions (CNVRs) using CNVRuler [44]. Genomic areas with density <10% were excluded ("recurrence 0.1"). The recurrence trims a CNVR based on its frequency to avoid false positive predictions. Additionally, the option "*Gain/Loss separated regions*" was used to compile CNVRs based on their genotype (gain or loss). Overlapping "gain" and "loss" CNVRs were merged into single regions to account for genomic regions in which both events can occur ("both" CNVRs).

Validation by Qpcr

Quantitative PCR was carried out for nine CNVRs in all 30 sires of the population. The CNVRs were selected because they contain SNPs that are associated with MT or are located near to MT-related QTLs and genes. Additionally, CNVR323 was validated by qPCR in 31 non-sire samples as this CNV was not present in the sires. To ensure specificity of the validation, we only considered 6 of 31 samples that were predicted in 3n state in the region covered by the primer amplicon. Primers were designed using Primer3plus [115] and quality testing was performed with NetPrimer [http://www.premierbiosoft.com/netprimer]. In the genomic region represented by the primer, the presence of SNPs from SNPdb [116] was checked against Ensembl-Biomart (http://www.ensembl.org/biomart/martview) [117]. To ensure qPCR accuracy, only primers that did not contain SNPs from SNPdb were synthetized.

All primers were tested in serial dilutions of pooled genomic DNA to achieve optimal qPCR conditions. A qPCR solution of 10 µl was used consisting of 5.0 µl SYBR Green 2x (Roche®), 0.5 µl forward primer (5 mM), 0.5 µl reverse primer (5 mM) and 4.0 µl of genomic DNA (2.5 ng/µl). The qPCR steps were as follows: 1) 95°C for 5 min, 2) 40 cycles of 95°C and 51–60°C (primer-dependent), 3) 72°C for 10–15 seconds (primer dependent). Each animal was tested in technical triplicates carried out in Light cycle 480 (Roche®).

Cycle thresholds (log2 Ct) were corrected by primer mean efficiency as calculated with LinReg [118]. ΔCt denotes Ct of targeted region minus Ct of control region (primer pair targeting the *BTF3* gene, which has been used before as a control gene for CNV studies in cattle [19,26]). ΔΔCt was calculated as ΔCt from the animal to be tested minus ΔCt of a diploid (2n) control [119]. The control value was estimated based on the average value of ΔCt from all 2n animals (estimated with PennCNV).

Copy number was estimated from the normalized ratio (NR): $2x2^{-(\Delta\Delta Ct)}$ [26]. Copy number states were categorized as 1n (partial deletion), 2n (normal state), 3n (one copy gain), 4n (two copy gain) and 5n (gain of more than two copies) based on the geometric average of two states [119]. Lack of amplification was considered as 0n (complete deletion). In PennCNV, each CNV is typically assigned to one of four possible states: 0n (complete deletion), 1n (partial deletion), 2n (normal state), 3n (one copy gain), and 4n (two copy gains). As qPCR can result in >4n copies (see above), a frequently used simplification is to separate all states in just two CNV types: gain or loss [26,56]. Accurate CNV detection can be biased due to biological or technical variation [57,68,69]. Moreover, qPCR amplicons usually represent only a fragment of the CNVR. Hence, to consider a sample as concordant, we required it to be detected by PennCNV (≥1 CNV call) as well as qPCR (≥1 CNVR primer pair) in the same CNV type (loss or gain). CNVRs with ≥1 concordant sample were considered as validated.

Annotation (QTLs, genes and enrichment)

CNVRs were screened for overlap with cattle QTLs from cattleQTLdb [45], specific MT-QTLs in Nelore [63], genes annotated in Biomart-Ensembl [120], and 25,620 predicted target transcripts of 676 miRNAs listed in the Microcosm Targets database, version 5 [121].

Genes, which overlapped CNVRs, were subjected to an enrichment analysis with DAVID [62] to identify metabolic pathways that are predominantly associated with genes prone to CNV. The enrichment analysis was carried out for the subset of regions in which CNVs occurred in >1% of the population, defined as polymorphic CNVRs.

Comparison with Previous Studies

The detected CNVRs (UMD_3.1 genomic coordinates) were compared to CNVRs from 14 previous cattle studies [19,21–26,56–59,114,122,123] and structural variation data for bovine from Ensembl [117]. Genomic coordinates of 11 studies were converted from Btau_4.0 to UMD_3.1 using liftOver [https://genome.ucsc.edu/cgi-bin/hgLiftOver]. CNVRs from different studies that overlapped >70% with each other were considered as detected by both studies. CNVRs from external sources are denoted as known CNVRs in this article.

Overlap Statistics

Overlap analysis was carried out using the Bioconductor package regioneR [124]. The package implements a general framework for testing overlaps of genomic regions based on permutation sampling. We repeatedly sampled random regions ($N = 1000$ permutations) from the UMD_3.1 genome assembly matching size and chromosomal distribution of the detected CNVRs (including all 2649 CNVRs and, separately, the 521 polymorphic CNVRs). CNVRs >1 Mb were excluded as they represented extreme outliers. In every permutation, the overlap was recomputed with a) cattle QTLs from QTLdb, b) UMD_3.1 genes, c) Nelore-specific QTLs, and d) known CNVRs. For robustness, we checked two overlap measures: (i) number of random regions that overlapped a target region, and (ii) total genomic size in bp that was overlapped.

Genome-wide association study of CNV state with SF-EBV

The tenderness evaluation was performed by measuring the shear force (SF) [7] on day 7 and day 14 after slaughtering. The SF values were used to estimate genetic breeding values for meat tenderness (SF-EBV, detailed procedure in [109]). A total of 250 animals with SF-EBV values were used herein.

Association analysis was carried out as follows: **1)** Probes within CNVs were assigned to the corresponding number of copies (0n, 1n, 2n, 3n, or 4n). **2)** As common CN polymorphisms (CNPs, allele frequency ≥5%) can be represented as SNPs of equal frequency [112], we selected all genotyped probes that deviated from the normal state (2n) in ≥5% of the population. **3)** Genome-wide association analysis with the SF-EBVs was carried out using PLINK 1.07, using the corresponding functionality for CNP data encoded in a probe-by-probe fashion (http://pngu.mgh.harvard.edu/~purcell/plink/gvar.shtml, [125]). **4)** Assuming subsequent probes to frequently measure the same CNV event [66], we concatenated subsequent probes with identical genotype in ≥95% of our population to CNV segments. It should be noted here, that they are thus identical by state (IBS) and not necessarily by descent (IBD) as parental information was not available. **5)** The probe with most significant *p*-value was selected as representative for the segment. **6)** CNV segment *p*-values were corrected for multiple testing by controlling the false discovery rate [64].

ACKNOWLEDGMENTS

The authors are grateful to Dr. Sónia Andrade and Dr. Aline Silva Mello Cesar from Luiz de Queiroz College of Agriculture (ESALQ/USP) for helpful suggestions, and to the BeefQuality project team for providing samples of the experimental animals. We are also thankful to the editor and anonymous reviewers which helped to considerably improve the manuscript.

AUTHOR CONTRIBUTIONS

Conceived and designed the experiments: LCAR LLC VHS. Performed the experiments: VHS RB FP NSM LCAR. Analyzed the data: VHS PFG LG RZ RB. Contributed reagents/materials/analysis tools: VHS LCAR LLC FP. Wrote the paper: VHS LG RZ FP. Contributed substantially in data interpretation: LG RZ.

REFERENCES

1. Herath S, Dobson H, Bryant CE, Sheldon IM. Use of the cow as a large animal model of uterine infection and immunity. J Reprod Immunol. 2006;69: 13–22. doi: 10. 1016/j. jri. 2005. 09. 007. pmid:16386311
2. Elsik CG, Tellam RL, Worley KC, Gibbs RA, Muzny DM, Weinstock GM, et al. The genome sequence of taurine cattle: a window to ruminant biology and evolution. Science. 2009;324: 522–528. doi: 10. 1126/science. 1169588. pmid:19390049
3. Beef and Veal Meat Exports by Country in 1000 MT CWE. Index mundi. 2014. Available: http://www.indexmundi. com/agriculture/?commodity=beef-and-veal-meat&graph=exports.
4. Produção da pecuária municipal. ISSN-0101-4234: 61. Instituto Brasileiro de Geografia e Estatística (IBGE). 2010. Available:http://www. ibge. gov. br/home/estatistica/economia/ppm/2010/ppm2010. pdf.
5. Oliveira PS, Cesar AS, Nascimento ML, Chaves AS, Tizioto PC, Tullio RR, et al. Identification of genomic regions associated with feed efficiency in Nelore cattle. BMC Genet. 2014;15: 100. doi: 10. 1186/s12863-014-0100-0. pmid:25257854
6. Connor SFO, Tatum JD, Wulf DM, Green RD, Smith GC. Genetic effects on beef tenderness in Bos indicus composite and Bos taurus cattle. J Anim Sci. 1997;75: 1822–1830. pmid:9222838
7. Wheeler TL, Shackelford SD, Koohmaraie M. Standardizing Collection and Interpretation of Warner-Bratzler Shear Force and Sensory Tenderness Data. Proceedings 50th Annual Reciprocal Meat Conference. 1997;50: 68–77.
8. Lomiwes D, Farouk MM, Wiklund E, Young OA. Small heat shock proteins and their role in meat tenderness: a review. Meat Sci. 2014;96: 26–40. doi: 10. 1016/j. meatsci. 2013. 06. 008. pmid:23896134
9. Purslow PP, Archile-Contreras AC, Cha MC. Meat Science and Muscle Biology Symposium: manipulating meat tenderness by increasing the turnover of intramuscular connective tissue. J Anim Sci. 2012;90: 950–959. doi: 10. 2527/jas. 2011-4448. pmid:21890505
10. Weston AR, Rogers RW, Althen TG. REVIEW: The role of Collagen in Meat Tenderness. Prof Anim Sci. 2002;18: 107–111. doi: 10. 15232/s1080-7446(15)31497-2

11. Ouali A, Gagaoua M, Boudida Y, Becila S, Boudjellal A, Herrera-Mendez CH, et al. Biomarkers of meat tenderness: present knowledge and perspectives in regards to our current understanding of the mechanisms involved. Meat Sci. 2013;95: 854–870. doi: 10. 1016/j. meatsci. 2013. 05. 010. pmid:23790743

12. Nonneman DJ, Shackelford SD, King DA, Wheeler TL, Wiedmann RT, Snelling WM, et al. Genome-wide association of meat quality traits and tenderness in swine. J Anim Sci. 2013;91: 4043–4050. doi: 10. 2527/ jas. 2013-6255. pmid:23942702

13. Wang D, Zhou L, Zhou H, Hou G, Shi L, Li M, et al. Effects of nutritional level of concentrate-based diets on meat quality and expression levels of genes related to meat quality in Hainan black goats. Anim Sci J. 2014. doi: 10. 1111/asj. 12251.

14. Hocquette J, Renand G, Levéziel H, Picard B, Cassar-malek I. The potential benefits of genetics and genomics to improve beef quality–a review. Anim Sci Pap Reports. 2006;24: 173–189.

15. Coró FAG, Youssef EY, Shimokomaki M. Carne do Zebu: o que está atrás da sua textura? Rev Nac da Carne. 1999;17: 28–34.

16. Whipple G, Koohmaraie M, Dikeman ME, Crouse JD. Predicting beef-longissimus tenderness from various biochemical and histological muscle traits. J Anim Sci. 1990;68: 4193–4199. pmid:2286561

17. Zhao C, Tian F, Yu Y, Luo J, Mitra A, Zhan F, et al. Functional genomic analysis of variation on beef tenderness induced by acute stress in angus cattle. Comp Funct Genomics. 2012. http://dx. doi. org/10. 1155/2012/756284.

18. Clop a, Vidal O, Amills M. Copy number variation in the genomes of domestic animals. Anim Genet. 2012;43: 503–517. doi: 10. 1111/j. 1365-2052. 2012. 02317. x. pmid:22497594

19. Bae JS, Cheong HS, Kim LH, NamGung S, Park TJ, Chun J-Y, et al. Identification of copy number variations and common deletion polymorphisms in cattle. BMC Genomics. 2010;11: 232. doi: 10. 1186/1471-2164-11-232. pmid:20377913

20. Cicconardi F, Chillemi G, Tramontano A, Marchitelli C, Valentini A, Ajmone-Marsan P, et al. Massive screening of copy number population-scale variation in Bos taurus genome. BMC Genomics. 2013;14: 124. doi: 10. 1186/1471-2164-14-124. pmid:23442185

21. Hou Y, Liu GE, Bickhart DM, Cardone MF, Wang K, Kim E, et al. Genomic characteristics of cattle copy number variations. BMC Genomics. 2011;12: 127. doi: 10. 1186/1471-2164-12-127. pmid:21345189

22. Hou Y, Bickhart DM, Chung H, Hutchison JL, Norman HD, Connor EE, et al. Analysis of copy number variations in Holstein cows identify potential mechanisms contributing to differences in residual feed intake. Funct Integr Genomics. 2012;12: 717–723. doi: 10. 1007/s10142-012-0295-y. pmid:22991089

23. Hou Y, Bickhart DM, Hvinden ML, Li C, Song J, Boichard DA, et al. Fine mapping of copy number variations on two cattle genome assemblies using high density SNP array. BMC Genomics. 2012;13: 376. doi: 10. 1186/1471-2164-13-376. pmid:22866901

24. Hou Y, Liu GE, Bickhart DM, Matukumalli LK, Li C, Song J, et al. Genomic regions showing copy number variations associate with resistance or susceptibility to gastrointestinal nematodes in Angus cattle. Funct Integr Genomics. 2012;12: 81–92. doi: 10. 1007/s10142-011-0252-1. pmid:21928070

25. Jiang L, Jiang J, Wang J, Ding X, Liu J, Zhang Q. Genome-wide identification of copy number variations in Chinese Holstein. PLoS One. 2012;7: e48732. doi: 10. 1371/journal. pone. 0048732. pmid:23144949

26. Jiang L, Jiang J, Yang J, Liu X, Wang J, Wang H, et al. Genome-wide detection of copy number variations using high-density SNP genotyping platforms in Holsteins. BMC Genomics. 2013;14: 131. doi: 10. 1186/1471-2164-14-131. pmid:23442346

27. Matukumalli LK, Lawley CT, Schnabel RD, Taylor JF, Allan MF, Heaton MP, et al. Development and Characterization of a High Density SNP Genotyping Assay for Cattle. PLoS One. 2009;4: 13. doi: 10. 1371/ journal. pone. 0005350

28. Seroussi E, Glick G, Shirak A, Yakobson E, Weller JI, Ezra E, et al. Analysis of copy loss and gain variations in Holstein cattle autosomes using BeadChip SNPs. BMC Genomics. 2010;11: 673. doi: 10. 1186/1471-2164-11-673. pmid:21114805

29. Liu GE, Bickhart DM. Copy number variation in the cattle genome. Funct Integr Genomics. 2012;12: 609–624. doi: 10. 1007/s10142-012-0289-9. pmid:22790923

30. Vincent AK, Noor A, Janson A, Minassian BA, Ayub M, Vincent JB, et al. Identification of genomic deletions spanning the PCDH19 gene in two unrelated girls with intellectual disability and seizures. Clin Genet. 2011. doi: 10. 1111/j. 1399-0004. 2011. 01812. x.

31. Pfundt R, Veltman JA. Structural genomic variation in intellectual disability. Methods Mol Biol. 2012;838: 77–95. doi: 10. 1007/978-1-61779-507-7_3. pmid:22228007

32. Kohler LH, Mireskandari M, Knösel T, Altendorf-Hofmann A, Kunze A, Schmidt A, et al. FGFR1 expression and gene copy numbers in human lung cancer. Virchows Arch. 2012. doi: 10. 1007/s00428-012-1250-y.

33. Norris BJ, Whan VA. A gene duplication affecting expression of the ovine ASIP gene is responsible for white and black sheep. Genome Res. 2008;18: 1282–1293. doi: 10. 1101/gr. 072090. 107. pmid:18493018

34. Pailhoux E, Vigier B, Chaffaux S, Servel N, Taourit S, Furet JP, et al. A 11. 7-kb deletion triggers intersexuality and polledness in goats. Nat Genet. 2001;29: 453–458. pmid:11726932 doi: 10. 1038/ng769

35. Elferink MG, Vallée AAA, Jungerius AP, Crooijmans RPMA, Groenen MAM. Partial duplication of the PRLR and SPEF2 genes at the late feathering locus in chicken. BMC Genomics. 2008;9: 391. doi: 10. 1186/1471-2164-9-391. pmid:18713476

36. Hirano T. Null Mutation of PCLN-1/Claudin-16 Results in Bovine Chronic Interstitial Nephritis. Genome Res. 2000;10: 659–663. pmid:10810088 doi: 10. 1101/gr. 10.

5. 659

37. Meyers SN, McDaneld TG, Swist SL, Marron BM, Steffen DJ, O'Toole D, et al. A deletion mutation in bovine SLC4A2 is associated with osteopetrosis in Red Angus cattle. BMC Genomics. 2010;11: 337. doi: 10. 1186/1471-2164-11-337. pmid:20507629

38. Flisikowski K, Venhoranta H, Nowacka-Woszuk J, McKay SD, Flyckt A, Taponen J, et al. A novel mutation in the maternally imprinted PEG3 domain results in a loss of MIMT1 expression and causes abortions and stillbirths in cattle (Bos taurus). PLoS One. 2010;5: e15116. doi: 10. 1371/journal. pone. 0015116. pmid:21152099

39. Xu Y, Zhang L, Shi T, Zhou Y, Cai H, Lan X, et al. Copy number variations of MICAL-L2 shaping gene expression contribute to different phenotypes of cattle. Mamm Genome. 2013;24: 508–516. doi: 10. 1007/ s00335-013-9483-x. pmid:24196410

40. Pinkel D, Segraves R, Sudar D, Clark S, Poole I, Kowbel D, et al. High resolution analysis of DNA copy number variation using comparative genomic hybridization to microarrays. Nat Genet. 1998;20: 207–211. pmid:9771718

41. Koboldt DC, Larson DE, Chen K, Ding L, Wilson RK. Massively parallel sequencing approaches for characterization of structural variation. Methods Mol Biol. 2012;838: 369–384. doi: 10. 1007/978-1-61779-507-7_18. pmid:22228022

42. Alkan C, Coe BP, Eichler EE. Genome structural variation discovery and genotyping. Nat Rev Genet. 2011;12: 363–376. doi: 10. 1038/nrg2958. pmid:21358748

43. Wang K, Li M, Hadley D, Liu R, Glessner J, Grant SF, et al. PennCNV: an integrated hidden Markov model designed for high-resolution copy number variation detection in whole-genome SNP genotyping data. Genome Res. 2007;17: 1665–1674. pmid:17921354 doi: 10. 1101/gr. 6861907

44. Kim J-H, Hu H-J, Yim S-H, Bae JS, Kim S-Y, Chung YJ. CNVRuler: a copy number variation-based case-control association analysis tool. Bioinformatics. 2012;28: 1790–1792. doi: 10. 1093/bioinformatics/bts239. pmid:22539667

45. Hu Z-L, Park CA, Wu X-L, Reecy JM. Animal QTLdb: an improved database tool for livestock animal QTL/ association data dissemination in the post-genome era. Nucleic Acids Res. 2013;41: D871–D879. doi: 10. 1093/ nar/gks1150. pmid:23180796

46. Nicholas TJ, Baker C, Eichler EE, Akey JM. A high-resolution integrated map of copy number polymorphisms within and between breeds of the modern domesticated dog. BMC Genomics. 2011;12: 414. doi: 10. 1186/1471-2164-12-414. pmid:21846351

47. Wang J, Jiang J, Wang H, Kang H, Zhang Q, Liu JF. Enhancing Genome-Wide Copy Number Variation Identification by High Density Array CGH Using Diverse Resources of Pig Breeds. PLoS One. 2014;9: e87571. doi: 10. 1371/journal. pone. 0087571. pmid:24475311

48. Agam A, Yalcin B, Bhomra A, Cubin M, Webber C, Holmes C, et al. Elusive copy number variation in the mouse genome. PLoS One. 2010;5: e12839. doi: 10. 1371/journal. pone. 0012839. pmid:20877625

49. Tuzun E, Sharp AJ, Bailey J a, Kaul R, Morrison VA, Pertz LM, et al. Fine-scale structural variation of the human genome. Nat Genet. 2005;37: 727–732. pmid:15895083 doi: 10. 1038/ng1562

50. Schaschl H, Aitman TJ, Vyse TJ. Copy number variation in the human genome and its implication in autoimmunity. Clin Exp Immunol. 2009;156: 12–16. doi: 10. 1111/j. 1365-2249. 2008. 03865. x. pmid:19220326

51. Aitman TJ, Dong R, Vyse TJ, Norsworthy PJ, Johnson MD, Smith J, et al. Copy number polymorphism in Fcgr3 predisposes to glomerulonephritis in rats and humans. Nature. 2006;439: 851–855. pmid:16482158 doi: 10. 1038/nature04489

52. Young JM, Endicott RM, Parghi SS, Walker M, Kidd JM, Trask BJ. Extensive copy-number variation of the human olfactory receptor gene family. Am J Hum Genet. 2008;83: 228–242. doi: 10. 1016/j. ajhg. 2008. 07. 005. pmid:18674749

53. Nozawa M, Kawahara Y, Nei M. Genomic drift and copy number variation of sensory receptor genes in humans. Proc Natl Acad Sci U S A. 2007;104: 20421–20426. pmid:18077390 doi: 10. 1073/pnas. 0709956104

54. Hasin Y, Olender T, Khen M, Gonzaga-Jauregui C, Kim PM, Urban AE, et al. High-resolution copy-number variation map reflects human olfactory receptor diversity and evolution. PLoS Genet. 2008;4: e1000249. doi: 10. 1371/journal. pgen. 1000249. pmid:18989455

55. Graubert TA, Cahan P, Edwin D, Selzer RR, Richmond TA, Eis PS, et al. A high-resolution map of segmental DNA copy number variation in the mouse genome. PLoS Genet. 2007;3: e3. pmid:17206864 doi: 10. 1371/ journal. pgen. 0030003

56. Liu GE, Hou Y, Zhu B, Cardone MF, Jiang L, Cellamare A, et al. Analysis of copy number variations among diverse cattle breeds. Genome Res. 2010;20: 693–703. doi: 10. 1101/gr. 105403. 110. pmid:20212021

57. Fadista J, Thomsen B, Holm L-E, Bendixen C. Copy number variation in the bovine genome. BMC Genomics. 2010;11: 284. doi: 10. 1186/1471-2164-11-284. pmid:20459598

58. Bickhart DM, Hou Y, Schroeder SG, Alkan C, Cardone MF, Matukumalli LK, et al. Copy number variation of individual cattle genomes using next-generation sequencing. Genome Res. 2012;22: 778–790. doi: 10. 1101/gr. 133967. 111. pmid:22300768

59. Choi J-W, Lee K-T, Liao X, Stothard P, An H-S, Ahn S, et al. Genome-wide copy number variation in Hanwoo, Black Angus, and Holstein cattle. Mamm Genome. 2013;24: 151–163. doi: 10. 1007/s00335-013-9449-z. pmid:23543395

60. Zhao C, Tian F, Yu Y, Luo J, Hu Q, Bequette BJ, et al. Muscle transcriptomic analyses in Angus cattle with divergent tenderness. Mol Biol Rep. 2012;39: 4185–4193. doi: 10. 1007/s11033-011-1203-6. pmid:21901422

61. Fontanesi L, Martelli PL, Beretti F, Riggio V, Dall'Olio S, Colombo M, et al. An initial comparative map of copy

number variations in the goat (Capra hircus) genome. BMC Genomics. 2010;11: 639. doi: 10. 1186/1471-2164-11-639. pmid:21083884

62. Huang DW, Sherman BT, Lempicki RA. Systematic and integrative analysis of large gene lists using DAVID bioinformatics resources. Nat Protoc. 2009;4: 44–57. doi: 10. 1038/nprot. 2008. 211. pmid:19131956

63. Tizioto PC, Decker JE, Taylor JF, Schnabel RD, Mudadu M, Silva FL, et al. Genome scan for meat quality traits in Nelore beef cattle. Physiol Genomics. 2013;45: 1012–1020. doi: 10. 1152/physiolgenomics. 00066. 2013. pmid:24022219

64. Benjamini Y, Hochberg Y. Controlling the false discovery rate: a practical and powerful approach to multiple testing. J R Stat Soc. 1995;57: 289–300.

65. Henrichsen CN, Vinckenbosch N, Zöllner S, Chaignat E, Pradervand S, Schütz F, et al. Segmental copy number variation shapes tissue transcriptomes. Nat Genet. 2009;41: 424–429. doi: 10. 1038/ng. 345. pmid:19270705

66. Yau C, Holmes CC. CNV discovery using SNP genotyping arrays. Cytogenet Genome Res. 2008;123: 307–312. doi: 10. 1159/000184722. pmid:19287169

67. Su S-Y, Asher JE, Jarvelin M-R, Froguel P, Blakemore AIF, Balding DJ, et al. Inferring combined CNV/SNP haplotypes from genotype data. Bioinformatics. 2010;26: 1437–1445. doi: 10. 1093/bioinformatics/btq157. pmid:20406911

68. Redon R, Ishikawa S, Fitch KR, Feuk L, Perry GH, Andrews TD, et al. Global variation in copy number in the human genome. Nature. 2006;444: 444–454. pmid:17122850 doi: 10. 1038/nature05329

69. Cahan P, Li Y, Izumi M, Graubert TA. The impact of copy number variation on local gene expression in mouse hematopoietic stem and progenitor cells. Nat Genet. 2009;41: 430–437. doi: 10. 1038/ng. 350. pmid:19270704

70. Kuhn RM, Haussler D, Kent WJ. The UCSC genome browser and associated tools. Brief Bioinform. 2013;14: 144–161. doi: 10. 1093/bib/bbs038. pmid:22908213

71. Sentandreu M. , Coulis G, Ouali A. Role of muscle endopeptidases and their inhibitors in meat tenderness. Trends Food Sci Technol. 2002;13: 400–421. doi: 10. 1016/s0924-2244(02)00188-7

72. Gardberg M, Talvinen K, Kaipio K, Iljin K, Kampf C, Uhlen M, et al. Characterization of Diaphanous-related formin FMNL2 in human tissues. BMC Cell Biol. 2010;11: 55. doi: 10. 1186/1471-2121-11-55. pmid:20633255

73. Li S, Xu X, Zhou G. The roles of the actin-myosin interaction and proteolysis in tenderization during the aging of chicken muscle. Poult Sci. 2012;91: 150–160. doi: 10. 3382/ps. 2011-01484. pmid:22184440

74. Olson MF. GTPase Signalling: New Functions for Diaphanous-Related Formins. Curr Biol. 2003;13: R360–R362. pmid:12725753 doi: 10. 1016/s0960-9822(03)00277-x

75. Romberg CF, Beqollari D, Meza U, Bannister RA. RGK protein-mediated impairment of slow depolarization-dependent Ca2+ entry into developing myotubes. Channels. 2014;8: 243–248. pmid:24476902 doi: 10. 4161/chan. 27686

76. Bannister RA, Colecraft HM, Beam KG. Rem inhibits skeletal muscle EC coupling by reducing the number of functional L-type Ca2+ channels. Biophys J. 2008;94: 2631–2638. doi: 10. 1529/biophysj. 107. 116467. pmid:18192376

77. Suzuki J, Kaziro Y, Koide H. Positive regulation of skeletal myogenesis by R-Ras. Oncogene. 2000;19: 1138–1146. pmid:10713701 doi: 10. 1038/sj. onc. 1203402

78. Hawke TJ, Kanatous SB, Martin CM, Goetsch SC, Garry DJ. Rad is temporally regulated within myogenic progenitor cells during skeletal muscle regeneration. Am J Physiol Cell Physiol. 2006;290: C379–C387. pmid:16221735 doi: 10. 1152/ajpcell. 00270. 2005

79. Tizioto PC, Gromboni CF, Nogueira ARDA, de Souza MM, Mudadu MDA, Tholon P, et al. Calcium and potassium content in beef: influences on tenderness and associations with molecular markers in Nellore cattle. Meat Sci. 2014;96: 436–440. doi: 10. 1016/j. meatsci. 2013. 08. 001. pmid:23995697

80. Geesink GH, Koohmaraie M. Effect of calpastatin on degradation of myofibrillar proteins by mu-calpain under postmortem conditions. J Anim Sci. 1999;77: 2685–2692. pmid:10521028

81. Monsón F, Sañudo C, Sierra I. Influence of cattle breed and ageing time on textural meat quality. Meat Sci. 2004;68: 595–602. doi: 10. 1016/j. meatsci. 2004. 05. 011. pmid:22062536

82. Pierzchala M, Hoekman AJW, Urbanski P, Kruijt L, Kristensen L, Young JF, et al. Validation of biomarkers for loin meat quality (M. longissimus) of pigs. J Anim Breed Genet. 2014;131: 258–270. doi: 10. 1111/jbg. 12081. pmid:24506540

83. Zhao C, Tian F, Yu Y, Liu G, Zan L, Updike MS, et al. miRNA-dysregulation associated with tenderness variation induced by acute stress in Angus cattle. J Anim Sci Biotechnol. 2012;3: 12. doi: 10. 1186/2049-1891-3-12. pmid:22958451

84. Felekkis K, Voskarides K, Dweep H, Sticht C, Gretz N, Deltas C. Increased number of microRNA target sites in genes encoded in CNV regions. Evidence for an evolutionary genomic interaction. Mol Biol Evol. 2011;28: 2421–2424. doi: 10. 1093/molbev/msr078. pmid:21441354

85. Bachmann VA, Riml A, Huber RG, Baillie GS, Liedl KR, Valovka T, et al. Reciprocal regulation of PKA and Rac signaling. Proc Natl Acad Sci U S A. 2013;110: 8531–8536. doi: 10. 1073/pnas. 1215902110. pmid:23657011

86. Verde I, Pahlke G, Salanova M, Zhang G, Wang S, Coletti D, et al. Myomegalin is a Novel Protein of the Golgi/Centrosome That Interacts with a Cyclic Nucleotide Phosphodiesterase. J Biol Chem. 2001;276: 11189–11198. pmid:11134006 doi: 10. 1074/jbc. m006546200

87. Kobzik L, Reid MB, Bredt DS, Stamler JS. Nitric oxide in skeletal muscle. Nature. 1994;372: 546–548.

pmid:7527495 doi: 10. 1038/372546a0

88. Berdeaux R, Stewart R. cAMP signaling in skeletal muscle adaptation: hypertrophy, metabolism, and regeneration. AJP Endocrinol Metab. 2012;303: E1–E17. doi: 10. 1152/ajpendo. 00555. 2011.

89. Mickelson JR, Beaudry TM, Louis CF. Regulation of skeletal muscle sarcolemmal ATP-dependent calcium transport by calmodulin and cAMP-dependent protein kinase. Arch Biochem Biophys. 1985;242: 127–136. doi: 10. 1016/0003-9861(85)90486-2. pmid:3931553

90. Cong M, Thompson VF, Goll DE, Antin PB. The bovine calpastatin gene promoter and a new N-terminal region of the protein are targets for cAMP-dependent protein kinase activity. J Biol Chem. 1998;273: 660–666. doi: 10. 1074/jbc. 273. 1. 660. pmid:9417129

91. Nowak D. Enzymes in Tenderization of Meat—The System of Calpains and Other Systems—a Review. Polish J Food Nutr Sci. 2011;61: 231–237. doi: 10. 2478/v10222-011-0025-5.

92. Abraham NG, Drummond G. CD163-mediated hemoglobin-heme uptake activates macrophage HO-1, providing an antiinflammatory function. Circ Res. 2006;99: 911–914. doi: 10. 1161/01. RES. 0000249616. 10603. d6. pmid:17068296

93. Cesari M, Penninx BWJH, Lauretani F, Russo CR, Carter C, Bandinelli S, et al. Hemoglobin levels and skeletal muscle: results from the InCHIANTI study. J Gerontol A Biol Sci Med Sci. 2004;59: 249–254. doi: 10. 1093/gerona/59. 3. M249. pmid:15031309

94. Schaer CA, Schoedon G, Imhof A, Kurrer MO, Schaer DJ. Constitutive endocytosis of CD163 mediates hemoglobin-heme uptake and determines the noninflammatory and protective transcriptional response of macrophages to hemoglobin. Circ Res. 2006;99: 943–950. doi: 10. 1161/01. RES. 0000247067. 34173. 1b. pmid:17008602

95. Abraham NG, Kappas A. Heme oxygenase and the cardiovascular-renal system. Free Radic Biol Med. 2005;39: 1–25. doi: 10. 1016/j. freeradbiomed. 2005. 03. 010. pmid:15925276

96. Zhou Y, Harrison DE, Love-Myers K, Chen Y, Grider A, Wickwire K, et al. Genetic analysis of tissue glutathione concentrations and redox balance. Free Radic Biol Med. 2014;71: 157–164. doi: 10. 1016/j. freeradbiomed. 2014. 02. 027. pmid:24613380

97. D'Alessandro A, Rinalducci S, Marrocco C, Zolla V, Napolitano F, Zolla L. Love me tender: an Omics window on the bovine meat tenderness network. J Proteomics. 2012;75: 4360–4380. doi: 10. 1016/j. jprot. 2012. 02. 013. pmid:22361340

98. Zhu X, Topouzis S, Liang L-F, Stotish RL. Myostatin signaling through Smad2, Smad3 and Smad4 is regulated by the inhibitory Smad7 by a negative feedback mechanism. Cytokine. 2004;26: 262–272. pmid:15183844 doi: 10. 1016/j. cyto. 2004. 03. 007

99. Spiess C, Meyer AS, Reissmann S, Frydman J. Mechanism of the eukaryotic chaperonin: protein folding in the chamber of secrets. Trends Cell Biol. 2004;14: 598–604. pmid:15519848 doi: 10. 1016/j. tcb. 2004. 09. 015

100. Florini JR, Ewton DZ, Coolican SA. Growth hormone and the insulin-like growth factor system in myogenesis. Endocr Rev. 1996;17: 481–517. pmid:8897022 doi: 10. 1210/edrv-17-5-481

101. Florini JR, Magri K a, Ewton DZ, James PL, Grindstaff K, Rotwein PS. "Spontaneous" differentiation of skeletal myoblasts is dependent upon autocrine secretion of insulin-like growth factor-II. J Biol Chem. 1991;266: 15917–15923. pmid:1651927

102. Alzhanov DT, McInerney SF, Rotwein P. Long range interactions regulate Igf2 gene transcription during skeletal muscle differentiation. J Biol Chem. 2010;285: 38969–38977. doi: 10. 1074/jbc. M110. 160986. pmid:20937833

103. Rehfeldt C, Stabenow B, Pfuhl R, Block J, Nürnberg G, Otten W, et al. Effects of limited and excess protein intakes of pregnant gilts on carcass quality and cellular properties of skeletal muscle and subcutaneous adipose tissue in fattening pigs. J Anim Sci. 2012;90: 184–196. doi: 10. 2527/jas. 2011-4234. pmid:21890499

104. Casas E, White SN, Riley DG, Smith TPL, Brenneman RA, Olson TA, et al. Assessment of single nucleotide polymorphisms in genes residing on chromosomes 14 and 29 for association with carcass composition traits in Bos indicus cattle. J Anim Sci. 2005;83: 13–19. pmid:15583037

105. Tait RG, Shackelford SD, Wheeler TL, King DA, Casas E, Thallman RM, et al. μ-Calpain, calpastatin, and growth hormone receptor genetic effects on pre-weaning performance, carcass quality traits, and residual variance of tenderness in Angus cattle selected to increase minor haplotype and allele frequencies. J Anim Sci. 2014;92: 456–466. doi: 10. 2527/jas. 2013-7075. pmid:24398843

106. Pinto LFB, Ferraz JBS, Meirelles FV, Eler JP, Rezende FM, Carvalho ME, et al. Association of SNPs on CAPN1 and CAST genes with tenderness in Nellore cattle. Genet Mol Res. 2010;9: 1431–1442. pmid:20662159 doi: 10. 4238/vol9-3gmr881

107. Curi RA, Chardulo LAL, Mason MC, Arrigoni MDB, Silveira AC, de Oliveira HN. Effect of single nucleotide polymorphisms of CAPN1 and CAST genes on meat traits in Nellore beef cattle (Bos indicus) and in their crosses with Bos taurus. Anim Genet. 2009;40: 456–462. doi: 10. 1111/j. 1365-2052. 2009. 01859. x. pmid:19392828

108. D'Alessandro A, Marrocco C, Rinalducci S, Mirasole C, Failla S, Zolla L. Chianina beef tenderness investigated through integrated Omics. J Proteomics. 2012;75: 4381–4398. doi: 10. 1016/j. jprot. 2012. 03. 052. pmid:22510581

109. Carvalho ME, Gasparin G, Poleti MD, Rosa AF, Balieiro JCC, Labate CA, et al. Heat shock and structural proteins associated with meat tenderness in Nellore beef cattle, a Bos indicus breed. Meat Sci. 2014;96: 1318–1324. doi: 10. 1016/j. meatsci. 2013. 11. 014. pmid:24342181

110. Diskin SJ, Li M, Hou C, Yang S, Glessner J, Hakonarson H, et al. Adjustment of genomic waves in signal intensities from whole-genome SNP genotyping platforms. Nucleic Acids Res. 2008;36: e126. doi: 10. 1093/nar/gkn556.

pmid:18784189

111. Zhang X, Du R, Li S, Zhang F, Jin L, Wang H. Evaluation of copy number variation detection for a SNP array platform. BMC Bioinformatics. 2014;15: 50. doi: 10. 1186/1471-2105-15-50. pmid:24555668

112. McCarroll SA, Kuruvilla FG, Korn JM, Cawley S, Nemesh J, Wysoker A, et al. Integrated detection and population-genetic analysis of SNPs and copy number variation. Nat Genet. 2008;40: 1166–1174. doi: 10. 1038/ng. 238. pmid:18776908

113. Zheng X, Shaffer JR, McHugh CP, Laurie CC, Feenstra B, Melbye M, et al. Using family data as a verification standard to evaluate copy number variation calling strategies for genetic association studies. Genet Epidemiol. 2012;36: 253–262. doi: 10. 1002/gepi. 21618. pmid:22714937

114. Zhan B, Fadista J, Thomsen B, Hedegaard J, Panitz F, Bendixen C. Global assessment of genomic variation in cattle by genome resequencing and high-throughput genotyping. BMC Genomics. 2011;12: 557. doi: 10. 1186/1471-2164-12-557. pmid:22082336

115. Untergasser A, Nijveen H, Rao X, Bisseling T, Geurts R, Leunissen JA. Primer3Plus, an enhanced web interface to Primer3. Nucleic Acids Res. 2007;35: W71–W74. pmid:17485472 doi: 10. 1093/nar/gkm306

116. Sherry ST. dbSNP: the NCBI database of genetic variation. Nucleic Acids Res. 2001;29: 308–311. pmid:11125122 doi: 10. 1093/nar/29. 1. 308

117. Kinsella RJ, Kähäri A, Haider S, Zamora J, Proctor G, Spudich G, et al. Ensembl BioMarts: a hub for data retrieval across taxonomic space. Database (Oxford). 2011. doi: 10. 1093/database/bar030.

118. Ramakers C, Ruijter JM, Deprez RHL, Moorman AFM. Assumption-free analysis of quantitative real-time polymerase chain reaction (PCR) data. Neurosci Lett. 2003;339: 62–66. pmid:12618301 doi: 10. 1016/s0304-3940(02)01423-4

119. D'haene B, Vandesompele J, Hellemans J. Accurate and objective copy number profiling using real-time quantitative PCR. Methods. 2010;50: 262–270. doi: 10. 1016/j. ymeth. 2009. 12. 007. pmid:20060046

120. Guberman JM, Ai J, Arnaiz O, Baran J, Blake A, Baldock R, et al. BioMart Central Portal: an open database network for the biological community. Database (Oxford). 2011. doi: 10. 1093/database/bar041.

121. Griffiths-Jones S, Grocock RJ, van Dongen S, Bateman A, Enright AJ. miRBase: microRNA sequences, targets and gene nomenclature. Nucleic Acids Res. 2006;34: D140–D144. pmid:16381832 doi: 10. 1093/nar/gkj112

122. Stothard P, Choi J-W, Basu U, Sumner-Thomson JM, Meng Y, Liao X, et al. Whole genome resequencing of black Angus and Holstein cattle for SNP and CNV discovery. BMC Genomics. 2011;12: 559. doi: 10. 1186/1471-2164-12-559. pmid:22085807

123. Kijas JW, Barendse W, Barris W, Harrison B, McCulloch R, McWilliam S, et al. Analysis of copy number variants in the cattle genome. Gene. 2011;482: 73–77. doi: 10. 1016/j. gene. 2011. 04. 011. pmid:21620936

124. Gel B, Díez-Villanueva A, Serra E, Buschbeck M, Peinado MA, Malinverni R. regioneR: an R/Bioconductor package for the association analysis of genomic regions based on permutation tests. Bioinformatics. 2015. doi: 10. 1093/bioinformatics/btv562.

125. Purcell S, Neale B, Todd-Brown K, Thomas L, Ferreira MAR, Bender D, et al. PLINK: a tool set for whole-genome association and population-based linkage analyses. Am J Hum Genet. 2007;81: 559–575. doi: 10. 1086/519795. pmid:17701901

CHAPTER 8 GENOME-WIDE ASSOCIATION FOR MILK PRODUCTION

Shadi Nayeri, Mehdi Sargolzaei, Mohammed K. Abo-Ismail, Natalie May, Stephen P. Miller, Flavio Schenkel, Stephen S. Moore and Paul Stothard

Keywords: Genome-wide association analysis, Functional analysis, Holstein cattle, Female fertility and milk production

BACKGROUND

Milk production and fertility are two economically important traits affecting profitability in dairy cattle. These traits are polygenic, affected by many genes and variants, each with small effects on the observed phenotype [1]. Improvements in management and nutrition, along with intense genetic selection have increased milk production in recent decades. However, selection has also changed the reproductive physiology of the cow and led to a decrease in reproductive efficiency [2]. For example, time in estrus has been reduced to less than 8 hours in lactating dairy cows [3], pregnancy rate has decreased, days open and services per conception has increased [2]. In the last decade, advances in genome sequencing technologies and availability of a tremendous number of genetic variants in the form of single nucleotide polymorphisms (SNP) (Bovine HapMap Consortium, 2009) have led to the application of genomic selection (GS) [4]. Genomic selection is based on linkage disequilibrium (LD) of unknown functional mutations and SNP genotypes that are spread out across the whole genome [4]. Incorporation of functional mutations into genotyping panels could increase GS accuracy and applicability across populations [1, 5].

Many quantitative regions and candidate genes associated with milk production and fertility have been identified by means of genome-wide association analysis studies (GWAS) [6, 7, 8]. Several QTL regions and genes associated with milk yield, fat yield, protein yield, fat deviation and protein deviation have been reported in previous studies [9, 10, 11, 12]. Earlier studies have also identified strong functional candidate genes that affect milk production traits such as *DGAT1* and *GHR* [13, 14]. Similarly, important genomic associations for fertility traits were found in previous GWAS studies, including significant QTLs for calving to first service interval [15, 16, 17], days open [18], cow non-return rate [15, 19], heifer non-return rate [20], daughter pregnancy rate [12, 21], age at puberty [22] and interval from first service to last service for cows and heifers [15].

Most of the previously described genetic variants, however, are not causal, but rather are in linkage disequilibrium with the functional mutation. The level of LD is a limiting factor for the precision of QTL location detection in dairy cattle populations [23]. This is because even SNPs at long distances from the QTL may show associations with the phenotypic trait of interest due to extended LD [24]. Identifying pathways and genes that are associated with significant SNPs can give us a deeper biological insight into expression mechanisms of the trait under study [1, 25]. Refining the position of QTL regions harbouring candidate genes, and identifying causal mutations underlying variation in complex traits, can lead to an increase in accuracy of selection for these traits [26].

The goals of the current study were to identify or refine the position of QTL regions for milk production (MILK), fat production (FAT), protein production (PROT), fat deviation

Citation: Shadi Nayeri, Mehdi Sargolzaei, Mohammed K. Abo-Ismail, Natalie May, Stephen P. Miller, Flavio Schenkel, Stephen S. Moore and Paul Stothard, Genome-wide association for milk production, DOI: 10.1186/s12863-016-0386-1.

(FATD), protein deviation (PROTD), heifer first service to calving interval (FSTCh), calving to first service interval (CTFS), daughter fertility (DF), and days open (DO) in Canadian Holstein dairy cattle. Additionally we performed an enrichment analysis to test for overrepresentation of significant SNPs in biological pathways.

RESULTS AND DISCUSSION

Association Analysis

Association analysis identified strong associations for most of the production traits and some of the fertility traits in this study. Representative Manhattan plots are shown in Fig. 1, for milk production (MILK) and calving to first service interval (CTFS) traits. No significant associations were detected for FSTCh and DF. For those traits yielding significant associations, the number of significant SNPs identified at a genome-wise FDR of 5 % varied from 1,416 for PROTD to 8 for DO (Table 1). Q-Q plots comparing the observed distribution of –log (P-value) to the expectation under null hypothesis: Figure S1 E-I (milk production traits): Figure S2 D-G (fertility traits). The plots show a distribution close to the expected distribution line for CTFS (λ_{median} = 1.0567), DO (λ_{median} = 1.0115) and some production traits (MILK λ_{median} = 1.0056; FAT λ_{median} = 0.9558; PROT λ_{median} = 1.05), whereas there were strong deviations from expectation for FATD (λ_{median} = 0.7844) and PROTD (λ_{median} = 0.8969). When a high-density marker panel is used in single maker association analysis potentially a large number of markers in linkage disequilibrium may display association (and similar low P-values) with the same QTL region. This yields many significant tests that are not independent and, therefore, deviate from the expected distribution of test statistics. However, this does not imply in an overall inflation of P-values. The deviations from expectations observed for milk, fat, protein and fat deviations appear to be due largely to the strong effect of the *DGAT1* gene and the many SNPs in linkage disequilibrium that show some degree of association with it. When BTA14 was excluded from the analyses to assess this possibility, the λ_{median} values for milk production (MILK), fat production (FAT), fat deviation (FATD) and protein deviation (PROTD) were 1.0605, 1.0896, 1.1005 and 0.9709, respectively, much closer to the expectation.

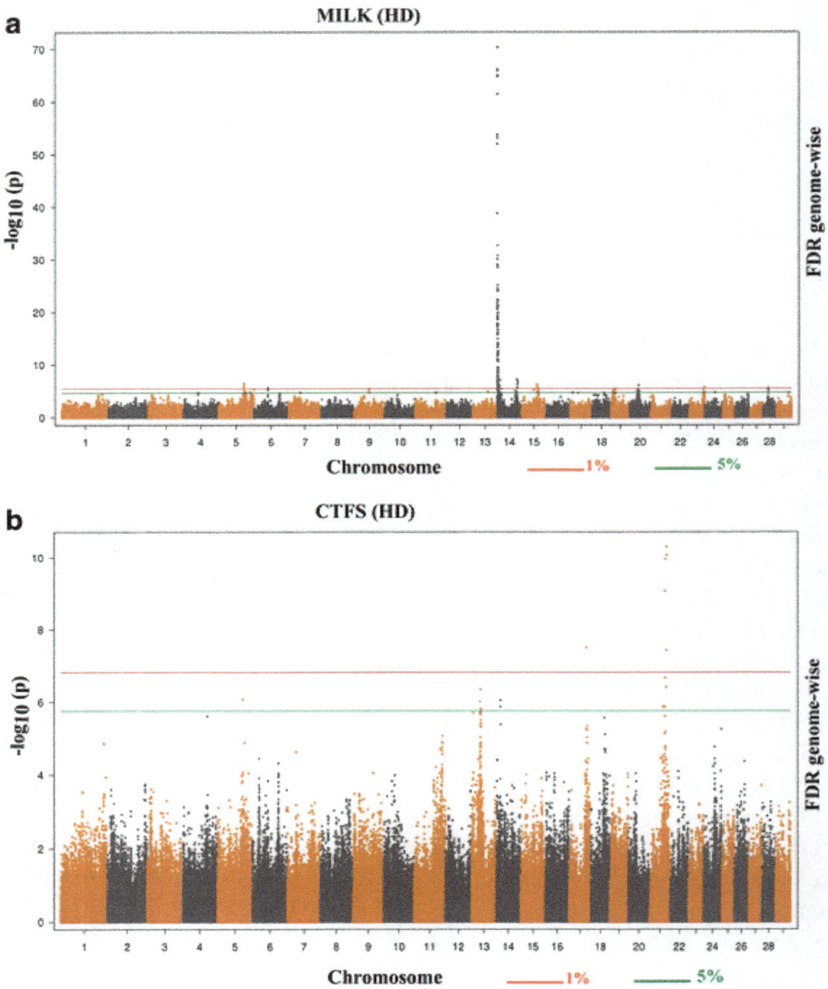

Figure 1: The $-\log_{10}$ of the P-values for association with SNPs is plotted for **a**. milk production (MILK) and **b**. calving to first service intervals (CTFS) in Holstein dairy cattle. Chromosome number is indicated on the horizontal axis. The red line is the threshold for significant SNPs at 1 % FDR. The green line is the threshold for significant SNPs at 5 % FDR.

Table 1: The number of significant SNPs (5 % and 1 % genome-wise FDR) using single SNP regression mixed linear model on imputed BovineHD (777 k) genotypes in Holstein dairy cattle

Trait	Heritability	No. of sig SNPs (1 % FDR)	No. of sig. SNPs (5 % FDR)
Milk production	0.410	221	292
Fat production	0.340	595	813
Protein production	0.370	41	87
Fat deviation (%)	0.370	998	1,230
Protein deviation (%)	0.370	912	1,416
Daughter fertility	0.070	0	0
First service to calving interval (heifer)	0.033	0	0
Calving to first service interval	0.072	8	20
Days open	0.102	8	0

Production Traits

Milk Production (MILK)

In total 292 SNPs were found to be significant for milk production at a genome-wise FDR of 5 %. Highly significant SNPs (genome-wise 1 % FDR) were mostly localized on BTA 5, 6, 14, 15, 20 (Fig. 1a).

Many of the strong associations detected in this study support previously reported regions for the same or correlated traits. For example, a region containing numerous highly significant SNPs on BTA14 includes*DGAT1*, a gene with a major effect on milk fat content [13] and several other production traits [6, 11, 27, 28,29]. A peak on BTA20 overlaps with growth hormone receptor (*GHR*) [14, 30, 31] and a nearby peak at 31–32 Mb contains several genes previously identified as potentially influencing milk production, milk fat percentage, lean meat yield, carcass weight, residual feed intake, age at puberty and male fertility traits in beef cattle [32, 33, 34, 35, 36]. We have also detected regions that were previously identified on BTA5 at position 87 and 107 Mb for milk yield [37]. The presence of a QTL at confidence interval of 45–52 Mb on BTA6 was confirmed by previously reported significant QTLs found for clinical mastitis in Norwegian Red dairy cattle [38]. A significant peak on BTA15 at the genomic interval of 54 to 58 Mb in this study is close to a significant SNP that was reported to be associated with persistency of milk production in dairy cattle by Kolbehdari et al. [9], which is located in an intronic region of the *CD44* gene.

The most significant SNP (BovineHD1400000216: rs134432442) for this trait was located within the *CPSF1* gene on BTA14. Cochran et al. [39] has reported this SNP to be significantly associated with fat yield and fat percentage in Holstein cattle. Other highly significant (genome-wise FDR≤1 %) SNPs in this study are located within genes *TONSL* (BovineHD1400000206: rs137472016), *CYHR1* (BovineHD1400000204: rs137727465),*FOXH1* and *PPP1R16A* (ARS-BFGL-NGS-57820: rs109146371).

Of the 7,586 SNPs that were introduced as candidate SNPs (P-value<0.01) to SNP2GO for enrichment analysis, a total of 1,576 were enriched in biological pathways that may provide better insight into key pathways and genes associated with milk production. Enrichment analysis found 545 significant GO terms (genome-wise FDR≤1 %) with a minimum of 10 genes associated with each GO term. Included among the top 10 most significant GO terms (FDR≤1 %) are those with clear relevance to the trait, such as GO:0007595 (lactation) and GO:0043627 (response to estrogen) whereas for others the relevance is less clear, for example GO:0003785 (actin monomer binding) and GO:0010569 (regulation of double-strand break repair via homologous recombination).

Fat Production (FAT) and Fat Deviation (FATD)

Significant SNPs identified for fat production (FAT) were located on BTA5 and 14. Similarly, GWAS identified highly significant peaks for fat deviation (FATD) on BTA5, 6, 14 and 20. Significant SNPs were mostly located on BTA14 for these two traits.

This association analysis supports the presence of QTLs on BTA5 (76 to 107 Mb), BTA14 (0 to 20 Mb and 31 to 63 Mb), BTA20 (22 Mb and 31 to 33 Mb) related to milk production traits reported in multiple studies [9, 10, 36,39, 40, 41]. In addition to a significant peak at 0 to 20 Mb on BTA14, the location of *DGAT1*, we identified another peak with highly significant SNPs at a genomic interval of 60 to 70 Mb associated with FATD and MILK. The presence of significant SNPs associated with milk production has been reported for this region at 63 Mb by Kolbehdari et al. [9]. Heyen et al. (1999) and Ashwell et al. [11] have also reported three regions on BTA14 at 55 and 69 Mb (between markers D14S55-ILSTS39 and D14S31-CSSM66) and 69 to 79 Mb (between markers BM4305-IRNA100), with a significant effect on fat yield and fat percentage, milk yield and protein yield respectively. The common highly significant SNPs among FAT and FATD were mostly associated with BTA14 and assigned to several known and some newly identified genes including*ARHGAP39* (BovineHD1400000188: rs134892687), *PPP1R16A* (BovineHD1400000199: rs134839376), *GRINA*(BovineHD1400000275: rs133271979), *SMPD5* (BovineHD1400000262: rs135549651) and *MROH1*(BovineHD1400000243: rs133119726).

Chromosome 5 contained the second highest number of significant SNPs for both FAT and FATD (135 and 185, respectively; genome-wise FDR≤1 %). A significant peak on this chromosome was detected extending between 87 to 100 Mb in this study. This region was reported to be associated with a SNP (ss117963826) in the*GABARAPL1* gene with an antagonistic effect on milk yield and fat percentage [37]. Kolbehdari et al. [9] also reported a SNP (rs41592948) in this region linked to the *GABARAPL1* gene with an effect on dairy strength. Furthermore, we identified five significant SNPs (genome-wise FDR≤5 %) within previously known QTL regions (BovineHD0500025075: rs137830740; BovineHD0500025146: rs133732696; BovineHD0500025147: rs42406616; BovineHD0500025415: rs109234621; BovineHD0500025488: rs109374096) on BTA5 for FAT trait. These SNPs are all intron variants and are located within genes *ST8SIA1*, *ABCC9*,*SLO1C1* and *PDE3A*. The QTL region that these SNPs are located in were found associated with C22:1 fatty acid content, milk fat yield and sire conception rate in Angus, Brown Swiss and Holstein cattle.

Several SNPs within the identified significant peak on BTA20 for FATD in this study (31–32 Mb) are located within growth hormone receptor (*GHR*) gene including BovineHD2000009236: rs109719726; ARS-BFGL-NGS-118998: rs110482506 and UA-IFASA-7069: rs41639261. Association of significant SNPs affecting milk fat content within *GHR* gene was already reported in German Holstein-Friesian cattle [44]. This region was also reported to be associated with early embryonic survival, sire conception rate [45] and carcass weight in Holstein-Friesian cattle [33]. Another highly significant peak region for FATD was located on BTA6 and within a known QTL region (at 37–47 Mb) [46]. This region has also been reported significant for milk protein content, carcass weight and fat production in US cattle breeds [46,47].

In the candidate SNP enrichment analysis, 1,958 SNPs were enriched in biological pathways for FAT (from 7,024 candidate SNPs P-value<0.01). These SNPs were overrepresented in 330 significant GO terms. Included among the top 10 most significant GO terms the most relevant ones with the trait are (genome-wise FDR≤1 %) GO:0051057 (positive regulation of small GTPase mediated signal transduction) and GO:0030513 (positive regulation of BMP signaling pathway). For other terms the relevance is less clear such as GO:0003281 (ventricular septum development), GO:0003279 (cardiac septum development) and GO:0003215 (cardiac right ventricle morphogenesis). A total of 5,290 candidate SNPs (*P*-value<0.01) associated with FATD were overrepresented in 445 enriched GO terms. Among the top seven most significant GO terms the ones with clear relevance to the FATD trait are GO:0042403 (thyroid hormone metabolic process) and GO:0035357 (peroxisome proliferator activated receptor signaling pathway), whereas the relevance is less clear for other GO terms including GO:0033158 (regulation of protein import into nucleus, translocation).

Protein production (PROT) and protein deviation (PROTD)

Association analysis detected significant SNPs on BTA 5, 9 and 14. The largest number of significant SNPs for protein production (PROT) was located on chromosome 14, with 41 significant SNPs (genome-wise 1 % FDR) (Additional file 1: Figure S1 C). Significant regions detected for protein deviation (PROTD) were identified on chromosomes 3, 5, 6, 10, 14, 15, 20, 26 and 29 (Additional file 3: Table S1) with the majority of significant SNPs located on chromosomes 3, 6, 14 and 20

Our study detected strong associations on chromosome 20 from 28 to 38 Mb for PROTD. These results are consistent with previous studies reporting QTL regions surrounding *GHR* on chromosome 20 for milk yield and milk composition trait [14, 27, 41]. SNPs affecting protein yield and protein percentage on BTA20 associated with growth hormone receptor gene has been reported for several breeds including Holstein [14, 48], and Ayrshire dairy cattle [30]. We also identified highly significant SNPs including (ARS-BFGL-NGS-118998: rs110482506) in our study at 32 Mb on BTA20 which supports a QTL reported by Wang et al. [44] with a SNP downstream of the *GHR* gene.

Significant peaks were detected on chromosome 6, spanning 24 – 40 and 80 to 90 Mb for PROTD. The presence of QTLs on BTA6 in dairy cattle affecting milk production traits, near the casein gene cluster (around 87 Mb) has been reported in multiple breeds and populations including Dutch, US Holstein cattle for milk fat and milk protein yield [49, 50, 51] German Holsteins [52, 53] and Brazilian Holstein cattle [54] for milk yield and fat yield. Several highly significant SNPs (FDR≤5 %) on BTA6 in this study associated with PROT and PROTD (including ARS-BFGL-NGS-42501: rs110388088; BovineHD0600006866:rs109858710;BovineHD0600009641: rs135525961; BovineHD0600009643: rs135142364; BovineHD0600009650: rs137464778) are located within a known QTL region associated with milk whey protein in dairy cattle [47]. Also two highly significant SNPs (genome-wise FDR≤1 %) (Hapmap24324-BTC-062449 and BTA-121739-no-rs) on BTA6 for PROTD in this study were reported to be very close to *ABCG2*gene (in a distance of 100 kb) [55, 56]. Variants close and within this gene were associated with protein percentage in Chinese dairy cattle [55, 56].

The peak detected for PROTD on chromosome 3 at 10 to 34 Mb is supported by previous investigations [9, 36, 37, 57]. For example, four highly significant SNPs (genome-wise FDR≤1 %) within this region (at 11 Mb) in this study (BovineHD0300003802: rs109857972; BovineHD0300003805:rs136467848;BovineHD0300003811: rs132784836 and BovineHD0300003779: rs110122034) are located within a reported QTL region associated with protein percentage in German Holstein cattle [37]. Two other SNPs in our study (BTB-00604223: rs41769311) at 52 Mb and (ARS-BFGL-NGS-4613: rs110428369) at 54 Mb on BTA15 were also found to be associated with protein percentage in Irish Holstein cattle [27].

We identified five significant SNPs in this study for PROT on BTA5 at 88 Mb including BovineHD0500025150: rs42406611; BovineHD0500025181: rs109795387 and BovineHD0500025189: rs136903701 within the gene*ABCC9*. This region was detected to be significantly associated with protein yield (at 75 to 110 Mb) in Cochran et al. [39] study.

Enrichment analysis of the candidate SNPs for PROT found 2,280 SNPs overrepresented in 658 GO terms (FDR≤1 %). Included among the top 10 most highly significant GO terms (FDR≤1 %) are those GO terms with a more clear relevance to the trait such as GO:0040037 (negative regulation of fibroblast growth factor receptor signaling pathway) and GO:0021903 (rostrocaudal neural tube patterning), whereas for others the relevance is less clear such as GO:0097228 (sperm principal piece). Candidate SNP enrichment analysis for PROTD identified 447 significant GO terms potentially involved in pathways affecting protein production deviation. These enriched GO terms were overrepresented by a total number of 2,557 highly significant SNPs. The most relevance GO terms among the top 10 most significant GO terms (FDR≤1 %) are GO:0007595 (lactation), GO:1902742 (apoptotic process involved in development), GO:0010257 (NADH dehydrogenase complex assembly) and GO:0033108 (mitochondrial respiratory chain complex assembly).

Fertility Traits

In balanced genomic selection programs, one aim is to select bulls which sire daughters with higher reproductive performance (daughters showing an early heat in the mating period with a high probability of conception) [58]. To examine fertility, we first focused on daughter performance traits including daughter fertility and then examined the regions associated with success of insemination traits such as calving interval and days open. Association analysis did not detect any significant SNP for daughter fertility or heifer first service to calving interval. The identified chromosomes and significant regions for CTFS and DO are discussed in detail below. In order to detect the regions with potential significant pleiotropic effects on both production and fertility traits, we also investigated the overlapping regions between these two traits. No overlapping region was detected between production and fertility traits in this study.

Calving to First Service Interval (CTFS)

A total of 20 SNPs (genome-wise FDR≤5 %) were found to be associated with calving to first service interval (Fig. 1b). These SNPs were mostly located on BTA13 and 21. The most significant SNP (genome-wise 1 % FDR) was BovineHD2100017054: rs136777407 (on BTA21), which is an intronic variant within the gene *FAM181A*.

Little is known about the function of this gene, although it was shown that methylation of this gene increases during mid-secretory phase of progesterone (P4) hormone in human endometrium [59]; This region (at 47–59 Mb) on BTA21 has not been previously reported to be associated with fertility traits in cattle or other species. The region also includes the *ASB2* gene. Therefore, these two genes can be considered as potential candidate genes affecting CTFS and calving interval associated traits in dairy cattle. The two other highly significant SNPs (genome-wise FDR≤5 %) in this study on BTA21 at the same region (47–57 Mb) were BovineHD2100016620: rs136994701, within the *SLC24A4* gene, and BovineHD2100013476: rs136407309 a 3-prime UTR variant within the *NKX2-1* gene. The *SLC24A4* gene encodes a member of potassium-dependent sodium or calcium exchanger protein family and has been identified to be associated with hair color, skin pigmentation [60] and eye color in humans [61]. The gene *NKX2-1* encodes a transcription factor that regulates the expression of thyroid-specific genes (Gene ID: 7080) as well as genes that are involved in morphogenesis (Gene ID: 7080). However, no association of these two genes with fertility trait has been previously reported.

We also detected two significant SNPs associated with CTFS at 30–32 Mb on BTA13; this region has previously been reported by Sahana et al. (2010) as associated with interval from calving to first insemination and fertility index traits [17]. These two SNPs on BTA13 (BovineHD1300008936: rs136296491 and BovineHD1300009502: rs109815929) are located within candidate genes *FAM188A* and *MRC1*, respectively. The location of these SNPs is within known QTL regions associated with Inhibin level, shear force and lean meat yield in several cattle breeds. The single significant SNP on BTA5 (BovineHD0500025143: rs42718239) is located within gene *ABCC9* at 88 Mb. This region previously has been validated for calving to first insemination (ICF) trait in Nordic Holstein, Nordic Red and Jersey cattle breed but the gene *ABCC9* was not assigned to any of the associated SNPs [24]. This SNP is also located within a known QTL region associated with sire conception rate in Holstein dairy cattle [62]. The protein coded by *ABCC9* gene is thought to form ATP-sensitive potassium channels in cardiac, skeletal, vascular and non-vascular smooth muscle (Gene ID: 10060). ATP-sensitive potassium channels are expressed in many tissues and regulate different cellular functions by coupling cell metabolism with membrane potential [63]. The opening of these channels results in a flow of K^+ions and thus, reducing cellular excitability and contractility [64]. The K^+ channel opening components are reported as potent inhibitors of human myometrial contractility [63]. It has been indicated that down-regulation of ATP-sensitive potassium channel (K_{ATP} channel) subunits may facilitate myometrial function during late pregnancy in humans [63]. In another study in humans, it has been revealed that estrogen may induce the activation of K_{ATP} channels to promote cell proliferation in the myometrial [65]. Therefore, the*ABCC9* gene can be considered as a potential candidate gene in dairy cattle and may have a role in cell proliferation of myometrial cells and resuming the reproduction cycle after calving.

Of 8,834 SNPs that were introduced to SNP2GO (P-value<0.01), 2,330 SNPs were overrepresented in biological pathways and molecular functions for CTFS. These SNPs were associated with 459 significant GO terms with FDR≤1 %. The GO terms with the clearest relevance for CTFS trait among the top nine most significant GO terms (FDR≤1 %) are GO:0040019 (positive regulation of embryonic development) and GO:0045663 (positive regulation of myoblast differentiation), while for others the relevance is less clear, for example GO:0045836 (positive regulation of mitotic nuclear division).

Days Open (DO)

Association analysis identified a total of eight highly significant SNPs (genome-wise 1 % FDR) on BTA21 associated with DO. These significant SNPs span the 53–59 Mb region. The peak on chromosome 21 overlaps with the one detected for CTFS, and has not been previously reported for this trait. As noted above, this peak includes gene *FAM181A*. Schulman et al. (2008) reported a QTL associated with days open EBV on chromosomes 1, 2, 5, and 25 for Finnish Ayrshire dairy cattle. These regions were not detected in our study.

The number of candidate SNPs overrepresented in the enrichment analysis was 1,897 (P-value≤0.01). These SNPs were involved in 381 significant GO terms for days open. Among the top 10 most significant GO terms, the ones with clear relevance to the trait are GO:0043268 (positive regulation of potassium ion transport), GO:0071353 (cellular response to interleukin-4). Additional GO terms, for which the relationship to the trait is less clear, include GO:0000786 (nucleosome) and GO:0005070 (SH3/SH2 adaptor activity).

Overlapping Regions among Milk Production Traits

The only overlapping peak among all of the production traits (MILK, FAT, FATD, PROT and PROTD) was the region identified on BTA14. This region spans around 1.4 to 2.9 Mb and includes 74 highly significant SNPs (genome-wise FDR≤1 %).

Overlapping Regions among Fertility Traits

Investigating overlapping regions among fertility traits has resulted in identifying eight significant SNPs on BTA21 at 53–59 Mb between two fertility traits CTFS and DO in this study. The overlapping region among these fertility traits was not reported to be associated with days open or calving to first interval traits in previous GWAS.

CONCLUSION

Genome wide association analysis in this study detected several regions associated with milk production and female fertility in Canadian Holstein cattle. Most of the regions in this study were identified in other independent studies. However, novel regions of association were detected. Our result shows a novel significant region on chromosome 21 (at 47–59 Mb), which overlaps among CTFS and DO, which was not reported for fertility traits in previous association studies. This region includes several genes including*FAM181A*, *SLC24A4* and *NKX2-1*. The inclusion of several traits in one study allowed us to more easily compare overlaps that might, for example, highlight regions with pleiotropic effects. Although overlaps were observed within production and fertility traits, we did not see any overlap between production and fertility in this study. GO term enrichment analysis of the GWAS results identified terms consistent with the known physiology of the traits as well as novel or unexpected terms. The chromosomal regions identified in this study confirm several previous findings affecting production traits. Our result could also highlight new regions and pathways that may contribute to variation in fertility trait in dairy cattle. These novel regions can be used for further functional analysis to identify genes, gene networks and variants that explain variation in these traits.

METHODS

Animals and Data

A population of 3,729 North American Holstein bulls was used in this study, which examined nine production and fertility traits: milk production (MILK), fat production (FAT), protein production (PROT), fat and protein deviation (FATD and PROTD respectively), daughter fertility (DF), first service to calving interval (FSTCh), calving to first service interval (CTFS), and days open (DO). The Canadian Dairy Network (CDN) provided available pedigree, genotypes and official evaluations for proven bulls born between 1956 and 2009. Individuals were genotyped using the BovineSNP50K (50 k) panel (3,729 bulls) (Illumina, San Diego, CA) or the high density (HD, 777 k) SNP panel (2,387 bulls), respectively. In this work only autosomal SNPs were included. For 50 k panel, SNP list used for official genomic evaluation by CDN was considered. These SNPs have passed standard quality control measures used by CDN. Quality control (QC) was performed on the HD genotyping data using snp1101 software (Mehdi Sargolzaei, personal communication) and 116,619 SNPs were excluded on the basis of Mendelian error rate higher>0.05 (3,566 SNPs), low call rate<0.9 (6,446 SNPs), with low MAF (1e-06 – 0.5; 61,577 SNPs), excess of heterozygosity>0.15 (90 SNPs) and excluded by user (46,433 SNPs excluded from sex chromosome and misplaced SNPs). The number of SNPs kept in the analyses before imputation was 40,666 and 657,986 SNPs for 50 k and HD panels, respectively.

Genotypes of 3,729 50 k animals were imputed to HD with a reference population of 2,387 HD individuals using FImpute V2.2 software [66]. Quality control (QC) was performed on the imputed data as well. SNPs with minor allele frequency (MAF) of less than 1 % (55,817 SNPs) and high Mendelian error rate more than 5 % (74 SNPs) were excluded. After quality control, 602,095 SNPs remained for use in the subsequent association analysis.

Calculating De-regressed Proofs

In this study de-regressed Holstein bull proofs were used as pseudo phenotypes. A bull's published estimated breeding value (EBV) is a weighted mean of his daughters deviations (DD) and his parental average (PA) [67]. The de-regressed bull proofs were computed by CDN as shown below [67]:

$$DE_{prg} = \frac{Rel_{EBV}}{1\text{-}Rel_{EBV}} - \frac{Rel_{PA}}{1\text{-}Rel_{PA}}$$

$$Rel_{DD} = \frac{DE_{prg}}{DE_{prg}+1}$$

$$DEBV = PA + \frac{(EBV\text{-}PA)}{Rel_{DD}}$$

Where, DE_{prg} is the daughter equivalent from progeny information, Rel_{EBV} and Rel_{PA} are the reliabilities of EBV and PA, respectively, Rel_{DD} is the reliability of DD, and DEBV is the de-regressed bull proof.

Genome wide Association Analysis (GWAS)

Association analysis was performed using a single SNP regression mixed linear model implemented in the snp1101 software [68]. The mixed linear model was:

$$Y_i = \mu + \beta g_i + a_i + e_i$$

where Y_i is pseudo phenotype of the ith bull (de-regressed bull proofs, DEBV); μ is the overall mean; β is the linear regression coefficient (allele substitution effect) of the SNP; g_i is the SNP genotype of the ith bull, which was coded as 0, 1 and 2 for SNP genotypes BB, AB and AA, respectively; α_i is the random additive polygenic effect of the ith bull and e_i is the random error term.

Assumptions for the model are $a_i : a \sim N(0, G\sigma_a^2)$ where G is the genomic relationship matrix [68] and σ_a^2 is the polygenic additive genetic variance; $e_i : e \sim N(0, R\sigma_e^2)$ where σ_e^2 is the residual variance. R is a diagonal matrix containing weights for the residual variance based on the reliabilities of the de-regressed bull proofs [68].

To account for multiple tests, 5 % and 1 % genome-wise false discovery rate (FDR) were used to identify significant and highly significant associations, respectively. The inflation factor λ [69] and quantile-quantile (Q-Q) plots were calculated to compare observed distributions of –log (*P*-value) to the expected distribution under the no association model for each trait.

Candidate SNP enrichment analysis

The SNP2GO R package was used for functional analysis of the genome-wide association results [25]. Genomic annotations, associated Gene Ontology terms (GO terms) and the list of significant (*P*-value < 0.01) and non-significant SNPs (termed "candidate" and "non-candidate" SNPs in the SNP2GO documentation) for each trait were provided as input to SNP2GO, which reports biological pathways or processes that are enriched for significant SNPs. For this analysis, the Ensembl version 78 genomic annotation file for *Bos taurus* UMD 3.1 assembly was used in conjunction with the Ensembl gene ID file from Ensembl version 78, which contains gene ID to GO term associations. The SNP2GO "extension" value was set to 50 nucleotides, which expands the gene region (by 50 nucleotides upstream and 50 nucleotides downstream) when identifying overlaps between genes and markers. The "runs" parameter was set to 100,000 and a false discovery rate (FDR) of 1 % was used to correct for multiple testing.

ABBREVIATIONS

BTA, Bos taurus autosome; CTFS, calving to first service interval; DF, daughter fertility; DO, days open; EBV, estimated breeding value; FAT, fat production; FATD, fat deviation; FDR, false discovery rate; FSTCh, heifer first service to calving interval;GO, gene ontology; GS, genomic selection; GWAS, genome-wide association study; LD, linkage disequilibrium; MILK, milk production; PROT, protein production; PROTD, protein deviation; QC, quality control; QTL, quantitative trait loci; SNP, single nucleotide polymorphism

DECLARATIONS

Acknowledgements

The authors acknowledge the financial contributions of L'Alliance Boviteq (LAB) and the Natural Sciences and Engineering Research Council (NSERC) of Canada. We thank the Canadian Dairy Network for providing some of the data used in this study.

Funding

This project was supported by Natural Sciences and Engineering Research Council (NSERC) of Canada and L'Alliance Boviteq (LAB) Grant no. CRDPJ 396290–10.

Availability of data and material

Relevant information supporting the results not already presented in the article is given in additional files. The estimated breeding values and pedigree information are publicly available at CDN website (www.cdn.ca). The bulls' genotypes cannot be made publicly available, because they are property of the bulls' owners.

Authors' contributions

PS, SPM, FS, SSM and SN developed the project, assisted with data analysis and edited the manuscript. SN performed data analysis and drafted the manuscript. NM helped to draft the manuscript and assisted with the interpretation of results. MS and MKA assisted with data analysis and edited the manuscript. All authors contributed to the manuscript and approved the final version of the manuscript.

AUTHORS' INFORMATION

Not applicable.

Competing interests

The authors declare that they have no competing interest.

REFERENCES

1. Snelling WM, Cushman R a, Keele JW, Maltecca C, Thomas MG, Fortes MRS, Reverter A. Networks and pathways to guide genomic selection. J Anim Sci. 2013;91:537–52.
2. Lucy MC. Reproductive loss in high-producing dairy cattle: where will it end? J Dairy Sci. 2001;84:1277–93.
3. Nebel R, Dransfield M, Jobst S, Bame J. Automated electronic systems for the detection of oestrus and timing of AI in cattle. Anim Reprod Sci. 2000;60–61:713–23.
4. Meuwissen THE, Hayes BJ, Goddard ME. Prediction of total genetic value using genome-wide dense marker maps. Genetics. 2001;157:1819–29.
5. Meuwissen T, Goddard M. Accurate prediction of genetic values for complex traits by whole-genome resequencing. Genetics. 2010;185:623–31.
6. Pimentel EC, Bauersachs S, Tietze M, Simianer H, Tetens J, Thaller G, Reinhardt F, Wolf E, König S. Exploration of relationships between production and fertility traits in dairy cattle via association studies of SNPs within candidate genes derived by expression profiling. Anim Genet. 2011;42:251–62.
7. Reverter A, Fortes MRS. Building single nucleotide polymorphism-derived gene regulatory networks: Towards functional genomewide association studies. J Anim Sci. 2012;91:530–6.
8. Höglund JK, Guldbrandtsen B, Su G, Thomsen B, Lund MS. Genome scan detects quantitative trait loci affecting female fertility traits in Danish and Swedish Holstein cattle. J Dairy Sci. 2009;92:2136–43.
9. Kolbehdari D, Wang Z, Grant JR, Murdoch B, Prasad

A, Xiu Z, Marques E, Stothard P, Moore SS. A whole genome scan to map QTL for milk production traits and somatic cell score in Canadian Holstein bulls. J Anim Breed Genet. 2009;126:216–27.

10. Boichard DB, Grohs C, Bourgeois F, Cerqueira F, Faugeras R, Neau A, Rupp R, Amigues Y, Boscher MY, Leveziel H. Detection of genes influencing economic traits in three French dairy cattle breeds. Genet Sel Evol. 2003;35:77–101.

11. Ashwell MS, Heyen DW, Sonstegard TS, Van Tassell CP, Da Y, VanRaden PM, Ron M, Weller JI, Lewin HA. Detection of quantitative trait loci affecting milk production, health, and reproductive traits in Holstein cattle. J Dairy Sci. 2004;87:468–75.

12. Schnabel RD, Sonstegard TS, Taylor JF, Ashwell MS. Whole-genome scan to detect QTL for milk production, conformation, fertility and functional traits in two US Holstein families. Anim Genet. 2005;36:408–16.

13. Grisart B, Farnir F, Karim L, Cambisano N, Kim J, Kvasz A, Mni M, Simon P, Frere J-M, Coppieters W, Georges M. Genetic and functional confirmation of the causality of the DGAT1 K232A quantitative trait nucleotide in affecting milk yield and composition. Proc Natl Acad Sci U S A. 2004;101:2398–403.

14. Blott S, Kim J, Moisio S, Schmidt-Küntzel A, Cornet A, Berzi P, Cambisano N, Ford C, Grisart B, Johnson D, Karim L, Simon P, Snell R, Spelman R, Wong J, Vilkki J, Georges M, Farnir F, Coppieters W. Molecular dissection of a quantitative trait locus: a phenylalanine-to-tyrosine substitution in the transmembrane domain of the bovine growth hormone receptor is associated with a major effect on milk yield and composition. Genetics. 2003;163:253–66.

15. Höglund JK, Buitenhuis AJ, Guldbrandtsen B, Su G, Thomsen B, Lund MS. Overlapping chromosomal regions for fertility traits and production traits in the Danish Holstein population. J Dairy Sci. 2009;92:5712–9.

16. Daetwyler HD, Schenkel FS, Sargolzaei M, Robinson JAB. A genome scan to detect quantitative trait loci for economically important traits in Holstein cattle using two methods and a dense single nucleotide polymorphism map. J Dairy Sci. 2008;91:3225–36.

17. Sahana G, Guldbrandtsen B, Bendixen C, Lund MS. Genome-wide association mapping for female fertility traits in Danish and Swedish Holstein cattle. Anim Genet. 2010;41:579–88.

18. Schulman NF, Sahana G, Lund MS, Viitala SM, Vilkki JH. Quantitative trait loci for fertility traits in Finnish Ayrshire cattle. Genet Sel Evol. 2008;40:195–214.

19. Olsen HG, Hayes BJ, Kent MP, Nome T, Svendsen M, Larsgard AG, Lien S. Genome-wide association mapping in Norwegian Red cattle identifies quantitative trait loci for fertility and milk production on BTA12. Anim Genet. 2011;42:466–74.

20. Holmberg M, Andersson-Eklund L. Quantitative trait loci affecting fertility and calving traits in Swedish dairy cattle. J Dairy Sci. 2006;89:3664–71.

21. Cole JB, Wiggans GR, Ma L, Sonstegard TS, Lawlor TJ, Crooker BA, Van Tassell CP, Yang J, Wang S, Matukumalli LK, Da Y. Genome-wide association analysis of thirty one production, health, reproduction and body conformation traits in contemporary U.S. Holstein cows. BMC Genomics. 2011;12:408.

22. Hawken RJ, Zhang YD, Fortes MRS, Collis E, Barris WC, Corbet NJ, Williams PJ, Fordyce G, Holroyd RG, Walkley JRW, Barendse W, Johnston DJ, Prayaga KC, Tier B, Reverter A, Lehnert SA. Genome - wide association studies of female reproduction in tropically adapted beef cattle. J Anim Sci. 2012;90:1398–410.

23. Pryce JE, Bolormaa S, Chamberlain AJ, Bowman PJ, Savin K, Goddard ME, Hayes BJ. A validated genome-wide association study in 2 dairy cattle breeds for milk production and fertility traits using variable length haplotypes. J Dairy Sci. 2010;93:3331–45.

24. Höglund JK, Sahana G, Guldbrandtsen B, Lund MS. Validation of associations for female fertility traits in Nordic Holstein, Nordic Red and Jersey dairy cattle. BMC Genet. 2014;15:8.

25. Szkiba D, Kapun M, von Haeseler A, Gallach M. SNP2GO: functional analysis of genome-wide association studies. Genetics. 2014;197:285–9.

26. Hirschhorn JN, Daly MJ. Genome-wide association studies for common diseases and complex traits. Nat Rev Genet. 2005;6:95–108.

27. Meredith BK, Kearney FJ, Finlay EK, Bradley DG, Fahey AG, Berry DP, Lynn DJ. Genome-wide associations for milk production and somatic cell score in Holstein-Friesian cattle in Ireland. BMC Genet. 2012;13:21.

28. Maxa J, Neuditschko M, Russ I, Förster M, Medugorac I. Genome-wide association mapping of milk production traits in Braunvieh cattle. J Dairy Sci. 2012;95:5357–64.

29. Xie Y, Yang S, Cui X, Jiang L, Zhang S, Zhang Q, Zhang Y, Sun D. Identification and expression pattern of two novel alternative splicing variants of EEF1D gene of dairy cattle. Gene. 2014;534:189–96.

30. Viitala S, Szyda J, Blott S, Schulman N, Lidauer M, Mäki-Tanila A, Georges M, Vilkki J. The role of the bovine growth hormone receptor and prolactin receptor genes in milk, fat and protein production in Finnish Ayrshire dairy cattle. Genetics. 2006;173:2151–64.

31. Flori L, Fritz S, Jaffrézic F, Boussaha M, Gut I, Heath S, Foulley J-L, Gautier M. The genome response to artificial selection: a case study in dairy cattle. PLoS One. 2009;4:e6595.

32. Blaschek M, Kaya A, Zwald N, Memili E, Kirkpatrick BW. A whole-genome association analysis of noncompensatory fertility in Holstein bulls. J Dairy Sci. 2011;94:4695–9.

33. Doran AG, Berry DP, Creevey CJ. Whole genome association study identifies regions of the bovine genome and biological pathways involved in carcass trait performance in Holstein-Friesian cattle. BMC Genomics. 2014;15:837.

34. Fortes MRS, Deatley KL, Lehnert SA, Burns BM, Reverter A, Hawken RJ, Boe-Hansen G, Moore SS, Thomas MG. Genomic regions associated with

fertility traits in male and female cattle: advances from microsatellites to high-density chips and beyond. Anim Reprod Sci. 2013;141:1–19.

35. Rolf MM, Taylor JF, Schnabel RD, McKay SD, McClure MC, Northcutt SL, Kerley MS, Weaber RL. Genome-wide association analysis for feed efficiency in Angus cattle. Anim Genet. 2012;43:367–74.

36. Jiang L, Liu X, Yang J, Wang H, Jiang J, Liu L, He S, Ding X, Liu J, Zhang Q. Targeted resequencing of GWAS loci reveals novel genetic variants for milk production traits. BMC Genomics. 2014;15:1105.

37. Pimentel ECG, Bauersachs S, Tietze M, Simianer H, Tetens J, Thaller G, Reinhardt F, Wolf E, König S. Exploration of relationships between production and fertility traits in dairy cattle via association studies of SNPs within candidate genes derived by expression profiling. Anim Genet. 2011;42:251–62.

38. Sodeland M, Kent MP, Olsen HG, Opsal M a, Svendsen M, Sehested E, Hayes BJ, Lien S. Quantitative trait loci for clinical mastitis on chromosomes 2, 6, 14 and 20 in Norwegian Red cattle. Anim Genet. 2011;42:457–65.

39. Cochran SD, Cole JB, Null DJ, Hansen PJ. Discovery of single nucleotide polymorphisms in candidate genes associated with fertility and production traits in Holstein cattle. BMC Genet. 2013;14:49.

40. Jiang L, Liu J, Sun D, Ma P, Ding X, Yu Y, Zhang Q. Genome wide association studies for milk production traits in Chinese Holstein population. PLoS One. 2010;5:e13661.

41. Raven L-A, Cocks BG, Hayes BJ. Multibreed genome wide association can improve precision of mapping causative variants underlying milk production in dairy cattle. BMC Genomics. 2014;15:62.

42. Saatchi M, Garrick DJ, Tait RG, Mayes MS, Drewnoski M, Schoonmaker J, Diaz C, Beitz DC, Reecy JM. Genome-wide association and prediction of direct genomic breeding values for composition of fatty acids in Angus beef cattle. BMC Genomics. 2013;14:730.

43. Guo J, Jorjani H, Carlborg Ö. A genome-wide association study using international breeding-evaluation data identifies major loci affecting production traits and stature in the Brown Swiss cattle breed. BMC Genet. 2012;13:82.

44. Wang X, Wurmser C, Pausch H, Jung S, Reinhardt F, Tetens J, Thaller G, Fries R. Identification and dissection of four major QTL affecting milk fat content in the German Holstein-Friesian population. PLoS One. 2012;7:e40711.

45. Ma L, Wiggans GR, Wang S, Sonstegard TS, Yang J, Crooker BA, Cole JB, Van Tassell CP, Lawlor TJ, Da Y. Effect of sample stratification on dairy GWAS results. BMC Genomics. 2012;13:536.

46. Saatchi M, Schnabel RD, Taylor JF, Garrick DJ. Large-effect pleiotropic or closely linked QTL segregate within and across ten US cattle breeds. BMC Genomics. 2014;15:442.

47. Huang W, Peñagaricano F, Ahmad KR, Lucey JA, Weigel KA, Khatib H. Association between milk protein gene variants and protein composition traits in dairy cattle. J Dairy Sci. 2012;95:440–9.

48. Chamberlain AJ, Hayes BJ, Savin K, Bolormaa S, McPartlan HC, Bowman PJ, Van der Jagt C, MacEachern S, Goddard ME. Validation of single nucleotide polymorphisms associated with milk production traits in dairy cattle. J Dairy Sci. 2012;95:864–75.

49. Georges M, Nielsen D, Mackinnon M, Mishra A, Okimoto R, Pasquino AT, Sargeant LS, Sorensen A, Steele MR, Womack JE, Hoeschele I,. Mapping quantitative trait loci controlling milk production in dairy cattle by exploiting progeny testing. Genet Soc Am. 1995;139:907–20.

50. Zhang Q, Boichard D, Hoeschele I, Ernst C, Eggen A, Murkve B, Pfister-Genskow M, Witte LA, Grignola FE, Uimari P, Thaller G, Bishop MD. Mapping quantitative trait loci for milk production and health of dairy cattle in a large outbred pedigree. Genetics. 1998;149:1959–73.

51. Bovenhuis H, Van Arendonk JA, Korver S. Associations between milk protein polymorphisms and milk production traits. J Dairy Sci. 1992;75:2549–59.

52. Kühn C, Freyer G, Weikard R, Goldammer T, Schwerin M. Detection of QTL for milk production traits in cattle by application of a specifically developed marker map of BTA6. Anim Genet. 1999;30:333–40.

53. Freyer G, Sørensen P, Kühn C, Weikard R, Hoeschele I. Search for pleiotropic QTL on chromosome BTA6 affecting yield traits of milk production. J Dairy Sci. 2003;86:999–1008.

54. Silva AA, Azevedo ALS, Verneque RS, Gasparini K, Peixoto MGCD, da Silva MVGB, Lopes PS, Guimarães SEF, Machado MA. Quantitative trait loci affecting milk production traits on bovine chromosome 6 in zebuine Gyr breed. J Dairy Sci. 2011;94:971–80.

55. Fang M, Fu W, Jiang D, Zhang Q, Sun D, Ding X, Liu J. A Multiple-SNP Approach for Genome-Wide Association Study of Milk Production Traits in Chinese Holstein Cattle. PLoS One. 2014;9:e99544.

56. Cohen-Zinder M, Seroussi E, Larkin DM, Loor JJ, Everts-Van Der Wind A, Lee JH, Drackley JK, Band MR, Hernandez AG, Shani M, Lewin HA, Weller JI, Ron M. Identification of a missense mutation in the bovine ABCG2 gene with a major effect on the QTL on chromosome 6 affecting milk yield and composition in Holstein cattle. Genome Res. 2005;15:936–44.

57. Lipkin E, Mosig MO, Darvasi A, Ezra E, Shalom A, Friedmann A, Soller M. Quantitative trait locus mapping in dairy cattle by means of selective milk DNA pooling using dinucleotide microsatellite markers: analysis of milk protein percentage. Genetics. 1998;149:1557–67.

58. Haile-Mariam M, Morton JM, Goddard ME. Estimates of genetic parameters for fertility traits of Australian Holstein-Friesian cattle. Anim Sci. 2003;76:35–42.

59. Houshdaran S, Zelenko Z, Irwin JC, Giudice LC. Human Endometrial DNA Methylome is cycle-dependent and is associated with gene expression regulation. Mol Endocrinol. 2014;28:1118–35.

60. Han J, Kraft P, Nan H, Guo Q, Chen C, Qureshi A,

Hankinson SE, Hu FB, Duffy DL, Zhao ZZ, Martin NG, Montgomery GW, Hayward NK, Thomas G, Hoover RN, Chanock S, Hunter DJ. A genome-wide association study identifies novel alleles associated with hair color and skin pigmentation. PLoS Genet. 2008;4:e1000074.

61. Liu F, Wollstein A, Hysi PG, Ankra-Badu GA, Spector TD, Park D, Zhu G, Larsson M, Duffy DL, Montgomery GW, Mackey DA, Walsh S, Lao O, Hofman A, Rivadeneira F, Vingerling JR, Uitterlinden AG, Martin NG, Hammond CJ, Kayser M. Digital Quantification of Human Eye Color Highlights Genetic Association of Three New Loci. PLoS Genet. 2010;6:e1000934.

62. Peñagaricano F, Weigel KA, Khatib H. Genome-wide association study identifies candidate markers for bull fertility in Holstein dairy cattle. Anim Genet. 2012;43 Suppl 1:65–71.

63. Curley M, Cairns MT, Friel AM, McMeel OM, Morrison JJ, Smith TJ. Expression of mRNA transcripts for ATP-sensitive potassium channels in human myometrium. Mol Hum Reprod. 2002;8:941–5.

64. Khan RN, Matharoo-Ball B, Arulkumaran S, Ashford ML. Potassium channels in the human myometrium. Exp Physiol. 2001;86:255–64.

65. Park S-H, Ramachandran S, Kwon S-H, Cha S-D, Seo EW, Bae I, Cho C, Song D-K. Upregulation of ATP-sensitive potassium channels for estrogen-mediated cell proliferation in human uterine leiomyoma cells. Gynecol Endocrinol. 2008;24:250–6.

66. Sargolzaei M, Chesnais JP, Schenkel FS. A new approach for efficient genotype imputation using information from relatives. BMC Genomics. 2014;15:478.

67. VanRaden PM, Van Tassell CP, Wiggans GR, Sonstegard TS, Schnabel RD, Taylor JF, Schenkel FS. Invited review: reliability of genomic predictions for North American Holstein bulls. J Dairy Sci. 2009;92:16–24.

68. VanRaden PM. Efficient methods to compute genomic predictions. J Dairy Sci. 2008;91:4414–23.

69. Devlin B, Roeder K. Genomic control for association studies. Biometrics. 1999;55:997–1004.

Some of the QTL for lodging co-localized with QTL for either plant height, grain length, or grain length-breadth ratio. In contrast to the above cases, the correlations of these traits were either absent or weak ($r^2 = 0.24$ for plant height and lodging, 0.02 for grain length and lodging, and −0.03 for length-breadth ratio and lodging in the dry season), yet in addition to the QTL described above, a single QTL for grain length and lodging was identified on chromosome 6 (*qGrL-6-1*, *qLG-6-2*) and a QTL for grain length-breadth ratio and lodging was identified on chromosome 12 (*qLBR-12*, *qLG-12*) (Figs. 2–3). These results would indicate that these traits are truly quantitative and teh variance explained by the shared QTL is less significant than that explained by loci elsewhere in the genome that segregate independently.

Application to Breeding

Performing association mapping on a panel of adapted breeding lines rather than on a diversity panel provides the opportunity to apply the results directly to breeding programs. Unlike the results from studies using diversity panels, our association mapping results can be readily used to identify favorable or unfavorable haplotypes that are currently segregating in our elite breeding material. These haplotypes could be used to determine the most suitable parents for crossing in order to exploit transgressive segregation and/or to increase the frequency with which favorable haplotypes appear in the progeny. MAS for favorable haplotypes among the progeny would allow us to increase breeding efficiency and decrease cost by reducing the number of plants advanced to the next generation of breeding or that need to be phenotyped. In this way, we aim to increase the rate of genetic improvement by increasing gain from selection.

MAS based on the haplotypes identified in an association mapping study such as ours necessitates a strong association between that haplotype and the phenotype-of-interest, and, ideally, a large effect of the haplotype on the phenotypic variation of the trait. When the association of haplotype and phenotype is non-perfect, as would be expected in many cases, there will be an r^2 value at or above which marker assisted selection would be more efficient than phenotypic selection, i.e., at or above this value, the increased efficiency of marker based selection would be expected to outweigh the drag on gain from selection that would result from mistakenly eliminating favorable individuals, or vice versa. Below this threshold, phenotypic selection would be preferable. Breeders generally expect marker assisted selection accuracy to be high (> = 95%) for traits with major genes or that are easy to phenotype. However, when a trait is laborious, time consuming, or technically difficult and/or expensive to phenotype, a lower accuracy ~70% could be acceptable to breeders [18].

We used PLINK to identify haplotypes that were associated with the significant peaks identified in our GWAS for three of our phenotypes that are routinely collected by breeders—flowering time, plant height, and yield. In many breeding programs in Southeast Asia, short stature, early flowering time (to allow for plant of multiple crops per year), and, of course, high yields are desirable breeding goals [19], so we therefore focus on performing selection to increase the frequency of favorable alleles at QTLs associated with these traits in our population.

Flowering Time

For flowering time in the dry season, 17 significant haplotypes were identified, all spanning the region of the large-effect QTL on chromosome three, *qFL-3*. The haplotypes ranged from 2–124 SNPs in length, and spanned regions ranging from 233 bp—900.4 Kb. The early-flowering variant of haplotype H1 was the single most significant haplotype identified, and explained 34% of the phenotypic variance. Haplotype H1 spanned 58 SNPs and 202.9 Kb. The most significant haplotype associated with late-flowering was the late variant of haplotype H3, which consisted of 8 SNPs spanning 49.3 Kb, and explained ~33% of the phenotypic variance. In the wet season, 22 significant haplotypes were identified, all of which also co-localized with *qFL-3*.

Of the 5% earliest flowering individuals in the dry season (17/342), 12 were confirmed to carry the early H1 haplotype. Only 1 out of the 17 individuals was confirmed *not* to have this haplotype. The other four individuals had missing data in the region of the haplotype which did not preclude the presence of the early H1 haplotype, but prevented confirmation of haplotype presence. For the 5% latest flowering individuals in the dry season, 6/17 were confirmed to have late haplotype H3, and 5/17 were confirmed to have other haplotypes (Fig. 4). Due to this high degree of correlation and the large effect of these haplotypes, a breeder could potentially perform either positive selection for early flowering or negative selection against late flowering using a set of linked SNPs derived from either of these haplotypes.

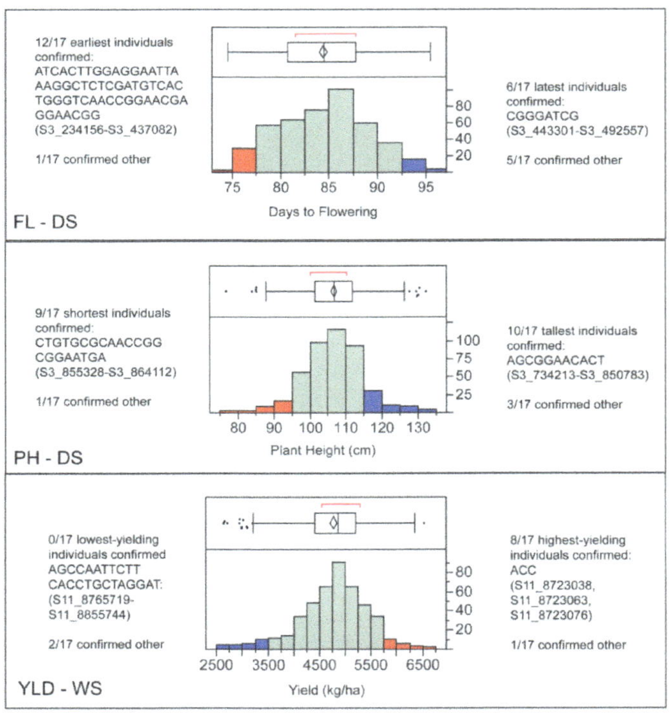

Figure 4. Phenotypic distributions for flowering time in the dry season (top), plant height in the dry season (middle), and yield in the wet season (bottom), showing the most desirable 5% of individuals (red) and least desirable 5% of individuals (blue). The most significant haplotypes associated with each end of the distribution are shown to the left and right of the histogram with the number of individuals in the best or worst 5% that carry the respective haplotypes. "Confirmed other" refers to individuals that were known NOT to carry the most significant haplotype. Individuals that were neither confirmed to carry

the significant haplotype or confirmed to carry other haplotypes had missing data at one or more SNPs that did not preclude the possibility of the individual carrying the significant haplotype.

Plant Height

For plant height in the dry season, 10 haplotypes were identified, all of which co-localized with *qPH-3* and *qFL-3*. The haplotypes ranged from 3–35 SNPs in length, and spanned between 4.2 and 416.2 Kb. The short stature variant of haplotype H2 was the most significant haplotype associated with shortness, and was 22 SNPs in length, spanned 8.8 Kb, and explained ~12% of the phenotypic variance. The tall-stature variant of haplotype H1 was the most significant for tallness. H1 was 11 SNPs in length, spanned 116.6 Kb, and explained ~10% of the phenotypic variance . For the 5% shortest individuals, 9/17 were confirmed to carry the H2 short haplotype, and only 1 was confirmed to carry a different haplotype. For the 5% of tallest individuals, 10/17 were confirmed to carry the H1 tall haplotype, and only 3/17 were confirmed to carry a different haplotype (Fig. 4). These results for the tall haplotype, in particular, suggest that markers linked to haplotype H1 could be developed by breeders to cull undesirably tall individuals from their program.

For the wet season, only one haplotype (H1) was identified that co-localized with qPH-2, the plant height QTL identified in the WS GWAS. H1 was 3 SNPs in length, spanned 7 bp, and explained about 7% of the phenotypic variance. All 12 of the shortest 5% of individuals that did not have missing data at the three SNPs that make up the haplotype were confirmed to carry the short H1 haplotype. Of the 5% tallest individuals, 4/17 had the tall stature H1 haplotype, and the 8/17 had the short haplotype. These data suggest that this haplotype would also be most useful for performing negative selection, i.e., the tall variant of H1 could be used to select against tall individuals in the wet season (Fig. 4).

Yield

Two significant haplotypes were identified each for the dry season and the wet season for grain yield. In the dry season, both significant haplotypes were associated with lower yield and were located at the top of chromosome six, co-localizing with a GWAS peak that fell just below our significance threshold. The most significant of these two haplotypes, H2, was 8 SNPs long, spanned a region of 47.9 Kb, and explained ~7% of the phenotypic variance. Of the 5% lowest yielding individuals, 3/17 were confirmed to carry this haplotype, and 7/17 were confirmed to carry a different haplotype. Of the 5% best yielding, individuals, however, only 1/17 individuals carried this haplotype, while 13/17 were confirmed to carry other haplotypes.

The results are more interesting for the wet season. Two significant haplotypes (H1 and H3) were identified on chromosome 11, one of which, H1, was associated with higher yields. The H1 high-yield haplotype consisted of only 3 SNPs, spanned 38 bp, and explained ~9% of the phenotypic variance. Of the 5% highest yielding individuals in the wet season, 8/17 were confirmed to carry high-yielding haplotype H1, and only 1/17 individuals were confirmed to carry a different haplotype, suggesting this could be a particularly useful haplotype for performing positive selection for high yield in the wet season (Fig. 4). Low yielding haplotype H3 spanned 24 SNPs and ~90 Kb, and explained ~6% of the phenotypic variance, however it was not possible to confirm its presence among any of the 5% lowest yielding individuals due to missing data at the relevant loci. 2/17 of the lowest yielding individuals were confirmed to carry alternative haplotypes.

Candidate Genes

To identify candidate genes underlying the above QTLs/haplotypes of interest, we searched rice genome browsers and existing literature at regions that were either near the most significant GWAS SNPs reported, or in the regions spanned by the significant haplotypes under selection (i.e., early flowering, short stature, high yields) reported in Tables (materials and methods). For flowering time, the large effect QTL identified on chromosome 3 (*qFL-3*) corresponds to the location of the QTL *Hd9* identified in early QTL mapping studies [20]. While there were a number of possible candidate genes underlying*qFL-3* , the *OsMADS50* gene (LOC_Os03g03100), an important activator of flowering in rice, and a homolog of *Arabidopsis* gene the *SUPPRESSOR OF OVEREXPRESSION OF CONSTANS1* (*SOC1*) was identified as the best candidate. QTL for plant height (*qPH-3*) and culm length (*qCulmL-3*) also co-localized with *qFL-3*, suggesting that *OsMADS50* may also play a role in the expression of these traits as well as for flowering time. Indeed, rice plants in which this gene was knocked down using RNAi exhibited elongated internodes, suggesting true pleiotropy may underlie these traits in rice [21].

OsMADS50 co-localizes with both *Hd9* and our peak association marker, but it does not fall within the most significant flowering time haplotype [21,22,23,24]. This suggests that a second genetic feature located ~800 Kb downstream of*OsMADS50* is a target of selection in this breeding population. No obvious candidate genes for flowering time were identified within this haplotype but the data suggest this is potentially a novel flowering time-related QTL.

Several known, important, rice flowering time genes also underlie the flowering time QTL identified on chromosome 6 (*qFL-6*), including *Hd1*, *RFT1* (a.k.a. *FT-3*), and *FT-2*. *Hd1*, however, is ~6Mb away from the peak QTL marker, and thus is not the best candidate [25,26,27]. *RFT1*, a homolog of an Arabidopsis florigen gene, is located directly beneath the association peak [25,26], while *FT-2* is located nearby ~11 Kb away. The flowering time QTL on chromosome 8 (*qFL-8*) also corresponded to a previously mapped QTL, *Hd5/DTH8*, which encodes a *HAP3* subunit of the CCAAT-box-binding transcription factor and has been shown to play an important role in early flowering [28,29,30]. Another possible candidate gene that encodes a putative cullin protein lies nearby; cullins are F-box protein subunits with the potential to regulate flowering time[31]. Additional candidate genes for the flowering time QTL identified in the wet season only (*qFL-2*, *qFL-9*).

OsMADS50 is also the best candidate gene for the plant height QTL on chromosome 3 (*qPH-3*), as mentioned previously, although several other candidates were also identified [21]. *qPH-6* and *qPH-8* were both were very near putative zinc finger domain containing proteins that provide likely candidates. *qPH-2* was near a putative GA encoding gene, which makes a good candidate for this wet season QTL.

No best single candidate gene was identified for the wet season yield QTL on chromosome 11 (*qYLD-11*), but several putative stress tolerance genes were identified in the vicinity of the significant haplotype which warrant further investigation.

Toward Genomic Selection for rice improvement

The lack of perfect association between the haplotypes described above and their respective phenotypes, as well as the limited number of QTL identified for many of the phenotypes in this study, highlight the complexity of the genetic architecture underlying many agronomic traits in rice. Even the haplotypes for flowering time, for which we identified a large effect QTL, did not explain 100% of the phenotypic variance. This poses a problem for the implementation of MAS in rice breeding programs. While genetic gain could potentially be increased over phenotypic selection alone, the probability of eliminating favorable individuals or selecting unfavorable individuals will limit overall breeding progress.

Genomic selection (GS), first described by Meuwissen et al., in 2001 could solve this problem [32]. Instead of making selections based on a subset of previously identified significant QTL, in a genomic selection breeding scheme, selections are based on the output of statistical models that are fit using all available high density genotyping data [15,16,32]. Our recently published study in PLoS Genetics explores the efficacy of performing genomic selection in this same population [33], but the GWAS results presented here by themselves also suggest that GS could be an effective strategy for rice improvement. The GWAS results are also informative for thinking about how GS could best be implemented for this population. The lack of any large-effect QTL for many of the agronomic traits studied here suggest that for some agronomic traits, linear, additive statistical models such as RR-BLUP could be the most effective means of predicting breeding value. The presence of large-effect QTL for other traits such as flowering time, however, suggest that for other traits in rice, non-additive genomic selection models and/or models that can account for differences in marker variance may be more accurate than the simple RR-BLUP models. The results also suggest that a model in which the large effect QTL identified from this GWAS are fit as fixed effects could further improve accuracy. Simulation experiments and a preliminary studies in wheat and cattle have shown this to be the case for high heritability traits with large fixed effects [14,17,34].

GWAS have been used to identify many genes that underlie a variety of agronomic traits in rice [8,9,10,11,12,13,35,36]. Here we performed a GWAS on a breeding population of elite inbred rice lines for the IRRI irrigated rice breeding program in order to a.) more directly apply the results to MAS in this breeding population, and b.) inform the genomic selection results presented in the companion paper. The successful identification of haplotypes for flowering time, plant height, and grain yield, as well as the elucidation of the genetic architecture underlying a variety of the agronomic traits in this population suggests that this can be a powerful strategy for subsequent rice improvement.

MATERIALS AND METHODS

Plant Material

363 elite breeding lines were selected for genotyping from the International Rice Research Institute (IRRI) irrigated rice breeding program based on the planned inclusion of the lines in the 2011 Multi-Environment Testing Program and in the 2011 and 2012 Replicated Yield Trials (RYT) at IRRI (Los Baños). Approximately half of the lines were also included in the 2009–2010 RYTs at IRRI. The other lines were promoted from the observational yield trial (OYT) to the RYT in 2011.

Phenotyping

A total of 19 agronomic, morphological, grain and yield-related traits were evaluated in the panel in 2012 (dry and wet seasons) as described in Table 1. A detailed analysis of these traits will be reported elsewhere. Briefly, all measurements were performed using two replications with 5 plants per plot, except plot yield, which was measured using a 6 m^2harvest area. All measurements were made on a quantitative scale, with the exception of lodging score and panicle exertion rate, which were nominal. Methods used followed IRRI's standard evaluation system (SES) or other routinely used protocols. Predicted means and variance components were calculated using linear mixed models in Genstat v. 16. Entries were considered as fixed effects and replication and entry by replication interactions were considered as random effects. A detailed trait analysis will be reported elsewhere.

Genotyping

DNA Extraction

Young leaf tissue was collected from each of the 369 breeding lines from plants grown in Gutterman Greenhouse in Ithaca, NY. DNA was extracted using the Qiagen 96-plex DNeasy kit as per the Qiagen fresh leaf tissue 96-plex protocol (www.qiagen.com/HB/DNeasy96Plant).

Library Preparation

384-plex genotyping-by-sequencing (GBS) libraries were prepared using the protocol by Elshire *et al.* 2011 [37], as described previously in Spindel and Wright *et al* 2013 [38].

GBS Data Analysis

SNPs were discovered and called from the raw 384-plex GBS data using the TASSEL3.0 GBS pipeline with physical alignment to the MSU version 6.0 Nipponbare rice reference genome using Bowtie2, as described in Spindel and Wright *et al* 2013 [38,39,40]. The IRRI breeding materials genotyped here are a collection of multi-parent related and unrelated inbred lines, so the GBS-PLAID algorithm for imputation, which was developed specifically for imputation of biparental rice mapping populations, was not useful [38]. Imputation of missing data was instead performed using the TASSEL3.0 FastImputationBitFixedWindow plugin with default settings [41]. The algorithm works by dividing the entire SNP dataset into small SNP windows, then identifying the most similar inbred line within each window to fill the missing data.

The algorithm takes advantage of small IBD regions shared between pairs of inbred lines in the collection; if the window from the closest neighbor has more than 5% difference from the line being imputed, the data point is left as missing [41]. The imputation error rate using this algorithm was estimated for each chromosome in our dataset by masking a fraction of the un-imputed allele calls and comparing the imputed and actual calls. The average imputation error rate across the twelve rice chromosomes was estimated in this way to be less than 1%.

SNPs that still had 10% or more data missing after imputation (or call rates of < 90%) were removed from the dataset along with all monomorphic SNPs, for a total SNP set of 73,147 SNPs. A second genotype dataset was also obtained in which we removed all monomorphic SNPs, SNPs with call rates < 75%, and SNPs with minor allele frequencies (MAF) < 0.05 for a total of 71,710 SNPs. Individuals with more than 60% missing data were dropped from both datasets, which resulted in the removal of six individuals that failed sequencing for the total of 363 genotyped lines used throughout the study.

Both final datasets were then transformed from nucleotide genotype coding (i.e., 'A', 'C', 'T', 'G') to numeric coding (1, 0, -1 for class I homozygotes, heterozygotes, and class II homozygotes, respectively) in order to facilitate statistical analysis. The minimal remaining missing data were filled using the numeric genotype means of each line in order to perform PCA and GWAS.

Subpopulation Structure Analysis

The majority of the 363 lines were characterized apriori from pedigree records to belong to the *indica* or *indica-admixed* subpopulation groups. In order to identify outlier individuals belonging to the *japonica* or *japonica-admixed* groups, principle components analysis (PCA) was performed in R (version 3.0.1, function prcomp) using the imputed 73,147 SNP dataset. The first principle component of high density SNP data in rice can separate the*indica* and *japonica* subgroups [11], so by plotting the first four principle components using JMP Pro 10, 13 *japonica* outliers were identified as pulled in a tight cluster apart from the rest of the 350 lines. These 13 lines were removed from the dataset, and a second PCA was performed using the same methodology as the first to identify any admixed outliers, i.e, outlier lines containing greater percentages of *japonica* derived SNPs. By plotting the first four principle components of the second PCA, another 18 lines were judged to be outliers and removed from the dataset, leaving a total of 332 lines to be used for the cross-validation experiments. A third PCA was performed using the remaining 332 to confirm that there were no additional subpopulation outliers. Percent variance explained by the first four principle components for each PCA was produced from the R function output.

Association Mapping

Association mapping was performed in R using the GEMMA implementation of the standard linear mixed model $y = W\alpha + X\beta + u + \varepsilon$, where y is a vector of phenotypes, $W = (w1,..wc)$ is an $n \times c$ matrix of covariates (fixed effect), α is a c-vector of the corresponding coefficients, β is the marker effect size, X is a matrix of allele dosages for the imputed variants, u is a vector of random effects and ε is a vector of errors. u is multivariate normal, $\sim$ MVNn(0, $\lambda\tau^{-1}$K), with τ^{-1} as the variance of the residual errors, λ the ratio between the two variance components, and K the $m \times m$ genomic relationship matrix. ε is $\sim$ MVNn(0, $\lambda\tau^{-1}$In) where In is an $n \times n$ identity matrix [42]. K was also calculated using GEMMA (parameter-gk = 2). For additional details on the GEMMA implementation of the mixed model, see Zhao et al., 2012 [43].

GWAS models that included the first principle component (PC) as a covariate were compared to models that did not include the first PC using TASSEL v. 4.3.1, and were found not to differ significantly, therefore, we did not include the first PC as a covariate when building the GEMMA model. This makes sense given the small percent variance explained by any of the principle components after removal of outliers (see results). The QQ plots also indicate that the GEMMA model fit the data well. The significance of the GWAS results were determined using a Wald test, and the significance threshold determined by a false discovery rate of 0.1 using the R function p.adjust (method = BH) [44]. Allele effects were calculated as the difference between the average trait value for all lines that were homozygous for the major allele (AA) and the average trait value for all lines that were homozygous for the minor allele (BB) for a given SNP. The percent variance explained by all significant SNPs discovered in each season was output from GEMMA. The percent variance explained by each individual significant SNP was calculated as the squared correlation between the phenotype and genotype of the SNP [45]. QQ plots and Manhattan plots were generated using R. all other plots were produced using the program JMP v. 10.0.

Haplotype Analysis

Haplotype blocks were calculated for flowering time, plant height, and yield for both seasons of data in PLINK (command—blocks). For all phenotypes except yield in the dry season, the input genotype files contained all SNPs that passed the GWAS significance threshold. For yield in the dry season, the input genotype file contained all SNPs with p-value < = .0001. The resulting PLINK blocks file was then used to calculate the haplotype frequencies in PLINK using the—hap-freq command. To test haplotypes for significant association with phenotype, we used the PLINK—hap-assoc command.

For each trait, the most significant haplotype associated with either low phenotypic value or high phenotypic value (e.g. early flowering and late flowering) was identified. The proportion of either the individuals with the 5% lowest phenotypic values or the 5% highest phenotypic values that carried the most significant haplotypes associated with the respective high/low phenotype were calculated.

Candidate Gene Analysis

Candidate genes were identified for the flowering time, plant height and yield QTLs, three of the most important traits to breeders. Previously identified QTL that co-localized with the QTL mapped in this study were identified using the Gramene QTL data archive (http://archive.gramene.org/qtl/). To ascertain whether QTL had been cloned, literature searches were performed. Regions covered by significant haplotypes and SNPs were also searched for candidates using the MSU v.7 rice genome browser (http://rice.plantbiology.msu.edu/cgi-bin/gbrowse/rice/).

ACKNOWLEDGMENTS

We thank Pavel Korniliev, Francisco Agosto-Perez, Genevieve DeClerck, and Anthony Greenberg for statistical and computational support.

AUTHOR CONTRIBUTIONS

Conceived and designed the experiments: SM BC PV JS. Performed the experiments: HB JS. Analyzed the data: HB JS. Wrote the paper: JS BC SM. Advised HB on project completion/design: AL TB GG JH.

REFERENCES

1. Gregorio GB IM, Vergara GV, Thirumeni S (2013) Recent advances in rice science to design salinity and other abiotic stress tolerant rice varieties. SABRAO Journal of Breeding and Genetics 45: 31–41.
2. Seck PA DA, Mohanty S, Wopereis MCS (2012) Crops that feed the world 7: Rice. Food Security 4: 7–24. doi: 10. 1007/s12571-012-0168-1
3. Varshney RK, Terauchi R, McCouch SR (2014) Harvesting the Promising Fruits of Genomics: Applying Genome Sequencing Technologies to Crop Breeding. PLoS Biol 12: e1001883. doi: 10. 1371/journal. pbio. 1001883. pmid:24914810
4. Walsh MLaB (1998) Genetics and Analysis of Quantitative Traits. Sunderland, MA: Sinauer Associates Inc. .
5. Yu J, Holland JB, McMullen MD, Buckler ES (2008) Genetic Design and Statistical Power of Nested Association Mapping in Maize. Genetics 178: 539–551. doi: 10. 1534/genetics. 107. 074245. pmid:18202393
6. Zhu C, Gore M, Buckler ES, Yu J (2008) Status and Prospects of Association Mapping in Plants. The Plant Genome Journal 1: 5. doi: 10. 3835/plantgenome2008. 02. 0089
7. Flint-Garcia SA, Thornsberry JM, Buckler ESt (2003) Structure of linkage disequilibrium in plants. Annual review of plant biology 54: 357–374. pmid:14502995 doi: 10. 1146/annurev. arplant. 54. 031902. 134907
8. Famoso AN, Zhao K, Clark RT, Tung CW, Wright MH, Bustamante C, et al. (2011) Genetic architecture of aluminum tolerance in rice (Oryza sativa) determined through genome-wide association analysis and QTL mapping. PLoS genetics 7: e1002221. doi: 10. 1371/journal. pgen. 1002221. pmid:21829395
9. Huang X, Wei X, Sang T, Zhao Q, Feng Q, Zhao Y, et al. (2010) Genome-wide association studies of 14 agronomic traits in rice landraces. Nature genetics 42: 961–967. doi: 10. 1038/ng. 695. pmid:20972439
10. Huang X, Zhao Y, Wei X, Li C, Wang A, Zhao Q, et al. (2012) Genome-wide association study of flowering time and grain yield traits in a worldwide collection of rice germplasm. Nature genetics 44: 32–39. doi: 10. 1093/jncimonographs/lgs007. pmid:22623593
11. Zhao K, Tung C-W, Eizenga GC, Wright MH, Ali ML, Price AH, et al. (2011) Genome-wide association mapping reveals a rich genetic architecture of complex traits in Oryza sativa. Nat Commun 2: 467. doi: 10. 1038/ncomms1467. pmid:21915109
12. Huang X, Han B (2014) Natural Variations and Genome-Wide Association Studies in Crop Plants. Annual review of plant biology 65: 531–551. doi: 10. 1146/annurev-arplant-050213-035715. pmid:24274033
13. Yang W, Guo Z, Huang C, Duan L, Chen G, Jiang N, et al. (2014) Combining high-throughput phenotyping and genome-wide association studies to reveal natural genetic variation in rice. Nat Commun 5. doi: 10. 1038/ncomms6087
14. Zhang Z, Ober U, Erbe M, Zhang H, Gao N, He JL, et al. (2014) Improving the Accuracy of Whole Genome Prediction for Complex Traits Using the Results of Genome Wide Association Studies. PloS one 9. doi: 10. 1371/journal. pone. 0093017
15. Heffner EL, Sorrells ME, Jannink JL (2009) Genomic Selection for Crop Improvement. Crop Science 49: 1–12. doi: 10. 2135/cropsci2008. 08. 0512
16. Lorenz AJ, Chao S, Asoro FG, Heffner EL, Hayashi T, Iwata H, et al. (2011) Genomic Selection in Plant Breeding. 110: 77–123. doi: 10. 1016/b978-0-12-385531-2. 00002-5
17. Bernardo R (2014) Genomewide Selection when Major Genes Are Known. Crop Science 54: 68. doi: 10. 2135/cropsci2013. 05. 0315
18. Collard BC, Mackill DJ (2008) Marker-assisted selection: an approach for precision plant breeding in the twenty-first century. Philosophical transactions of the Royal Society of London Series B, Biological sciences 363: 557–572. pmid:17715053 doi: 10. 1098/rstb. 2007. 2170
19. Dalrymple D (1986) Development and spread of high yielding rice varieties in developing countries. Bureau for Science and Technology, Agency for International Development Washington, DC.
20. Lin H, Ashikari M, Yamanouchi U, Sasaki T, Yano M (2002) Identification and Characterization of a Quantitative Trait Locus, Hd9, Controlling Heading Date in Rice. Breeding Science 52: 35–41. doi: 10. 1270/jsbbs. 52. 35
21. Lee SY, Kim J, Han JJ, Han MJ, An GH (2004) Functional analyses of the flowering time gene OsMADS50, the putative SUPPRESSOR OF OVEREXPRESSION OF CO 1/AGAMOUS-LIKE 20 (SOC1/AGL20) ortholog in rice. Plant Journal 38: 754–764. pmid:15144377 doi: 10. 1111/j. 1365-313x. 2004. 02082. x
22. Rao NN, Prasad K, Kumar PR, Vijayraghavan U (2008) Distinct regulatory role for RFL, the rice LFY homolog, in determining flowering time and plant architecture. Proceedings of the National Academy of Sciences of the United States of America 105: 3646–3651. doi: 10. 1073/pnas. 0709059105. pmid:18305171
23. Ryu CH, Lee S, Cho LH, Kim SL, Lee YS, Choi SC, et al. (2009) OsMADS50 and OsMADS56 function antagonistically in regulating long day (LD)-dependent flowering in rice. Plant Cell and Environment 32: 1412–1427. doi: 10. 1111/j. 1365-3040. 2009. 02008. x. pmid:19558411

24. Yoo SK, Chung KS, Kim J, Lee JH, Hong SM, Yoo SJ, et al. (2005) CONSTANS activates SUPPRESSOR OF OVEREXPRESSION OF CONSTANS 1 through FLOWERING LOCUS T to promote flowering in Arabidopsis. Plant Physiology 139: 770–778. pmid:16183837 doi: 10. 1104/pp. 105. 066928
25. Kojima S, Takahashi Y, Kobayashi Y, Monna L, Sasaki T, Araki T, et al. (2002) Hd3a, a rice ortholog of the Arabidopsis FT gene, promotes transition to flowering downstream of Hd1 under short-day conditions. Plant and Cell Physiology 43: 1096–1105. pmid:12407188 doi: 10. 1093/pcp/pcf156
26. Komiya R, Ikegami A, Tamaki S, Yokoi S, Shimamoto K (2008) Hd3a and RFT1 are essential for flowering in rice. Development 135: 767–774. doi: 10. 1242/dev. 008631. pmid:18223202
27. Yano M, Katayose Y, Ashikari M, Yamanouchi U, Monna L, Fuse T, et al. (2000) Hd1, a major photoperiod sensitivity quantitative trait locus in rice, is closely related to the arabidopsis flowering time gene CONSTANS. Plant Cell 12: 2473–2483. pmid:11148291 doi: 10. 2307/3871242
28. Fujino K, Yamanouchi U, Yano M (2013) Roles of the Hd5 gene controlling heading date for adaptation to the northern limits of rice cultivation. Theoretical and Applied Genetics 126: 611–618. doi: 10. 1007/s00122-012-2005-5. pmid:23090144
29. Lin H, Liang Z-W, Sasaki T, Yano M (2003) Fine Mapping and Characterization of Quantitative Trait Loci Hd4 and Hd5 Controlling Heading Date in Rice. Breeding Science 53: 51–59. doi: 10. 1270/jsbbs. 53. 51
30. Wei X, Xu J, Guo H, Jiang L, Chen S, Yu C, et al. (2010) DTH8 Suppresses Flowering in Rice, Influencing Plant Height and Yield Potential Simultaneously. Plant Physiology 153: 1747–1758. doi: 10. 1104/pp. 110. 156943. pmid:20566706
31. Jain M, Nijhawan A, Arora R, Agarwal P, Ray S, Sharma P, et al. (2007) F-Box Proteins in Rice. Genome-Wide Analysis, Classification, Temporal and Spatial Gene Expression during Panicle and Seed Development, and Regulation by Light and Abiotic Stress. Plant Physiology 143: 1467–1483. pmid:17293439 doi: 10. 1104/pp. 106. 091900
32. Meuwissen THE, Hayes BJ, Goddard ME (2001) Prediction of total genetic value using genome-wide dense marker maps. Genetics 157: 1819–1829. pmid:11290733
33. Spindel J, Begum, H. , Akdemir, D. , Virk, P. , Collard, B. , Redoña, E. , Atlin, G. , Jannink, J. , McCouch, S. (2014 *In Press*) Genomic Selection and Association Mapping in rice (Oryza sativa): Effect of trait genetic architecture, training population composition, marker number and statistical model on accuracy of rice genomic selection in elite, tropical rice breeding lines. PLoS genetics.
34. Zhao Y, Mette MF, Gowda M, Longin CFH, Reif JC (2014) Bridging the gap between marker-assisted and genomic selection of heading time and plant height in hybrid wheat. Heredity 112: 638–645. doi: 10. 1038/hdy. 2014. 1. pmid:24518889
35. Xu X, Liu X, Ge S, Jensen JD, Hu FY, Li X, et al. (2012) Resequencing 50 accessions of cultivated and wild rice yields markers for identifying agronomically important genes. Nature Biotechnology 30: 105–U157. doi: 10. 1038/nbt. 2050. pmid:22158310
36. Yonemaru J, Mizobuchi R, Kato H, Yamamoto T, Yamamoto E, Matsubara K, et al. (2014) Genomic regions involved in yield potential detected by genome-wide association analysis in Japanese high-yielding rice cultivars. Bmc Genomics 15. doi: 10. 1186/1471-2164-15-346
37. Elshire RJ, Glaubitz JC, Sun Q, Poland JA, Kawamoto K, Buckler ES, et al. (2011) A robust, simple genotyping-by-sequencing (GBS) approach for high diversity species. PloS one 6: e19379. doi: 10. 1371/journal. pone. 0019379. pmid:21573248
38. Spindel J, Wright M, Chen C, Cobb J, Gage J, Harrington S, et al. (2013) Bridging the genotyping gap: using genotyping by sequencing (GBS) to add high-density SNP markers and new value to traditional bi-parental mapping and breeding populations. Theoretical and Applied Genetics: 1–18.
39. Bradbury PJ, Zhang Zhiwu, Dallas E. Kroon, Casstevens Terry M. , Ramdoss Yogesh, Buckler Edward S. (2007) TASSEL: software for association mapping of complex traits in diverse samples. Bioinformatics 23: 2633–2635. pmid:17586829 doi: 10. 1093/bioinformatics/btm308
40. Langmead B, Salzberg SL (2012) Fast gapped-read alignment with Bowtie 2. Nat Meth 9: 357–359. doi: 10. 1038/nmeth. 1923
41. Romay MC, Millard MJ, Glaubitz JC, Peiffer JA, Swarts KL, Casstevens TM, et al. (2013) Comprehensive genotyping of the USA national maize inbred seed bank. Genome Biol 14: R55. doi: 10. 1186/gb-2013-14-6-r55. pmid:23759205
42. Zhou X, Stephens M (2012) Genome-wide efficient mixed-model analysis for association studies. Nature genetics 44: 821–824. doi: 10. 1038/ng. 2310. pmid:22706312
43. Zhou X, Stephens M (2012) Genome-wide efficient mixed-model analysis for association studies. Nature genetics 44: 821–824. doi: 10. 1038/ng. 2310. pmid:22706312
44. Benjamini Y, Hochberg Y (1995) Controlling the false discovery rate: a practical and powerful approach to multiple testing. Journal of the Royal Statistical Society Series B (Methodological): 289–300.
45. Faraway JJ (2002) Practical Regression and Anova using R. r-project.

CHAPTER 10

ANALYSIS OF THE GENETIC DIVERSITY AND STRUCTURE ACROSS A WIDE RANGE OF GERMPLASM REVEALS PROMINENT GENE FLOW IN APPLE AT THE EUROPEAN LEVEL

Jorge Urrestarazu, Caroline Denancé, Elisa Ravon, Arnaud Guyader, Rémi Guisnel, Laurence Feugey, Charles Poncet, Marc Lateur, Patrick Houben, Matthew Ordidge, Felicidad Fernandez-Fernandez, Kate M. Evans, Frantisek Paprstein, Jiri Sedlak, Hilde Nybom, Larisa Garkava-Gustavsson, Carlos Miranda, Jennifer Gassmann, Markus Kellerhals, Ivan Suprun, Anna V. Pikunova, Nina G. Krasova, Elnura Torutaeva, Luca Dondini , Stefano Tartarini, François Laurens and Charles-Eric Durel

Keywords: *Malus* x *domestica* Borkh., Genetic resources, Population structure, Variability, SSR markers, Differentiation, Parentage analysis

BACKGROUND

Cultivated apple (*Malus* x *domestica* Borkh.) is one of the most important fruit crops grown in temperate zones and the most important in the *Rosaceae* family [1]. Although there are more than 10,000 documented apple cultivars worldwide and the apple production area is widespread geographically, the global production is dominated by relatively few cultivars, many of which are closely related [2, 3]. Moreover, in the last century, despite the existence of a large number of apple breeding programs worldwide, only a few well-adapted genotypes (e.g., 'Red Delicious', 'Golden Delicious', 'Jonathan', 'McIntosh' or 'Cox´s Orange Pippin') were extensively used in apple breeding to release new varieties with desirable traits [2, 4, 5]. The additional release of clonal selections of the most popular and widely grown varieties has further contributed towards the uniformity of commercial apple orchards [6, 7, 8]. The gradual replacement of the traditional and locally well-adapted cultivars by a few wide-spread modern varieties has led to a dramatic loss of genetic diversity in the orchards and may also hamper future plant breeding.

The recognition of this situation has encouraged the establishment of action towards the preservation of apple genetic resources worldwide. Multiple apple collections are presently maintained in Europe, preserving mainly old cultivars which have been grown traditionally in their respective regions, but also other cultivars with diverse geographic origins introduced a long time ago, that represent elite selections from before the time of formal breeding. Most of these existing collections were established before molecular identification became available, and in the absence of marker data, the criteria used in the past for selecting the germplasm to be preserved in collections focused mainly on morphology (pomology), eco-geography and/or passport information [9]. The effectiveness of these conservation approaches depends upon the criteria

Citation: Jorge Urrestarazu, Caroline Denancé, Elisa Ravon, Arnaud Guyader, Rémi Guisnel, Laurence Feugey, Charles Poncet, Marc Lateur, Patrick Houben, Matthew Ordidge, Felicidad Fernandez-Fernandez, Kate M. Evans, Frantisek Paprstein, Jiri Sedlak, Hilde Nybom, Larisa Garkava-Gustavsson, Carlos Miranda, Jennifer Gassmann, Markus Kellerhals, Ivan Suprun, Anna V. Pikunova, Nina G. Krasova, Elnura Torutaeva, Luca Dondini, Stefano Tartarini, François Laurens and Charles-Eric Durel, Analysis of the genetic diversity and structure across a wide range of germplasm reveals prominent gene flow in apple at the European level, DOI: 10.1186/s12870-016-0818-0.

used for selecting germplasm and it has been suggested that genetic diversity may not always be optimal in these, or equivalent collections in other crops [10, 11], and therefore, unintended internal redundancies are expected. Assessment of the genetic diversity in fruit tree species is nowadays mainly performed by marker genotyping techniques [12]. Molecular markers have therefore become an indispensable tool in the management of germplasm collections, and their use is widely applied in characterization to assist and complement phenotypic assessments and to re-examine the composition of the collections [11, 13, 14, 15, 16]. The use of molecular markers has not only important implications with regard to the efficiency of the management of the genetic resources, but constitutes a key instrument to evaluate diversity, to elucidate the underlying genetic structure of the germplasm and to quantify relatedness and differentiation between populations among other multiple applications [17, 18, 19, 20]. Such knowledge is of high relevance since the conservation of plant genetic resources only fulfills its full potential when they are used effectively, which requires knowledge of the extent and structure of the variation occurring within the material preserved [21].

Until now, the studies of diversity and genetic structure conducted in European apple have been based on the analyses of material from limited geographic areas (mostly nation-scale) [11, 14, 22, 23, 24, 25, 26]. By contrast, the extent and structure of the apple genetic diversity conserved at a European level have remained largely unknown. The main obstacle is the different sets of SSR markers used in the different European collections preventing an overall comparison [27]. Thus, in the frame of the EU-FruitBreedomics project [28] a single set of 16 SSR markers was used in a very broad set of apple germplasm (~2440 accessions, mostly of dessert use) preserved in collections located in eleven countries and representing three broad European geographical regions (North+East, West and South) in order to determine the diversity in apple collections at a European scale, to evaluate gene flow in cultivated apple across Europe, as well as to elucidate the stratification of germplasm into population subdivisions and finally, to perform parentage analysis. This is the largest study of apple genetic resources at the pan-European level.

RESULTS

SSR Polymorphism – Identification and Redundancy

Among the 2,446 accessions, ten accessions did not show clear PCR amplifications and were discarded from the analysis. Pairwise comparison of multilocus profiles revealed 219 groups of redundancies, leading to the removal of 405 redundant accessions before further analyses (16 % of redundancy). The number of accessions in each of these identical SSR profile groups varied from two to nine. The cumulative probability of identity (P_{ID}) was extremely low: $P_{ID} = 1.3 \times 10^{-22}$, thus highlighting the low risk of erroneous attribution of accessions to duplicate groups. Redundancies were found both within and between collections, leading to the confirmation of numerous previously documented synonyms (e.g., 'Papirovka' and 'White Transparent', 'London Pippin' and 'Calville du Roi', or 'Président van Dievoet' and 'Cabarette') and allowing the putative identification of numerous unknown synonyms or mutant groups (e.g., 'Gloria Mundi'='Mela Zamboni'='Audiena de Oroz'='Belle Louronnaise', 'Court-Pendu Plat/Doux/Gris'='Krátkostopka královská', 'Reinette de Champagne'='Maestro Sagarra' or 'Reinette Simirenko'='Renetta Walder'='Burdinche'). Redundancy groups also supported the notion of several national/local name translations such as the English cultivar 'Cornish Gilliflower' translated into 'Cornwallské hřebíčkové' (i.e., 'Cornish clove'), or 'White Transparent' and 'Skleněné žluté' (i.e., 'yellow glass') in Czech and 'Transparente Blanca' in Spanish, the Russian cultivar 'Korichnoe polosatoe' translated into 'Kaneläpple' in Swedish (i.e., 'cinnamon apple'), or the cultivar 'La Paix' translated into 'Matčino' (i.e., 'Mother', a synonym of 'La Paix') in Czech. Several cases of homonymy (i.e., accessions with the same name but different SSR profiles) were also found, e.g., three different SSR profiles for the same accession names 'Pomme Citron' or 'Charles Ross'. Data allowed identifying some obvious labeling errors, e.g., X2698 'Court Pendu Plat' which was shown to be the rootstock 'MM106', or CRAW-0362 'Transparente de Croncels' which was found likely to actually be 'Filippa'. Following these observations, the apple germplasm dataset was reduced to 2,031 unique genotypes (i.e., exhibiting distinct SSR profiles). Among these individuals, 162 (8 % of the different genotypes) were removed since they had a putative triploid profile, while another ten were discarded because of too much missing SSR data, or because further identified as rootstock or outliers in a preliminary Principal Coordinate Analysis. The final number of unique diploid genotypes further analyzed was therefore 1,859. Using passport data and other accessible information, it was possible to attribute geographical regions of origin (either for three broad designated European regions or, when possible, specific countries) for a large part of the unique genotypes. Roughly 89 % (1,653) of these genotypes could be geographically assigned, with 261, 1,074 and 318 genotypes assigned to Northern+Eastern, Western and Southern historical regions of origin, respectively. In brief, the Northern+Eastern region was composed of germplasm originating in Nordic European countries plus Russia, the Western region was composed of germplasm originating in Western and Central European countries and the Southern region was composed of germplasm from Spain and Italy (see Methods for more details). The remaining 11 % consisted of either genotypes lacking passport information or genotypes with contradictory information in passport data from different origins. Similarly, the specific country of origin could be attributed to 1,550 genotypes out of the 1,653 geographically assigned. It is important to note that the European region or country of origin assigned to a genotype was independent from the location of the collection where the sampled accession was maintained, since many collections contained accessions from various origins.

Genetic Diversity Across and within European Regional Groups

The 16 SSR markers amplified a total of 369 alleles across the 1,859 apple accessions used for diversity analysis, ranging from 17 (CH02c09 and CH05f06) to 35 (CH02c06) alleles per locus. The average number of alleles per locus was 23.06, whereas the mean effective number of alleles per locus was

6.59 (Table 1). High average number of alleles per locus and almost identical mean effective number of alleles per locus were noted for the three geographical regions of origin of the germplasm. Allelic richness was normalized to the smallest group (i.e., North+East) to avoid a group size-dependent bias of results. Overall, the results obtained for the material of the three designated regions of origin suggested the existence of a high and relatively homogeneous allelic diversity across Europe (Table 1). Within the 369 alleles identified in the overall set (i.e., across Europe), 73.4 % and 52.0 % were found at frequencies below 5 % and 1 %, respectively (Table 1; data not shown for 1 %). A similar proportion of rare alleles was obtained for the material from the three designated geographical regions of origin, with the exception of alleles detected at a frequency<1 % with Northern+Eastern and Southern European origins, for which slightly lower percentages were identified (≈38 %). Almost identical mean *He* values were obtained for the overall dataset (0.83) and for the germplasm from each of the three geographical groups (Table 1). Cross-comparison of the allelic composition for the accessions classified into geographic categories showed that 221 out of the 362 alleles (seven alleles appeared only in accessions that could not be classified into geographic groups) were detected in all three geographical groups, 59 alleles (16.3 %) were identified in two geographic groups only, whereas 82 alleles (22.6 %) were specifically found only in one geographic group (i.e., private alleles). At the national level (i.e., countries of origin of the unique genotypes), some countries exhibited a higher rate of private alleles than others: especially, genotypes assigned to Switzerland, Italy and Russia harboured 15, 14 and 14 private alleles (respectively), genotypes from Spain and France harboured 7 private alleles each, whereas genotypes from the Netherlands, Belgium, Great Britain or Sweden had a maximum of one private allele. The pattern of distribution of the frequent alleles (frequency>0.05) between Southern, Northern+Eastern and Western germplasm was analyzed for each locus separately using Chi^2 tests. Highly significant differences in the allelic distributions ($P<0.001$) were found between all the geographic groups for all markers except for the CH-Vf1 locus when comparing Southern and Western germplasm (data not shown).

Table 1: Average measures of genetic diversity at two different levels: overall set of accessions and according to the three geographical regions of origin (Northern+Eastern, Southern and Western)

Material	N_A	N_B [a]	N_E	A_R [b]	Ho	He
Overall set (1859 genotypes)	23.06	16.94	6.59	-	0.81	0.83
European regions of origin						
Northern + Eastern Europe	16.75	10.87	6.24	16.57	0.83	0,82
Southern Europe	17.50	11.87	6.29	16.95	0.81	0.82
Western Europe	20.31	13.94	6.18	16.36	0.81	0.82

[a]Rare alleles were considered if they appeared in a frequency below 5 %

[b]For the geographical European regions of origin, allelic richness was computed after normalization according to the smallest population size (i.e., Northern+Eastern Europe). Number of alleles per locus (N_A), number of rare alleles (N_B), effective number of alleles (N_E), allelic richness (A_R), and observed (Ho) and expected (He) heterozygosity are included

Genetic Structure and Differentiation

A Bayesian model-based clustering method was applied to the 1,859 unique diploid genotypes in order to elucidate the underlying genetic structure at a European scale. The analysis of Evanno's ΔK statistic indicated unambiguously $K=3$ as the most likely level of population stratification (Fig. 1 a1). The mean proportion of ancestry of the genotypes to the inferred groups was 0.81. Using the threshold of $qI \geq 0.80$ to define strong assignments to groups, 1,175 genotypes (63 %) were identified as strongly associated to a group. This partitioning level corresponded to an asymmetric division of the material into three groups: K1 composed of 506 genotypes, K2 containing 401 genotypes, and K3, the largest group, comprising 952 genotypes. Diversity estimates revealed high levels of allelic variation within each group, with allelic richness ranging between 16.0 (K3) and 18.6 (K1) (Table 2). Genetic discrimination between the three groups was confirmed through a multivariate Principal Coordinate Analysis (PCoA) (Fig. 2). In the bi-dimensional plot, K1 was located mostly to the left of the Y axis, and K2 mostly below the X axis, while K3 occurred to the right of the Y axis and mostly above the X axis. A Neighbor-joining tree also showed three different main clusters (Fig. 3), supporting the identification of the three groups by the Bayesian method.

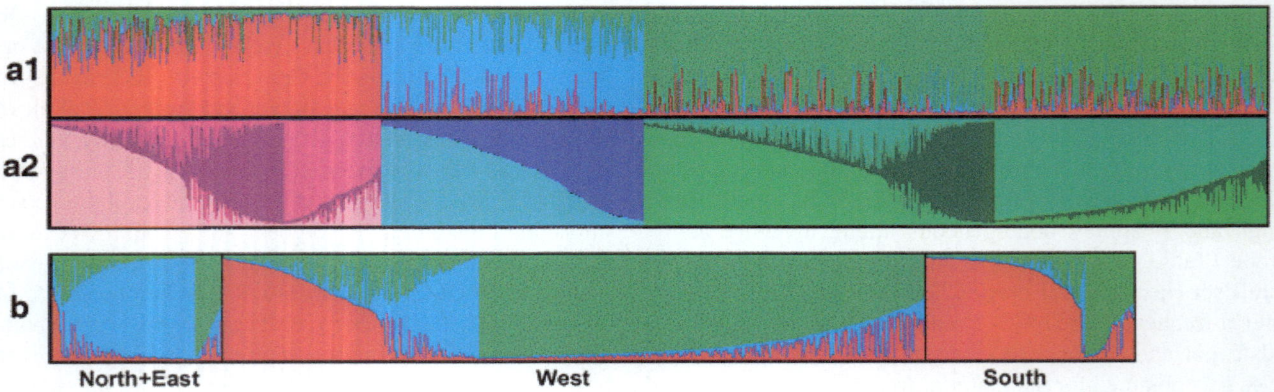

Figure 1: Graphical display of the results of the Structure analyses. **a1**) Proportions of ancestry of 1859 unique diploid apple genotypes for $K=3$ groups inferred with Structure v.2.3.4 software [67]. Each genotype is represented by a vertical bar partitioned into $K=3$ segments representing the estimated membership fraction in three groups. The three groups are depicted using the following color codes: Red=group K1; Blue=group K2; Green=group K3. **a2**) Proportions of ancestry of the same 1859 genotypes following a nested Structure analysis within each previously defined group. For K1 and K3 three subgroups are shown and for K2 two subgroups are shown. Each genotype is represented by a vertical bar partitioned into $K=2$ or 3 subgroups representing the estimated membership fraction in each subgroup. Genotypes are presented in the same order than in a1. The subgroups are depicted using the following color codes: light Pink=K1.1; Purple=K1.2; dark Pink=K1.3; light Blue=K2.1; dark Blue=K2.2; fluorescent Green=K3.1; dark Green=K3.2; light Green=K3.3. **b**) Proportions of ancestry of 1653 unique diploid apple genotypes with known European region of origin for $K=3$ groups inferred with the same Structure analysis as in a. The genotypes are sorted according to their European region of origin (North+East, West, and South)

Table 2: Descriptive information for each of the three major groups and eight subgroups of genotypes identified by the Bayesian model-based clustering method

	Number of genotypes in the group/ subgroup		**He**	**Number of alleles**				**Allelic richness**
Group/Subgroup	**Number Genotypes**	**Frequency of genotypes with $qI \geq 0.8$**		**Total**	**Private**	**Unique**	**A**	
K1	506	60 %	0.823	307	34	16	19.19	18.63
K2	401	57 %	0.816	287	23	15	17.94	17.76
K3	952	67 %	0.801	294	22	14	18.36	15.99
K1.1	209	42 %	0.842	282	17	12	17.63	16.38
K1.2	149	54 %	0.789	215	3	1	13.44	13.20
K1.3	148	36 %	0.761	228	6	3	14.25	13.86
K2.1	244	48 %	0.818	268	14	11	16.75	14.73
K2.2	157	53 %	0.778	211	5	4	13.19	12.67
K3.1	375	41 %	0.775	242	7	6	15.13	12.32
K3.2	162	57 %	0.760	171	0	0	10.69	10.31
K3.3	415	51 %	0.809	255	14	8	15.94	13.43

Summary statistics include the partitioning of number of individuals in each group, expected heterozygosity (He), total, private, unique, and average number of alleles (A). Allelic richness is scaled to the smallest group (K2; $N=401$) or subgroup (K1.3; $N=148$)

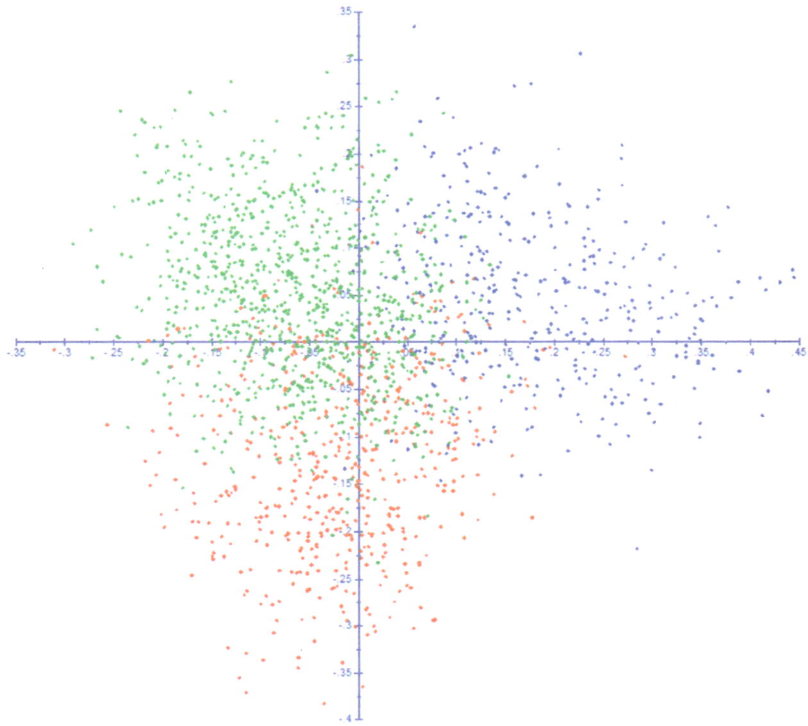

Figure 2: Scatter plot of the Principal Coordinate Analysis (PCoA) of the 1859 apple accessions based on the 16 SSR data. The three groups are depicted using the following color codes: Red=group K1; Blue=group K2; Green=group K3.

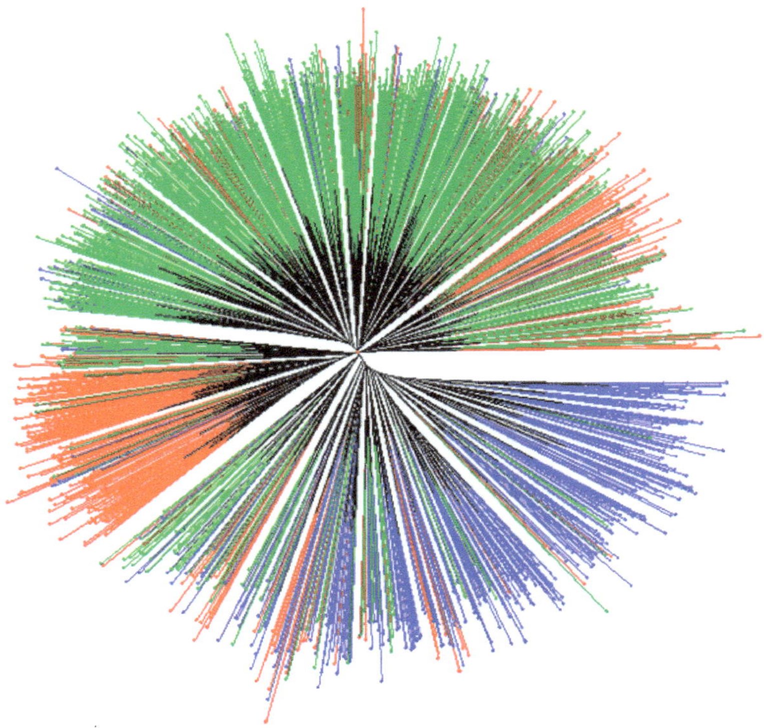

Figure 3: Neighbor-joining dendrogram based on simple matching dissimilarity matrix calculated from the dataset of 16 SSR markers for the 1859 genotypes clustered in the three groups revealed by the Bayesian model-based clustering method. The three groups are depicted using the following color codes: Red=group K1; Blue=group K2; Green=group K3.

The genetic differentiation between the three designated geographic regions of origin was low ($F_{ST}=0.021$, $P<0.001$, Table 3), suggesting a weak genetic structure for this crop at a European scale in terms of geographical origin. The level of genetic differentiation between the three groups inferred by Structure was only slightly higher ($F_{ST}=0.031$, $P<0.001$). The largest differentiation between pairs of groups was found between Northern+Eastern and Southern germplasm ($F_{ST}=0.042$, $P<0.001$), whereas much lower F_{ST} values were found between the Western and each of the Northern+Eastern ($F_{ST}=0.023$, $P<0.001$) and Southern ($F_{ST}=0.015$, $P<0.001$) materials.

Table 3: Analysis of molecular variance (AMOVA) based on the 16 SSR loci of the apple germplasm evaluated in this study corresponding to three regions of origin (Northern+Eastern, Southern and Western Europe) and groups and subgroups defined by Structure analysis.

Populations	*df* [a]		Variance components (%)		
	W [b]	*A* [c]	*W*	*A*	*p* value
3, geographic origins	1653	2	97.9	2.1	0.001
3, groups defined by Structure	1859	2	96.9	3.1	0.001
3, subgroups of K1	506	2	96.3	3.7	0.001
2, subgroups of K2	401	1	96.6	3.4	0.001
3, subgroups of K3	952	2	97.3	2.7	0.001
8, subgroups (K1+ K2+ K3)	1859	7	95.6	4.4	0.001

[a] *df* degrees of freedom

[b] *W* within populations

[c] *A* among populations

The relationship between membership of accessions within the three groups defined by Structure and their geographical regions of origin was also analyzed. 80 % and 75 % of the accessions from Northern+Eastern and Southern Europe clustered in K2 and K1 respectively. The relationship between the material with Western European origin and the third group (K3) was less evident (63 %), but still visible by comparison (Fig. 1b). Although the genetic differentiation revealed between the three groups defined by Structure was not very high, the existence of a relationship between the grouping by geographical regions of origin of the accessions and the three inferred groups is noteworthy. Furthermore, when considering the specific country of origin attributed to the cultivars, the distribution within the three Structure-defined groups appears to follow a clear gradient from North(East) to South of Europe (Fig. 4); the cultivars from Northern Europe and Russia were mainly assigned to the K2 group and the Spanish and Italian cultivars were mainly assigned to the K1 group, with intermediate patterns found for those countries located at the interfaces of the broad regions.

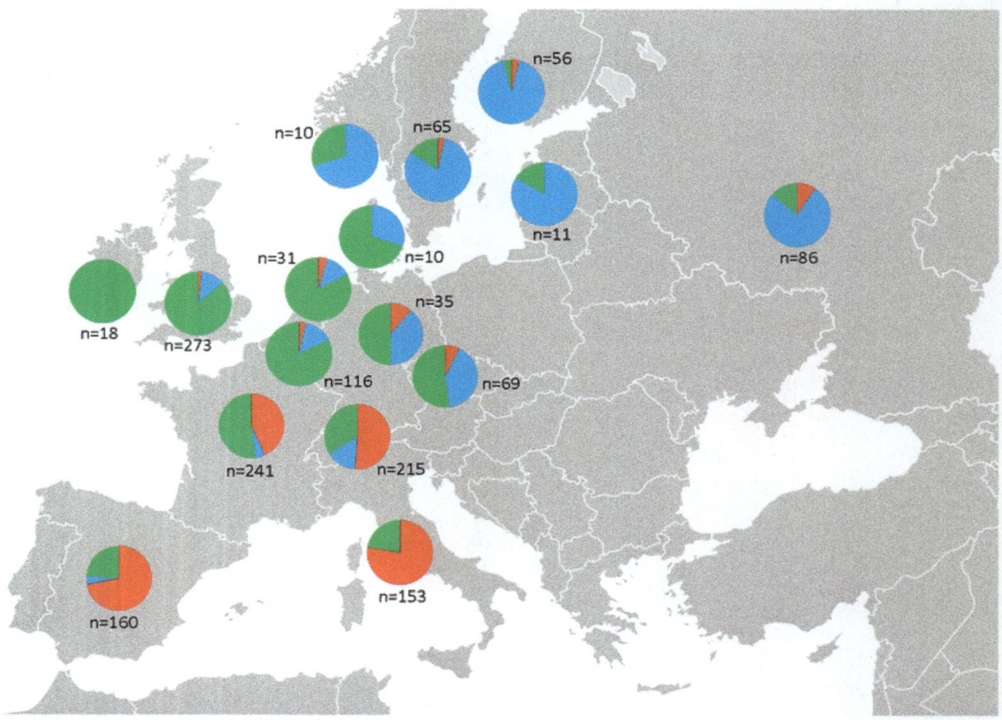

Figure 4: Genetic composition of the groups of cultivars clustered by country of origin for $K=3$ groups inferred with Structure. For the detailed country list. The pies represent the proportion of each group in each country; color codes are as per Fig. 1 a1.

Within the admixed accessions (i.e., $qI<0.8$) for which the geographical regions of origin (Northern+Eastern, Southern and Western) was known, we defined a membership coefficient threshold ($qI<0.55$) with the aim of identifying genotypes unambiguously in admixis, in order to examine whether a supplemental relationship could be found between geographical region and grouping by Structure for the admixed material. For the unambiguously admixed material (i.e., $qI<0.55$) of Southern European origin, the average proportion of ancestry (qI) was 0.45 to K1 (the group mostly associated with material from Southern Europe), followed by 0.42 to K3 and 0.13 to K2, the groups mostly composed by material from Western and Northern+Eastern European origins, respectively (data not shown); a slightly less pronounced, but complementary, pattern was observed for the unambiguously admixed germplasm (i.e., $qI<0.55$) of Northern+Eastern Europe with average proportions of ancestry of 0.43, 0.35 and 0.22 to K2, K3 and K1, respectively. For the unambiguously admixed material (i.e., $qI<0.55$) of Western origin the average proportion of ancestry to each of these three groups was almost identical (approximately 1/3). This result was in line with the lower F_{ST} values found between the groups K1/K3 ($F_{ST}=0.024$, $P<0.001$) in comparison with the slightly higher differentiation between the groups K1/K2 ($F_{ST}=0.039$, $P<0.001$) and K2/K3 ($F_{ST}=0.036$, $P<0.001$). The dispersion of the three groups in the PCoA plot was also in agreement with these results, showing the highest overlap between K1 and K3 followed by K2 and K3.

Nested-Bayesian Clustering Approach: Substructuring of the Diversity

In order to investigate the substructuring of the diversity within each of the three groups identified in the initial analysis we used a nested application of the Structure software. To do this, the three groups were analyzed independently. To evaluate the strength of the hypothetical subdivisions (i.e., subgroups) within each group, simulations for each K value were examined, paying attention to the internal consistency between the runs, the mean proportion of ancestry of accessions within each subgroup, and the proportion of accessions unequivocally assigned ($qI\geq 0.80$).

The analysis of the relationships between K and ΔK for K1 suggested a probable subdivision of this material into three subgroups and the assignment of genotypes was well correlated between runs. The average proportion of ancestry for the accessions clustered in the three subgroups of K1 was 0.75, with 44 % of the accessions showing strong assignments. Two subgroups for K2 and three for K3 were similarly established. In both cases, the assignment of genotypes was well correlated between runs, and almost identical average proportions of ancestry to those for the subgroups of K1 were obtained with slightly higher proportions of strongly assigned accessions (47 % and 50 % respectively). Secondary peaks at other K values were also explored but these subdivisions had less statistical support (data not shown). Therefore, we adopted eight subgroups as the most suitable partitioning degree of substructuring (Fig. 1 a2). For these eight subgroups the affinity of almost half of the individuals (47 %) to their respective subgroups was strong and the assignment of admixed accessions was consistent between runs. The examination of the eight subgroups showed considerable differences in size, ranging from 148 (K1.3) to 415 (K3.3) genotypes, and variable proportion of accessions strongly assigned to the inferred subgroups (Table 2). K3.2 was the subgroup with the highest proportion of strongly assigned genotypes (57 %), whereas K1.3 had the highest proportion of admixed accessions. The proportion of accessions unambiguously assigned for the remaining six subgroups ranged from 41 % to 54 %, whereas the mean proportion of ancestry for the accessions clustered in each one of the eight subgroups was very stable (≈0.75).

The analysis of the relationship between the different subgroups and the putative countries of origin of the germplasm indicated potentially interesting correlations, especially for groups K1 and K3. About 70 % of the subgroup K1.2 consisted of germplasm originating from Spain. Similarly, 46 % of the subgroup K1.1 and 50 % of the subgroup K1.3 consisted of germplasm originating from Switzerland and Italy, respectively; the latter subgroup was also composed of a further 39 % of the cultivars with a French origin and interestingly, a significant proportion of these were attributed to Southeastern France (data not shown). The disentangling of the substructuring pattern therefore allowed not only the dissection of the internal distribution of the diversity within group K1, but also the detection of three subgroups strongly associated with some particular countries of origin. With respect to the collections from the Northern+Eastern part of Europe (Sweden, Finland and Russia), no clear differentiation of the germplasm in the two subgroups of K2 was observed. For the subgroup K3.1, about half of the germplasm consisted of cultivars from either the United Kingdom or France. All of the 40 cultivars selected in the French collection as being recently bred, clustered in a single small subgroup (K3.2) which was mostly composed of English, US and, perhaps more surprisingly, Spanish cultivars. Major standard cultivars such as 'Golden Delicious', 'Red Delicious', 'Jonathan' and 'Ingrid Marie' were also assigned to this subgroup, as well as 'Cox's Orange Pippin' and 'James Grieve'. Interestingly, most of the 40 cider apple cultivars (87 %) were assigned to one subgroup (K3.3) which was mostly composed of French, English, and Swiss cultivars. The other standard cultivars were assigned to the latter subgroup and to subgroup K3.1.

Genetic diversity estimates were calculated for all the subgroups obtained by the nested Bayesian model-based clustering (Table 2). While He ranged from 0.76 (K1.3 and K3.2) to 0.84 (K1.1), indicating a high level of heterozygosity contained in all the subgroups, the percentage of alleles represented in each one of the eight subgroups was very variable, ranging from 46 % (K3.2) to 76 % (K3.1). Some private alleles were identified in all subgroups except for K3.2. They were most abundant in K1.1, but a considerable number of them were found also in K3.3 and K2.1. Most of the private alleles (approx. 72 %) were also unique as they were identified in only one accession. To properly evaluate the allelic diversity between the eight subgroups, we applied a rarefaction approach to compensate for the differences in subgroup size. The allelic richness obtained for the eight subgroups supported the previous results, confirming the highest diversity in K1.1 and the lowest diversity in K3.2.

Estimates of genetic differentiation showed that only 3.7 % (K1) and 3.4 % (K2) accounted for variation among subgroups within groups (Table 3). The genetic differentiation between the subgroups into which K3 was subdivided was considerably lower (Table 3). Considering the eight subgroups obtained by the overall Nested Bayesian approach, the results showed that variation among subgroups accounted for 4.4 % of the total variation. Regarding the F_{ST} pairwise tests between subgroups (Table 4), irrespective of whether they belonged to the same group or not, the highest F_{ST} corresponded to the pair K1.3/K2.2 ($F_{ST}=0.087$, $P<0.001$), followed by K1.3/K2.1 ($F_{ST}=0.077$, $P<0.001$), and the lowest to the pairs K1.1/K3.3 ($F_{ST}=0.016$, $P<0.001$) and K3.1/K3.3 ($F_{ST}=0.023$, $P<0.001$).

Table 4: Pairwise estimates of F_{ST} among the eight subgroups obtained by the nested Bayesian clustering approach

Subgroup	K1.1	K1.2	K1.3	K2.1	K2.2	K3.1	K3.2	K3.3
K1.1	—							
K1.2	0.030	—						
K1.3	0.035	0.051	—					
K2.1	0.028	0.067	0.077	—				
K2.2	0.049	0.076	0.087	0.035	—			
K3.1	0.034	0.051	0.061	0.061	0.055	—		
K3.2	0.051	0.065	0.070	0.070	0.058	0.029	—	
K3.3	0.016	0.042	0.060	0.038	0.051	0.023	0.038	—

All the estimates were highly significant ($P<0.001$)

Parentage Reconstruction

Two-parents-offspring relationships within the 1,859 diploid genotypes were explored using CERVUS software. A total of 46 putative trios (offspring and two inferred parents) were identified with high (95 %) confidence level. These consisted of two already documented trios, ('Calville Rouge du Mont Dore' and 'Belle de Mleiev' and their parents; [23]), as well as another 10 recent and 34 old cultivars (Table 5). The two parents of the 10 modern cultivars, for which full parentage was already documented were correctly inferred (e.g., 'Heta', 'Jaspi' and 'Pirkko'='Lobo' x 'Huvitus', 'Pirja'='Huvitus' x 'Melba', or 'Mio'='Worcester Pearmain' x 'Oranie'). In most cases, the two parents of the older cultivars were not known and thus newly inferred (Table 5). Inferred parentage was found for old cultivars from various European countries (6x for Italy; 4x for Great Britain, Switzerland, Czech Republic, and Sweden; 3x for Germany and Spain; 2x for Belgium). Perhaps unsurprisingly, some accessions were more frequently inferred as parents, such as the two French cultivars 'Reine des Reinettes' (= 'King of the Pippins') or 'Transparente de Croncels' which were each identified three times. Geographic convergence of parentage was frequently observed (e.g., 'Kramforsäpple'='Sävstaholm' x 'Åkerö', all three from Sweden; 'Beauty of Moray'='Keswick Codlin' x 'Stirling Castle', all three from Great Britain; 'Roja de Guipuzcoa'='Urte Sagarra' x 'Maxel Gorri', all three from Spain; or 'Scodellino'='Abbondanza' x 'Decio', all three from Italy). But hybridizations between cultivars from distant countries were also observed (e.g., 'Rotwiler' presumably from Switzerland='King of the Pippins' x 'Alexander' from France and Ukraine, respectively; or 'Godelieve Hegmans' from Belgium='Red Astrakan' x 'Transparente de Croncels' from Russia and France, respectively). It should be noted that the female and male status of the inferred parents could not be specified from the available SSR markers.

Table 5: Full parentages of 46 apple cultivars inferred within the set of the 1859 apple unique accessions using 16 SSR markers with their accession codes, accession names (AcceNumber), their duplicate codes according to the SSR profile (FBUNQ) and their putative country of origin (OriginHist)

Offspring ID	Accename	FBUNQ	Orig-inHist	First candidate ID	Accename	FBUNQ	OriginHist	Second candi-date ID	Accename	FBUNQ	Origin-Hist	Status[j]
X1618	Calville Rouge du Mont Dore	963	FRA	BAL086	Alexander	30	UKR	DCA_I05	Mele Ubriache[a]	361	FRA	doc.
X1846	Belle de Mleiev	1563	-	X0557	Mc Intosh	508	CAN	1957218	King of the Pippins	37	FRA	doc.
BAL035	Heta	1774	FIN	CRAW-0433	Lobo	788	CAN	FIN09	Huvitus	4922	FIN	recent
BAL039	Jaspi	1776	FIN	CRAW-0433	Lobo	788	CAN	FIN09	Huvitus	4922	FIN	recent
FIN18	Pirkko	4930	FIN	CRAW-0433	Lobo	788	CAN	FIN09	Huvitus	4922	FIN	recent
BAL010	Rödluvan	107	SWE	CRAW-0433	Lobo	788	CAN	BAL023	Barchatnoje	1768	RUS	recent
BAL109	Arona	1819	LVA	CRAW-0433	Lobo	788	CAN	BAL112	Iedzenu	1822	LVA	recent
BAL176	Nyckelby	1861	SWE?	CRAW-0433	Lobo	788	CAN	1957188	Cox's Pomona	2033	GBR	recent?
BAL059	Pirja	444	FIN	FIN09	Huvitus	4922	FIN	CRAW-0836	Melba	167	CAN	recent
FIN43	Pirkkala	4949	FIN	BAL042	Kaneläpple	512	RUS	FIN14	Lavia	4926	FIN	recent
BAL154	Mio	543	SWE	CZ_G2D_0045	Worcester parména	550	GBR	BAL056	Oranie	48	SWE	recent
BAL052	Oberle	1784	CAN	BAL027	Early Red Bird	236	CAN	CRAW-0266	Stark Earliest	468	USA	old
BAL091	Förlovningsäpple	1804	SWE	CHE0893	Heuapfel	1248	CHE	X1646	Saint Germain	31	-	old
BAL167	Valldaäpple	1853	SWE	CHE0893	Heuapfel	1248	CHE	BAL179	Göteborgs Flick-äpple	1863	SWE	old
BAL099	Kramforsäpple	1811	SWE	BAL161	Sävstaholm	573	SWE	BAL195	Åkerö	308	SWE	old
BAL158	Stenkyrke	463	SWE	BAL171	Fullerö	1857	SWE	CZ_LJ_0045	Malinové podzimní[b]	722	POL	old
FIN07	Finne	4920	FIN	BAL161	Sävstaholm	573	SWE	FIN08	Grenman	4921	FIN	old

1942035	Beauty of Moray	1925	GBR	2000053	Keswick Codlin	1438	GBR	2000090	Stirling Castle	2103	GBR	old
1951242	Brighton	2011	NZL?	X4915	Red Dougherty	939	NZL	CZ_LC_0411	Hlaváčkovo[c]	23	USA	old
1957208	Ben's Red	2035	GBR	CRAW-0020	Devonshire Quarrenden	622	GBR	1955077	Box Apple	2025	GBR	old
1965004	Fred Webb	2054	GBR	1946088	Winter Marigold	324	GBR	1957181	Gascoyne's Scarlet	45	GBR	old
2000083	Rivers' Early Peach	2099	GBR	2000051	Irish Peach	2093	IRL	BAL169	Aspa	1855	SWE	old
BMN0011	Roja de Guipuzcoa	3854	ESP	BMN0017	Urte Sagarra	956	ESP	BMN0171	Maxel Gorri	3896	ESP	old
BMZ016	Cella	3935	ESP	BMN0022	Erreka Sagarra	957	ESP	X5102	Bisquet	535	FRA	old
BMN0070	Madotz-01	3869	ESP	1957218	King of the Pippins	37	FRA	X7201	Transparente de Croncels	62	FRA	old
CHE1322	Rotwiler	1271	CHE?	1957218	King of the Pippins	37	FRA	BAL086	Alexander	30	UKR	old
CHE1788	Roseneggler	3718	CHE	1957218	King of the Pippins	37	FRA	CZ_BoN_0429	Trat. Laze	2284	CZE	old
CHE0032	Ernst Bosch	1003	DEU	1947074	Ananas Reinette	69	NLD	CZ_GF_0415	Evino[d]	7	GBR	old
CHE0168	Eibner	3258	CHE	CRAW-0836	Melba	167	CAN	CZ_BoN_0424	Trevínské červené[e]	71	USA	old
CHE1390	Klefeler	3589	CHE	KRAS123	Papirovka	25	RUS	X7199	Rose de Berne	83	CHE	old
CRAW-0226	Laubain n°1	2126	BEL	CRAW-0086	Bismarck	3	AUS	CZ_GS_0478	Ušlechtilé žluté[f]	90	GBR	old
CRAW-0105	Godelieve Hegmans	2116	BEL	BAL175	Röd Astrakan	82	RUS	X7201	Transparente de Croncels	62	FRA	old
CZ_BB_0442	Nathusiovo	2268	DEU	CZ_GL_0464	Bláhovo Libovické	2311	CZE	X7201	Transparente de Croncels	62	FRA	old
CZ_BB_0434	Panenské veliké	2265	CZE	CZ_GP_0469	Panenské české	1529	CZE	X1344	Reinette de Landsberg	61	DEU	old
CZ_GK_0412	Proche	2308	CZE	CRAW-0425	Calville Rouge d'Automne	13	FRA	X1344	Reinette de Landsberg	61	DEU	old
CZ_BoN_0421	Moravcovo	2283	CZE	CZ_GP_0469	Panenské české	1529	CZE	CHE0269	Pomme Bölleöpfel	1377	-	old
CZ_GL_0456	Bláhův poklad	694	-	CZ_GL_0464	Bláhovo Libovické	2311	CZE	CZ_GG_0438	Malinové hornokrajské[g]	47	NLD	old
CZ_GP_0473	Petr Broich	2321	DEU	1957175	Annie Elizabeth	15	GBR	2000075	Peasgood's Nonsuch	51	GBR	old
CZ_BB_0458	Šarlatová parména	2269	CZE	CZ_GG_0442	Malinové holovouské	452	CZE	X8233	Petite Madeleine	24	-	old
CZ_BB_0466	Podzvičinské[h]	231	-	X0691	Boiken	108	DEU	X1071	Reinette de Caux	629	NLD	old
DCA_017	S.Giuseppe	1646	ITA	DCA_090	Abbondanza	327	ITA	DCA_C44	Rambour Frank (MI)	493	FRA	old
DCA_H03	Scodellino	1642	ITA	DCA_090	Abbondanza	327	ITA	DCA_E52	Decio	397	ITA	old
DCA_E72	Gelato Cola	330	ITA	DCA_E69	Gelato (CT)	780	-	DCA_F74	Limoncella (TN)[i]	708	ITA	old
DCA_H62	Liscio di Cumiana	1713	ITA	DCA_H29	Carla	114	-	DCA_C21	Renetta di Grenoble	263	ITA	old
DCA_I96	Ros Magior	1658	ITA	DCA_I80	Rus d' Muslot	321	-	X1115	Rome Beauty	334	USA	old
DCA_F47	Mela Golden Simile di Villa Collemandina	1692	ITA	DCA_A20	Rosa Mantovana (TN)	101	ITA	CRAW-0025	Yellow Bellflower	77	USA	old

[a]DCA_I05 'Mele Ubriache' duplicate with 'Calville Rouge d'Hiver' [23]

[b]based on 11 SSR [64] the accession CZ_LJ_0045 'Malinové podzimní' was shown to be duplicated with 'Danziger Kantapfel'

[c]based on 11 SSR [64] the accession CZ_LC_0411 'Hlaváčkovo' duplicate with 'Northen Spy'

[d]based on 11 SSR [64] the accession CZ_GF_0415 'Evino' duplicate with 'Mank's Codlin'

[e]based on 11 SSR [64] the accession CZ_BoN_0424 'Trevínské červené' duplicate with 'King David'

[f] based on 11 SSR [64] the accession CZ_GS_0478 'Ušlechtilé žluté' duplicate with 'Golden Noble'

[g]based on 11 SSR [64] the accession CZ_GG_0438 'Malinové hornokrajské' duplicate with 'Framboise'

[h]based on 11 SSR [64] and on 13 SSR [14] the accession CZ_BB_0466 'Podzvičinské' duplicate with 'Altlander Pfannkuchenapfel' and 'Thurgauer Kent'

[i]based on 11 SSR [64] the accession DCA_F74 'Limoncella' (TN) duplicate with 'Cola'

[j]recent or old cultivars ; doc. = inferred parentage already documented in [23]

DISCUSSION

Identification and Redundancy

The exchange of genotyping data between research units has increased considerably in recent years, with the aim to investigate the extent and distribution of diversity for specific crops at a wide geographic scale. In this study, the application of a common set of 16 SSR markers on a wide set of dessert apple cultivars distributed across three broad European regions allowed the detection of redundant accessions and duplicated genotypes between and within collections, and the description of the structuration of a significant part of the European apple diversity. Cross-comparison of SSR data in attempts to combine datasets from multiple sources has often been problematic due to challenges in harmonizing the allelic sizes between different laboratories [18, 29, 30]. By combining existing data over numerous shared reference accessions in our collections with the re-genotyping of a subset of the accessions, we were able to strongly secure the SSR allele adjustment over sites. This dataset represents a highly valuable resource for the comparison of apple germplasm collections throughout Europe and the rest of the world. Taking into consideration the rich allelic diversity present in the European apple germplasm, it would be useful to identify a relatively small set of varieties that offer a good representation of the allelic variability identified in this germplasm to act as an internal control (i.e., a reference set) between laboratories for future use.

Interestingly, duplicate groups involving accessions from different collections underlined some putative drift in the cultivar denomination. Some good examples were 'Pott's seedling' and 'Pottovo' (FBUNQ14), or 'Signe Tillish' and 'Signatillis' (FBUNQ34). In addition, 'sports' are often given derivative names (e.g., 'Crimson Peasgood' as a sport of 'Peasgood's Nonsuch') but the current analysis was not set up to distinguish between clones and 'sports' of cultivars with potential morphological differences. Many likely errors in denomination of genotypes were also detected when multiple representatives of a given cultivar were detected within a group, but a single supposed representative was obviously outside of the group and was often associated with representatives of a different cultivar. For example, 'Drap d'Or' and 'Chailleux' (FBUNQ92) are known to be synonyms used in France for the same cultivar, and accession DCA_D35 'Drap Dore', which was found to belong to the group FBUNQ50, was most likely a denomination error since almost all other members of this group were 'Winter Banana'. In other cases, accessions with uncertain denomination could be resolved, such as CRAW-1858 'Reinette Baumann?' (FBUNQ21) and accession CRAW-1108 'Peasgood Nonsuch?' (FBUNQ51) for which the molecular analyses confirmed that they were most likely 'true-to-type' cultivars. The question of 'trueness-to-type' is a major issue in apple germplasm management where extensive budwood exchange between regions and countries has occurred for centuries. Indeed, an erroneously denominated accession can be transmitted from collection to collection for years, such that a large number of representatives within a duplicate group (as per the present study) should not always be considered definitive proof of the trueness-to-type of accessions but this objective evidence is extremely valuable in highlighting issues to resolve. Since genebank curators have often collected material of old cultivars from private gardens or from tree pasture orchards, unidentified or misidentified material can later be detected either by classical phenotypic characters and/or by using genetic markers. As an example, this study showed that an old so called local cultivar 'Madame Colard' (CRAW-0365 – FBUNQ72), described to have been raised in 1910 by the nurseryman Joseph Colart at Bastogne (Belgium), exhibited the same SSR profile as the old English apple cultivar called 'Royal Jubilee' (UK-NFC 2000085) raised already in 1888. Further comparison with historical descriptions could conclude that they are the same cultivar. Additional insights from the passport data of accessions would be needed to help in tracing the transmission of the material from collection to collection and pomological characterization will be required to compare accessions to published descriptions of the variety. This will remain a task for the curators of collections, in order to improve curation of germplasm in a coordinated way.

It is important to note that the criteria used to select the accessions at the country-level were not always the same. For instance, the INRA and UNIBO material corresponded to former "core collections" built to encompass a large variability not restricted to the national/local accessions [23, 24]. Conversely, the UK-NFC and FRUCTUS material was restricted to older diploid accessions considered to derive from UK and Switzerland, respectively. A similar, despite less stringent situation was applied also for CRA-W, RBIPH, SLU, and the Spanish accessions (UPNA, UDL and EEAD). For MTT, NCRRI, VNIISPK, and KNAU, the national representativeness was more limited and strictly restricted to accessions considered to be emblematic landrace cultivars. The germplasm was thus somewhat heterogeneous in nature, but still allowed a broad examination of the European dessert apple diversity. In the future, it will be useful to enlarge the dataset to include additional accessions from the collections considered here as well as other European collections [11,31] or collections from other regions worldwide [32, 33, 34] to provide a wider perspective on genetic resource conservation of apple worldwide.

Genetic Diversity

The high level of diversity and heterozygosity in apple germplasm at a European level agreed with previous results obtained at collection-scale in several European countries, e.g., Italy [24], Spain [26], France [23], Sweden [22], Czech Republic [25] or Switzerland [14]. The large diversity found is consistent with the weak bottleneck effect reported in connection with the domestication of this species [35, 36, 37]. Probably a combination of factors are involved: i) vegetative propagation methods that have been adopted since ancient times favoring the dispersal of cultivars across geographic regions [38, 39], ii) forced allogamy due to the self-incompatibility system of *Malus* × *domestica* [40], iii) multiple hybridization events at each geographical region combined with human activities, e.g., selection and breeding [36, 37] and, iv) diversifying selection associated with adaptive criteria for the subsistence in diverse agricultural environments [41, 42]. Interestingly, the distribution of private SSR alleles over the countries of origin of the unique genotypes was somewhat unbalanced at the European level with much higher occurrences in genotypes assigned to Switzerland, Italy or Russia than in genotypes originating from Northern-

Western Europe. Whilst these findings should be considered with caution because of possible biases linked to the initial sampling or to the size differences of the genotype sets, this study underlines that accessions originating from Southern Europe and Russia could be expected to bring original genetic diversity into modern breeding programs especially for traits related to more extreme climate adaptation. Overall, the highly diverse germplasm studied here contains much more genetic variation than do modern apple cultivars, many of which having been selected for optimal performance within a narrow range of environmental conditions [5, 37, 42].

Coordinated Actions: a Key Point for Better Knowledge of the Resources Conserved

This large-scale analysis in apple germplasm constitutes a good example of the efficiency and value of coordinated international actions to enhance the knowledge of diversity conserved at a European level. The results obtained offer a valuable step to undertake actions to coordinate European resources towards optimizing the management of apple germplasm across Europe in line with the aspirations of the European Cooperative program on Plant Genetic Resources (ECPGR). The results also offer a potential starting point that may open new opportunities for apple breeding in the near future. All breeding advances are built upon the diversity available, and a key role of the germplasm collections is to help safeguard natural forms of genetic variation and to make them accessible to plant biologists, breeders, and other key users [15]. The extensive germplasm evaluated in this study consisted mainly of old and/or locally grown accessions across Europe, many of which remain underutilized in cultivation or breeding programs. The preservation of traditional cultivars in living germplasm collections must be regarded as an invaluable reservoir of insufficiently explored genetic diversity that may become useful for apple breeding in a near future, and the establishment of coordinated genetic data is hoped to increase the accessibility of this material to breeding programs. From the perspective of modern-day fruit production, most of these old varieties would now be considered as obsolete since they are not particularly well-adapted to current agricultural practices and marketing. Nevertheless, this material should be considered as a reservoir of potentially interesting genes to be used for further improvement. This is particularly relevant in a crop like apple, for which the current production is highly dependent on a very limited number of cultivars with a narrow genetic basis for the bulk of current production [5]. As an example, it can be mentioned that 50 % of the commercially marketed apple production in the European Union consists of only four cultivars, 'Golden Delicious', 'Gala', 'Idared' and 'Red Delicious' [43]. The low diversity of the subset of elite cultivars used for commercial production during recent decades is likely to result in a bottleneck hampering future genetic improvement [37]. The recognition of this situation should encourage the establishment of coordinated actions across different levels (regional, national and international scales) to define strategies for the efficient conservation of the genetic resources of this species.

Genetic Structure: Major Divisions and Sub-structuring of the Diversity

The attribution of country of origin to traditional cultivars can be a matter of endless debate, especially for those dating back two-three centuries or more. Initial descriptions in pomologies and booklets can be subject to errors in denomination confused by historical distribution and renaming, resulting in synonymy, as well as the re-use of old names for more recent findings or misidentifications. This is less problematic for the better known old cultivars as many of them have been widely documented and monitored over years in several countries. However, for local cultivars and/or landraces where less information is available, the correct attribution can be complicated, especially between neighboring countries. It is also worthy of note that the 'country of origin' relies on a political construct, which can be prone to significant change within the potential lifetime of many varieties of apple (and other long lived perennial crops). Therefore, we first used a conservative approach and discussed our findings in terms of three broad European regions of origin. Then, we analyzed the structuration at a country-scale, but noting that the exact attribution of a given country to a genotype was not always unanimously agreed so that this finer level of analysis should be considered with an element of caution.

Using a Bayesian model-based clustering method we were able to initially discern the existence of three robust groups reflecting major divisions of the germplasm. These groups were linked with the three geographical regions of origin, although differentiated only to a low degree. This would reflect a situation whereby the cultivars from a given region were more frequently derived from crosses between parental cultivars from the same region than from cultivars from elsewhere. Nevertheless, the migration of the plant material associated to human movement together with hundreds of years of empirical selection may have caused a significant gene flow across Europe. This is clearly indicated by the low genetic differentiation between groups and has shaped the overall pattern of genetic diversity. A spatially and temporally dynamic process where seeds and mainly graftwood were exchanged between geographically distinct populations has contributed to the increase of the genetic diversity in each area through unintentional gene flow or human-mediated intentional crosses [35, 36, 44]. The background common to other long lived tree fruits, including factors such as multiple origins of cultivated populations, ongoing crop-wild species gene flow and clonal distribution of genotypes together with the features associated with fruit tree species (lengthy juvenile phase, extensive outcrossing, widespread hybridization or mechanisms to avoid selfing) has defined the way they evolve in nature and resulted in extensive population genetic variation, but limited population structure [44]. A possible cause of divergence between the three identified groups could be the differential adaptation to distinct environmental conditions as are the case between Southern, Western and Northern + Eastern Europe. A similar situation was postulated for grapevine cultivars where the genetic structure appeared to be strongly shaped by geographic origin and intentional selection [13]. But since selection causes differentiation in particular regions of the genome on which selection pressure is acting [45], another likely cause of the population structure is genetic drift (i.e., changes in allelic frequencies caused by chance events) as also shown in e.g., apricot [46]. Together with selection, migration and drift can shape the local adaptation of species [47].

Although there may have been some mistakes in attributing cultivars to country of origin, the genetic makeup of the cultivars at the European level clearly appeared to show a North-East to South gradient. Interestingly, some countries exhibit intermediate marker data patterns in consistency with their intermediate geographic positions. This was clearly manifested at the national scale for the German and Czech cultivars which were shared between K2 and K3 groups. Similarly, the French and Swiss cultivars were shared between K1 and K3 groups. By contrast, cultivars from Southern Europe (Spain and Italy), from Northwestern Europe (United Kingdom and Ireland, Belgium, the Netherlands), and from North+Eastern Europe (Sweden and Finland) and Russia were mostly assigned to a single group (K1, K3, and K2, respectively). For the admixed germplasm from Southern and Northern+Eastern European geographical regions of origin, a certain degree of introgression with the Western germplasm was also indicated in contrast to the low contribution of the Northern+Eastern germplasm into the Southern germplasm and viceversa. Thus, in agreement with the correspondence between clustering and regions or countries of origin of the germplasm, the geographical proximity appears to align with the patterning observed in the admixed accessions.

In cases demonstrating the presence of a significant hierarchical population structure as this study suggests, this method preferentially detects the uppermost level of structure [26, 48, 49, 50]. As a consequence, when large datasets in species with a complex background are analyzed, it is possible for an underlying substructure to remain undetected within the major divisions of the germplasm. In this context, the "*nested (or two-steps) Structure*" clustering method has been shown to be an efficient tool to delineate further levels of substructure in both apple and other plant species [10, 24, 26, 49, 50, 51, 52]. In this study, the three groups inferred from the first round of Structure analysis were used as the starting point for revealing internal substructuring. Eight subgroups were identified with remarkable differences in both allelic composition and richness, as well as a considerable number of private alleles associated to particular subgroups. Nevertheless, the relationship between the placement of the genotypes in the subgroups and their country of origin varied considerably between subgroups in contraposition to the clearer and more consistent clustering trend within the three groups. As discussed earlier, this stratification may reflect historical processes of selection and adaptation to local conditions that might suggest a "*fine-delineation*" of the intra-variation within each main geographical region of origin. This is most probably the case for the K1.1 subgroup which mainly consists of Spanish cultivars and could reflect a process of both local adaption and isolation by distance related to the Pyrenean barrier. For the K1.3 subgroup, local adaptation to the Southern region could be inferred together with a potential for more intense commercial exchange between Italy and Southern France. For other subgroups, the relationship with particular countries or small regions was not obvious, but some interesting associations between subgroups of group K3 and recent cultivars and some of their founders or particular usage (cider apple cultivars) could be noticed.

Relatedness and Family Relationships

The previously reported parentage of 10 recent cultivars was correctly inferred in all cases. These results served as a control and validated the parentage assignment obtained with the CERVUS software [53] indicating that the number and informativeness of SSR markers were sufficient at least for these cultivars. The 16 SSR markers were nevertheless limited in their ability to infer parentages, and additional cases might have been detected with a larger number of SSR markers. In a recent paper [54], it was suggested that the number of 27 SSR loci used in that study was a minimum to be utilized for full parentage reconstruction. Basically, the LOD score tests used in the CERVUS software are computed according to the SSR allelic frequencies, and thus, parentages involving common alleles are more difficult to detect. By contrast, parentages involving low frequency and rare alleles are more easily detected. On that basis, it is worthy to note that the more frequently detected parents (i.e., 'Reine des Reinettes'='King of the Pippins', and 'Transparente de Croncels') are possibly representing a biased view of the frequently involved parents, as they most probably carry rare or low frequency alleles in at least some SSR loci. Putative parents present in the dataset but carrying more common alleles may have been hidden because of the statistical limits of their detection with 16 SSR markers. A similar situation was observed by [23] with the frequent appearance of 'Reine des Reinettes' as a parent of four old cultivars out of 28, using 21 SSR markers. In the near future, medium and high density SNP arrays [55,56, 57] will provide much more power to infer parentages.

The parentage of some old cultivars was either confirmed, in the case of 'Ernst Bosch'='Ananas Reinette' x 'Mank's Codlin' (synonym: 'Evino') or augmented, in the case of 'Ben's Red'='Devonshire Quarrenden' x 'Box Apple' (Table 5) where the second parent was initially hypothesized to be 'Farleigh Pippin' [58]. Distances between the geographic origins of the inferred parents (when known), ranged from crosses between geographically close cultivars to crosses between very distant cultivars, reflecting the large gene flow across Europe caused by, e.g., extensive exchange of budwood over centuries.

Some traditional folklore about the origination of old apple cultivars could be either substantiated or refuted by the SSR-based parentage information. As one example, the old Swedish cultivar 'Förlovningsäpple' is said to derive from a locally acquired seed in Northern Sweden where only a few cold-hardy apples can be grown. The two unknown parents were here inferred to be the Swiss cultivar 'Heuapfel' and the wide-spread cultivar 'Saint Germain' (X1646) also known as 'Vitgylling' in Sweden, a name used for a group of more or less similar, white-fruited, early-ripening and winter-hardy cultivars. Interestingly, the 'Vitgylling' accession included in the present study (BAL072) did not have the same SSR profile as 'Saint Germain', but they share one allele for all 16 SSR loci and may therefore be related. In two other cases, traditional Swedish folklore indicated that a sailor brought an exotic seed to the island of Gotland and to Kramfors in Northern Sweden, respectively, resulting in 'Stenkyrke' and 'Kramforsäpple'. For 'Stenkyrke', one parent is the Swedish 'Fullerö' and the second is the German cultivar 'Danziger Kantapfel' which has been much grown in Sweden. The origin of 'Stenkyrke' is thus probably much more local than anticipated. Similarly, the surmised American sailor origin of the seed giving rise to 'Kramforsäpple' is refuted by the fact that the parents of this cultivar are the Swedish 'Sävstaholm' and 'Åkerö'.

It is important to keep in mind that trueness-to-type of the accessions is not guaranteed, thus the labeling of the offspring or the parents can be erroneous in some cases. Conversely, the inferred parentages are robustly established so that the genetic relationships between the accessions are valid independently of their names. Crosses between the two inferred parents could be performed to reproduce the cross which gave birth to the offspring cultivar, especially if genetic analysis of some particular traits of the latter genotype indicates an interesting application in plant breeding.

CONCLUSIONS

The analysis of a large and representative set of *Malus* x *domestica* genotypes indicated that apple germplasm diversity reflects its origination within three main geographic regions of Europe, and that a weak genetic structure exists at the European level. This structuring of genetic variation in European dessert apple is caused by evolutionary processes relevant to the domestication of perennial fruit species with factors such as gene flow created by, e.g., ancient roads of commerce across the continent, other human activities like intentional selection and later breeding, and genetic drift. The remarkable differences in the allelic variation found at group and subgroup levels of germplasm stratification constitute a strong indication of that the diversity is hierarchically organized into three *genepools*, with consistent evidence of a pattern of internal substructure. The potential value for modern fruit production is mostly unknown since a majority of the accessions are poorly evaluated from an agronomic point of view. Thus, phenotypic data obtained with standardized methods is required to determine the commercial potential of the preserved material and to enable its use in new crosses to increase the genetic basis of the cultivated apple.

The integration of data for collections from different European geographic regions using standardized methods will undoubtedly form an important step in developing the European strategy for conservation of apple germplasm and constitute the starting point to define a European "apple core collection". This will constitute a decisive step in the field of conservation genetics, and may also have direct implications on the improvement of our understanding of the species, including i) the identification of genomic regions associated with commercially important horticultural traits, ii) the discovery of new germplasm features that may be taken advantage of for efficient breeding and iii) the analysis of genotype x environmental interactions for studying the stability of the most economically important traits for this species.

METHODS

Plant Material

Apple germplasm collections from nine European countries, plus Western part of Russia and Kyrgyzstan, were available for this study: France (INRA, Institut National de la Recherche Agronomique, 399 accessions), Italy (UNIBO, University of Bologna, 216 acc.), Belgium (CRA-W, Centre Wallon de Recherche Agronomique, 408 acc.), Czech Republic (RBIPH, Research and Breeding Institute of Pomology Holovousy, 263 acc.), United Kingdom (UK-NFC, University of Reading, 310 acc.), Sweden (SLU, Swedish University of Agricultural Sciences, 199 acc.), Finland (MTT Agrifood Research, 50 acc.), Spain (UPNA, Public University of Navarre, UDL, University of Lleida, and EEAD, Aula Dei Experimental Station, 269 acc.), Switzerland (FRUCTUS, Agroscope, 237 acc.), Russia (NCRRIHV, North Caucasian Regional Research Institute of Horticulture and Viticulture, and VNIISPK, The All Russian Research Institute of Horticultural Breeding, 83 acc.), and Kyrgyzstan (KNAU, Kyrgyz National Agrarian University, 12 acc.). In all countries, the accessions were mostly chosen as old local/national dessert cultivars (registered or at least known before 1950), but 12 standard dessert cultivars were also included to strengthen comparisons between collections, namely 'Golden Delicious', 'Red Delicious', 'McIntosh', 'Rome Beauty', 'Granny Smith', 'Jonathan', 'Winter Banana', 'Ingrid Marie', 'Ananas Reinette', 'Reinette de Champagne', 'Discovery' and 'Alkmene'. Moreover, 40 old cider apple cultivars and 40 recently-bred dessert cultivars were sampled in the INRA collection in order to investigate particular patterns. Altogether, 2,446 accessions were thus considered. Available collections were somewhat heterogeneous in nature as some of them corresponded to already established core collections (INRA and UNIBO) whereas others were selected for the present study thanks to available SSR marker data (UK-NFC and FRUCTUS, see below), or were chosen as a subset of mainly local cultivars (CRA-W, RBIPH, SLU, MTT, UPNA, UDL, EEAD, NCRRIHV, VNIISPK and KNAU). Cultivars that were known to be triploid or duplicated were avoided since this analysis was performed with an aim to subsequently use a major part of the material in a Genome Wide Association Study to be carried out within the EU FruitBreedomics project [28].

SSR Genotyping

A set of 16 SSR markers developed by different groups [59, 60, 61, 62] was used to genotype the 2,446 accessions. These SSR markers are distributed over 15 out of the 17 apple linkage groups, and 15 of them are included in a former list recommended by the ECPGR *Malus/Pyrus* working group [63]. The 16th marker of this list, NZ05g08, was replaced by the marker CH-Vf1 because the former showed either complex scoring pattern or low level of polymorphism in previous studies [23, 26]. SSR marker data were fully available for the collection from INRA [23]. SSR data were available (i.e., for some, but not all of the 16 SSR markers) for collections from UK-NFC [64], FRUCTUS [14], UPNA, UDL and EEAD [26], and UNIBO [24], so that only the missing SSR marker data were generated in the present study. Fully new SSR datasets were generated for collections from CRA-W, RBIPH, SLU, MTT, NCRRIHV, VNIISPK, and KNAU.

Forward primers were labeled with four different fluorescent dyes (6-FAM, VIC, NED, or PET) in order to be combined into four different multiplexed reactions. Polymerase chain reactions (PCR) for the four multiplex PCRs were performed in a final volume of 11 µL using 10 ng of DNA template, 0.18 µM of each primer, and 1× PCR Master mix of QIAGEN kit multiplex PCR (Qiagen, Hilden, Germany). PCR cycling conditions were as follows: pre-incubation for 15 min at 94°C, followed by 4 cycles using a touchdown amplification program with an annealing temperature reduced by 1°C per cycle from 60°C to 55°C, followed by 34 cycles, each consisting of 30 s denaturing at 94°C, 90 s annealing at 55°C, and 60 s elongation at 72°C, the last cycle ending with a final 15-min extension at 72°C. SSR amplification products were analyzed with an ABI3730 XL sequencing system (Applied Biosystems, Foster City, CA, USA). Fragment analysis and sizing were carried out using GeneMapper v.4.0 software (Applied Biosystems, Foster City, CA, USA); chromatograms were independently read by two operators. When SSR marker data were already available and obtained at different sites, SSR allele sizes were carefully adjusted between collections, both by use of reference accessions known to be in common between collections and by re-genotyping a subset of each collection with the full set of 16 SSR markers to confirm the allele adjustment.

Diversity Assessments

The multilocus SSR profiles were compared pairwise in order to establish the genetic uniqueness of each accession. Accessions were considered as duplicates if they had identical SSR fingerprints, or if they had one allelic difference for a maximum of two SSR loci thus making room for some genotyping errors and/or spontaneous SSR mutations. On this basis, redundant profiles were removed from the dataset to avoid bias in genetic analyses and duplicate groups were labeled with unique group ID codes (FBUNQ codes). An accession was declared as a putative triploid when at least three of the 16 SSR loci exhibited three distinct alleles. Analyses of descriptive diversity statistics were conducted at locus level. For each SSR marker, SPAGeDi v.1.3 software [65] was used to estimate the number of alleles (N_A), the number of alleles with a frequency below 5 % (N_B), the number of effective alleles (N_E), and the observed (*Ho*) and expected (*He*) heterozygosity. The probability of identity (P_{ID}) was calculated as follows [66]:

$$P_{ID} = \sum p_i^4 + \sum\sum \left(2p_ip_j\right)^2$$

where *pi* and *pj* are the frequencies of the ith and j[th] alleles and i≠j. The cumulative P_{ID} over the 16 SSR was computed as the product of the P_{ID} of each individual marker.

Determination of the Geographical Regions of Origin of the Unique Genotypes

Using passport data along with reviewing published records with a focus on old literature (national compilations/varietal catalogues/reports) and specialized websites we were able to discern the geographical regions of origin for a large part of the unique genotypes analyzed. This was further helped by the resolution of identified duplicates and comparison of accessions against additional SSR data of the whole UK-NFC apple collection kindly made available from the UK-NFC database [64] and of the whole FRUCTUS collection kindly made available by Agroscope [14]. We first decided to define three broad historical European regions of origin of the germplasm according to geographical proximity and traditional agricultural relations between them: North+East (Sweden, Norway, Finland, Denmark, Baltic countries, plus Russia, Ukraine and Kyrgyzstan), West (Ireland, United Kingdom, France, Belgium, the Netherlands, Switzerland, Germany, Czech Republic) and South (Spain and Italy). When available, countries of origin of the cultivars were also documented although, this information should be considered with caution since the information on the countries of origin was not always fully consistent within duplicates groups.

Analysis of the Genetic Structure

The software Structure v.2.3.4 [67] was used to estimate the number of hypothetical subpopulations (*K*) and to quantify the proportion of ancestry of each genotype to the inferred subpopulations. No prior information about the geographical origin of the accessions was considered in the analysis. Ten independent runs were carried out for *K* values ranging from two to 10 using 500,000 Markov Chain Monte Carlo (MCMC) iterations after a burn-in of 200,000 steps assuming an admixture model and allelic frequencies correlated. In order to assess the best *K* value supported for our dataset, the Δ*K* method [68] was used through the Structure harvester v.0.6.93 website [69] to examine the rate of change in successive posterior probabilities over the range of *K* values. When the results described above suggested additional substructuring of the diversity in subgroups, a second-level (nested) application of the Structure clustering method was carried out analyzing separately each of the *K* major groups previously obtained [10, 24, 26, 50, 51]. Genotypes were assigned to the group (or subgroup) for which they showed the highest membership coefficient, considering an accession strongly assigned to each partitioning level if its proportion of ancestry (*qI*) was ≥0.80 [70, 71, 72]; otherwise they were considered as "admixed". The placement of genotypes on groups (or subgroups) was determined using CLUMPP v.1.1 [73], which evaluates the similarity of outcomes between population structure runs. CLUMPP output was used directly as input for Distruct v1.1 [74] in order to graphically display the results.

To validate the genetic structure revealed by the Bayesian model-based clustering two complementary approaches using the Darwin software package v6.0.10 [75] were considered: i) an unweighted neighbor-joining tree constructed based on dissimilarities between the unique genotypes (using a Simple Matching coefficient), and ii) a multivariate Principal Coordinate Analysis (PCoA).

Genetic Differentiation

Population differentiation was estimated by analyses of molecular variance (AMOVA) through Genodive [76] under two scenarios: i) three broad European geographic regions of origin of the material (North+East, West and South); and ii) the major groups (and subgroups) defined by Structure. Pairwise F_{ST} estimates for the different partitioning levels considered in each case were also obtained using Genodive [76]. Descriptive statistics were calculated for the material clustered according to geographical regions of origin as well as for each group (or subgroup) identified by the Bayesian model-based clustering method, including *Ho* and *He*, number of total alleles, number of private alleles, i.e., those only found in one (sub)division level, and number of unique alleles, i.e., those only detected in one unique accession. The software FSTAT v.2.9.3.2 [77] was applied to compute the allelic richness after scaling down to the smallest partitioning level in the different scenarios considered.

Parentage Reconstruction and Relatedness Between the Accessions

On the basis of the SSR profiles of the unique genotypes, accessions were analyzed to infer possible parent-offspring relationships using Cervus v.3.0 software [53]. In order to reveal only robust parentages, we limited the study to the inferences of 'two-parents offspring' relationships and did not consider inferences of 'one-parent offspring' relationships where the lacking parent offers more flexibility but more speculative assignments as well, especially with only 16 SSR markers. Two criteria were considered to establish strict parentage relationships: i) a confidence level of the LOD score and ii) the Delta LOD value (defined as the difference in LOD scores between the first and second most likely two-candidate parents inferred) both higher than 95 %. Finally, an additional constraint was added to strengthen the results by limiting the maximum number of tolerated locus mismatches to only one in any inferred two-parents offspring trio, assuming that such a slight difference may be attributable to possible scoring errors, occurrence of null alleles or occasional mutational events [54, 78].

Abbreviations

ECPGR, European cooperative programme for plant genetic resources; F_{ST}, fixation index 'F-statistics'; LOD, logarithm of odds ratio; PCR, polymerase Chain reaction; SSR, simple sequence repeat.

DECLARATIONS

Acknowledgments

The staff at the ANAN genotyping platform of the SFR 149 QUASAV (Angers, France), at the GENTYANE genotyping platform (INRA, Clermont-Ferrand, France) and at the Unité Expérimentale Horticole (INRA, Angers, France) are warmly acknowledged for their help in producing the genotyping data and for maintaining the apple germplasm collection, respectively. The INRA MIGALE bioinformatic platform is also greatly acknowledged for having given support to intense data analyses. Natural Resources Institute, Finland, is gratefully acknowledged for providing access to 50 Finnish apple cultivars in their germplasm collections. UDL, University of Lleida, and EEAD, Aula Dei Experimental Station, are gratefully acknowledged for providing access to apple cultivars in their germplasm collections. Acknowledgements to Slepkov S. from Maykop experimental station of Vavilov's N.I. Vavilov Institute of Plant Industry (MOSVIR) for making available plant material of Russian apple cultivars included in the present analysis.

Funding

This work has been partly funded under the EU seventh Framework Programme by the FruitBreedomics project N°265582: "Integrated approach for increasing breeding efficiency in fruit tree crops". The views expressed in this work are the sole responsibility of the authors and do not necessarily reflect the views of the European Commission. Genotyping of the Spanish collection was partially funded by INIA, Instituto Nacional de Investigación y Tecnología Agraria y Alimentaria (project grant no RF2004-008-C03-00). Genotyping of the Swiss collection was funded by the Swiss Federal Office for Agriculture. Genotyping of the French collection was funded by the FRB, 'Fondation pour la Recherche sur la Biodiversité'. Initial genotyping of the UK-NFC material was funded by the UK Depart for the Environment Food and Rural Affairs (Defra), grant GC0140. Providing of VNIISPK material (DNA isolation and delivery) have been done with support of Russian Scientific Fund, Project 14-1600127. JU has been partially supported by an Early Stage Research Fellowship of the Institute of Advanced Studies (University of Bologna).

Availability of supporting data

The dataset supporting the conclusions of this article will be available in the Genome Database for Rosaceae (GDR) (https://www.rosaceae.org/).

Authors' contributions

JU and CD carried out the statistical analyses. CD and ER carried out the molecular analysis of the accessions not previously genotyped and CD performed the alignment of the SSR profiles of the whole dataset. CP coordinated the fragment analyses of the accessions not previously genotyped. ML, PH, MO, FP, JS, HN, LGG, CM, JG, IS, AVP, LD, and ST contributed in the classification of the plant material in the three broad European geographic regions and national origins, and in the interpretation of the SSR duplicated groups, synonyms and parentages inferred. AG, RG, LF, ML, PH, MO, FFF, KME, FP, JS, HN, LGG, CM, JG, MK, IS, AVP, NGK, ET, LD and ST provided plant material, DNA or SSR profiles of accessions preserved in the studied germplasm. CED conceived and coordinated the study. FL coordinated the EU FruitBreedomics project. JU and CED wrote the manuscript. ML, MO, HN, LGG, CM, MK, IS, AVP, and ST critically reviewed the manuscript. All authors read and approved the final manuscript.

Authors' information

Jorge Urrestarazu and Caroline Denancé share first authorship.

Competing interests

The authors declare that they have no competing interests.

REFERENCES

1. Food and Agriculture Organization of the United Nations. FAO statistics database on the World Wide Webhttp://faostat.fao.org/site/567/DesktopDefault.aspx?PageID0567#ancor. Accessed 27 July 2015.
2. Hokanson SC, Lamboy WF, Szewc-McFadden AK, McFerson JR. Microsatellite (SSR) variation in a collection of *Malus*(apple) species and hybrids. Euphytica. 2001;118:281–94.
3. Janick J, Moore JN. Fruit breeding. Volume I: tree and tropical fruits. New York: Wiley; 1996.
4. Laurens F, Durel CE, Patocchi A, Peil A, Salvi S,

Tartarini S, Velasco R, van de Weg WE. Review on apple genetics and breeding programs and presentation of a new initiative of a news European initiative to increase fruit breeding efficiency. J Fruit Sci. 2010;27:102–7.

5. Noiton DAM, Alspach PA. Founding clones, inbreeding, coancestry, and status number of modern apple cultivars. J Am Soc Hortic Sci. 1996;121:773–82.

6. Brooks RM, Olmo HP. Register of new fruit and nut varieties list 35. HortScience. 1991;26:951–78.

7. Brooks RM, Olmo HP. Register of new fruit and nut varieties list 36. HortScience. 1994;29:942–69.

8. Brooks RM, Olmo HP. Register of fruit and nut varieties. Alexandria: ASHS; 1997.

9. de Vicente. The evolving role of genebanks in the fast-developing field of molecular genetics. Ed. Issues in genetic resources No XI, August 2004. Rome: International Plant Genetic Resources Institute-IPGRI; 2004.

10. Jing RC, Vershinin A, Grzebyta J, Shaw P, Smykal P, Marshall D, Ambrose MJ, Ellis THN, Flavell AJ. The genetic diversity and evolution of field pea (*Pisum*) studied by high throughput retrotransposon based insertion polymorphism (RBIP) marker analysis. BMC Evol Biol. 2010;10:44.

11. van Treuren R, Kemp H, Ernsting G, Jongejans B, Houtman H, Visser L. Microsatellite genotyping of apple (*Malus* x*domestica* Borkh.) genetic resources in the Netherlands: application in collection management and variety identification. Genet Resour Crop Evol. 2010;57:853–65.

12. Nybom H, Weising K, Rotter B. DNA fingerprinting in botany: past, present, future. Investig Genet. 2014;5:1.

13. Bacilieri R, Lacombe T, Le Cunff L, Di Vecchi-Staraz M, Laucou V, Genna B, Péros JP, This P, Boursiquot JM. Genetic structure in cultivated grapevines is linked to geography and human selection. BMC Plant Biol. 2013;13:25.

14. Bühlmann A, Gassmann J, Ingenfeld A, Hunziker K, Kellerhals M, Frey JE. Molecular characterization of the Swiss fruit genetic resources. Erwerbs-Obstbau. 2015;57:29–34.

15. McCouch SR, McNally KL, Wang W, Sackville HR. Genomics of gene banks: a case study in rice. Am J Bot. 2012;99:407–23.

16. Wünsch A, Hormaza JI. Cultivar identification and genetic fingerprinting of temperate fruit tree species using DNA markers. Euphytica. 2002;125:59–67.

17. Ellegren H. Microsatellites: simple sequences with complex evolution. Nat Rev Genet. 2004;5:435–45.

18. Guichoux E, Lagache L, Wagner S, Chaumeil P, Léger P, Lepais O, Lepoittevin C, Malausa T, Revardel E, Salin F, Petit RJ. Current trends in microsatellite genotyping. Mol Ecol Resour. 2011;11:591–611.

19. Kalia RK, Rai MK, Kalia S, Singh R, Dhawan AK. Microsatellite markers: an overview of the recent progress in plants. Euphytica. 2011;177:309–34.

20. Sunnucks P. Efficient genetic markers for population biology. Trends Ecol Evol. 2000;15:199–03.

21. Urrestarazu J, Royo JB, Santesteban LG, Miranda CM. Evaluating the influence of the microsatellite marker set on the genetic structure inferred in *Pyrus communis* L. PLoS One. 2015;10:e0138417.

22. Garkava-Gustavsson L, Kolodinska Brantestam A, Sehic J, Nybom H. Molecular characterisation of indigenous Swedish apple cultivars based on SSR and S-allele analysis. Hereditas. 2008;145:99–112.

23. Lassois L, Denancé C, Ravon E, Guyader A, Guisnel R, Hibrand-Saint-Oyant L, Poncet C, Lasserre-Zuber P, Feugey L, Durel CE. Genetic diversity, population structure, parentage analysis and construction of core collections in the French apple germplasm based on SSR markers. Plant Mol Biol Rep. 2015. doi:10.1007/s11105-015-0966-7.

24. Liang W, Dondini L, De Franceschi P, Paris R, Sansavini S, Tartarini S. Genetic diversity, population structure and construction of a core collection of apple cultivars from Italian germplasm. Plant Mol Biol Rep. 2015;33:458–73.

25. Patzak J, Paprštein F, Henychová A, Sedlák J. Comparison of genetic diversity structure analyses of SSR molecular marker data within apple (*Malus* × *domestica*) genetic resources. Genome. 2012;55:647–65.

26. Urrestarazu J, Miranda C, Santesteban LG, Royo JB. Genetic diversity and structure of local apple cultivars from Northeastern Spain assessed by microsatellite markers. Tree Genet Genomes. 2012;8:1163–80.

27. Sehic J, Garkava-Gustavsson L, Nybom H. More harmonization needed for DNA-based identification of apple germplsm. Acta Hortic. 2013;976:277–83.

28. Laurens F, Aranzana MJ, Arús P, Bassi D, Bonany J, Corelli L, Durel CE, Mes J, Pascal T, Patocchi A, Peil A, Quilot B, Salvi S, Tartarini S, Troggio M, Vecchietti A, Velasco R, van de Weg WE. Review of fruit genetics and breeding programmes and a new European initiative to increase fruit breeding efficiency. Acta Hortic. 2012;929:95–102.

29. Morin PA, Manaster C, Mesnick SL, Holland R. Normalization and binning of historical and multi-source microsatellite data: overcoming the problems of allele size shift with ALLELOGRAM. Mol Ecol Resour. 2009;9:1451–5.

30. Putman AI, Carbone I. Challenges in analysis and interpretation of microsatellite data for population genetic studies. Ecol Evol. 2014;4:4399–28.

31. Gasi F, Simon S, Pojskic N, Kurtovic M, Pejic I. Genetic assessment of apple germplasm in Bosnia and Herzegovina using microsatellite and morphologic markers. Sci Hortic. 2010;126:164–71.

32. Gharghani A, Zamani Z, Talaie A, Oraguzie NC, Fatahi R, Hajnajari H, Wiedow C, Gardiner SE. Genetic identity and relationships of Iranian apple (*Malus* x *domestica* Borkh.) cultivars and landraces, wild *Malus* species and representative old apple cultivars based on simple sequence repeat (SSR) marker analysis. Genet Resour Crop Evol. 2009;56:829–42.

33. Gao Y, Liu F, Wang K, Wang D, Gong X, Liu L, Richards CM, Henk AD, Volk GM. Genetic diversity of *Malus* cultivars and wild relatives in the Chinese National

Repository of Apple Germplasm Resources. Tree Genet Genomes. 2015. doi:10.1007/s11295-015-0913-7.

34. Gross BL, Volk GM, Richards CM, Forsline CL, Fazio G, Chao CT. Identification of "duplicate" accessions within the USDA-ARS National Plant Germplasm System *Malus* Collection. J Am Soc Hortic Sci. 2012;137:333–42.

35. Cornille A, Gladieux P, Smulders MJM, Roldán-Ruiz I, Laurens F, Le Cam B, Nerseyan A, Clavel J, Olonova M, Feugey L, Gabrielyan I, Zhang XG, Tenaillon MI, Giraud. New insight into the history of domesticated apple: secondary contribution of the European wild apple to the genome of cultivated varieties. PLoS Genet. 2012;8:e1002703.

36. Cornille A, Giraud T, Smulders MJM, Roldán-Ruiz I, Gladieux P. The domestication and evolutionary ecology of apples. Trends Genet. 2014;30:57–65.

37. Gross BL, Henk AD, Richards CM, Fazio G, Volk GM. Genetic diversity in *Malus* x *domestica* (*Rosaceae*) through time in response to domestication. Am J Bot. 2014;101:1770–9.

38. Hartmann HT, Kester DE, Davies FT, Geneve RL. Plant propagation: principles and practices. 7th ed. Upper Saddle River: Prentice Hall; 2002.

39. Zohary D, Hopf D. Domestication of plants in the Old World: the origin and spread of cultivated plants in West Asia, Europe and the Nile Valley. Oxford: Oxford University Press; 2000.

40. De Franceschi P, Dondini L, Sanzol J. Molecular bases and evolutionary dynamics of self-incompatibility in the *Pyrinae*(*Rosaceae*). J Exp Bot. 2012;63:4015–32.

41. Knutson L, Stoner AK. Biotic diversity and germplasm preservation: global imperative. Boston: Kluwer Academic Publishers; 1989.

42. McCouch S. Diversifying selection in plant breeding. PLoS Biol. 2004;2:e347.

43. World Apple and Pear Association. European apple and pear crop forecast. Brussels: World Apple and Pear Association; 2013.

44. Miller AJ, Gross BL. Forest to field: perennial fruit crop domestication. Am J Bot. 2011;98:1389–414.

45. Biswas S, Akey JM. Genomic insights into positive selection. Trends Genet. 2006;22:437–46.

46. Bourguiba H, Audergon JM, Krichen L, Trifi-Farah N, Mamouni A, Trabelsi S, D'Onofrio C, Asma BM, Santoni S, Khadari B. Loss of genetic diversity as a signature of apricot domestication and diffusion into the Mediterranean Basin. BMC Plant Biol. 2012;12:49.

47. Blanquart F, Gandon S, Nuismer SL. The effects of migration and drift on local adaptation to a heterogeneous environment. J Evol Biol. 2012;25:1351–63.

48. Emanuelli F, Lorenzi S, Grzeskowiak L, Catalano V, Stefanini M, Troggio M, Myles S, Martinez-Zapater JM, Zyprian E, Moreira FM, Grando MS. Genetic diversity and population structure assessed by SSR and SNP markers in a large germplasm collection of grape. BMC Plant Biol. 2013;13:39.

49. Lia VV, Poggio L, Confalonieri VA. Microsatellite variation in maize landraces from Northwestern Argentina: genetic diversity, population structure and racial affiliations. Theor Appl Genet. 2009;119:1053–67.

50. Li XW, Meng XQ, Jia HJ, Yu ML, Ma RJ, Wang LR, Cao K, Shen ZJ, Niu L, Tian JB, Chen MJ, Xie M, Arus P, Gao ZS, Aranzana MJ. Peach genetic resources: diversity, population structure and linkage disequilibrium. BMC Genet. 2013;14:84.

51. Jacobs MMJ, Smulders MJM, van den Berg RG, Vosman B. What's in a name; genetic structure in *Solanum* section *Petota*studied using population-genetic tools. BMC Evol Biol. 2011;11:42.

52. Jing R, Ambrose MA, Knox MR, Smykal P, Hybl M, Ramos A, Caminero C, Burstin J, Duc G, van Soest LJM, Swiecicki WK, Pereira MG, Vishnyakova M, Davenport GF, Flavell AJ, Ellis THN. Genetic diversity in European *Pisum* germplasm collections. Theor Appl Genet. 2012;125:365–80.

53. Kalinowski ST, Taper ML, Marshall TC. Revising how the computer program cervus accommodates genotyping error increases success in paternity assignment. Mol Ecol. 2007;16:1099–2006.

54. Salvi S, Micheletti D, Magnago P, Fontanari M, Viola R, Pindo M, Velasco R. One-step reconstruction of multi-generation pedigree networks in apple (*Malus × domestica* Borkh.) and the parentage of Golden Delicious. Mol Breed. 2014;34:511–24.

55. Bianco L, Cestaro A, Sargent DJ, Banchi E, Derdak S, Di Guardo M, Salvi S, Jansen J, Viola R, Gut I, et al. Development and validation of a 20K single nucleotide polymorphism (SNP) whole genome genotyping array for apple (*Malus* x *domestica*Borkh). PloS ONE. 2014;9:e110377.

56. Bianco L, Cestaro A, Linsmith G, Muranty H, Denancé C, Théron A, Poncet C, Micheletti D, Kersshbamer E, Di Pierro EA, Larger S, Pindo M, van de Weg WE, Davassi A, Laurens F, Velasco R, Durel CE, Troggio M. Development and validation of the Axiom®Apple480K SNP genotyping array. Plant J. 2016;86:62–74.

57. Chagné D, Crowhurst RN, Troggio M, Davey MW, Gilmore B, Lawley C, Vanderzande S, Hellens RP, Kumar S, Cestaro A et al. Genome-wide SNP detection, validation, and development of an 8K SNP array for apple. PloS ONE. 2012;7:e31745.

58. Morgan J, Richards A, Dowle E. The new book of Apples: the definitive guide to apples, including over 2000 varieties. London: Ebury Press; 2002.

59. Hokanson SC, Szewc-McFadden AK, Lamboy WF, McFerson JR. Microsatellite (SSR) markers reveal genetic identities, genetic diversity and relationships in a *Malus* x *domestica* Borkh. core subset collection. Theor Appl Genet. 1998;97:671–83.

60. Liebhard R, Gianfranceschi L, Koller B, Ryder CD, Tarchini R, van de Weg E, Gessler C. Development and characterisation of 140 new microsatellites in apple (*Malus* x *domestica* Borkh.). Mol Breed. 2002;10:217–41.

61. Silfverberg-Dilworth E, Matasci CL, van de Weg WE, van Kaauwen MPW, Walser M, Kodde LP, Soglio V,

Gianfranceschi L, Durel CE, Costa F, Yamamoto T, Koller B, Gessler C, Patocchi A. Microsatellite markers spanning the apple (*Malus* x*domestica* Borkh.) genome. Tree Genet Genomes. 2006;2:202–24.

62. Vinatzer BA, Patocchi A, Tartarini S, Gianfranceschi L, Sansavini S, Gessler C. Isolation of two microsatellite markers from BAC clones of the Vf scab resistance region and molecular characterization of scab-resistant accessions in *Malus*germplasm. Plant Breed. 2004;123:321–6.

63. Evans KM, Fernández F, Govan C. Harmonising fingerprinting protocols to allow comparisons between germplasm collections - *Pyrus*. Acta Hortic. 2009;814:103–6.

64. Fernández-Fernández F. Fingerprinting the National apple and pear collections. Final report of DEFRA research project GC0140. 2010. p. 1–18. http://randd.defra.gov.uk/Document.aspx?Document=GC0140SID5FingerprintingFernandez.pdf.

65. Hardy OJ, Vekemans X. SPAGEDi: a versatile computer program to analyse spatial genetic structure at the individual or population levels. Mol Ecol Notes. 2002;2:618–20.

66. Waits LP, Luikart G, Taberlet P. Estimating the probability of identity among genotypes in natural populations: cautions and guidelines. Mol Ecol. 2001;10:249–56.

67. Pritchard JK, Stephens M, Donnelly P. Inference of population structure using multilocus genotype data. Genetics. 2000;155:945–59.

68. Evanno G, Regnaut S, Goudet J. Detecting the number of clusters of individuals using the software STRUCTURE: a simulation study. Mol Ecol. 2005;14:2611–20.

69. Earl DA, von Holdt BM. Structure harvester: a website and program for visualizing STRUCTURE output and implementing the Evanno method. Cons Genet Resour. 2012;4:359–61.

70. Breton C, Pinatel C, Médail F, Bonhomme F, Bervillé A. Comparison between classical and Bayesian methods to investigate the history of olive cultivars using SSR-polymorphisms. Plant Sci. 2008;175:524–32.

71. Miranda C, Urrestarazu J, Santesteban LG, Royo JB, Urbina V. Genetic diversity and structure in a collection of ancient Spanish pear cultivars assessed by microsatellite markers. J Am Soc Hortic Sci. 2010;135:428–37.

72. Marra FP, Caruso T, Costa F, Di Vaio C, Mafrica R, Marchese A. Genetic relationships, structure and parentage simulation among the olive tree (*Olea europaea* L. subsp. europaea) cultivated in Southern Italy revealed by SSR markers. Tree Genet Genomes. 2013;9:961–73.

73. Jakobsson M, Rosenberg NA. CLUMPP: a cluster matching and permutation program for dealing with label switching and multimodality in analysis of population structure. Bioinformatics. 2007;23:801–6.

74. Rosenberg NA. Distruct: a program for the graphical display of population structure. Mol Ecol Notes. 2004;4:137–8.

75. Perrier X, Jacquemoud-Collet JP. DARwin Software. 2006. http://darwin.cirad.fr/darwin.

76. Meirmans PG, van Tienderen PH. Genotype and Genodive: two programs for the analysis of genetic diversity of asexual organisms. Mol Ecol Notes. 2004;4:792–4.

77. Goudet J. FSTAT 2.9.3.2. 2002. http://www2.unil.ch/popgen/softwares/fstat.htm.

78. Lacombe T, Boursiquot JM, Laucou V, Di Vecchi-Staraz M, Péros JP. This P. Large-scale parentage analysis in an extended set of grapevine cultivars (*Vitis vinifera* L.). Theor Appl Genet. 2013;126:401–14.

CHAPTER 11 GENOME ENGINEERING AND PLANT BREEDING

Fabien Nogué, Kostlend Mara, Cécile Collonnier, Josep M. Casacuberta

Keywords: Genome engineering, Plant breeding, Genetic diversity, Site-directed nucleases, QTLs

GENETIC DIVERSITY, MUTATIONS AND PLANT BREEDING: A FRUITFUL "MÉNAGE À TROIS"

In spite of the remarkable fidelity of DNA replication and the high efficiency of genome repair and surveillance systems, mutations are incorporated into the DNA at a rate that makes virtually every organism unique. In plants, the mutation rate has been determined to be of 7×10^{-9} per site and generation in *Arabidopsis thaliana* (Ossowski et al. 2010), which means that even for this very compact genome, roughly one mutation is incorporated in each newborn individual. These mutations, which in some cases trigger phenotypic differences among individuals, are the raw material on which selection can operate making species adaptation and long-term evolution possible.

The most frequent types of genomic mutations are short insertions/deletions and single nucleotide changes leading to single nucleotide polymorphisms (SNPs). However, genomes also experience many drastic changes, including chromosome rearrangements, long deletions, duplications and insertions of long sequences, such as transposable elements. In fact, it is considered that the two major mechanisms for plant genome dynamics are the whole genome doubling (WGD) events (Wendel 2015) and the differential and lineage-specific proliferation (and deletion) of various types of transposable elements (TEs) (Lisch 2013). All these mechanisms endow plant genomes with a remarkable plasticity that allows plants to adapt to new environments and that has also been exploited for their domestication and breeding. The domestication of wild species and their breeding are a particular type of evolution directed mainly by human selection. However, the principles and mechanisms governing these processes are by no means different from those involved in general evolution. In fact, domestication and breeding constitute excellent models to study evolution at large (Meyer and Purugganan 2013; Olsen and Wendel 2013).

Domestication and Crop Breeding Alleles

During the last 10 years, an impressive amount of plant domestication and improvement alleles have been characterized and we can now start to pinpoint the genes and the mutations underlying the substantial phenotypic differences between crops and their wild relatives. There are a number of traits that have been repeatedly selected during plant domestication. Essentially, these are traits that facilitate crop planting and harvesting and are collectively referred to as the domestication syndrome (Olsen and Wendel 2013). A list of the traits already characterized can be found in recent reviews (Meyer and Purugganan 2013; Olsen and Wendel2013). These include loss of seed shattering and seed dormancy, uniformity in germination and erect growth allowing increased plant density in crop fields. Although the same traits have been repeatedly selected in domestication processes, it seems that the genes behind the phenotypic changes were not always the same, suggesting that this adaptive process is remarkably flexible (Gaut 2015). This is the case of the dual domestication of common bean, for which a population genetics analysis showed that among the genes presenting signs of having been selected, those that are common in both events are not different from a random expectation (Schmutz et al.2014).

On the other hand, it is also becoming increasingly clear that many domestication-related traits are not monogenic but involve many genes with small effects (Olsen and Wendel 2013).

However, even taking into account the complexity and the flexibility of the genomic changes associated to crop

Citation: Fabien Nogué, Kostlend Mara, Cécile Collonnier, Josep M. Casacuberta, Genome engineering and plant breeding, doi:10.1007/s00299-016-1993-z.

domestication, the analysis of the many relatively simple examples already characterized allows for deducing general trends of this process. In addition, these analyses provided us with a collection of genes and specific alleles behind important phenotypic transitions that can be of use for plant breeding. As already noticed in 2006 (Doebley et al.2006), changes in morphology or other complex phenotypes often involve transcriptional or development regulators, and rarely structural proteins and enzymes (Olsen and Wendel 2013).

Mutations behind the Domestication and Breeding Phenotypic Changes

The most frequent causal mutations of the domestication phenotypes are nonsense mutations, such as premature truncation of ORFs through induced frameshifts, introduction of STOP codons, changes in splicing signals, or AA changes leading to a loss of function of the protein. For example, the transition from the prostrate growth of ancestral wild rice to the erect growth of rice cultivars, which was one of the critical events in rice domestication, was the result of the selection of a single mutation in the *PROG1* (*PROSTRATE GROWTH 1*) gene. This mutation produces an AA substitution in the zinc-finger nuclear transcription factor encoded by *PROG1*leading to the loss of function (Jin et al. 2008; Tan et al. 2008).

However, mutations inducing changes in cis-regulatory sequences or changes in AAs leading to modified activity are also relatively frequent. For example, the key domestication change on plant architecture in maize (i.e., loss of auxiliary branches and increased apical dominance) was due to a mutation affecting the expression of the *tb1* (*teosinte branched 1*) gene (Wang et al.1999). In this case, the causal mutation has been shown to be the insertion of a *Hopscotch* retrotransposon, some 60 kb upstream of the coding region, that increases *tb1* expression (Studer et al. 2011). *tb1* encodes a transcription regulator that represses growth and its overexpression in the auxiliary branches inhibits their development. As for mutations that change the function of a protein, an interesting example is that of the *Flowering Locus T* (*FT*) paralogs in sunflower that played a role in the domestication of this species. It has been shown that a frameshift mutation in the *Helianthus annuus FT 1* (*HaFT1*) gene generates a dominant-negative allele that interacts and blocks the action of another FT paralog resulting in delayed flowering (Blackman et al. 2010).

In addition to traits that can be considered as linked to the domestication process, crops have also been intensively modified in other traits to improve their cultivation and the quality and the diversity of their derived products. Here again, spontaneous mutations through the panoply of already mentioned mechanisms caused null alleles but also modified the expression or the coding capacity of genes. An interesting example that illustrates the high plasticity of plant genomes is the mutation associated to the *SUN* gene, linked to the elongated phenotype of tomato, which consists of a retrotransposon-mediated gene duplication that placed the duplicated gene under the control of a different promoter in the genome (Xiao et al. 2008).

In summary, plant domestication and breeding led to the profound modification of plant phenotypes through the recurrent selection of spontaneous mutations present in plant populations. These mutations encompass a diversity of genome changes, from SNPs and small indels to chromosomal rearrangements, gene duplications and TE insertion/deletion events that have modified an important number of genes, often regulators of key processes in the plant.

Introducing New Mutations

Plant breeding has been done for the most part of its 10,000 years of history in an unconscious way. However, the progress of the scientific knowledge has an immediate impact in these practices throughout history. Understanding the key role of genetic variability and selection had a profound impact on plant breeders and the discovery that physical and chemical agents induce mutations in the DNA (Muller 1928; Stadler 1928) prompted the breeders to start mutagenesis programs to increase genetic variability. Over the past 70 years, more than 3200 new crop varieties in over 200 plant species have been obtained through mutagenesis (IAEA/FAO mutant variety database, http://www-infocris.iaea.org/MVD/). This approach has proved to be extremely useful despite its limitations (Parry et al. 2009). Mutations introduced by these methods are of the same type as spontaneous mutations. Most mutations consist of SNPs and short indels but long insertion/deletions, chromosome rearrangements and gene duplications are also frequent (Bolon et al. 2014).

The most important limitation of chemical and physical mutagenesis is its stochastic nature. Generating mutants for specific traits requires the generation of extensive mutant collections and the development of powerful forward genetic screens. A refinement of the strategy of induced mutagenesis is TILLING (which stands for Targeting Induced Local Lesions In Genomes) that links mutagenesis to a DNA-analysis method that selects from thousands of mutagenized plants those that display the mutations in the desired gene. However, in most cases, the outcome of these approaches is null mutations and obtaining particular alleles known to confer certain phenotypes is, in most cases, virtually impossible.

In the last few years, a new generation of mutagenic agents, based on site-directed nucleases (SDNs) [or site-specific nucleases (SSNs)], such as zinc-finger nucleases, TALENs or the more recent CRISPR/Cas9, has been developed and very rapidly been incorporated into the plant breeders toolkit (Podevin et al. 2013). SDNs allow the generation of all the types of mutations that have been introduced during the domestication and breeding of crops with an unprecedented precision and extent. Different genes can be the targets of these SDNs, and interesting alleles can be discovered through forward and reverse genetic screens, for simple traits. However, for more complex traits, the use of these new techniques will require the identification of the genomic regions involved by the quantitative traits loci (QTL) analysis.

GENETIC DIVERSITY CAPTURED BY BREEDERS THROUGH THE IDENTIFICATION OF QTL

New Techniques for the Discovery of QTLs

Quantitative traits loci mapping has so far been very efficient for identifying the genetic regions linked to quantitative agronomic traits. It was supported by the recent development of new means of high-throughput genotyping and phenotyping. However, it is still a challenge to identify genomic regions associated to phenotypes that could result from different stresses and to show overlapping responses in plants. This is particularly the case for abiotic stresses, where repeated evaluations to confirm the value of the detected QTLs are needed (Takeda and Matsuoka 2008). Moreover, QTL mapping is limited to the detection of genetic variants in populations segregated from bi-parental crosses. Some other techniques that are now used in routine for the detection of alleles of interest, such as association mapping and selection screening, have been developed (Takeda and Matsuoka 2008). Association mapping correlates genetic markers to a given phenotype based on the analysis of populations derived from large germplasms. It allows for a large number of alleles for each tested locus to be taken into account, instead of the two parental alleles tested with classical QTL mapping (Takeda and Matsuoka 2008). A key factor in this approach is the linkage disequilibrium (LD) that determines the marker density and the experimental design. For species, such as *A. thaliana*and maize, that exhibit a high recombination rate and a rapid decay of LD (<1 kb), this approach is successful (Yu and Buckler 2006). In self-crossing crops, such as rice, soybean, sorghum and barley, the relatively large scale of LD leads to a lower association mapping resolution. An alternative approach, selection screening, is based on the theory that, after strong selection, the selected loci present a decrease in nucleotide diversity and an increase in LD (Takeda and Matsuoka 2008). This approach is not suitable for all crops, as for sorghum, for example, where no selection was found after the screening of 445 loci (Hamblin et al. 2006). On the contrary, it is particularly well suited for species, such as maize, for which the demographic history has been well studied. However, the choice between the different approaches by the breeders will influence the strategy and the success rate of their breeding plans.

Exploitation of the QTLs Necessitates Long and Cost Effective Breeding Plans

In the last two decades, numerous efforts have been made to identify molecular markers and QTLs linked to important agronomic traits for a wide range of crop species. Marker assisted selection (MAS) is one of the most used techniques to select for QTLs in various plant species. An alternative approach that has found wide application is known as QTL pyramiding. Here, combinations of two or more QTLs identified in different varieties or compatible species and responsible for different traits are introgressed into the same elite line (Ashikari and Matsuoka2006). However, contrasting results were reported regarding the successful rate of the QTL introgression, especially when five or more QTLs for certain traits are introgressed into a given elite line (Semagn et al. 2006). Among the various factors that negatively impact the achievement of successful results of MAS, the low repeatability of QTLs through different genetic backgrounds, interactions between QTLs, and recombination between genes that might be present in the same chromosomal segment within the QTL are the most reported ones (Semagn et al. 2010). Another limit of working with QTLs is that their exploitation requires long and expensive breeding plans. This is particularly true when working with forest tree species. Indeed, most forest trees are clonally propagated resulting in narrow genetic diversity and small segregating populations. In addition, they have large genomes, grow under highly heterogenic environmental conditions and, most importantly, their generation cycles are very long (Lidder and Sonnino 2012). Technological advances, such as genome editing, will provide new perspectives for more efficient breeding strategies.

Characterization of the Genes Underlying QTLS Could Be an Exceptional Source of Targets for Genome Engineering

Once QTLs involved in a desired trait have been identified unequivocally, all the genes present in those loci need to be characterized to look for polymorphic genes linked to the desired phenotype. When an annotated version of the whole genome sequence is available, the putative function of the genes might point towards particular ones (candidate genes). This approach is more complicated for species, whose genomes are still not available, or for genes, whose function is not known. The availability and continuing cost decrease of molecular markers and the development of new genomic techniques, such as NGS, might enhance efficiency in identifying and characterizing candidate genes for various species (Pérez-de-Castro et al. 2012; Lim et al. 2014). The validation of the candidate gene is the key step of the whole process. For that matter, one possible strategy consists of the analysis of the genetic profile of the candidate gene with the aim of finding some nucleotide variations (i.e., SNPs, premature stop codons, etc.) that are absent in another variety of the same species which does not present that phenotype/trait. In the case, the comparison with other varieties of the same species is not possible, and a study of the importance of those particular mutations can be carried out in a model plant, where the presence of a reference genome and availability of obtaining mutants are more feasible than in other plant species. For this purpose, genetic engineering techniques can help to introduce point mutations. Once the link of that particular nucleotidic variation with the wanted phenotype/trait is certified, the same kind of variation can be introduced in the desired variety of interest or elite line. Direct modification of the variety of interest/elite line genome through genetic engineering techniques can result in restoring the desired trait and reducing the time of introgression of the allele by avoiding backcrossing of a large number of inbred lines. In the conventional breeding, a minimum of five-six backcrossing generations is necessary to transfer the gene of interest from a donor to a recipient line (Lidder and Sonnino2012). This process requires a lot of time and effort, especially when more than one allele needs to be transferred.

Role of Model Plants in the Validation of the Candidate Genes

The use of model plants, with a well-characterized genome, and ample genetic and molecular resources, is of high value in the validation of candidate genes for interesting phenotypes. The confirmation of the role of candidate genes in model species is a good indicator of their potential implication in the related crop species. For the final validation in crops, again, academic laboratories could provide assistance to breeders for the rapid production of the desired genetic profiles by genome engineering. Academic labs would benefit from their implication in such studies by deepening their overall knowledge of crop genomics and enlarging the field of application of their discoveries. Thus, cooperation between breeders/farmers and public research institutions could be very fruitful for the genome engineering of elite lines and should be enhanced by appropriate innovation programs.

SHOWCASING OF THE CANDIDATE GENES THROUGH GENOME ENGINEERING OF ELITE LINES

Precise Modification of Elite Line Genomes

Once an allele responsible for a particular desired trait has been identified, its transfer to an agronomically interesting variety can sometimes be a challenge. For that matter, genome engineering can provide means of inserting or modifying it into elite lines. Because of the new SDNs capable of inducing mutagenesis or gene replacement at specific targets, new alleles can be engineered very precisely at endogenous loci and with a very limited impact on the rest of the genome. Different types of mutations (deletions, insertions, and substitutions) can be generated randomly via error-prone non-homologous end-joining (NHEJ). In addition, it seems that by targeting a locus presenting micro-homologies, the output of the repair, that could be mediated via micro-homology mediated end-joining (MMEJ), can be predicted (Butler et al. 2015; Collonnier et al. 2016).

In addition to the introduction of targeted mutations in a particular gene, SDNs can be used to induce mutagenesis in different genes simultaneously, which can be particularly interesting for the modification of loci underlying quantitative traits. For that matter, several SDNs targeting different loci can be co-transformed (Ma et al. 2015; Shan et al. 2015; Lowder et al. 2015), but a single SDN targeting a sequence common to a defined set of genes can also be used. This latter approach is very useful to knock-out TAGs (tandemly arrayed genes) which are prevalent in higher eukaryotic genomes (e.g., ~17 % in *A. thaliana*) and to trigger gene clusters deletions (Christian et al. 2013; Qi et al. 2013; Zhou et al. 2014). Multiplexing can also lead to chromosome rearrangements and create novel chimeric genes (Qi et al. 2013). Inducing multi-KOs is particularly interesting for the creation of new phenotypes depending on paralogous genes in diploid species or on multiple homeologous genes in polyploid species (Wang et al.2014). Thus, SDNs are particularly amenable to engineer complex plant genomes and to confer valuable traits to crops with high ploidy and genome size (Gil-Humanes and Voytas 2014).

Knock-Out and Knock-in Strategies

Knock-out of specific candidate genes through imprecise NHEJ-mediated repair of SDN-induced DSBs offers a lot of opportunities for the creation of valuable agronomic traits in crops. Some innovative processing traits were developed in potato lines using TALENs targeting the three alleles of the *VInv* gene (vacuolar invertase). The resulting mutations limit the accumulation of reducing sugars in tubers during cold storage and minimize the production of acrylamide during frying (Clasen et al. 2016). Different quality traits were also produced via the knock-out strategy. Fragrant rice was created by targeted knock-out of the *OsBADH2* gene using TALENs (Shan et al. 2015). Tomatoes with a modified fruit ripening pattern were produced by CRISPR/Cas9-mediated mutagenesis of the *RIN* locus (Ito et al. 2015). Soybean lines with improved oil quality were obtained by TALEN-induced targeted mutagenesis of the fatty acid desaturase 2 gene families (Haun et al. 2014). Knocking-out host susceptibility genes is also a good way to develop a new resistance to diseases. In bread wheat, TALEN-induced mutation in three homeologous genes (*TaMLO*) conferred resistance to powdery mildew (Wang et al. 2014). TALENs and CRISPRs were also used to engineer resistance to bacterial blight in rice by targeting the binding sites of bacterial transcription activator-like effectors (Li et al.2012; Zhou et al. 2015a). Provided that candidate genes are identified, and other traits, such as tolerance to abiotic stresses (drought), resistance to diseases (such as sharka in *Prunus*) or better nutritional quality (lesser anti-nutritional compounds and healthier fatty acids) might be within reach via the knock-out strategy. However, knock-in of new alleles into the genome by homologous recombination-mediated targeted gene replacement offers an even wider range of potentialities and will probably be the method of choice in the future. So far, very few agronomic traits have been engineered by gene targeting in crops. Resistance to herbicides was, for example, conferred to tobacco lines by ZFN-stimulated allele replacement of two endogenous genes (Townsend et al. 2009). One technical way of supporting the soaring of the knock-in strategy could be to develop methodologies facilitating the proper and stable integration of the donor template. New tools, such as the recently identified nucleases Cpf1, which trigger sticky ends DSBs instead of blunt cuts like the Cas9 protein (Fagerlund et al.2015), have to be explored.

Will Genome Editing Progressively Take Over Marker Assisted Selection (MAS)?

The association of defined regions of the genome to traits of interest relies on diagnostic markers identified through the detection of classical QTLs by bi-parental crossing, or via association mapping. In MAS, these markers are then used to predict the genetic value of the individuals that are kept in the successive generations of breeding (Das and Rao 2015; Hasan et al. 2015; Pradhan et al. 2015). As proposed before, genome editing could help to speed up the breeding process of QTLs normally followed by MAS technology. Provided the number and the size of the genomic regions associated with the desired traits are not too high, one could even imagine that, in a near future, it could replace MAS in some cases. Because of the possibility of targeting different loci at the same time, the modifications could be done over very few generations, depending on the total number of loci to target (the number of SDNs capable of working simultaneously in a given cell being high but not unlimited) (Farré et al. 2014) and on the type of transformation used (some leading to homogenous and stable transformants in T0, T1 or T2 generation) (Feng et al. 2014). There is then no need then for labor intensive, costly and time-consuming backcrossing to perform QTL pyramiding. This approach is particularly indicated for mono- or oligo-genic traits and for polygenic traits depending on few genes with strong effects. For quantitative traits depending on many genes with small effects, whether the candidate genes are transferred into elite lines by hybridization or by genome engineering, the value of the phenotype may not be exactly the one expected, partly due to the modification of genetic backgrounds and the resulting loss of the relevant genetic interactions (Takeda and Matsuoka 2008). Genome editing could be very useful to evaluate and validate the strength of the predictive value of a given candidate gene by easily transferring its best alleles into a set of different genetic backgrounds representative of the diversity of the genetic material used in the selection schemes. As multiple genes can be individually engineered at the same time, genome editing also provides a way to modify linked genes or QTLs that are usually difficult to segregate due to the limits of meiotic recombination (Flavell 2010).

Beyond the Constraints of Elitism: Increasing Transformation and Mutation Frequencies

The promise of totipotency in the context of transformation has not yet been fulfilled in all crops. Genetic modifications were preferably realized in varieties amenable to transformation and regeneration protocols, that were, once transformed, backcrossed to elite lines. The optimization and the industrialization of certain critical steps of these protocols now allow for the acceleration the breeding process for many crops by directly transforming elite lines. For these crops, candidate genes can be engineered in elite lines without difficulty. For the others, genome editing relies on the development of robust and possibly genotype-independent methods of transformation and regeneration adapted to these highly selected materials. Most of the time, regeneration is the bottleneck and a better understanding of the genetics of this trait would be determinant. For the moment, more immediate solutions consist in trying to increase transformation efficiency and mutation frequency. One promising way is to use engineered plant virus expression vectors, such as modified geminivirus replicons, to express SDNs and bring the donor DNA (Butler et al. 2015; Čermák et al. 2015; Yin et al. 2015; Baltes et al. 2015). This approach also presents the advantage to produce edited plants without stable integration of the SDNs-coding transgenes that are only transiently expressed and do not need to be segregated in the progeny (Čermák et al. 2015). It is particularly interesting for species with high degrees of heterozygosis and long juvenile stages, such as fruit trees, for which genetic segregation would be long and costly (Ilardi and Tavazza 2015). This applies also to vegetatively propagated species, such as potato or grapes, for which segregation has a strong genetic cost. Very efficient mutation frequencies were also obtained in certain species by transforming nucleases directly as proteins (Luo et al. 2015; Woo et al. 2015). Finally, mutation efficiency can be optimized using specific promoters favoring the expression of SDN-coding genes in reproductive cells, the ratio of mutated T1 plants being sometimes 20-fold higher than with constitutive promoters (Wang et al. 2015).

No-one Left Behind: Breeding to Support Development

Genome editing could contribute to support the breeding of orphan crops that are of key economic importance in certain local environments, in particular in developing countries. These crops regroup species, such as sorghum and millets, groundnut, cowpea, common bean, chickpea, pigeonpea, cassava, yam and sweet potato (Varshney et al. 2012). They do not benefit from large investments from both public research institutes and private breeding companies. As a consequence, very little genomic information is available on them, which makes difficult the mapping of regions involved in traits of interest, as well as the testing of candidate genes. MAS can also be complicated by the use of landraces or wild germplasm as the source of favorable alleles (Varshney et al. 2012). By allowing very precise modification of genomic regions and targeted introgression of potential candidate genes coming from any source, genome editing can help lift some of these difficulties. In addition, by making breeding of new varieties less labor intensive and less costly, it could not only facilitate the work of local research teams, but also make private investment into small programs dedicated to this type of crops more appealing.

Several initiatives aimed at developing GM traits to answer health and agronomic issues in developing countries have been conducted in the world. They have been long tested without full success due to societal and governmental opposition to the arrival onto the market of seeds modified via classical transgenesis. The Golden Rice case (Potrykus 2015) and the attempts to grow virus-resistant papaya in Asia (Davidson 2012; Gonsalves 2015) illustrate very well this problem.

Back to the Future: Looking for Candidate Genes in Wild Species

As presented before, the search for the QTLs and identification of the genes that are beyond is a major source of information for the development of SDNs strategies. Most of the time, the choice of the material used for the QTLs identification is driven by the capacity of the parent lines to be cross-fertile with elite lines that will be the final recipient of the allele(s) of interest. In general, the genetic bottlenecks imposed on crops during domestication and through modern breeding practices have greatly reduced the genetic variability that can be used for breeding. For this reason, wild relatives or local landraces are a particularly useful source for allele mining (Tanksley and McCouch 1997). As a consequence, there are now more and more reports proposing to look for genetic diversity in wild species or local landraces through selection screens (Zhou et al. 2015b). However, crosses are not always easy or even feasible between domesticated varieties and their wild relatives. This barrier of sexual compatibility limits the gene pools accessible to the breeders (Michelmore 2003). On the contrary, in principle, the sole limitation for the application of the SDNs technology to a specific gene is the presence of its homolog in a donor organism, which could be a cross-incompatible wild relative or even more distant species, including non-plant species (e.g., bacteria). In addition, SDNs can also be used for what some authors have called a "rewilding" of actual crops through GE to get rid of mutations that compromise the hereditary basis of crop survival during environmental stresses and are rarely counter-selected (Palmgren et al. 2015). Thus, the search for superior alleles of a given agronomically important gene should no longer be restricted any more to related genotypes but should be extended to divergent ones.

GENOME TUNING THROUGH SDNS TO GO A STEP FURTHER THAN CLASSICAL PLANT BREEDING

Inventing New Candidate Genes

As already explained, the search for new alleles for plant improvement is the basis of plant breeding. However, this search for superior alleles should not be restricted to natural variations but should also include artificial design or induced genetic variations leading to the discovery of alleles otherwise not present in the available genetic diversity. Indeed, in addition to the screening of natural variation, several technologies can be employed for the improvement of a gene product, including protein-structure-based rational design (Lutz 2011), random mutagenesis and the more recent strategy of directed evolution. Directed evolution of proteins integrate random and focused mutagenesis using different strategies, such as error-prone PCR, DNA shuffling (in vivo or in vitro), and screening for better alleles in microbe systems (Minshull and Stemmer 2001). These strategies have been very useful for improving or altering the activity of biomolecules for industrial, research and therapeutic applications (Packer and Liu 2015) and, although they are still in their infancy for plant proteins (Lassner and Bedbrook 2001), they have a great potential for the optimization of some agronomically important gene products. However, we must keep in mind that these screenings in microbial systems will be essentially valid for enzymes and that the optimization achieved in a bacterial system will not necessarily be maintained in the plant. One way to avoid these two limitations would be to implement the directed evolution strategy in the plant itself.

Targeted Induced Genetic Variation (TIGV) an Alternative to TILLING?

Mutations, in combination with recombination, are the sources from which plant breeders are able to produce new varieties. As stated before, one way to increase the genetic variability available for breeding is to induce mutations and characterize them using the TILLING technique. However, the use of the newly produced allele in breeding plans is still time consuming and necessitates multiple cycles of crossing for introgression in an elite line. One way to avoid or limit these time-consuming steps of introgression could be the use of SDNs for Targeted Induced Genetic Variation (TIGV) (Fig. 1). CRISPR/Cas9, and in general SDNs, can be used to introduce specific or random mutations at a precise location in the genome depending on the use or not of a template DNA to repair the introduced DSB. In addition, a collection of SDNs can be engineered to introduce a combination of DSBs in the same gene. In the case of CRISPR/Cas9, this can be easily achieved using a collection of guide RNAs instead of a single one. This approach could be used to mutate a gene in multiple sites also presenting different modifications, allowing to generate a population of alleles of a particular gene. This would strongly increase the chance to find, for a given gene product, a superior allele that would result from a combination of specific mutations in multiple domains of the protein. One obvious target for TIGV is resistance genes, such as the NBS-LRR gene family, as the evolution of these genes is thought to be facilitated by recombination and sequence exchange, leading to variation in the recognition patterns of plants to pathogen elicitors (Joshi and Nayak 2013). Last but not least, the use of TIGV on different genes in the same plant would permit the generation of a population of mutants that would contain different combinations of alleles at different loci. Strategies consisting in the targeting of multiple genes in the same plant (multiplex editing) have been proven functional in crops (Xie et al. 2015). The production of such targeted mutagenized populations and their screening in different conditions of stresses would represent a very innovative solution to tackle the question of multiple and combined stresses in crop production (Kissoudis et al. 2014). Moreover, because stress conditions, such as salinity and drought, vary depending on local climates and geographical features, site-specific screening of these populations of mutants should permit the selection of superior gene alleles for very specific agricultural conditions leading to custom-made breeding. Nevertheless, the TIGV strategy has limitations. The first one is the capacity to transform efficiently elite lines for a given crop. Moreover, as the strategy is based on screening for mutants in natural conditions of growth, a second limitation is the possibility to grow stably transformed GM plants (i.e., mutated plants where the CAS9/gRNA module has not been segregated) in field trials.

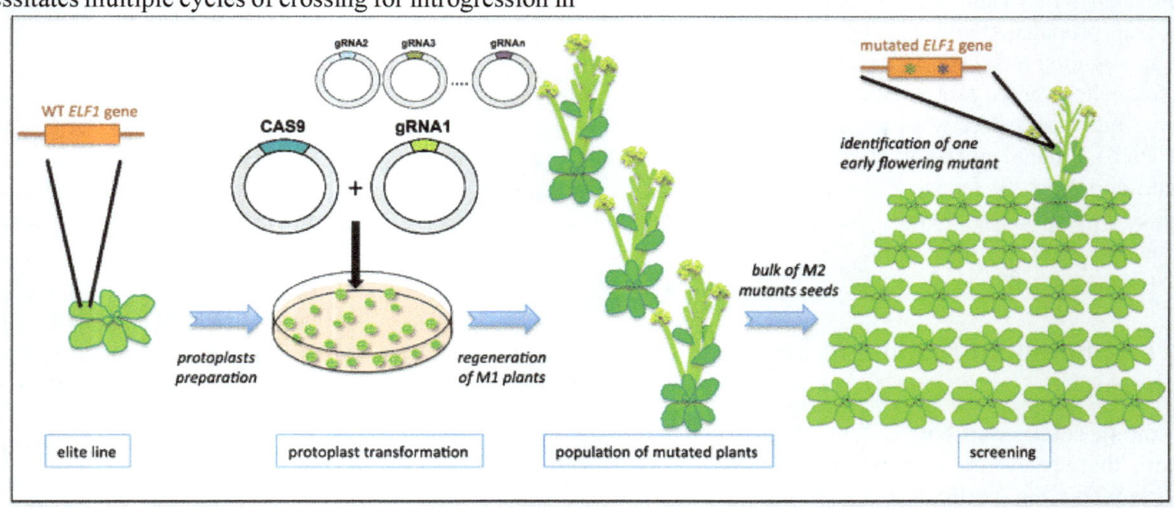

Figure 1: Principle of targeted induced genetic variation (TIGV). The aim of TIGV is to produce genetic variability at one or more loci in the genome and screen for individuals that would have acquired, at the targeted loci, more favorable alleles for one or more agronomic traits.

In this case, the desired trait is a short-cycle rapeseed line. For this purpose, the *ELF1* gene that has been shown to be involved in the transition from the vegetative to reproductive growth will be targeted using the CRISPR/Cas9 strategy. Wild type plants are stably or transiently (in this case via protoplasts PEG fusion) transformed with the CAS9 nuclease and one or more gRNAs targeting different sites in the *ELF1*gene. M1 plants are regenerated and selfed. M2 mutants seeds are then collected and sown in field conditions for screening of early flowering plants.

"Genome Remodeling", Crop Genome Optimization through SDNs

The use of SDNs is not restricted to the modification of coding sequences but can also be used to delete large chromosomal segments (up to 170 kb) or target repeated sequences, as already demonstrated using the CRISPR/Cas9 strategy in rice, for example, (Endo et al. 2015; Zhou et al. 2014). As already commented, these types of genomic structure-associated mutations, such as insertions, deletions, duplications, inversions and translocations, are part of the genomic differences within a species that are not limited to single nucleotide. These large polymorphisms are named structural variants (SVs) (Saxena et al. 2014). Some SVs have been clearly associated with specific phenotypes, such as aluminum-tolerance in maize that is due to an increase in the copy number of the *MATE1* gene (Maron et al. 2013). However, the SVs are not always associated with a gain of function and some of them can be detrimental to fitness. In addition, the proliferation of repetitive sequences can also be detrimental to the genome. For example, the banana genome is invaded by numerous badnavirus sequences, including those of banana streak viruses (BSVs) and some of these named endogenous BSVs (eBSVs), can release infectious pararetrovirus following stresses (Chabannes et al. 2013). Potential awakening of eBSVs has become the major constraint to banana breeding programs worldwide (Chabannes et al. 2013). One way to get rid of this risk could be the specific knock-out or, even better, the deletion of the BSVs sequences in the banana genome using the CRISPR/Cas9 strategy.

Mimic the CRISPR/Cas9 Defense System in Plants

Some of these new tools of genome engineering can also be used for purposes other than modifying the plant genome. The CRISPR-Cas system originates from bacteria and is an adaptive immune system to fight against foreign DNA, such as phages and, in some aspects, could be compared to the eukaryotic, RNA interference (RNAi) pathway, even if these two systems are not homologous (Marraffini and Sontheimer 2010). Recently, different groups have tried to transpose this bacterial immune system to plants. For this purpose, the CAS9 nuclease and single guide RNA targeting different sequences in the genome of geminiviruses have been expressed in *Nicotiana benthamiana* and Arabidopsis (Ali et al. 2015; Baltes et al. 2015; Ji et al.2015) and shown to induce mutations in the viral DNA sequence and to decrease the copy number of replicating virus. Interestingly, the co-expression of two sgRNAs in the plant increases the efficiency of reduction of viral genome copy number and targeting a common sequence in different viruses allows multiple virus resistance (Ali et al. 2015; Baltes et al. 2015). Thus, this strategy, due to its flexibility and ease of implementation, presents a great potential to protect plants against viruses, although this needs to be confirmed in field conditions. Moreover, like all the strategies to fight against viruses, the CRISPR/Cas9 also has potential limitations. First, because resistant virus strains could emerge due to mutations appearing in the targeted sequences and preventing the cleavage, or due to the production by the virus of suppressors of the CRISPR/Cas9 system, as it is already the case for some bacteriophages (van der Oost and Brouns 2015). Second, the constitutive expression of a CAS9/sgRNA module into the plant raises the question of the potential risk of unintended mutagenic effects (due to direct or off-target interaction) on the plant genome and/or on organisms that would be in contact with the plant.

CRISPR/Cas9 and Friends to Control Meiotic Recombination

The improvement of crop species relies on the possibility to produce new allele combinations, and, as mentioned before, new techniques, such as MAS, have greatly helped the breeders to do so. However, if the breeders can now more easily identify good allele combinations, they are still limited by the possibility of actually combining them in the plant. Indeed, this depends on the number and localization of recombination events that occur between homologous chromosomes during meiosis, which ensures the reshuffling of genetic diversity for evolution and adaptation of species. This is especially true when crossing is done between different species where sequence divergence in homeologous regions can impair cross-over formation during meiosis (Canady et al. 2006). Mastering the recombination at the crossing stage would, therefore, be a wonderful tool to control the combination of the chromosomal regions that contain the favorable alleles while avoiding the presence of unfavorable alleles that can be linked to them. A key step of meiotic recombination is the formation of double strand breaks (DSBs) into the chromosomes that subsequently define loci of mutual genetic exchange. To manipulate meiotic recombination in crop plants, one could consider targeting the meiotic double strand breaks to specific regions to induce site-specific cross-over, where genetic exchange is needed (Wijnker and de Jong 2008). The SPO11 protein and its partners are the key actors of the meiotic-programmed DSBs. In yeast, it has been shown that a fusion of the SPO11 protein to the DNA-binding domain of the transcriptional activator Gal4 can stimulate DSB formation, with associated recombination, near Gal4-consensus-binding sites (Peciña et al. 2002). Based on these data, it would be tempting to test in plants whether SPO11 fusions with a variety of different DNA-binding domains, such as TALE effectors or dead CAS9 associated to specific sgRNA, could target DSBs to specific regions of the genome. Such a tool would be very useful to breeders to accelerate the combination of favorable alleles.

CONCLUSION

Plant evolution relies on spontaneous genome mutations potentially resulting in new traits fixed by natural selection. Plant breeding also relies on natural genetic variability but, in addition, breeders have increased it using random mutagenesis. Genome editing now provides means to introduce almost any type of mutation and chromosome rearrangements in a very precise way. This not only empowers the breeders to accelerate and direct crop selection in an unprecedented way; it also opens up the door to an almost unlimited range of possibilities in terms of the combination of new alleles by erasing sexual barriers. These new tools could be integrated in breeding schemes very rapidly in the upcoming years. From a scientific point of view, the main limiting factor is the reliable and efficient identification of the genes underlying traits of interest and the evaluation of their combination on the value of these traits. For that matter, developing efficient gene-function analysis tools and precise high-throughput phenotyping methods are essential. However, the main uncertainty on the future use of these techniques for plant breeding is the regulatory framework that will be applied to their commercial products. As discussed in chapter 10, whether these plants and their products are considered under the GMO legislation and risk assessed as GMOs, or whether they are submitted to an alleviated legal framework or are completely deregulated, will have a profound impact on the development and the use of these techniques in plant breeding. The costs and delays associated to the GMO approval process would probably block the use of these techniques for most crops and traits, and would make it impossible for small breeders and seed companies to engage in the development of new varieties using genome editing, as it has already happened with conventional GMOs.

Author contribution statement

All authors wrote and approved the final manuscript.

ACKNOWLEDGMENTS

K.M. was supported by the Agence National de la Recherche (Program ANR-11-BTBR-0001-GENIUS). Work done at CRAG is funded by a grant from Ministerio de economía y competitividad (AGL2013-43244-R). The authors would like to thank Samantha Mara for careful reading of the manuscript.

CONFLICT OF INTEREST

The authors declare that they have no conflict of interest.

REFERENCES

1. Ali Z, Abulfaraj A, Idris A et al (2015) CRISPR/Cas9-mediated viral interference in plants. Genome Biol 16:238. doi:10.1186/s13059-015-0799-6
2. Ashikari M, Matsuoka M (2006) Identification, isolation and pyramiding of quantitative trait loci for rice breeding. Trends Plant Sci 11:344–350. doi:10.1016/j.tplants.2006.05.008
3. Baltes NJ, Hummel AW, Konecna E et al (2015) Conferring resistance to geminiviruses with the CRISPR–Cas prokaryotic immune system. Nat Plants 1:15145. doi:10.1038/nplants.2015.145
4. Blackman BK, Strasburg JL, Raduski AR et al (2010) The role of recently derived FT paralogs in sunflower domestication. Curr Biol 20:629–635. doi:10.1016/j.cub.2010.01.059
5. Bolon Y-T, Stec AO, Michno J-M et al (2014) Genome resilience and prevalence of segmental duplications following fast neutron irradiation of soybean. Genetics 198:967–981. doi:10.1534/genetics.114.170340
6. Butler NM, Atkins PA, Voytas DF, Douches DS (2015) Generation and inheritance of targeted mutations in potato (*Solanum tuberosum* L.) using the CRISPR/Cas system. PLoS One 10:e0144591. doi:10.1371/journal.pone.0144591
7. Canady MA, Ji YF, Chetelat RT (2006) Homeologous recombination in *Solanum lycopersicoides* introgression lines of cultivated tomato. Genetics 174:1775–1788. doi:10.1534/genetics.106.065144
8. Čermák T, Baltes NJ, Čegan R et al (2015) High-frequency, precise modification of the tomato genome. Genome Biol 16:232. doi:10.1186/s13059-015-0796-9
9. Chabannes M, Baurens F-C, Duroy P-O et al (2013) Three infectious viral species lying in wait in the banana genome. J Virol 87:8624–8637. doi:10.1128/JVI.00899-13
10. Christian M, Qi Y, Zhang Y, Voytas DF (2013) Targeted mutagenesis of *Arabidopsis thaliana*using engineered TAL effector nucleases. G3 (Bethesda) 3:1697–1705. doi:10.1534/g3.113.007104
11. Clasen BM, Stoddard TJ, Luo S et al (2016) Improving cold storage and processing traits in potato through targeted gene knockout. Plant Biotechnol J 14:169–176. doi:10.1111/pbi.12370
12. Collonnier C, Nogué F, Casacuberta JM (2016) Targeted genetic modification in crops using site-directed nucleases. In: Watson RR, Preedy VR (eds) Genetically modified organisms in food. Elsevier, London, pp 133–145
13. Das G, Rao GJN (2015) Molecular marker assisted gene stacking for biotic and abiotic stress resistance genes in an elite rice cultivar. Front Plant Sci 6:1–18. doi:10.3389/fpls.2015.00698
14. Davidson SN (2012) Power, progress and prevarication: local knowledge and GE papaya in Thailand. GM Crop Food 3:104–110. doi:10.4161/gmcr.19331
15. Doebley JF, Gaut BS, Smith BD (2006) The molecular genetics of crop domestication. Cell 127:1309–1321. doi:10.1016/j.cell.2006.12.006
16. Endo M, Mikami M, Toki S (2015) Multigene knockout utilizing off-target mutations of the CRISPR/Cas9 system in rice. Plant Cell Physiol 56:41–47. doi:10.1093/pcp/pcu154
17. Fagerlund RD, Staals RHJ, Fineran PC (2015) The Cpf1 CRISPR-Cas protein expands genome-editing tools. Genome Biol 16:251. doi:10.1186/s13059-015-0824-9
18. Farré G, Blancquaert D, Capell T et al (2014) Engineering complex metabolic pathways in plants. Annu Rev Plant Biol 65:187–223. doi:10.1146/annurev-arplant-050213-035825
19. Feng Z, Mao Y, Xu N et al (2014) Multigeneration analysis reveals the inheritance, specificity, and patterns of CRISPR/Cas-induced gene modifications in Arabidopsis. Proc Natl Acad Sci USA 111:4632–4637. doi:10.1073/pnas.1400822111
20. Flavell R (2010) From genomics to crop breeding. Nat Biotechnol 28:144–145. doi:10.1038/nbt0210-144

21. Gaut BS (2015) Evolution is an experiment: assessing parallelism in crop domestication and experimental evolution. Mol Biol Evol 32:1661–1671. doi:10.1093/molbev/msv105
22. Gil-Humanes J, Voytas DF (2014) Wheat rescued from fungal disease. Nat Biotechnol 32:886–887. doi:10.1038/nbt.3013
23. Gonsalves D (2015) The wayward Hawaiian boy returns home. Annu Rev Phytopathol 53:1–17. doi:10.1146/annurev-phyto-080614-120314
24. Hamblin MT, Casa AM, Sun H et al (2006) Challenges of detecting directional selection after a bottleneck: lessons from *Sorghum bicolor*. Genetics 173:953–964. doi:10.1534/genetics.105.054312
25. Hasan MM, Rafii MY, Ismail MR et al (2015) Marker-assisted backcrossing: a useful method for rice improvement. Biotechnol Biotechnol Equip 29:237–254. doi:10.1080/13102818.2014.995920
26. Haun W, Coffman A, Clasen BM et al (2014) Improved soybean oil quality by targeted mutagenesis of the fatty acid desaturase 2 gene family. Plant Biotechnol J 12:1–7. doi:10.1111/pbi.12201
27. Ilardi V, Tavazza M (2015) Biotechnological strategies and tools for Plum pox virus resistance: trans-, intra-, cis-genesis, and beyond. Front Plant Sci 6:379. doi:10.3389/fpls.2015.00379
28. Ito Y, Nishizawa-Yokoi A, Endo M et al (2015) CRISPR/Cas9-mediated mutagenesis of the RIN locus that regulates tomato fruit ripening. Biochem Biophys Res Commun 467:76–82. doi:10.1016/j.bbrc.2015.09.117
29. Ji X, Zhang H, Zhang Y et al (2015) Establishing a CRISPR–Cas-like immune system conferring DNA virus resistance in plants. Nat Plants 1:15144. doi:10.1038/nplants.2015.144
30. Jin J, Huang W, Gao J-P et al (2008) Genetic control of rice plant architecture under domestication. Nat Genet 40:1365–1369. doi:10.1038/ng.247
31. Joshi RK, Nayak S (2013) Perspectives of genomic diversification and molecular recombination towards R-gene evolution in plants. Physiol Mol Biol Plants 19:1–9. doi:10.1007/s12298-012-0138-2
32. Kissoudis C, van de Wiel C, Visser RGF, van der Linden G (2014) Enhancing crop resilience to combined abiotic and biotic stress through the dissection of physiological and molecular crosstalk. Front Plant Sci 5:1–20. doi:10.3389/fpls.2014.00207
33. Lassner M, Bedbrook J (2001) Directed molecular evolution in plant improvement. Curr Opin Plant Biol 4:152–156. doi:10.1016/S1369-5266(00)00152-7
34. Li T, Liu B, Spalding MH et al (2012) High-efficiency TALEN-based gene editing produces disease-resistant rice. Nat Biotechnol 30:390–392. doi:10.1038/nbt.2199
35. Lidder P, Sonnino A (2012) Biotechnologies for the management of genetic resources for food and agriculture. Adv Genet 78:1–167. doi:10.1016/B978-0-12-394394-1.00001-8
36. Lim J-H, Yang H-J, Jung K-H et al (2014) Quantitative trait locus mapping and candidate gene analysis for plant architecture traits using whole genome re-sequencing in rice. Mol Cells 37:149–160. doi:10.14348/molcells.2014.2336
37. Lisch D (2013) How important are transposons for plant evolution? Nat Rev Genet 14:49–61. doi:10.1038/nrg3374
38. Lowder LG, Zhang D, Baltes NJ et al (2015) A CRISPR/Cas9 toolbox for multiplexed plant genome editing and transcriptional regulation. Plant Physiol 169:971–985. doi:10.1104/pp.15.00636
39. Luo S, Li J, Stoddard TJ et al (2015) Non-transgenic plant genome editing using purified sequence-specific nucleases. Mol Plant 8:1425–1427. doi:10.1016/j.molp.2015.05.012
40. Lutz S (2011) Beyond directed evolution—semi-rational protein engineering and design. Curr Opin Biotechnol 21:734–743. doi:10.1016/j.copbio.2010.08.011.Beyond
41. Ma X, Zhang Q, Zhu Q et al (2015) A robust CRISPR/Cas9 system for convenient, high-efficiency multiplex genome editing in monocot and dicot plants. Mol Plant 8:1274–1284. doi:10.1016/j.molp.2015.04.007
42. Maron LG, Guimarães CT, Kirst M et al (2013) Aluminum tolerance in maize is associated with higher MATE1 gene copy number. Proc Natl Acad Sci USA 110:5241–5246. doi:10.1073/pnas.1220766110
43. Marraffini LA, Sontheimer EJ (2010) CRISPR interference: RNA-directed adaptive immunity in bacteria and archaea. Nat Rev Genet 11:181–190. doi:10.1038/nrg2749
44. Meyer RS, Purugganan MD (2013) Evolution of crop species: genetics of domestication and diversification. Nat Rev Genet 14:840–852. doi:10.1038/nrg3605
45. Michelmore RW (2003) The impact zone: genomics and breeding for durable disease resistance. Curr Opin Plant Biol 6:397–404. doi:10.1016/S1369-5266(03)00067-0
46. Minshull J, Stemmer WPC (2001) Molecular breeding: the natural approach to protein design. Adv Protein Chem 55:261–292. doi:10.1016/S0065-3233(01)55006-8
47. Muller HJ (1928) The production of mutations by X-rays. Proc Natl Acad Sci 14:714–726. doi:10.1126/science.68.1751.59
48. Olsen KM, Wendel JF (2013) A bountiful harvest: genomic insights into crop domestication phenotypes. Annu Rev Plant Biol 64:47–70. doi:10.1146/annurev-arplant-050312-120048
49. Ossowski S, Schneeberger K, Lucas-Lledó JI et al (2010) The rate and molecular spectrum of spontaneous mutations in *Arabidopsis thaliana*. Science 327:92–94. doi:10.1126/science.1180677
50. Packer MS, Liu DR (2015) Methods for the directed evolution of proteins. Nat Rev Genet 16:379–394. doi:10.1038/nrg3927
51. Palmgren MG, Edenbrandt AK, Vedel SE et al (2015) Are we ready for back-to-nature crop breeding? Trends Plant Sci 20:155–164. doi:10.1016/j.tplants.2014.11.003
52. Parry MAJ, Madgwick PJ, Bayon C et al (2009) Mutation discovery for crop improvement. J Exp Bot

60:2817–2825. doi:10.1093/jxb/erp189

53. Peciña A, Smith KN, Mézard C et al (2002) Targeted stimulation of meiotic recombination. Cell 111:173–184

54. Pérez-de-Castro AM, Vilanova S, Cañizares J et al (2012) Application of genomic tools in plant breeding. Curr Genomics 13:179–195. doi:10.2174/138920212800543084

55. Podevin N, Davies HV, Hartung F et al (2013) Site-directed nucleases: a paradigm shift in predictable, knowledge-based plant breeding. Trends Biotechnol 31:375–383. doi:10.1016/j.tibtech.2013.03.004

56. Potrykus I (2015) From the concept of totipotency to biofortified cereals. Annu Rev Plant Biol 66:1–22. doi:10.1146/annurev-arplant-043014-114734

57. Pradhan SK, Nayak DK, Mohanty S et al (2015) Pyramiding of three bacterial blight resistance genes for broad-spectrum resistance in deepwater rice variety, Jalmagna. Rice (N Y) 8:51. doi:10.1186/s12284-015-0051-8

58. Qi Y, Li X, Zhang Y et al (2013) Targeted deletion and inversion of tandemly arrayed genes in*Arabidopsis thaliana* using zinc finger nucleases. G3 (Bethesda) 3:1707–1715. doi:10.1534/g3.113.006270

59. Saxena RK, Edwards D, Varshney RK (2014) Structural variations in plant genomes. Brief Funct Genomics 13:296–307. doi:10.1093/bfgp/elu016

60. Schmutz J, McClean PE, Mamidi S et al (2014) A reference genome for common bean and genome-wide analysis of dual domestications. Nat Genet 46:707–713. doi:10.1038/ng.3008

61. Semagn K, Bjørnstad Å, Ndjiondjop MN (2006) Progress and prospects of marker assisted backcrossing as a tool in crop breeding programs. J Biotechnol 5:2588–2603

62. Semagn K, Bjørnstad Å, Xu Y (2010) The genetic dissection of quantitative traits in crops. Electron J Biotechnol 13:16–17. doi:10.2225/vol13-issue5-fulltext-14

63. Shan Q, Zhang Y, Chen K et al (2015) Creation of fragrant rice by targeted knockout of the OsBADH2 gene using TALEN technology. Plant Biotechnol J 13:791–800. doi:10.1111/pbi.12312

64. Stadler LJ (1928) Mutations in barley induced by X-rays and radium. Science 68:186–187. doi:10.1126/science.68.1756.186

65. Studer A, Zhao Q, Ross-Ibarra J, Doebley J (2011) Identification of a functional transposon insertion in the maize domestication gene tb1. Nat Genet 43:1160–1163

66. Takeda S, Matsuoka M (2008) Genetic approaches to crop improvement: responding to environmental and population changes. Nat Rev Genet 9:444–457. doi:10.1038/nrg2342

67. Tan L, Li X, Liu F et al (2008) Control of a key transition from prostrate to erect growth in rice domestication. Nat Genet 40:1360–1364. doi:10.1038/ng.197

68. Tanksley SD, McCouch SR (1997) Seed banks and molecular maps: unlocking genetic potential from the wild. Science 277:1063–1066. doi:10.1126/science.277.5329.1063

69. Townsend JA, Wright DA, Winfrey RJ et al (2009) High-frequency modification of plant genes using engineered zinc-finger nucleases. Nature 459:442–445. doi:10.1038/nature07845

70. van der Oost J, Brouns SJJ (2015) CRISPR sabotage. Genome Biol 16:248. doi:10.1186/s13059-015-0820-0

71. Varshney RK, Ribaut J-M, Buckler ES et al (2012) Can genomics boost productivity of orphan crops? Nat Biotechnol 30:1172–1176. doi:10.1038/nbt.2440

72. Wang RL, Stec A, Hey J et al (1999) The limits of selection during maize domestication. Nature 398:236–239. doi:10.1038/18435

73. Wang Y, Cheng X, Shan Q et al (2014) Simultaneous editing of three homoeoalleles in hexaploid bread wheat confers heritable resistance to powdery mildew. Nat Biotechnol Adv 32:947–951. doi:10.1038/nbt.2969

74. Wang Z-P, Xing H, Dong L et al (2015) Egg cell-specific promoter-controlled CRISPR/Cas9 efficiently generates homozygous mutants for multiple target genes in Arabidopsis in a single generation. Genome Biol 16:144. doi:10.1186/s13059-015-0715-0

75. Wendel JF (2015) The wondrous cycles of polyploidy in plants. Am J Bot. doi:10.3732/ajb.1500320

76. Wijnker E, de Jong H (2008) Managing meiotic recombination in plant breeding. Trends Plant Sci 13:640–646. doi:10.1016/j.tplants.2008.09.004

77. Woo JW, Kim J, Kwon SI et al (2015) DNA-free genome editing in plants with preassembled CRISPR-Cas9 ribonucleoproteins. Nat Biotechnol 33:1162–1164. doi:10.1038/nbt.3389

78. Xiao H, Jiang N, Schaffner E et al (2008) A retrotransposon-mediated gene duplication underlies morphological variation of tomato fruit. Science 319:1527–1530. doi:10.1126/science.1153040

79. Xie K, Minkenberg B, Yang Y (2015) Boosting CRISPR/Cas9 multiplex editing capability with the endogenous tRNA-processing system. Proc Natl Acad Sci USA 112:3570–3575. doi:10.1073/pnas.1420294112

80. Yin K, Han T, Liu G et al (2015) A geminivirus-based guide RNA delivery system for CRISPR/Cas9 mediated plant genome editing. Sci Rep 5:14926. doi:10.1038/srep14926

81. Yu J, Buckler ES (2006) Genetic association mapping and genome organization of maize. Curr Opin Biotechnol 17:155–160. doi:10.1016/j.copbio.2006.02.003

82. Zhou H, Liu B, Weeks DP et al (2014) Large chromosomal deletions and heritable small genetic changes induced by CRISPR/Cas9 in rice. Nucleic Acids Res 42:10903–10914. doi:10.1093/nar/gku806

83. Zhou J, Peng Z, Long J et al (2015a) Gene targeting by the TAL effector PthXo2 reveals cryptic resistance gene for bacterial blight of rice. Plant J 82:632–643. doi:10.1111/tpj.12838

84. Zhou Z, Jiang Y, Wang Z et al (2015b) Resequencing 302 wild and cultivated accessions identifies genes related to domestication and improvement in soybean. Nat Biotechnol 33:408–414. doi:10.1038/nbt.3096

CHAPTER 12

TARGETED DELETION AND INVERSION OF TANDEMLY ARRAYED GENES IN ARABIDOPSIS THALIANA USING ZINC FINGER NUCLEASES

Yiping Qi, Xiaohong Li, Yong Zhang, Colby G. Starker, Nicholas J. Baltes, Feng Zhang, Jeffry D. Sander, Deepak Reyon, J. Keith Joung, and Daniel F. Voytas

Genome sequences of many plant species have been completed, and all contain a significant number of tandemly arrayed genes (TAGs). For example, approximately 14% of rice genes (International Rice Genome Sequencing Project 2005), 17% of Arabidopsis genes (Arabidopsis Genome Initiative 2000), and 35% of maize genes (Messing *et al.* 2004) are organized in tandem. The genetic redundancy resulting from TAGs presents a challenge for reverse genetics (Jander and Barth 2007). It is difficult, if not impossible, to eliminate TAG expression with the use of methods such as ethyl methanesulfonate mutagenesis and targeting-induced local lesions in genomes, *i.e.*, TILLING (Koornneef *et al.* 1982; McCallum *et al.* 2000), T-DNA and transposon insertional mutagenesis (Alonso *et al.* 2003; Raina *et al.* 2002; Rosso *et al.*2003; Sessions *et al.* 2002; Woody *et al.* 2007), or RNA interference and miRNA-based gene silencing (Abbott *et al.* 2002; Alvarez *et al.* 2006; Hamilton and Baulcombe 1999; Schwab *et al.* 2006). One promising approach for studying TAGs is to make chromosomal deletions. Ionizing radiation, however, acts randomly (Li *et al.* 2001), making it difficult to recover the desired deletion. Although the Cre-Lox system has proven effective for making deletions, it relies on large LoxP T-DNA insertion populations (Zhang *et al.*2003), which currently are unavailable for most plant species.

An alternative approach to make targeted genome deletions is to use sequence-specific nucleases. These proteins, which include zinc finger nucleases (ZFNs), transcription activator-like effector nucleases (TALENs), and meganucleases, make site-specific DNA double-strand breaks (DSBs) at a locus of interest (Christian *et al.* 2010; Kim *et al.* 1996; Smith *et al.* 2006). Repair of DSBs occurs by two pathways, namely nonhomologous end-joining (NHEJ) and homologous recombination (HR) (Puchta 2005; Puchta *et al.* 1996). NHEJ is error-prone and typically leads to insertions, deletions (indels), and substitutions at the cleavage site. In contrast, repair by HR is typically error-free because it uses a DNA template to correct the break. Of the three nuclease platforms, ZFNs have been most widely used in plants. ZFNs have been successfully used for targeted mutagenesis by NHEJ in Arabidopsis (Osakabe *et al.* 2010; Zhang *et al.* 2010) and soybean (Curtin *et al.* 2011), as well as for gene targeting by HR in tobacco (Townsend *et al.* 2009) and maize (Shukla *et al.* 2009). In addition, Petolino *et al.* (2010) reported that when a 4.3-kb beta-glucuronidase transgene was flanked by two ZFN sites, it could be efficiently deleted from the tobacco genome, thus demonstrating that ZFNs can induce chromosomal deletions of transgenes in plants. Although ZFN-mediated deletion, inversion, and duplication of endogenous chromosomal DNAs has been achieved in human cells (Lee *et al.* 2010, 2011), none of these chromosome rearrangements have yet to be demonstrated in plant cells by the use of sequence-specific nucleases.

In Arabidopsis, the receptor-like kinase (*RLK*) and the nucleotide-binding and leucine-rich repeat resistance (*R*) gene families are large and have ~600 and ~150 gene members, respectively (Meyers *et al.* 2003; Shiu *et al.* 2004). Both gene families play important roles in plant development and immunity. For example, many plant hormone receptors and almost all plant immune receptors are members of these two families. Genes in both families are organized in tandem throughout the genome. In this study, we sought to delete endogenous TAGs by using ZFNs that target three *RLK* gene clusters and one large *R* gene cluster. We successfully demonstrated targeted deletions, inversions, and duplications of multiple gene clusters as well as large chromosomal deletions exceeding 9 Mb.

Citation: Yiping Qi, Xiaohong Li, Yong Zhang, Colby G. Starker, Nicholas J. Baltes, Feng Zhang,, Jeffry D. Sander, Deepak Reyon, J. Keith Joung and Daniel F. Voytas, Targeted Deletion and Inversion of Tandemly Arrayed Genes in Arabidopsis thaliana Using Zinc Finger Nucleases, doi: 10.1534/g3.113.006270.

MATERIALS AND METHODS

ZFN assembly

Genomic DNA sequences of target genes were analyzed with the software ZiFiT Targeter (version 3.3) to identify ZFN sites for which ZFNs could be engineered using the Context-Dependent Assembly (CoDA) method (Curtin *et al.* 2011;Sander *et al.* 2011). DNA sequences encoding ZFNs of choice (Supporting Information, Figure S1 and Table S1) were assembled by mutagenesis and overlapping polymerase chain reaction (PCR) using standard molecular cloning procedures. For each ZFN, ZF arrays were first cloned into the yeast expression vectors pCP3 and pCP4 using available *Xba*I and *Bam*HI sites (Zhang *et al.*2010). Then, DNA sequences for the left and right ZF arrays were excised from the yeast expression vectors with *Xba*I and *Bam*HI and moved into the pZHY013 entry clone using the *Xba*I-*Bam*HI and *Nhe*I-*Bgl*II sites, respectively (Figure S2). pZHY013 contains an obligate *Fok*I heterodimer architecture (Miller *et al.* 2007), and the ZFNs are linked by a T2A translational skipping sequence. The plant ZFN expression vectors were made using a Gateway LR reaction between the aforementioned entry clones and the pFZ19 destination vector (Zhang *et al.*2010).

Transgenic Plants and Expression of ZFNs

Agrobacterium tumefaciens GV3101/pMP90 was transformed with pFZ19 plasmids containing the ZFNs. The transformed *A. tumefaciens* strain was then used to transform Arabidopsis Col-0 (wild-type) plants using the floral dip method (Clough and Bent 1998). T1 transgenic plants were selected by growing the sterilized seeds on 0.5× MS solid medium (0.8% agar) that contained 100 μg/mL timentin (PlantMedia) and 20 μg/mL hygromycin B (Roche). For inducing ZFN expression, 20 μM β-estradiol (Sigma-Aldrich) was included in the medium.

ZFN Activity Measurement

One-week-old seedlings grown on MS medium with estradiol were harvested for DNA extraction with the CTAB DNA isolation method (Stewart and Via 1993). Eight T2 transgenic plants from the same T1 parent were bulked to represent each sample, whereas eight wild-type plants were bulked as the negative control.

To detect ZFN activity, an enrichment PCR procedure was used. To summarize, ~500 ng of genomic DNA from each sample was digested overnight (16 hr) with 1 μL of *Dde*I (for At1g53-ZFN), *Afl*II (for At1g70-ZFN), *Bfa*I (for At4g16-ZFN), *Bmg* BI (for At3g21-ZFN), or *Ear*I (for At5g01-ZFN) in a 20-μL reaction volume. Four microliters of digested genomic DNA was used for PCR amplification of the corresponding ZFN target sites in a 25-μL reaction volume. Ten microliters of unpurified PCR product was then digested with 1 μL of the same restriction enzyme in a 40-μL reaction volume for 12–16 hr. Digested products were resolved by electrophoresis in 1.5% agarose gels; mutations created by ZFNs were evidenced as undigested PCR products.

An alternative method to detect and measure ZFN activity by restriction digestion is similar to enrichment PCR except that the genomic DNA digestion step is omitted. Four microliters of 50 ng/μL genomic DNA was directly used for PCR and subsequent digestion by the corresponding restriction enzyme. The frequency of ZFN-mediated mutagenesis was measured by quantifying the percentage of undigested PCR product that was resolved by electrophoresis on a 1.5% agarose gel.

ZFN activity also was measured by the Surveyor assay in which T7 endonuclease was substituted for Cel-I (Guschin *et al.* 2010). In summary, PCR products amplified from genomic DNA templates were purified with a QIAquick PCR purification kit. For each sample, ~500 ng of purified PCR product was mixed with NEB buffer 2 in a 30-μL volume. To promote heteroduplex formation, PCR amplicons were denatured and reannealed using the following regime: 95° for 5 min, 95° to 85° at −1.5°/sec, and 85° to 25° at −0.1°/sec. One microliter of T7 endonuclease (NEB) was added to each sample for digestion at 37° for 1 hr. The digested products were resolved by electrophoresis in 1.5% agarose gels, and the frequency of ZFN-mediated mutagenesis was quantified as described previously (Guschin *et al.* 2010).

Sequence Confirmation of Mutagenesis and Deletions

Both undigested PCR products (for assessing mutagenesis) and PCR products (for detecting deletions, inversions, and duplications) were purified with the QIAquick Gel Extraction Kit. The purified DNA products were then cloned using either the pCR8/GW/TOPO TA Cloning Kit or the pCR2.1 Original TA Cloning Kit (Invitrogen). Multiple clones for each experiment were randomly picked and subjected to DNA sequencing.

Deletion Frequency Measurements

A method similar to digital PCR analysis (Lee *et al.* 2010) was used to estimate deletion frequencies. Genomic DNA samples were serially diluted (in a 3× gradient) in distilled water, and they were used for PCR via the use of deletion-specific primer pairs (Table S2). The same DNA samples were used to amplify a fragment of the *ADH1* gene as a genomic DNA copy number control. In each case, the difference of dilution factors for both deletion PCR and control PCR was used to calculate deletion frequency.

RESULTS

Strategy for Targeted Deletion of TAGs

Chromosomal deletions can be stimulated by two coordinated DSBs (Petolino *et al.* 2010). In our strategy, we used a single pair of ZFNs, which because of the high sequence similarity among TAGs, created two or more DSBs. Five ZFNs were engineered using the CoDA method to target three *RLK* gene clusters, the*RPP4 R* gene cluster, and the *ASK8* gene cluster (Figure 1 and Figure S3). These five ZFNs were expected to induce a total of 13 DNA DSBs in targeted exons of the Arabidopsis genome. The resulting size of predicted deletions ranged from a few to more than 50 kb. In addition to inducing deletions, ligation of broken chromosomes can create novel gene fusions. However, because NHEJ is error-prone, we anticipated that indels would be introduced at the cleavage site, thus rendering some of the deleted TAGs nonfunctional.

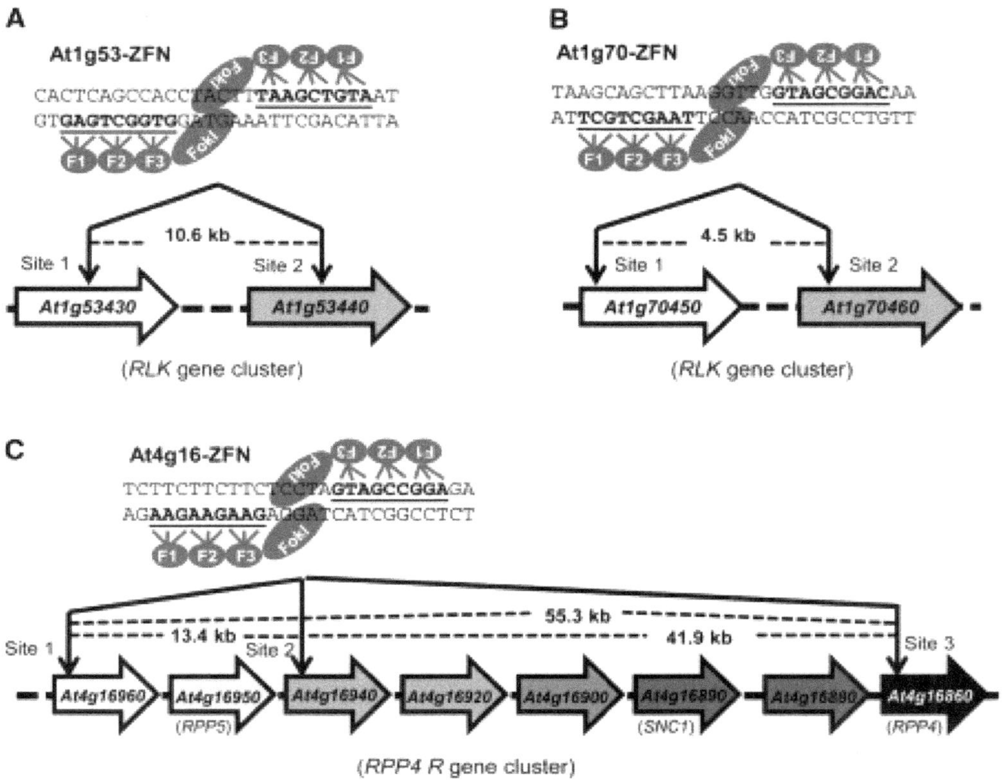

Figure 1: Schematic of target genes and ZFN sites. (A) The At1g53-ZFN targets both*At1g53430* and *At1g53440*in the 14th exon of each gene. (B) The At1g70-ZFN targets *At1g70450* in the 1st exon and *At70460* in the 2nd exon. (C) The At4g16-ZFN targets three sites in the*RPP4* R gene cluster: the 1st exon of *At4g16960*, the 5′ UTR of *At4g16940*, and the 1st exon of *At4g16860*. Illustration of ZFN pairs depicts the DNA recognition triplets for each zinc finger. The zinc finger binding sequences are underlined, and the distance between cleavage sites is shown.

Detection of ZFN-induced Mutagenesis

For all ZFNs used in this study, both the left and right ZFNs were expressed from an estradiol-inducible promoter and separated by a "self-cleaving" T2A peptide, which promotes production of two proteins from one mRNA by a translational skipping mechanism (Szymczak *et al.* 2004). T1 transgenic plants were screened on MS medium containing both hygromycin (to identify transgenic plants) and β-estradiol, which allowed for induction of ZFN transgenes. Detection of ZFN-induced mutations was performed by the use of enrichment PCR (Qi *et al.* 2013). Indels introduced by ZFNs frequently occur in the "spacer" region, where the *Fok*I nuclease domains dimerize and cleave the DNA. Thus, if there is a unique restriction enzyme site in the spacer, this restriction enzyme site will likely be destroyed through ZFN-mediated mutagenesis and thus render the DNA uncuttable by the restriction enzyme.

For each locus evaluated, T1 plants in which ZFNs were induced by estradiol were pooled together and genomic DNA was extracted. For the At1g53-ZFN, which targets the *At1g53430-At1g53440* gene cluster, ZFN activity was detected at the *At1g53430* target site (Figure S4A). Similarly, the At1g70-ZFN and At4g16-ZFN were found to be active at the *At1g70450-At1g70460* gene cluster and the *RPP4* gene cluster, respectively (Figure S4, B and C). We did not detect ZFN activity for the At3g21-ZFN and the At5g01-ZFN (Figure S3, data not shown). These data suggest that three of the five ZFN pairs are functional, and their activity can be temporally controlled by the estradiol-inducible promoter (Figure 1). We thus focused on the three active ZFNs for further study.

Quantification of ZFN Activity at Seven Endogenous Loci

We screened multiple T2 populations of Arabidopsis plants transformed with each ZFN pair and selected two independent lines showing high ZFN activity. These transgenic plants were used to quantify ZFN activity based on the level of NHEJ mutagenesis in whole seedlings. Note that the frequency of mutation observed by enrichment PCR is influenced by the position of the restriction enzyme site. The *Afl*II and *Bfa*I sites are very close to the middle of the spacer in the At1g70-ZFN and At4g16-ZFN recognition sites, respectively. In these cases, we reasoned the percentage of uncut DNA by these restriction enzymes should approximate the ZFN-mediated mutagenesis frequency. On the other hand, the *Dde*I site is farther away from the spacer in the At1g53-ZFN target site. Because only a fraction of mutagenesis events are likely assessed by measuring loss of the *Dde*I site, the T7 endonuclease assay was instead used to measure activity of the At1g53-ZFN. Activity of the three ZFNs at seven endogenous loci ranged from a few percent to 20% (Figure 2, A–C), with the At4g16-ZFN having the greatest activity (13.4–20.1%). Differences observed in ZFN activity among different transgenic plants is likely due to different expression levels of the ZFN transgenes, depending on the chromosomal site of T-DNA integration.

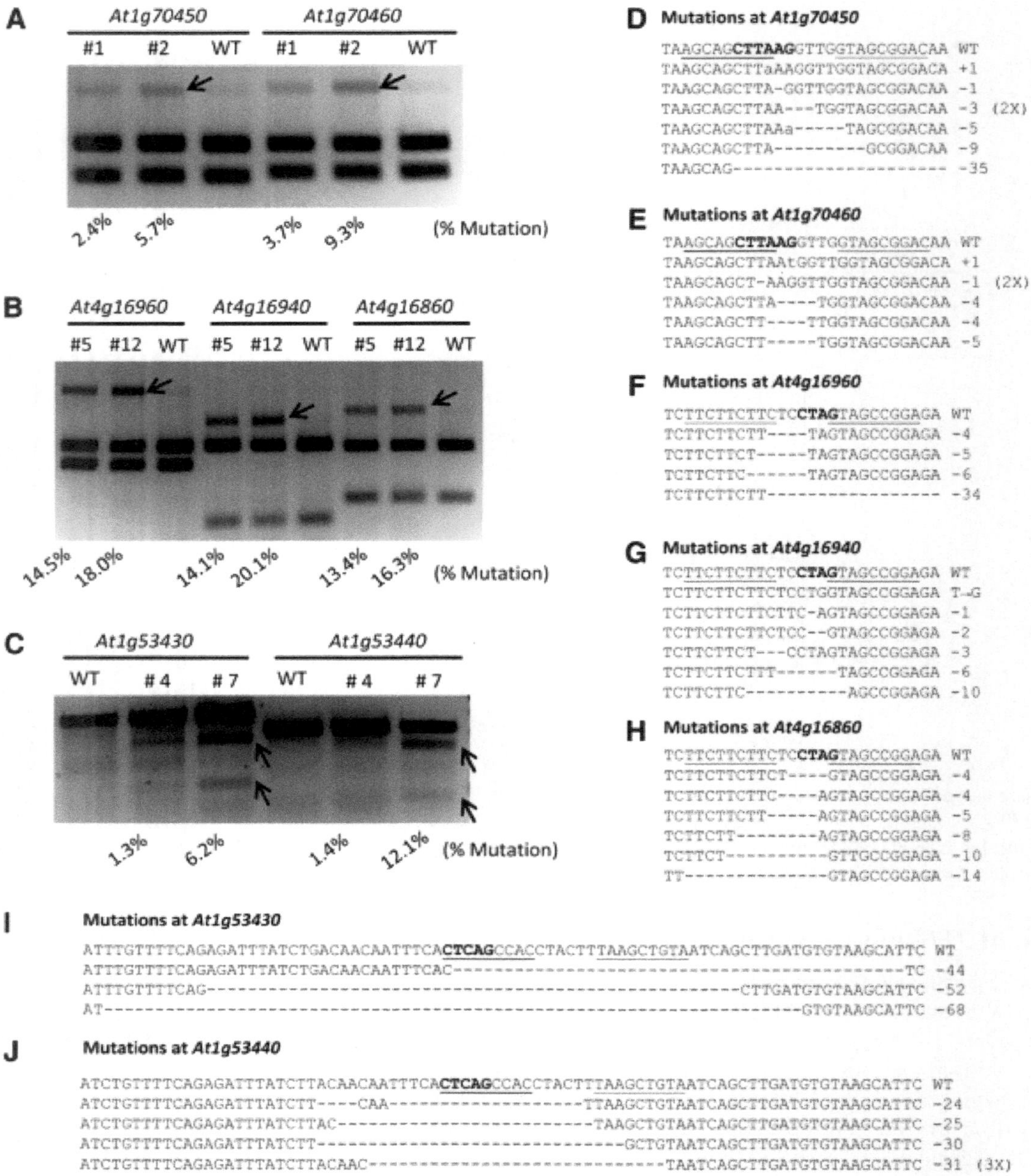

Figure 2: ZFN activity at seven endogenous loci. (A) At1g70-ZFN activity at target sites in *At1g70450*and *At1g70460*. PCR products were digested with*Afl*II. (B) At4g16-ZFN activity at target sites in *At4g16960,At4g16940*, and *At4g16860*. PCR products were digested with *Bfa*I. (C) At1g53-ZFN activity at target sites in*At1g53430* and *At1g53440*as measured by the T7 endonuclease assay. For both restriction digestion assays (A and B) and the T7 endonuclease assay (C), mutagenesis frequencies (shown at the bottom of the figures) were determined by measuring the signal intensity of each band using the Labworks analysis software. (D-J) ZFN-induced mutations at seven endogenous target sites with different mutation types indicated. The restriction enzyme sites used for activity measurement are marked in bold letters. The uncut PCR products (from A and B, and Figure S2) were cloned and sequenced to reveal ZFN-induced mutations at each target site.

To further confirm activity of each ZFN, DNA fragments resistant to restriction enzyme digestion were cloned and sequenced (Figure 2, A and B). The sequencing results revealed ZFN-induced mutations at all seven target loci, with small deletions (1–10 bp) being the most prevalent (Figure 2, D–J). Because the *Dde*I restriction site is farther away from the spacer, deletions recovered at *At1g53430* and *At1g53440* were typically larger (24–68 bp;Figure 2, I and J and Figure S5).

Deletion of TAGs by ZFNs

The observed mutations at all seven target sites indicate that ZFN-induced DSBs were generated. To examine deletion of TAGs as an outcome to ZFN activity, we conducted PCR using specific primers flanking all ZFN target sites, such that the production of PCR products would indicate the presence of deletions (Figure 3, A–C). With this strategy, we detected deletions of the*At1g53430* gene cluster (Figure 3A) and the *At1g70450* gene cluster (Figure 3B), as well as three different types of deletions at the *RPP4* gene cluster (Figure

3C). We further confirmed these deletions by cloning the deletion-specific PCR products and sequencing randomly selected clones (Figure 3, A–C, lower panels). Not only did we observe indels at the site of deletions, we also observed perfect ligations of two sticky ends derived from ZFN-cleaved DNA (Figure 3, A–C). These events are most likely due to the presence of compatible overhangs. Taken together, these data suggest that deletions of up to 55 kb can be made by ZFNs on different Arabidopsis chromosomes.

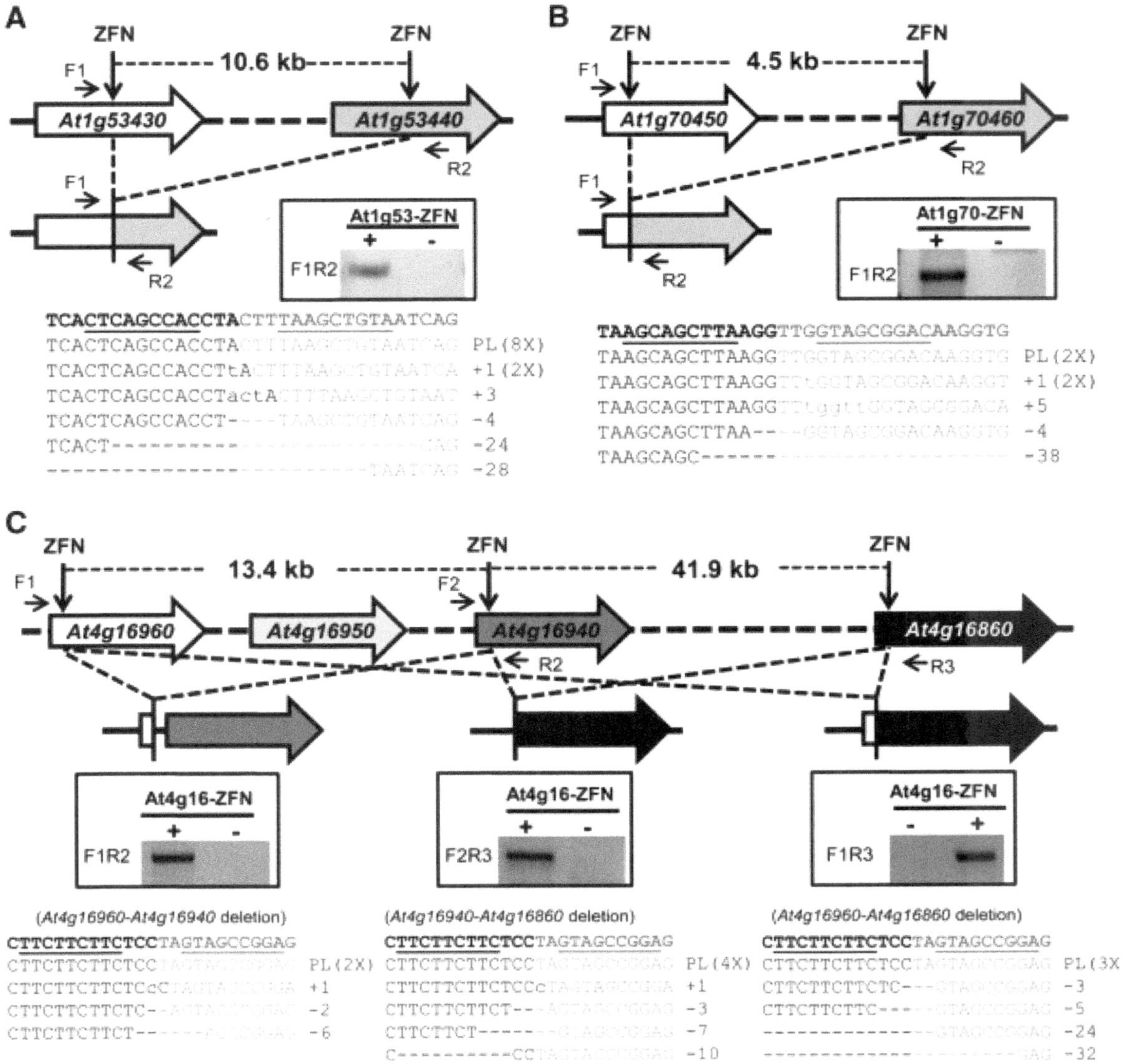

Figure 3: Deletion of gene clusters by ZFNs. (A) Deletion of the*At1g53430-At1g53440*cluster by the At1g53-ZFN. (B) Deletion of the-*At1g70450-At1g70460*cluster by the At1g70-ZFN. (C) Three types of deletions at the *At4g16960-At4g16860* gene cluster generated by the At4g16-ZFN. The gene clusters and resulting deletions are depicted. Deletion events were confirmed by PCR, as shown in windows in the middle of each panel. The positions of PCR primers are indicated by arrows. PCR products were subsequently cloned and sequenced. The sequencing results shown in the lower panels confirmed perfect ligations after loss of the intervening DNA or ligations with mutations at the ZFN cleavage site.

Large chromosomal deletions by ZFNs

Because we achieved deletions of TAGs spanning up to 55 kb, we next tested whether much larger chromosomal deletions could be generated. To do so, we took advantage of an existing ZFN (ADH1-ZFN), which we previously engineered to target the *ADH1* gene (*At1g77120*) at the end of chromosome 1 (Zhang *et al.* 2010) (Figure 4A and Figure S6A). Simultaneous expression of At1g53-ZFN and ADH1-ZFN could potentially result in chromosomal deletions as large as 9 Mb, almost one-third the length of Arabidopsis chromosome 1 (Figure 4A). We first screened independent estradiol-inducible *ADH1-ZFN* lines and identified the *ADH1-ZFN* #3 line as showing strong estradiol-inducible mutagenesis activity (Figure S6B). We also mapped the *ADH-ZFN* transgene insertion site in this line to chromosome 2 to aid in genotyping (Figure S6C). We then obtained a homozygous T3 *ADH1-ZFN* #3 line for use in our experiments.

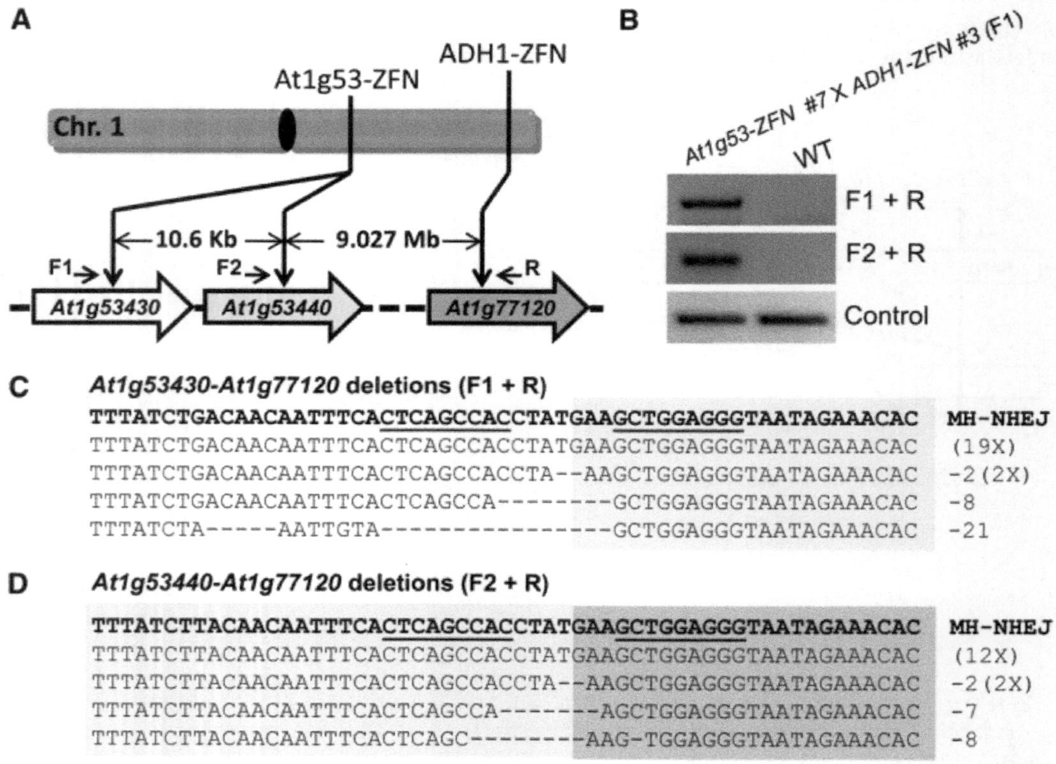

Figure 4: Large chromosomal deletions by ZFNs. (A) Schematic of ZFN targets on the right arm of Arabidopsis chromosome 1. The distance between the ZFN sites is shown and the positions of primers used to confirm large deletions are indicated. (B) PCR confirmation of large chromosomal deletions. The F1 and R primers amplify the junction fragment of the deletion of 9.037 Mb; primers F2 and R amplify the junction fragment of the deletion of 9.027 Mb (upper panels). PCR amplification of a part of the *ADH1* gene was used as a genomic DNA control (lower panel). F1 seedlings generated from the cross between At1g53-ZFN #7 line and ADH1ZFN #3 line were treated with estradiol, and the wild-type plants served as a negative control. (C) Sequenced clones indicative of large chromosomal deletions between *At1g53430* and*At1g77120*. (D) Sequenced clones indicative of large chromosomal deletions between *At1g53440* and *At1g77120*. The DNA sequences resulting from perfect ligation of DNA ends are shown in the first line of the text boxes; deletions with indels are shown below. ZFN binding sequences are underlined. MH-NHEJ, end-joining that appears to have been facilitated by microhomology.

To test for chromosomal deletions, we crossed the homozygous *At1g53-ZFN*#7 T2 line and the *ADH1-ZFN* #3 T3 line. F1 plants were obtained, and both ZFNs were induced by growing them on estradiol-containing MS medium. As with the gene cluster deletions, PCR was used to detect the large chromosomal deletions, and deletions were detected for both *At1g53430-At1g77120* and*At1g53440-At1g77120* (Figure 4B). Resulting PCR products were cloned and sequenced (Figure 4, C and D). Interestingly, the most prevalent product was a ligation of a 7-bp spacer sequence. We predict this product resulted from microhomology-based NHEJ using only 1 bp of microhomology (Figure S7). The resulting 7-bp spacer would be expected to be cut inefficiently by the hybrid ZFN, which contains At1g53-ZFN-left and ADH1-ZFN-right monomers (Figure S7). This may explain the PCR product's predominance.

Frequency of ZFN-induced Chromosomal Deletions

Having demonstrated ZFN-induced chromosomal deletions ranging from a few kilobases to 9 Mb, we next sought to estimate their frequency of occurrence. We adapted a digital PCR method used to measure deletion frequencies in human cells (Lee *et al.* 2010). In our case, deletion frequency was detected by PCR using a series of genomic DNA dilutions as templates. Amplification of*ADH1* was used as an internal control for DNA copy number. As summarized inTable 1, the frequency of gene cluster deletions was approximately 1%—positively correlated with ZFN activity and negatively correlated with the length of the deletions. As the length of deletion increased to 9 Mb, the frequency decreased to <0.1%.

Table 1: Frequency of ZFN-induced chromosomal deletions.

Transgenic Lines	At1g53430 to At1g53440 (deletion of 10.6 kb)	At1g70450 to At1g70460 (deletion of 4.5 kb)	At4g16940 to At4g16860 (deletion of 41.9 kb)	At4g16960 to At4g16860 (deletion of 55.3 kb)	At1g53440 to At1g77120 (deletion of 9.027 Mb)	At1g53430 to At1g77120 (deletion of 9.037 Mb)
A1g53-ZFN #4	0.3%					
At1g53-ZFN #7	1%					
At1g70-ZFN #1		3%				
At1g70-ZFN #2		3%				
At4g16-ZFN #5			1%	1%		
At4g16-ZFN #12			3%	3%		
ADH1-ZFN #3 X At1g53-ZFN #7[a]					0.046%	0.137%

ZFN, zinc finger nucleases.
[a] F1 plants.

Targeted Inversions of Gene Clusters

In addition to deletions, chromosomal DNA released by two DSBs could create inversions (illustrated in Figure 5A). Such inversions will have two novel junction sites, which can be detected with PCR using specific primer sets (Figure 5A). The predicted novel junctions were, in fact, detected at the At1g53430 gene cluster by PCR; DNA sequencing confirmed that the inversions occurred (Figure 5B). As anticipated, many inversion junctions had small deletions indicative of imprecise NHEJ (Figure 5C). We could also detect and confirm inversions at the *At1g70450* gene cluster (Figure S8), and these occurred at a frequency of ~0.05% as measured by digital PCR. Gene cluster inversions, therefore, appear to occur at a lower frequency than deletions.

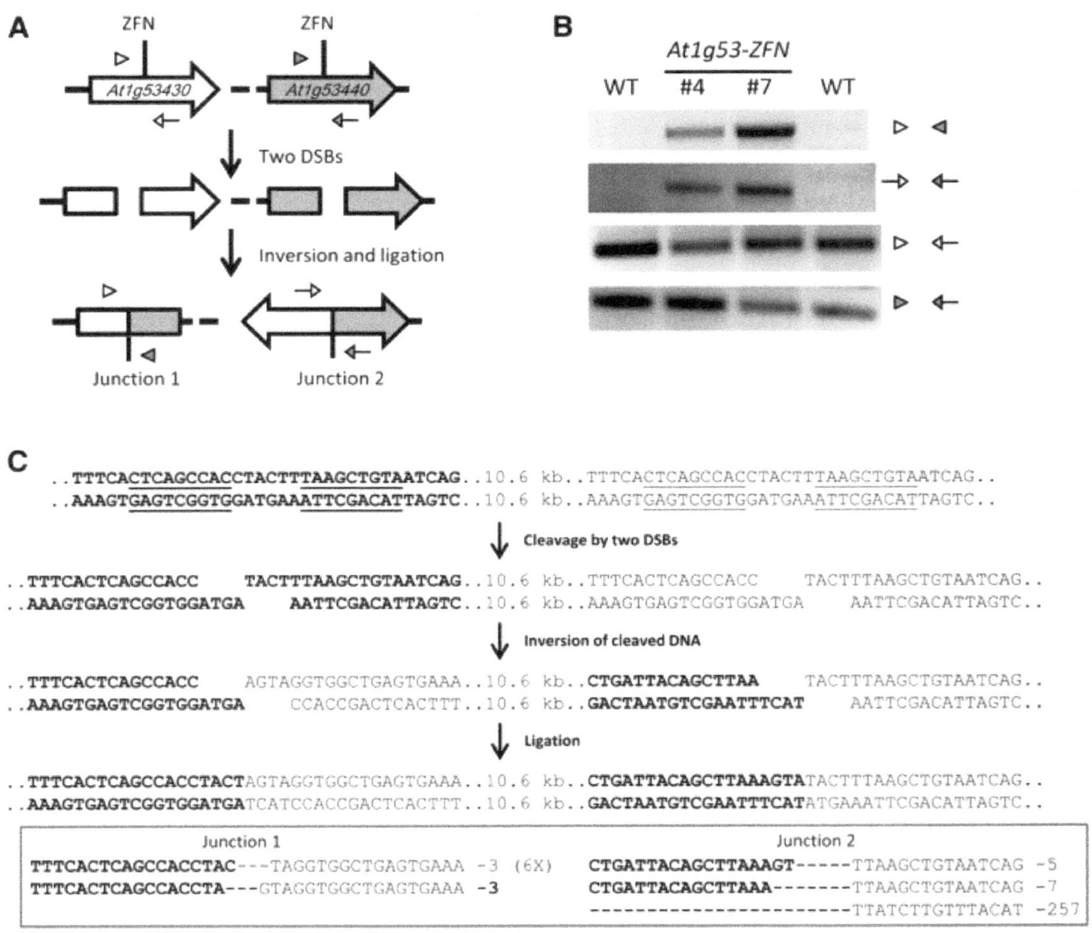

Figure 5: Inversion of the *At1g53430*gene cluster. (A) Schematic of the *At1g53430-At1g53440* gene cluster inversion. Positions of PCR primers for confirmation of inversions are indicated by empty or filled triangles and arrows. (B) PCR confirmation of gene cluster inversions. Two independent T2 lines were used to detect inversions; wild-type (WT) plants were used as the negative control. (C) Detailed depiction of the inversion event and DNA sequence confirmation of the inversions.

Possible Targeted Duplication of Gene Clusters

When two DSBs are induced in different TAGs on different chromosomes (either homologous chromosomes or sister chromatids), interchromosomal ligations of the broken DNA ends can result in deletions as well as gene cluster duplications (Figure S4A). For TAGs that only contain two genes, such as the*At1g70450* gene cluster, the duplication will create a hybrid gene (Figure S9A). We indeed detected specific products indicative of ZFN-induced duplications of the *At1g70450* gene cluster (Figure S9C), and these were also confirmed by DNA sequencing of the cloned PCR products (Figure S9D). However, if deleted DNA was circularized and retained in cells, it would give the same PCR products as depicted in Figure S9B. Because we cannot distinguish between these two outcomes, and because we do not know how long DNA circles persist in cells, it remains unclear how many of the events detected (Figure S9, C and D) truly reflect gene cluster duplications.

DISCUSSION

We previously reported engineering active ZFNs for six endogenous loci in Arabidopsis using the CoDA method (Sander *et al.* 2011). In this study, we successfully targeted seven additional loci with three ZFNs, each of which showed NHEJ mutagenesis activity. Two other ZFNs targeting different endogenous loci failed to show detectable activity in Arabidopsis. For these ZFNs, it remains possible that chromatin structure or epigenetic modifications impeded ZFN access to endogenous DNA targets. Our overall 60% success rate is in line with the success rate previously reported for CoDA, and our work further demonstrates the usefulness of the CoDA method for engineering active ZFNs.

We aimed to create deletions of TAGs in the Arabidopsis genome by using a single ZFN pair to target multiple genes within the cluster. However, not every gene cluster can be deleted with such a strategy. In cases in which the TAGs differ significantly in DNA sequence similarity, two different pairs of ZFNs might be required to create the desired deletion (Sollu *et al.* 2010). In addition, TALENs are good alternatives to ZFNs, because there seems to be less restriction in designing TALENs to target diverse DNA sequences (Bogdanove and Voytas 2011; Doyle *et al.* 2012; Reyon *et al.* 2012). The scale of deletions that we obtained (from a few kilobases to 9 Mb) suggests ZFNs are capable of creating very large deletions in plants. We recognize, however, that Arabidopsis plants are not likely to survive the loss of megabase pairs of chromosomal DNA. Targeted insertions created by HR were previously demonstrated by the use of ZFNs in plants (Cai *et al.* 2009). Generating deletions (such as in this study) and concomitantly creating insertions by HR may be another genome engineering approach of value for basic research and crop improvement.

Being highly homologous and repetitive, TAGs are particularly prone to change either through unequal crossover or gene conversion. This provides the opportunity to evolve new gene functions, some of which may be adaptive (Hanada *et al.* 2008; Lehti-Shiu *et al.* 2009). Our approach for making deletions and duplications in TAGs has demonstrated that we can now create novel chimeric genes which may otherwise not occur naturally. For example, we frequently detected hybrid genes due to perfect ligation of broken chromosomes after loss of the intervening DNA (Figure 3, A–C). In addition, duplication of gene clusters increases genetic redundancy and frees some gene members to evolve new functions. Tandem duplication in Arabidopsis has provided a means for adaptive evolution of *R* genes (Meyers *et al.* 2005). Our approach in creating rearrangements in the complex *RPP4* gene cluster and the two *RLK* gene clusters is thus a promising first step toward generating valuable genetic material for both molecular and evolutionary studies.

Gene cluster inversion is another consequence of our TAG-targeting approach. For TAGs encoding two genes, DNA inversion is likely to destroy the function of both gene targets. However, because inversions preserve DNA sequences (which would otherwise be lost in deletions), they may have unique applications, such as serving as templates for future genome evolution. Another application of inversions is that they may only knock out two target genes at both ends of the TAG while retaining the function of the genes in between. This is only true if there are more than two members in the TAG, such as the *RPP4* gene cluster evaluated in this study.

To recover plants with germline-transmitted deletions, we screened large T3 populations of At1g70-ZFN #2 (a total of 2539 plants) and At4g16-ZFN #12 (a total of 2322 plants) with no success (data not shown). Clearly, the frequencies of germline-transmitted deletions are much lower than the somatic frequencies. This observation is consistent with other ZFN-mediated mutagenesis studies we have conducted (unpublished data), where the frequency of somatic mutation did not directly reflect the frequency of germline mutation. Rather, we believe there is a threshold somatic mutation frequency that must be surpassed to ensure successful germline transmission. In previous work, we found that somatic mutagenesis frequencies in excess of 7% were sufficient to recover germinal mutations at high frequency (Zhang *et al.* 2010). The observed ~1% somatic deletion frequency observed here appears to be under this threshold.

It has been shown that stem cell niches in Arabidopsis are hypersensitive to DNA damage, and an even a low dose of DNA damage can trigger programmed cell death selectively in these stem cells (Fulcher and Sablowski 2009). Interestingly, programmed cell death in the shoot meristem was greatly suppressed when the DNA damage early response gene, *ATM*, was knocked out. Thus, it might be useful to use an *atm* mutant background to screen for germline deletions in plants. In addition, we recently reported that *smc6b*mutations promote NHEJ in Arabidopsis (Qi *et al.* 2013). As chromosomal breaks that lead to deletions are joined through NHEJ, it is likely that *smc6b*mutations will enhance deletion frequencies. It is also possible some deletions are deleterious to pollen and egg cells, and this impedes their transmission. Because whole plant regeneration routinely is performed from somatic cells for many species (*e.g.*, rice, maize, and tobacco), plants with deletions may be easier to achieve by regenerating somatic tissues that have been treated with sequence specific nucleases.

Note Added in Proof: See also Michelle Christian, Yiping Qi, Yong Zhang, and Daniel F. Voytas, 2013 Targeted Mutagenesis of *Arabidopsis thaliana* using Engineered TAL Effector Nucleases (TALENs) G3: Genes, Genomes, Genetics 3: 1697–1705.

ACKNOWLEDGMENTS

We thank all members of the Voytas lab for helpful discussions and suggestions. This work is supported by grants from the National Science Foundation (MCB 0209818 and DBI 0923827) to D.F.V. and a National Institutes of Health (NIH) Director's Pioneer Award DP1 GM105378 (J.K.J.) and NIH R01 GM088040 to J.K.J. D.F.V. has a financial interest in and serves as Chief Science Officer of Cellectis Plant Sciences. These interests have been reviewed and managed by the University of Minnesota in accordance with its Conflict of Interest policies. J.K.J. has a financial interest in Transposagen Biopharmaceuticals. J.K.J.'s interests were reviewed and are managed by Massachusetts General Hospital and Partners HealthCare in accordance with their conflict of interest policies.

LITERATURE CITED

1. Abbott, J. C., A. Barakate, G. Pincon, M. Legrand, C. Lapierre et al., 2002 Simultaneous suppression of multiple genes by single transgenes. Down-regulation of three unrelated lignin biosynthetic genes in tobacco. Plant Physiol. 128: 844–853.
2. Alvarez, J. P., I. Pekker, A. Goldshmidt, E. Blum, Z. Amsellem et al., 2006 Endogenous and synthetic microRNAs stimulate simultaneous, efficient, and localized regulation of multiple targets in diverse species. Plant Cell 18: 1134–1151.
3. Bogdanove, A. J., and D. F. Voytas, 2011 TAL effectors: customizable proteins for DNA targeting. Science 333: 1843–1846.
4. Cai, C. Q., Y. Doyon, W. M. Ainley, J. C. Miller, R. C. Dekelver et al., 2009 Targeted transgene integration in plant cells using designed zinc finger nucleases. Plant Mol. Biol. 69: 699–709.
5. Christian, M., T. Cermak, E. L. Doyle, C. Schmidt, F. Zhang et al., 2010 Targeting DNA double-strand breaks with TAL effector nucleases. Genetics 186: 757–761.
6. Doyle, E. L., N. J. Booher, D. S. Standage, D. F. Voytas, V. P. Brendel et al., 2012 TAL Effector-Nucleotide Targeter (TALE-NT) 2.0: tools for TAL effector design and target prediction. Nucleic Acids Res. 40: W117–122.
7. Fulcher, N., and R. Sablowski, 2009 Hypersensitivity to DNA damage in plant stem cell niches. Proc. Natl. Acad. Sci. USA 106: 20984–20988.
8. Guschin, D. Y., A. J. Waite, G. E. Katibah, J. C. Miller, M. C. Holmes et al., 2010 A rapid and general assay for monitoring endogenous gene modification. Methods Mol. Biol. 649: 247–256.
9. Hamilton, A. J., and D. C. Baulcombe, 1999 A species of small antisense RNA in posttranscriptional gene silencing in plants. Science 286: 950–952.
10. International Rice Genome Sequencing Project, 2005 The map-based sequence of the rice genome. Nature 436: 793–800.
11. Jander, G., and C. Barth, 2007 Tandem gene arrays: a challenge for functional genomics. Trends Plant Sci. 12: 203–210.
12. Kim, Y. G., J. Cha, and S. Chandrasegaran, 1996 Hybrid restriction enzymes: zinc finger fusions to Fok I cleavage domain. Proc. Natl. Acad. Sci. USA 93: 1156–1160.
13. Koornneef, M., L. W. Dellaert, and J. H. van der Veen, 1982 EMS- and radiation-induced mutation frequencies at individual loci in Arabidopsis thaliana (L.) Heynh. Mutat. Res. 93: 109–123.
14. Lee, H. J., E. Kim, and J. S. Kim, 2010 Targeted chromosomal deletions in human cells using zinc finger nucleases. Genome Res. 20: 81–89.
15. McCallum, C. M., L. Comai, E. A. Greene, and S. Henikoff, 2000 Targeting induced local lesions IN genomes (TILLING) for plant functional genomics. Plant Physiol. 123: 439–442.
16. Messing, J., A. K. Bharti, W. M. Karlowski, H. Gundlach, H. R. Kim et al., 2004 Sequence composition and genome organization of maize. Proc. Natl. Acad. Sci. USA 101: 14349–14354
17. Osakabe, K., Y. Osakabe, and S. Toki, 2010 Site-directed mutagenesis in Arabidopsis using custom-designed zinc finger nucleases. Proc. Natl. Acad. Sci. USA 107: 12034–12039.
18. Petolino, J. F., A. Worden, K. Curlee, J. Connell, T. L. Strange Moynahan et al., 2010 Zinc finger nuclease-mediated transgene deletion. Plant Mol. Biol. 73: 617–628.
19. Qi, Y., Y. Zhang, F. Zhang, J. A. Baller, S. C. Cleland et al., 2013 Increasing frequencies of site-specific mutagenesis and gene targeting in Arabidopsis by manipulating DNA repair pathways. Genome Res. 23: 547–554.
20. Raina, S., R. Mahalingam, F. Chen, and N. Fedoroff, 2002 A collection of sequenced and mapped Ds transposon insertion sites in Arabidopsis thaliana. Plant Mol. Biol. 50: 93–110.
21. Reyon, D., S. Q. Tsai, C. Khayter, J. A. Foden, J. D. Sander et al., 2012 FLASH assembly of TALENs for high-throughput genome editing. Nat. Biotechnol. 30: 460–465.
22. Sander, J. D., E. J. Dahlborg, M. J. Goodwin, L. Cade, F. Zhang et al., 2011 Selection-free zinc-finger-nuclease engineering by contextdependent assembly (CoDA). Nat. Methods 8: 67–69.
23. Schwab, R., S. Ossowski, M. Riester, N. Warthmann, and D. Weigel, 2006 Highly specific gene silencing by artificial microRNAs in Arabidopsis. Plant Cell 18: 1121–1133.
24. Townsend, J. A., D. A. Wright, R. J. Winfrey, F. Fu, M. L. Maeder et al., 2009 High-frequency modification of plant genes using engineered zinc-finger nucleases. Nature 459: 442–445.
25. Woody, S. T., S. Austin-Phillips, R. M. Amasino, and P. J. Krysan, 2007 The WiscDsLox T-DNA collection: an Arabidopsis community resource generated by using an improved high-throughput T-DNA sequencing pipeline. J. Plant Res. 120: 157–165.
26. Zhang, F., M. L. Maeder, E. Unger-Wallace, J. P. Hoshaw, D. Reyon et al., 2010 High frequency targeted

mutagenesis in Arabidopsis thaliana using zinc finger nucleases. Proc. Natl. Acad. Sci. USA 107: 12028–12033.

27. Zhang, S., S. Raina, H. Li, J. Li, E. Dec et al., 2003 Resources for targeted insertional and deletional mutagenesis in Arabidopsis. Plant Mol. Biol. 53: 133–150.

CHAPTER 13

GENETIC DIFFERENTIATION IN A PIONEER TROPICAL TREE

Antonio R. Castilla, Nathaniel Pope, Rodolfo Jaffé, Shalene Jha

INTRODUCTION

Forest ecosystem function is dependent on the colonization and establishment of native trees, which provide essential ecosystem services such as carbon sequestration, erosion control and the provisioning of food and nesting resources for humans and wildlife [1]. However, many of the world's forests are threatened by frequent and intense human-driven disturbance events, and rates of land conversion to agriculture, pasture, and urban areas are expected to further increase in response to a growing human population [2]. As vast regions of primary forest have already been cleared, recent conservation efforts are emphasizing the reforestation of degraded habitats as an additional strategy to safeguard ecosystem services in key areas of ecological and economic importance [3,4]. Specifically, there is growing interest from both conservation managers and landholders in the use of early successional native tree species for reforestation instead of fast-growing exotic timber species, as these native species can have beneficial impacts on both biodiversity conservation and rural livelihoods [5,6].

Reforestation implies the release of cultivated and translocated trees, which can have important genetic consequences for native populations such as alterations in their genetic diversity and gene flow dynamics and breakdowns of local adaptation [7–9]. Thus the design of effective restoration strategies using native trees requires detailed knowledge of genetic diversity and structuring within and among native populations of the selected species [10,11]. A first step in reforestation programs must therefore be the screening of native population genetic structure and levels of genetic diversity prior to implementation [10]. Furthermore, the successful establishment of translocated trees may depend on the species' breeding system (self- vs. out-crossed), the impact of inbreeding on the offspring, and the ongoing gene flow from native source populations [12]. Thus, understanding how native pioneer species have colonized areas in the past and how their genetic diversity is currently structured across the landscape is critical for the design of reforestation strategies within degraded areas.

The first and simplest hypothesis regarding gene flow across space is the isolation by distance pattern (IBD, hereafter). The IBD pattern is characterized by a declining probability of genetic identity with geographic distance [13] and has been reported for many living organisms, existing largely as a result of drift-dispersal processes (reviewed in [14]). Within the context of conservation genetics, IBD may suggest that optimal conservation of genetic diversity and potential adaptive genetic variation will require conservation of multiple populations across the species range [15,16]. Furthermore, IBD has frequently been reported for trees (e.g. [17,18]), although there are exceptions to this rule (e.g. [19]). Specifically, in some cases, topographic or geographic barriers, such as mountain ranges, may better explain divergence in genetic composition than geographic distance alone (e.g. [19]).

Thus, a second potential factor influencing gene flow among tree populations is elevation. As in any ecological gradient, many species have an ecological optimum along elevational gradients [20,21]. For instance, lowland plant species typically exhibit abundance peaks at low elevations and declines towards upper altitudinal limits as a consequence of suboptimal habitats. In this context, pollinators and frugivores can shift along elevational gradients to other more abundant resources, potentially compromising the dispersal service to lowland species at high elevation sites [22,23]. Furthermore, elevational gradients drive variation in environmental factors such as temperature or precipitation [24], which may affect flying conditions for insects, and potentially impact pollination [25–27]. Variation in the abiotic environment along elevational gradients may also lead to marked differences in the phenology of both flowering and fruiting events, also potentially altering pollination and seed dispersal [28,29]. Specifically, recent studies have suggested that elevation-mediated phenological

Citation: Castilla AR, Pope N, Jaffé R, Jha S (2016) Genetic Differentiation in a Pioneer Tropical Tree. PLoS ONE 11(6): e0156694. doi:10.1371/journal.pone.0156694.

time lags can impact both migration and selfing rates in tree species, potentially leading to long-term genetic structure across elevation gradients [30]. While a number of studies have found evidence of genetic differentiation for plant species living along elevational gradients [31], this past work has largely focused on small spatial scales (1–10km) and has not explicitly investigated species with short generation times and early-successional life history strategies.

Finally, forest fragmentation is expected to reduce genetic diversity, increase genetic differentiation, and potentially increase inbreeding in tree populations [32,33]. However, the empirical support to these theoretical expectation remains elusive [34,35]. A number of studies have indicated that the response of each species to habitat loss may differ greatly depending on their biological attributes, in particular animal vs. wind dispersal [36,37]. For instance, some tree species are able to ameliorate the genetic consequences of fragmentation through flexible mating systems and animal dispersers with high mobility [38–40]. For other tree species, though forest fragmentation may alter ecological processes, alterations may be challenging to detect if the process of fragmentation has occurred over a shorter period relative to the generation time of the species [35]. Therefore, it is important to examine trees of younger age-classes or progenies sired in those landscapes where deforestation has been recent [35].

Given the potentially major impacts of distance, elevation, and forest cover on plant gene flow, it is particularly urgent to investigate gene flow processes for tropical trees, as tropical regions exhibit substantial elevation gradients, greater proportions of animal-dispersed plant species, and rapid rates of deforestation. Specifically, species ranges in the tropics are frequently delimited by elevation, with distinct community assembly processes along elevation gradients [41] and potential implications for elevation-mediated gene flow. Second, most tropical tree species are animal-pollinated and self-incompatible and thus dependent on mobile animals for reproduction, making them particularly vulnerable to declines in disperser communities [42]. Finally, tropical biomes have the highest deforestation rates of any other biome, with an estimated ~2100 square kilometers of tropical forest destroyed every year [43]. Despite these factors, we still know very little about how tropical landscape composition and structure impacts plant dispersal and gene flow.

In this study, we use landscape genetic tools to assess how geographic distance, elevational gradients, and forest cover influence patterns of genetic structure and genetic diversity in a pioneer tree species across the Isthmus of Panama. This region is a global biodiversity hotspot that suffers from deforestation, biodiversity loss, and heavy erosion in streams and canals [44,45] and thus there is great need to address the potential of early successional native tree species for restoration programs. Our study model is *Miconia affinis*, a common early successional Neotropical tree that frequently colonizes forest gaps, riparian areas, and exposed hillsides [46]. *Miconia* represents one of the most abundant and diversified genera in the Neotropics [47,48], and because *M. affinis* is dispersed by a diverse pollinator and seed disperser community [46,49], it may be a particularly critical species for facilitating pollinator and frugivore community assembly in degraded regions. Specifically, we address how deforestation, elevation, and geographic distance influence the genetic diversity and differentiation of *M affinis* populations, and how gene flow has changed across historic and contemporary time periods. By examining genetic diversity, genetic differentiation, and migration rates within human-altered regions, this study determines the ability of a native species to maintain effective dispersal and colonization across both pristine and disturbed tropical landscapes. Further, we provide management suggestions that may be relevant for the many tree species with similar phenology and life histories.

MATERIALS AND METHODS

Study Species

Miconia affinis D.C. (Melastomataceae) is a common, early successional self-incompatible understory tree (3–6 m height) that is broadly distributed in the Neotropics, ranging from Mexico to Brazil [49,50]. The species blooms for ≈ 2 days at the onset of the first rains (January to June), and its flowers are visited by a large diversity of social and solitary bees [50]. Fruits develop in 3–4 months (May-July) and seeds are dispersed by a variety of birds and bats ([46] and references therein).

Previous studies have reported short lifespans (i.e., < 20 years) and early first flowering events (i.e., 4–5 years) for several *Miconia* species of similar size and habit [51–53]. To confirm this life history strategy for *M. affinis*, we estimated the annual growth rate of the species using data from the 50-ha Forest Dynamic Plot on Barro Colorado Island (BCI), Panama [54–56]. The plot consists of a standing number of over 350,000 mapped stems 10mm or above in diameter at breast height (DBH) and approx. 300 plant species (http://ctfs.arnarb.harvard.edu/webatlas/datasets/bci/). Censuses have been conducted since 1982. Specifically, we used data from the censuses conducted from 1990 to 2010 (http://www.ctfs.si.edu/site/Barro+Colorado+Island/abudance/) for trees that survived the whole 20-year period, as this was the period when trees were censused every 5 years (N = 32 trees)[57]. We estimated the annual growth rate as per Condit and colleagues [58], a method widely used for tropical tree species (e.g. [59]). For each tree, we estimated the annual growth rate as the difference in DBH between 2010 and 1990 divided by the number of years (i.e., 20 years). We posit that early growth in *M. affinis* (i.e., seedling and sapling stages) would not substantially alter our growth estimates given that congeneric species (e.g. *M. argentea)* exhibit growth from germination to the 1m tall size class in 1 to 6 years [60].

Using this information, we estimated the lifespan of the species by calculating the approximate number of years to reach the largest DBH. Specifically, we used data from five populations along the Panama Canal (CP, GB, PL, AG and RC; Fig 1) in which we surveyed all *M. affinis* trees ≥ 10 mm DBH in a plot of 3 km x 3 km per population (358, 349, 96, 316 and 370 individuals respectively). Using the largest DBH in each population, we calculated the mean largest DBH for *M. affinis*. We estimated the lifespan (LS) by dividing the mean largest DBH by the mean annual growth rate.

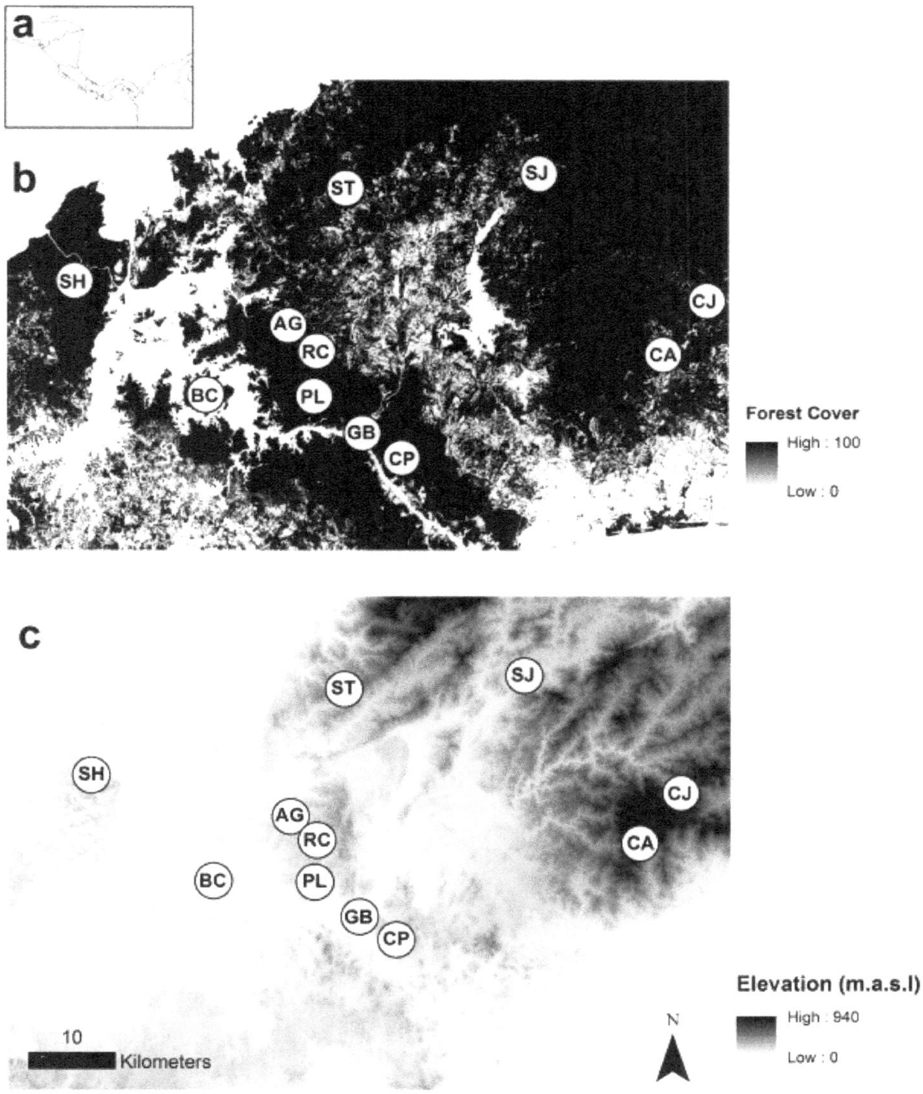

Figure 1. Study region. (a) Location of the study in Central America and (b & c) distribution of *M. affinis*populations across the Isthmus of Panama. (b) Forest cover, where increasing percentages of forest cover are represented by darker shades (0–100%) and (c) elevation, where increasing elevation is represented by darker shades (0–940 m.a.s.l.), and for both cases, darker shades represent higher resistance habitat. The location of the study populations in the map corresponds with the population centroids.

We also estimated the generation time (T_g) as the age at which members of a given cohort are expected to reproduce. Based on previous studies of the species, only *M. affinis* trees with diameter at breast height ≥ 10 mm produce flowers [61]. Therefore, we estimated the generation time (T_g) in years by dividing the DBH when the tree reproduces for the first time (i.e., 10 mm) by the mean annual growth rate.

Study Region

This study was conducted across the Panama Canal bioregion, an area that has lost more than 30% of its forested area to agro-pastoral development during the past 50 years [45]. We received permission for our field work and collection of samples from Panama's National Authority for the Environment (ANAM). Using reported localities for *M. affinis* from the University of Panama's Herbarium, we extensively surveyed the Isthmus of Panama locating eleven populations (Fig 1A–1C). The elevation of our study sites ranged from 26 to 804 m.a.s.l. (mean = 257.5 m.a.s.l.; Table 1), and the geographic distance between populations ranged from 4.2 to 67.9 km (mean = 28.5 km).

Table 1. Attributes of *M. affinis* **populations.** Population (Pop.) geographic coordinates (Lat. and Long.), elevation, number of trees (N), mean diameter at breast height (DBH), allelic richness (AR), private allelic richness estimated by rarefaction (PAR), and Nei's gene diversity (He).

Pop	Lat.	Long.	Elevation (m.a.s.l.)	N	DBH (mm)	AR	PAR	Unbiased He
					Mean (SD)	Mean (SD)	Mean (SD)	Mean (SD)
AG	9.2256	-79.766	120	29	27.85 (11.89)	3.83 (0.57)	0.19 (0.09)	0.635 (0.087)
BC	9.1554	-79.847	150	30	44.23 (21.30)	3.76 (0.46)	0.07 (0.03)	0.631 (0.075)
CA	9.1958	-79.400	628	30	53.60 (20.29)	2.92 (0.46)	0.08 (0.05)	0.505 (0.098)
CJ	9.2509	-79.357	804	30	48.53 (17.86)	3.78 (0.44)	0.23 (0.06)	0.616 (0.078)
CP	9.0910	-79.655	164	30	48.38 (19.89)	3.38 (0.46)	0.15 (0.07)	0.609 (0.066)
GB	9.1163	-79.694	26	28	43.31 (16.53)	3.68 (0.52)	0.12 (0.05)	0.616 (0.081)
PL	9.1540	-79.741	40	30	38.71 (12.95)	3.77 (0.56)	0.15 (0.06)	0.620 (0.093)
RC	9.1995	-79.738	183	30	32.64 (11.72)	3.42 (0.53)	0.08 (0.04)	0.597 (0.090)
SH	9.2706	-79.975	180	30	49.17 (20.74)	3.66 (0.63)	0.18 (0.09)	0.571 (0.100)
SJ	9.3787	-79.521	98	30	29.52 (13.07)	3.82 (0.59)	0.15 (0.07)	0.620 (0.104)
ST	9.3642	-79.710	440	25	35.47 (17.16)	3.53 (0.64)	0.19 (0.09)	0.563 (0.121)
Mean					*41.13 (18.76)*	*3.60 (0.08)*	*0.14 (0.02)*	*0.598(0.012)*

doi:10.1371/journal.pone.0156694.t001

We assessed the influence of elevation on genetic structure using a digital elevation map (DEM) of the study region with a resolution of 30 m x 30 m, available through STRI GIS online portal (http://strimaps.si.edu/portal/home/index.html). We examined the influence of contemporary deforestation patterns by utilizing a high-resolution global tree canopy cover map for the year 2000 with a resolution of 30 m x 30 m (University of Maryland:http://earthenginepartners.appspot.com/science-2013-global-forest). Forest cover represents the percentage of canopy closure for all vegetation taller than 5 meters in height, and each pixel is encoded with the % forest cover value, ranging from 0 to 100% in the study region (mean forest cover = 63.14%). In addition, we analyzed the influence of mean annual precipitation and mean annual temperature using the Worldclim database (http://www.worldclim.org/), at a resolution of 2.5 arc-min (roughly 5 km at the equator).

Sampling and Genotyping

We collected leaf tissue from 25–30 randomly chosen trees bearing inflorescences or infructescences (reproductive trees, hereafter) in each of the 11 populations (mean = 29 trees per population, total of 322 trees; Table 1). Tissue was preserved in liquid nitrogen and stored at -80°C until genotyping. Total genomic DNA was extracted from adult leaf tissue using the CTAB protocol [62]. All trees were genotyped for a set of nine highly polymorphic microsatellite loci [63,64]. Multiplex polymerase chain reactions (PCRs) using fluorescent dye-labeled selective primers were performed using a QIAGEN multiplex PCR kit (QIAGEN). Fragment separation and detection were conducted using an ABI 3730 DNA Sequencer, and alleles were scored manually using GENEMARKER® (Softgenetics).

Allelic Richness, Gene Diversity, and Population Structure

The probability of null alleles was calculated using the software Micro-Checker [65], and deviations from HWE and linkage disequilibrium (LD) were tested in GENEPOP v 4.0.10 [66] with 1000 dememorizations, 100 batches, and 1000 iterations per batch using the Markov chain approximation for the exact tests and likelihood-ratio tests, respectively. As Micro-Checker results indicated that one locus (B102) exhibited substantial evidence of null alleles (>70% of populations with evidence of null alleles), we chose to exclude this locus from subsequent analyses. One locus (Micaff 19) showed strong evidence of LD, so we also excluded it from subsequent analyses. The remaining seven loci exhibited deviations from HWE in less than 50% of populations, so we kept these seven loci for subsequent analyses.

Allelic richness (AR) and private allelic richness (PAR) per population were estimated using rarefaction with 10 genes as the sample size in HP-RARE [67], and unbiased Nei's gene diversity was calculated using GenAlEx 6.501 [68]. For each population, we calculated the geographic distance to the ten other populations (spatial isolation, hereafter) and the percentage of forest cover within a 3 km radius. We analyzed the effect of spatial isolation, elevation, and forest cover on our estimates of genetic diversity (i.e., AR, PAR and gene diversity) using generalized linear models. For each response variable, we built 7 a priori models comprised of various combinations of the three explanatory variables (spatial isolation, elevation, and forest cover) and used Akaike information criterion (hereafter, AIC) as a criterion for model selection. For the three explanatory variables, we estimated the model averaged parameter estimates (β) with standard errors based on models with ΔAIC of less than 5, using the package MuMIn for R [69].

A two-level hierarchical analysis of molecular variance (AMOVA; [70]) was run to analyze the partitioning of molecular variance among and within populations, using the program GenAlEx 6.501 [68]. We examined population genetic structuring using the Bayesian clustering method in STRUCTURE 2.3.3 [71]. The number of genetic groups was explored by performing 10 replicates of each simulation from K = 1 to K = 11, with a burn-in of 50,000 and MCMC of 100,000 assuming admixture and correlated frequencies [71]. We implemented CLUMPP using Structure Harvester [72] and we applied modal ΔK parameter as the choice criterion [73] to detect the true number of genetic groups. As a follow-up, we ran additional AMOVAs on the clusters assigned from the STRUCTURE analyses to compare the partitioning of variance among and within the assigned clusters.

Landscape Genetic Analysis

Using circuit theory [74], we examined if genetic differentiation among *M. affinis* populations is accounted for by three explanatory variables: geographic distance, elevation, and deforestation. We estimated the mean effective resistance between all pairs of sample sites (pairwise mode in the Circuitscape software; [74]) and define these values as Geographic Resistance Distance (RD), Elevation RD, and Deforestation RD. This software is based on electronic circuit theory and evaluates contributions of multiple dispersal pathways [74]. Geographic RD was analyzed by running a raster with equal resistances for all pixels. We did this by assigning all pixels a 0.5 resistance and then calculating resistance distances. Null resistance distances were log-transformed. Deforestation represents the percentage of open canopy for all vegetation taller than 5 meters in height; each pixel was encoded with the % deforestation, which we obtained by subtracting its forest cover percentage from 100%. Elevation ranged from 0 to 940 m.a.s.l. in the study region, and each pixel was encoded with the raw elevation value before resistance distance was calculated. Zero values in the rasters were replaced by 0.0001, since Circuitscape does not accept zeros. Finally, mean annual temperature and mean annual precipitation ranged from 22.3°C to 26.9°C and from 1754 mm to 3812 mm respectively in the study region, and each pixel was encoded with the raw environmental value before resistance distances were calculated.

We used maximum likelihood population effects (MLPE) models to test for the relationship between pairwise genetic distance [F_{ST} / (1-F_{ST})] and pairwise explanatory variables [75]. We also ran the same models using G_{ST} and D_{ST} instead of F_{ST} as estimates of the genetic differentiation among *M. affinis* populations. We assessed multi-collinearity between predictors before conducting our analyses using the variance inflation factor (VIF function, package car; [76]) and this prevented the incorporation of additional predictors such as Precipitation RD and Temperature RD, both of which correlated with Geographic RD and exhibited a VIF > 10. Among these three factors, we chose to include Geographic RD because the isolation by distance pattern is the most parsimonious explanation for genetic differentiation among plant populations [14]. Furthermore, elevation is often a driver of temperature and precipitation patterns [24], and thus inclusion of the Elevation RD allows us to indirectly explore the role of precipitation and temperature (see S1 Fig).

Overall, we examined a set of three explanatory variables, Geographic RD, Elevation RD, and Deforestation RD in our MLPE models (VIF < 2 for all variables). In our MLPE models, genetic distance was used as the response variable and resistance distances as predictors. We centered all explanatory variables around their mean and fitted MLPE models with Maximum Likelihood estimation using the "gls" function in the "nlme" R package [77]. The MLPE model uses a residual covariance structure to account for the non-independence of pairwise distances and is becoming a standard approach in landscape genetic studies since it accounts for the non-independence of pairwise distances within a likelihood framework and is compatible with AIC-based model selection [78]. Code implementing the MLPE correlation structure within the R package nlme is provided [79,80]. We built 7 a priori models comprised of various combinations of the three explanatory variables (geographic distance, elevation, and deforestation). For the three explanatory variables, we estimated the model averaged parameter estimates (β) with standard errors based on models with ΔAIC of less than 5 using the package MuMIn for R [69]. Finally, some studies have reported an increased abundance of pioneer tropical trees and resilient gene flow among their populations with forest fragmentation [81]. For this reason, we repeated our landscape genetic analysis running the same MLPE models but coding Deforestation as a conductance variable (i.e., Deforestation CD) and results were consistent using both approaches. Finally, we also ran the same MLPE models but used raw elevation differences between populations instead of Elevation RD.

Historical and Contemporary Gene Flow

To compare migration rates over contemporary and historical timescales, we used the programs BAYESASS and MIGRATE [82,83], respectively. Both programs generate parameters from which a comparable measure of gene flow can be inferred (*m*: proportion of population consisting of genetic migrants per generation). However, each program estimates this parameter over a different timescale. BAYESASS uses a Bayesian approach with MCMC sampling to estimate recent migration rates over the last five to seven generations (i.e., 25–35 years) given a generation time of 5 years for *M. affinis* (i.e., time to the first flowering event). This method does not assume the population to be in HWE or migration-drift equilibrium. The program was run for 5 x 10^6 iterations of which 500,000 were burn-in. We performed 3 runs with different seed numbers to ensure the convergence and consistency among runs. We used MIGRATE-n v. 3.6.11 to estimate levels of historical gene flow (approx. $4N_e$ generations in the past, approximately 1000's of years) and effective population size. MIGRATE uses a coalescent method with MCMC to estimate the mutation-scaled effective population size ($\theta = 4N_e\mu$, where N_e is effective population size and μ is mutation rate) and the mutation-scaled migration rate ($M = m/\mu$). All starting values were optimized by multiple test runs, adjusted for subsequent runs. In the final run, we used 10 short chains of 500,000 trees sampled and 5,000 trees recorded and 4 long chains of 5,000,000 trees sampled and 50,000 recorded. We used four heated chains with temperatures, 1.0, 1.5, 3.0 and 100,000.0, to ensure sufficient sampling of genealogical space. MIGRATE assumes that populations are in migration-drift equilibrium but does not require them to be in HWE.

To compare historical and contemporary migration rates, we used the values of *m* directly generated by BAYESASS and the estimated *m* from values of *M* (m/μ) generated by MIGRATE by dividing all *M* values by an estimated mutation rate of 5 x 10^{-4} [84]. Concordance of historical and contemporary migration rates was checked using a Mantel test with 10000 permutations using the function "mantel" in the "ecodist" R package [85].

Test of Population History

Finally, we evaluated which of two models of demographic history (gene flow versus genetic drift only) best described processes leading to the current genetic structure in the *M. affinis*populations using the approach developed by

Ciofi and colleagues [86] and implemented in the program 2MOD by MA Beaumont (software available fromhttp:// www.maths.bris.ac.uk/~mamab/software/). This program assesses the relative likelihoods of both demographic models using MCMC procedure. The gene flow model assumes populations are at drift-migration equilibrium and uses the allelic frequencies within populations to estimate the relative strength of drift versus gene flow for each population. The drift model assumes a historical panmictic population separated into many smaller populations that have since been evolving independently through genetic drift alone in the absence of gene flow. Both models assume that mutation rate is small and calculate a parameter F, the probability that two alleles in a given population share a common ancestor. We ran the program twice for each model with 500,000 MCMC updates, discarding the first 10% as burn-in. Results from the two runs were combined and the probability of each model was calculated as the number of draws for a given model out of the total draws.

RESULTS

Lifespan and Generation time of *M. affinis*

We found a mean annual growth rate of 1.75 ± 0.16 mm per year for *M. affinis* in the 50-ha BCI plot. Based on these data, we predict an average lifespan of 64.3 years and a generation time of 5.71 years for *M. affinis* (S1 Table). According to this estimated growth rate, 50 year-old trees would exhibit DBH > 87.5 mm. In the five intensively surveyed populations in this study (CP, GB, PL, AG and RC), the DBH ranged from 7.55 to 125.5 mm with a mean value of 37.29 mm. In our data set, only 2% of trees have DBH greater than 87.5 (older than ~50 years), 31% have DBH ranging from 43.75 to 87.5 (~ 25–50 year-old), while 67% of trees have DBH < 43.75 mm (i.e., 1–25 year-old).

Allelic Richness, Gene Diversity, and Population Structure

Average allelic richness per site based on rarefaction was 3.60 (±0.08) and average private allelic richness per site was 0.14 (±0.02). Average within-population unbiased Nei's gene diversity was 0.598 (±0.027) (Table 1). Neither spatial isolation, nor elevation were related to private allelic richness ($P = 0.41$ and $P = 0.49$ respectively). There was a marginally significant positive relationship between private allelic richness and forest cover ($z = 1.86$, $P = 0.06$). Allelic richness was not related to spatial isolation, elevation, or forest cover ($P = 0.85$, $P = 0.27$ and $P = 0.90$, respectively). The unbiased gene diversity was also not related to spatial isolation, elevation, or forest cover ($P = 0.28$, $P = 0.13$ and $P = 0.69$ respectively). We also ran the same models using the percentage of forest cover within a smaller 2 km and 1 km radius and obtained consistent results (see S2, S3 and S4 Tables in Supporting Information).

AMOVA revealed significant genetic differentiation among populations ($F_{ST} = 0.08$, $P < 0.001$), with 8% of molecular variance accounted for by differences among populations. Results of the STRUCTURE analysis revealed a modal maximum of ΔK at K = 2, although it was only two times higher than the ΔK at K = 3 (S2 Fig). At both values of K, the level of admixture was substantial (Fig 2 and S3 Fig). At K = 2, three eastern populations (i.e., CA, CJ and SJ) were differentiated from the remaining populations (Fig 2A and 2B). At K = 3, our results showed an additional third group consisting of ST, SH, and BC populations (S3 Fig). For K = 2 and K = 3, AMOVAs conducted on the STRUCTURE assigned clusters revealed 7% of the variation was due to differences among clusters in both analyses.

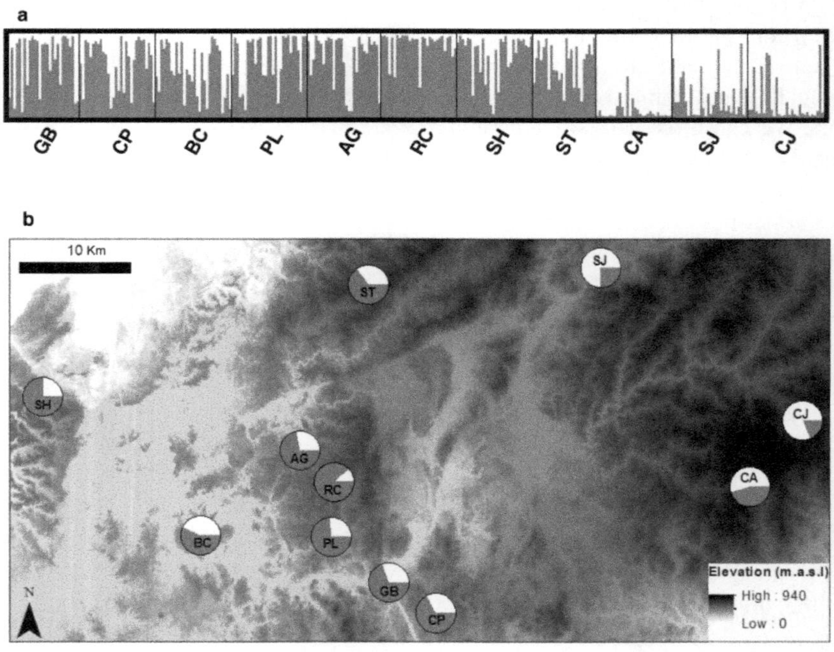

Figure 2. Genetic structure of *M. affinis* **populations based on STRUCTURE output for K = 2.** (a) Structure bar plot with individuals sorted by populations. Each vertical bar represents a single individual and color segments are proportional to its membership in two genetic clusters inferred from STRUCTURE analyses. From left to right, populations are sorted by increasing distance to the Panama Canal. (b) Populations represented by individual pie charts with the mean proportion of membership of each population for the inferred number of K = 2 genetic groups. In the background, higher elevation is represented by darker shades (0–940 m.a.s.l.).

Landscape Genetic Analysis

The first ranked model contained the explanatory variables Geographic RD and Elevation RD, with a 59% probability of being the best approximating model in the set (Table 2). Our results showed a positive relationship between genetic distance and Geographic RD (Table 3, Fig 3A). Our results also supported an isolation by elevation pattern with a positive relationship between the genetic distance and the Elevation RD (Table 3, Fig 3B). We obtained consistent results using G_{ST} and D_{ST} as response variables instead of F_{ST} and using raw elevation differences between populations instead of Elevation RD (S5–S10Tables).

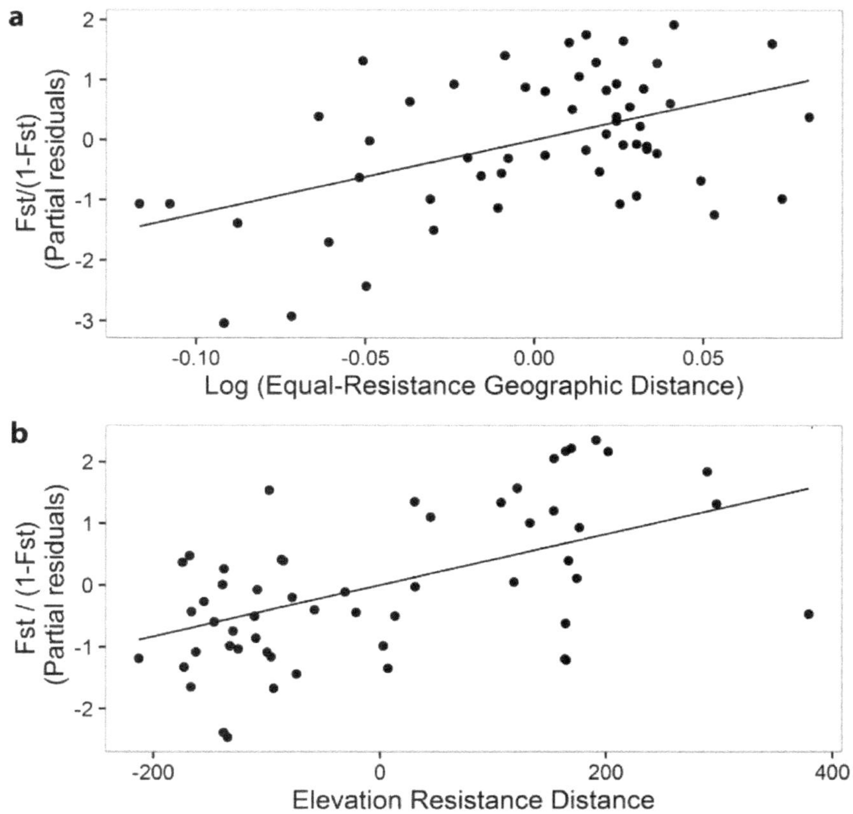

Figure 3. Isolation by Resistance Distance. (a) Isolation by Geographic Resistance Distance (RD) and (b) Elevation RD. Pairwise comparisons of genetic differentiation (F_{ST}) as a function of (a) Geographic RD and (b) Elevation RD. Explanatory variables are centered around their mean. We graphically represent the partial residuals of the dependent variable to remove the effects of other explanatory variables included in the model. Partial residuals were computed using the top-ranked model (ΔAIC; see Table 2).

Table 2. Ranked models explaining landscape effects on genetic differentiation (i.e., F_{ST}) among *M. affinis*'s populations. Models are ranked based upon the difference between Akaike's Information Criterion (AIC) in each individual model and the lowest AIC model (ΔAIC). Akaike's weights (ω_i) for each model are listed. The genetic differentiation between populations was calculated as $[F_{ST} / (1-F_{ST})]$ and Geographic refers to Geographic Resistance Distance (RD), Deforestation refers to Deforestation RD, and Elevation refers to Elevation RD.

Model statement	AIC	ΔAIC	ω_i
Geographic + Elevation	-240.66	0.00	0.588
Geographic + Elevation + Deforestation	-238.93	1.73	0.247
Geographic	-237.49	3.17	0.120
Geographic + Deforestation	-235.51	5.15	0.045
Elevation	-225.49	15.17	0.000
Elevation + Deforestation	-224.00	16.66	0.000
Deforestation	-216.16	24.50	0.000

doi:10.1371/journal.pone.0156694.t002

Table 3. Model averaged coefficients (β) and their standard errors (SE) calculated from the candidate model set (i.e. models with ΔAIC < 5). The genetic differentiation between populations was calculated as $[F_{ST} / (1-F_{ST})]$. Geographic refers to Geographic Resistance Distance (RD), Deforestation refers to Deforestation RD, and Elevation refers to Elevation RD.

	β	SE	Z value	P value
Geographic	0.40800	0.09091	4.390	<0.001*
Elevation	0.00013	0.00006	2.362	0.018*
Deforestation	-0.00144	0.00283	0.495	0.621

* Model averaged coefficients not overlapping with zero

doi:10.1371/journal.pone.0156694.t003

Historical and Contemporary Gene Flow

The historical migration estimates (M, hereafter) ranged from 0 to 59.6. Of the 55 pairwise estimates of M for all sets of populations, 32 were significantly different (non-overlapping 95% CI for pairwise migration). The number of migrants per generation, N_{em} (product of theta and M divided by four) ranged from 0 to 16.6. Multiple runs of BAYESASS revealed low contemporary immigration with the exception of CJ, AG, GB, and PL. Our analyses also reveal a source-sink relationship where CJ receives a substantial proportion of migrants from CA (m = 0.13 ± 0.05), while the expected proportion of migrants into CA from CJ is much smaller (m = 0.01 ± 0.01). RC also constitutes a source population for AG, PL, and GB, where these three populations all receive similar proportions of migrants from RC (0.19 ± 0.03, 0.20 ± 0.03 and 0.14 ± 0.04, respectively). Historical and contemporary migration estimates were not significantly correlated (r = 0.02, P < 0.88). There is a contemporary decrease in the migration between populations adjacent to the Panama Canal and the eastern populations (Fig 4).

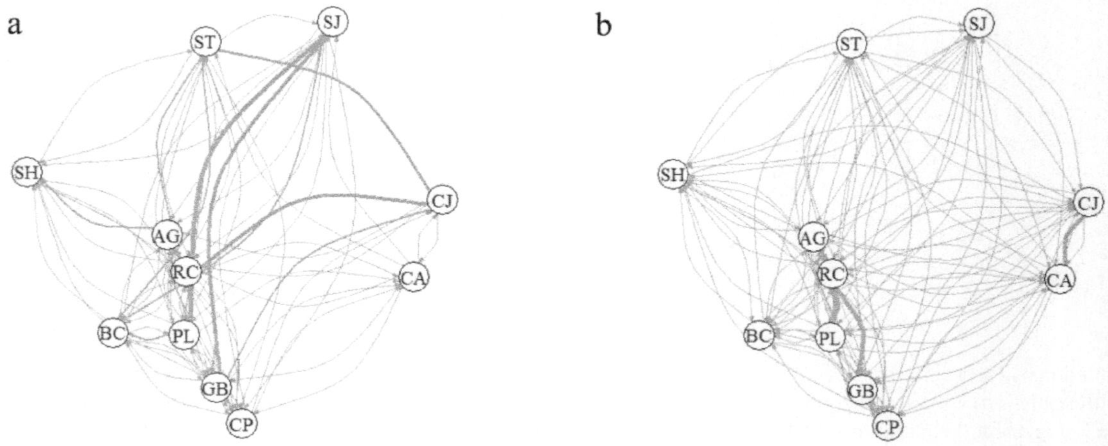

Figure 4. Historical and contemporary migration among *M. affinis populations.* (a) Historic mutation-scaled migration rates ($M = m/\mu$) approximately $4N_e$ generations in the past calculated with Migrate-n. (b) Contemporary migration rates (i.e., *m*) over the last 25–35 years estimated with Bayesass. The thickness of the arrows is proportional to *M* and *m* respectively. Only migration rates significantly different from zero are shown in both panels.

Test of Population History

The long-term gene flow population model was much more strongly supported than the pure drift model (Prob. of drift model = 0.00, Prob. of gene flow model = 1.00). This indicates that populations have been at drift-migration equilibrium for a substantial period of time. Overall, the effect of genetic drift was low in most of *M. affinis*' populations, with F values ranging from 0.04 to 0.22. ST and CA populations were the most affected by genetic drift (*F*= 0.14 and 0.22, respectively). Estimates of θ from MIGRATE were not correlated with F values from the 2MOD analysis (r_p = -0.05, *P* = 0.89).

DISCUSSION

Given global tropical forest declines and the importance of reforestation, there is increasing interest in the ecology and evolution of pioneer tropical trees. Specifically, due to their short generation times, these trees provide valuable information regarding the impacts of deforestation on both contemporary and historic gene flow processes, with implications for restoration species selection and translocation strategies. Using novel landscape genetic tools, we evaluate the genetic structure of an early successional species, and reveal that populations of *M. affinis* maintain moderately-high levels of genetic diversity and low levels of genetic differentiation despite contemporary deforestation patterns. Furthermore, we show that both elevation and geographic distance play a critical role in increasing genetic differentiation between populations. Thus, we highlight the importance of considering both topography and geographic distance when designing reforestation strategies aimed at preserving genetic diversity within tropical pioneer trees.

Regional Genetic Structure and Gene Flow

Our results indicate that *M. affinis* exhibits higher levels of within-population genetic diversity and lower levels of among-population genetic differentiation than many early successional plant species across the globe [87], but similar levels of within-population genetic diversity as other tropical pioneer species [88]. Interestingly, the level of among-population genetic differentiation for *M. affinis* was lower than the average reported for tropical trees [89], but similar to light-demanding tropical pioneer species with wind-mediated seed dispersal [88]. In addition, our Bayesian clustering analysis shows substantial admixture levels in *M. affinis* populations in the study region, suggesting that *M.affinis* experiences substantial dispersal across human-altered landscapes [46], leading to weak but significant genetic differentiation among populations. This high level of genetic diversity and low level of genetic differentiation among populations is likely due to the species' propensity to colonize recently deforested landscapes, leading to increases in the connectivity among populations throughout the region.

The genetic differentiation among *M. affinis* populations was significantly explained by two factors, geographic distance and elevation. First, we found a strong signal of isolation by distance across our study region, as documented for a number of tropical tree species [17,18]. Isolation by distance is frequently stronger in plants dispersed by animals with restricted mobility, such as understory tropical trees, which often have fleshy fruits that are dispersed by small-bodied birds [90]. Previous studies have reported shorter dispersal distances and a lower abundance of long-distance dispersal events for small bird species relative to larger-bodied species [91]. Because *Miconia* seeds are often dispersed by small-bodied bird species [46,92], the strong isolation by distance documented in this study likely reflects the restricted ability of these small-bodied animals to cover the long distances between populations.

We also detected significant isolation by elevation, suggesting that high-elevation populations of *M. affinis* are experiencing more restricted gene flow. These results, in line with other recent studies, document a relevant role of elevation as a factor influencing genetic differentiation in plant populations [21,30,93]. Harsher environmental conditions (e.g., lower temperature and higher precipitation) in high-elevation sites compared to low-elevation sites may affect flying conditions for insects and potentially impact pollen-mediated gene flow [25–27]. For instance, lower temperature and higher precipitation severely reduced the pollination service in high-elevation populations of another melastome, *Tibouchina pulchra* [94]. Additionally, elevation can result in reproductive isolation due to phenological shifts that cause temporal separation in the timing of flowering [28]. Specifically, our field research has revealed that populations located at the highest elevation, where there is an earlier onset of the first rains, exhibit a more advanced flowering and fruiting phenology than populations at the lowest elevations (A.C., personal observation), potentially making high-elevation populations more asynchronous in flowering and fruiting relative to other populations. This relationship between reproductive phenology and water availability during the dry season has been reported for other *Miconia*species [95] and may underlie reproductive isolation. This relationship could be strengthened by the dispersal patterns of frugivorous birds, given that understory fruit-eating birds often track changes in fruit abundance across altitudinal gradients within a single mountain side, exploiting higher elevation sites first and lowland sites later [96], potentially enhancing localized gene flow for each temporal fruiting period.

We did not find evidence of increased genetic differentiation with increased deforestation between *M. affinis* populations. While forest fragmentation is theorized to drive genetic differentiation among tropical tree populations [32,35], empirical support of this theory is still scarce. Recent reviews suggest that the genetic signature of deforestation may be more evident in younger age-classes or progenies sired in disturbed landscapes ([35] and references therein). Despite the short generation time and relatively young estimated age of the trees sampled in our focal populations, we found no evidence of genetic structuring in response to recent deforestation for *M. affinis*. One potential explanation is that the high dispersal and colonization ability of the study species promotes gene flow for *M. affinis*across the study region, regardless of forest cover. Past studies have documented that moderate levels of deforestation, whether human-induced or natural, can be well tolerated by disturbance-adapted tree species [97]. As documented in other studies, forest fragment boundaries may not represent mating boundaries due to long-distance dispersal events connecting populations [34]. Our results likewise suggest that the human-altered landscapes along the Panama Canal do not limit gene flow between isolated forest populations.

Other alternative explanations for the genetic structure observed in *M. affinis* populations could be related to the construction of the Panama Canal. The Panama Canal was completed in 1914 and led to extensive deforestation in the adjacent areas, except for Barro Colorado Island which, though disturbed, retained substantial old-growth forest [45]. Also, the forests east of Lake Alajuela were primarily undisturbed and retained large areas of near pristine old-growth forest. Our STRUCTURE analyses revealed substantial support for the clustering of *M. affinis* populations adjacent to the Panama Canal, excluding the Barro Colorado Island population. Furthermore, our estimates of historical and contemporary migration support a recent collapse of gene flow between populations located east of Lake Alajuela and populations in the Panama Canal Zone. These two analyses provide evidence that deforestation during the construction of the canal may have restricted gene flow; interestingly recovery of the forest in the past 100 years could be preventing detection of a relationship between current deforestation patterns and genetic differentiation.

Implications for Reforestation

In the past 50 years, the Republic of Panama has lost more than 30% of its forested area [45]. This high rate of deforestation is associated with high levels of erosion and sediment deposition in streams and canals. Given the socio-economic value of the Panama Canal and the large amount of water required for shipping traffic, there is great need to reduce sediment deposition and improve the water-holding capacity of the Panama Canal watershed via reforestation [44,98]. To date, however, the majority of the region's reforestation programs are dominated by fast-growing, exotic timber species, which support only low levels of plant diversity, often promote soil erosion, and provide only limited goods and services to local landholders [5]. Major efforts are being made in Panama to develop reforestation programs with native species that dually increase biodiversity and contribute to rural livelihoods [5], and recent studies have revealed that Panamanian rural farmers are highly interested in land restoration that promotes native trees [6].

However, one of the constraints to using native species in reforestation programs is the lack of ecological information about these species [99]. In this regard, the reproductive processes of tropical trees are particularly important to consider for the long-term viability of the restored population. Our study is one of the first to analyze the gene flow of an early successional tropical tree species using a spatially-explicit landscape genetics approach. Based on our results for *M. affinis*, we recommend the use of local genotypes for small-scale restoration, as they may maintain local phenological patterns and promote local adaptation to future land use and climate conditions. We also suggest utilizing species like*M. affinis* that demonstrate substantial gene flow at regional-scales, even across large tracts of deforested land. Finally, our results suggest prioritizing the conservation of geographically isolated or high-elevation populations, as these populations may face the greatest challenges of maintaining effective pollen- and seed-mediated gene flow.

Supporting Information

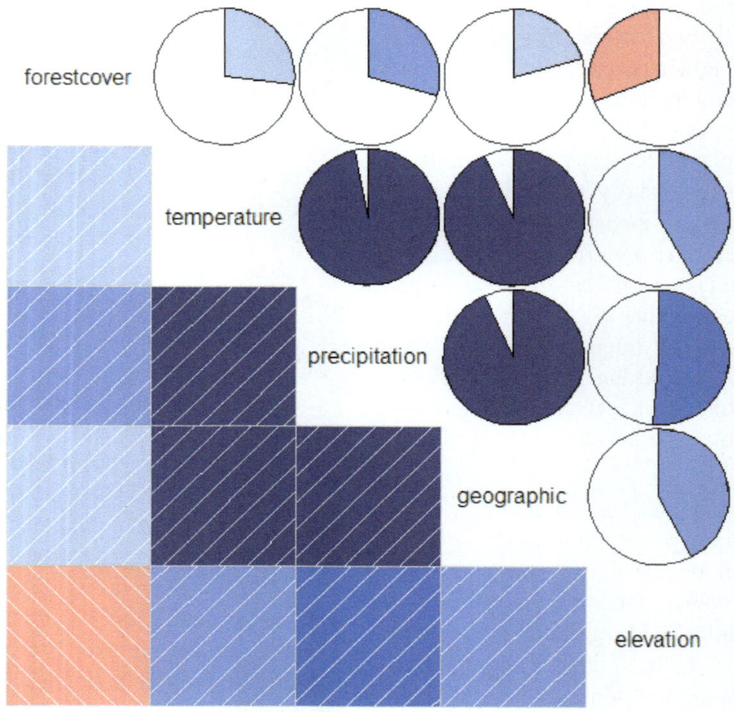

S1 Figure. Correlogram of the initial set of explanatory variables used the MLPE models (Geographic RD, Elevation RD, Forest Cover CD, Precipitation RD, and Temperature RD).

The color in both the filled portion of the pie and the shade squares indicates the sign of the correlation, with positive and negative values encoded by blue and red respectively. The intensity of the color increases uniformly as the correlation values move away from 0. There is a significant positive correlation between precipitation and elevation ($r_p = 0.51$, $P < 0.05$) and a marginally significant positive correlation between mean annual temperature and elevation ($r_p = 0.42$, $P = 0.08$), thus we opted to include only Elevation RD in our models.

doi:10.1371/journal.pone.0156694.s001

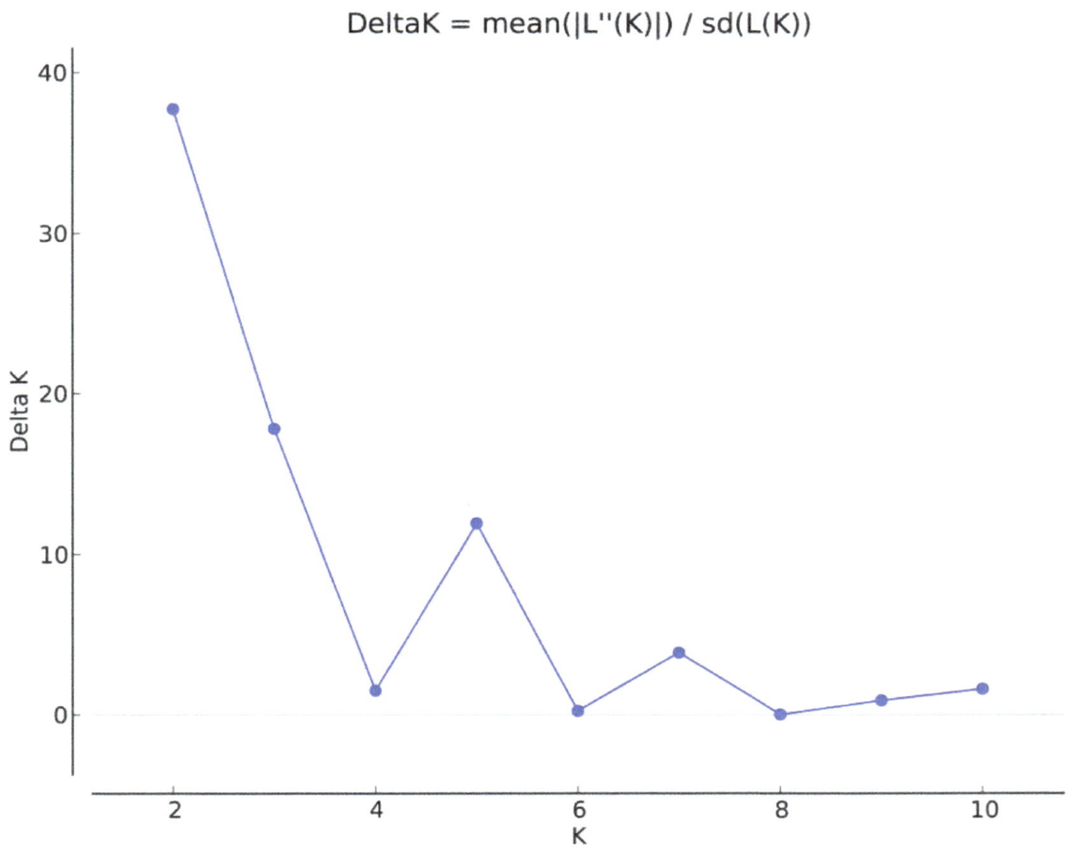

S2 Fig. Magnitude of ΔK as a function of K over 10 runs of STRUCTURE for each K value from 1 to 11.

Runs used a total number of iterations of 150,000 with a burn-in of 50,000. The admixture and the correlated frequency models were used.

doi:10.1371/journal.pone.0156694.s002

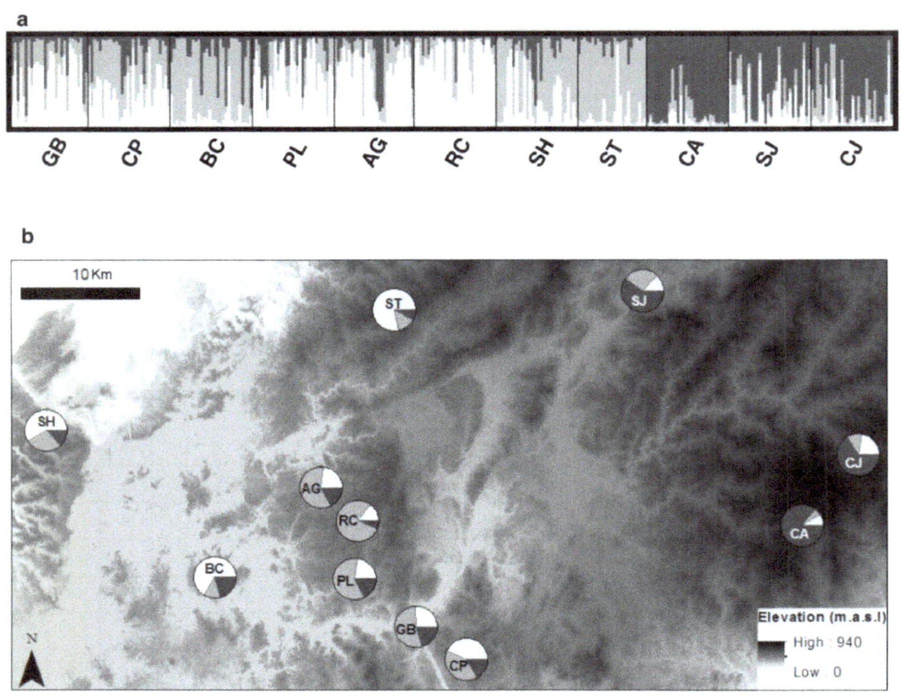

S3 Fig. Genetic structure of *M. affinis* populations based on STRUCTURE output for K = 3.

(a) Structure bar plot with individuals sorted by populations. Each vertical bar represents a single individual and color segments are proportional to its membership in three genetic clusters inferred from STRUCTURE analyses. From left to right, populations are sorted by increasing distance to the Panama Canal. (b) Populations represented by pie charts with the mean proportion of membership of each population for the inferred number of K = 3 genetic groups. In the background, higher elevation is represented by darker shades (0–940 m.a.s.l.).

doi:10.1371/journal.pone.0156694.s003

S1 File. Genotypes of *M. affinis* trees in the eleven study populations.

popu-lation	id_tree	Micaff 5a	Micaff 5b	B109a	B109b	micaff 14a	micaff 14b	micaff 16a	micaff 16b	micaff 7a	micaff 7b	micaff 8a	micaff 8b	B9a	B9b	B102a	B102b	Micaff 19a	Micaff 19b
AG	AG003	264	264	346	346	221	223	262	264	278	278	282	298	386	386	366	366	270	282
AG	AG033	264	268	342	354	221	221	252	266	280	280	290	298	376	386	364	370	272	282
AG	AG036	264	268	344	346	221	221	252	256	278	278	292	292	386	386	364	370	282	282
AG	AG056	268	272	338	354	221	221	254	262	278	282	292	298	386	386	366	370	272	272
AG	AG057	264	264	352	356	221	225	256	256	276	278	298	300	386	386	366	366	274	282
AG	AG062	264	268	354	354	221	221	252	254	278	280	0	0	386	386	370	370	272	272
AG	AG063	268	270	352	354	221	223	252	256	252	278	278	290	386	386	364	364	274	282
AG	AG072	264	268	354	354	221	223	252	254	278	280	292	300	386	386	366	366	282	282
AG	AG127	262	268	352	352	221	221	240	264	252	280	270	292	376	376	370	370	272	272
AG	AG181	264	264	346	354	221	221	262	262	252	280	298	298	376	386	366	370	272	272
AG	AG193	264	264	352	352	221	221	242	248	280	280	294	298	374	374	370	370	270	270
AG	AG209	264	268	342	352	221	221	254	262	278	278	262	270	376	376	366	370	272	272
AG	AG215	264	268	336	354	221	221	264	264	280	280	292	298	386	386	366	370	272	272
AG	AG248	264	264	346	352	221	221	262	262	278	280	298	298	386	386	366	370	272	272
AG	AG251	264	264	346	346	221	221	244	256	278	282	270	302	376	376	366	370	272	272
AG	AG254	264	268	346	352	221	221	252	266	280	280	290	292	386	386	364	370	282	282
AG	AG259	264	264	346	362	221	221	252	262	280	280	0	0	0	0	370	370	0	0
AG	AG266	264	264	344	362	221	221	252	262	282	282	0	0	0	0	370	370	0	0
AG	AG282	264	268	350	354	221	221	256	262	252	282	302	302	374	386	366	370	270	282
AG	AG296	264	264	344	344	221	221	264	264	280	282	282	292	376	376	370	370	272	272
AG	AG299	264	264	344	344	221	221	256	264	278	278	298	306	376	386	366	370	272	282
AG	AG301	264	268	344	344	221	223	262	262	278	280	270	292	376	386	366	366	272	282
AG	AG318	264	274	346	352	221	221	262	262	252	280	290	298	376	386	366	370	282	282
AG	AG329	264	268	346	346	221	221	254	262	252	278	290	292	376	386	364	370	272	282
AG	AG345	268	268	346	354	221	221	254	264	278	278	292	292	376	386	364	370	272	282
AG	AG361	268	270	352	354	221	221	262	262	278	280	282	292	386	386	366	366	282	282
AG	AG365	268	268	344	352	217	221	266	266	252	280	282	298	376	386	366	366	272	282
AG	AG366	264	268	352	352	221	221	262	262	280	280	292	298	386	386	366	366	282	282
AG	AG369	268	268	346	346	221	221	248	262	252	280	292	294	386	386	364	364	272	272
CA	CA001	264	266	364	364	221	221	262	262	280	280	290	292	374	386	366	366	270	270
CA	CA002	268	268	0	0	221	221	252	252	280	280	282	290	386	386	370	370	282	282
CA	CA003	264	268	364	364	221	221	0	0	280	280	290	294	374	386	370	370	270	272
CA	CA004	264	264	364	364	221	221	252	252	280	280	288	290	376	386	368	368	272	278
CA	CA005	264	264	364	364	221	221	252	264	280	280	294	294	374	376	370	370	270	278
CA	CA006	264	264	364	364	221	221	264	264	280	282	288	294	374	374	368	370	270	278
CA	CA007	264	268	364	364	221	221	252	252	280	280	292	292	386	386	370	370	274	274
CA	CA008	264	264	364	364	221	221	252	252	280	282	290	292	386	386	366	366	272	272
CA	CA009	268	268	354	358	221	221	262	262	280	280	290	294	376	386	370	370	272	282
CA	CA010	264	264	344	346	221	221	262	262	280	280	282	282	374	376	370	370	270	272
CA	CA011	264	264	354	364	221	221	264	264	280	280	290	292	374	386	370	370	270	278
CA	CA012	264	266	344	364	221	221	252	252	280	280	288	290	386	386	370	370	272	282
CA	CA013	264	268	346	360	221	221	0	0	280	280	282	292	376	386	370	370	272	272

CA	CA014	264	268	364	364	221	221	252	252	280	280	294	294	376	386	370	370	272	278
CA	CA015	264	268	0	0	221	221	0	0	280	280	282	290	386	386	370	370	274	274
CA	CA016	264	264	0	0	221	221	0	0	280	280	290	290	386	386	370	370	0	0
CA	CA017	264	264	0	0	221	221	0	0	280	280	288	290	374	386	370	370	270	270
CA	CA018	268	268	348	348	221	221	262	262	280	280	292	294	386	386	370	370	272	272
CA	CA019	264	266	362	364	221	221	0	0	280	282	288	292	386	386	366	366	270	270
CA	CA020	264	268	340	348	221	221	262	262	280	282	294	294	374	376	366	366	270	272
CA	CA021	264	268	340	346	221	221	0	0	280	280	290	294	376	386	370	370	272	272
CA	CA022	268	268	346	364	221	221	0	0	280	280	288	292	376	386	370	370	272	272
CA	CA023	268	268	348	364	221	221	0	0	280	280	288	288	386	386	366	366	270	270
CA	CA024	264	268	348	348	221	221	260	260	282	282	294	294	386	386	370	370	282	282
CA	CA025	268	268	346	346	221	221	0	0	252	280	294	294	386	386	366	366	274	282
CA	CA026	268	268	348	364	221	221	0	0	282	282	294	294	386	386	370	370	274	282
CA	CA027	264	268	340	364	221	221	262	262	282	282	294	294	386	386	366	366	282	282
CA	CA028	264	268	340	364	221	221	0	0	282	282	288	288	386	386	370	370	270	270
CA	CA029	268	270	340	364	221	221	252	252	280	280	294	294	386	386	370	370	270	270
CA	CA030	264	268	364	364	221	221	0	0	280	280	290	294	386	386	366	366	272	272
CJ	CJ001	264	264	344	352	221	221	262	262	280	280	282	294	374	386	366	366	270	270
CJ	CJ002	270	274	344	364	221	221	0	0	278	280	282	294	386	386	370	370	282	282
CJ	CJ003	264	264	346	350	221	221	258	262	280	280	282	300	386	386	366	366	282	282
CJ	CJ004	264	264	348	348	221	221	0	0	0	0	0	0	374	374	366	370	270	272
CJ	CJ005	264	270	340	348	221	221	0	0	278	282	290	294	374	376	366	366	270	272
CJ	CJ006	268	270	354	360	221	221	252	252	280	280	282	300	376	386	370	370	272	282
CJ	CJ007	264	264	348	348	221	221	0	0	282	282	0	0	0	0	366	370	0	0
CJ	CJ008	264	268	350	350	221	221	0	0	280	280	0	0	0	0	370	370	0	0
CJ	CJ009	264	270	344	344	217	221	256	256	280	280	0	0	386	386	370	370	0	0
CJ	CJ010	270	270	346	358	221	221	256	256	280	280	282	294	386	386	370	370	282	282
CJ	CJ011	264	266	340	364	221	223	260	262	280	280	294	294	386	386	370	370	282	282
CJ	CJ012	268	270	348	352	219	221	256	262	280	280	282	288	386	386	366	366	272	282
CJ	CJ013	264	270	346	352	221	221	0	0	282	282	294	294	386	386	370	370	272	282
CJ	CJ014	268	268	348	364	221	223	252	252	282	282	290	294	374	386	366	366	270	282
CJ	CJ015	264	264	344	364	221	221	258	262	252	282	290	294	386	386	366	370	280	282
CJ	CJ016	264	268	348	362	221	221	254	254	280	280	288	294	374	376	366	366	270	272
CJ	CJ017	264	264	344	364	221	221	262	262	280	280	294	294	386	386	366	366	282	282
CJ	CJ018	264	268	348	352	221	221	262	262	278	282	288	294	386	386	366	366	272	282
CJ	CJ019	268	274	348	352	221	221	262	262	282	282	294	294	374	386	366	370	270	270
CJ	CJ020	266	270	348	364	221	221	0	0	252	280	290	294	386	386	370	370	282	282
CJ	CJ021	264	268	344	348	221	221	264	264	278	280	290	294	386	386	366	370	270	282
CJ	CJ022	264	270	344	348	221	223	262	262	280	280	282	294	386	386	370	370	270	270
CJ	CJ023	264	268	352	364	221	227	256	258	280	280	288	294	386	386	370	370	282	282
CJ	CJ024	268	270	348	348	221	223	260	260	278	282	282	294	386	386	370	370	272	282
CJ	CJ025	264	266	346	348	221	221	0	0	280	282	294	310	374	386	366	370	270	282
CJ	CJ026	264	274	348	352	221	227	256	256	276	276	290	294	374	374	370	370	270	280
CJ	CJ027	264	270	364	364	221	221	0	0	280	280	282	310	386	386	364	370	272	282
CJ	CJ028	270	270	344	348	221	221	258	258	280	280	282	282	386	386	370	370	272	272
CJ	CJ029	264	270	344	352	221	221	256	256	280	280	282	298	376	386	366	366	272	282
CJ	CJ030	264	270	348	352	221	221	0	0	252	280	288	294	376	386	366	370	272	282
CP	CP006	270	270	352	352	221	221	264	264	280	282	284	290	374	374	366	370	270	270
CP	CP029	264	270	344	352	221	221	250	264	280	280	282	282	386	386	370	370	282	282
CP	CP045	264	268	344	346	221	221	262	262	278	280	284	298	374	386	370	370	270	282
CP	CP048	268	268	344	354	221	221	262	264	280	280	290	298	374	374	370	370	270	270
CP	CP060	268	268	352	354	221	221	262	264	280	280	282	288	386	386	370	370	270	282

CP	CP069	264	268	344	344	219	221	262	262	278	282	292	296	386	386	364	364	270	282
CP	CP077	264	270	344	352	219	221	264	266	280	280	288	298	374	378	370	370	270	274
CP	CP114	268	268	344	352	221	221	264	264	278	280	284	290	374	374	370	370	270	270
CP	CP122	264	268	344	346	221	223	262	262	278	278	282	298	374	386	370	370	270	282
CP	CP125	264	264	344	344	219	221	262	264	278	280	290	292	374	376	370	370	270	272
CP	CP137	264	264	344	344	221	221	262	264	278	280	284	290	374	386	364	364	270	270
CP	CP145	264	270	344	350	221	221	248	264	280	280	290	290	386	386	366	370	270	270
CP	CP162	264	270	344	344	221	221	262	262	280	280	292	294	374	386	364	370	270	270
CP	CP180	264	268	348	352	221	221	264	264	280	280	284	294	386	386	370	370	282	282
CP	CP184	264	268	346	362	221	221	264	264	280	280	288	298	376	386	370	370	272	272
CP	CP197	264	268	344	356	221	221	262	262	280	282	282	288	386	386	366	370	0	0
CP	CP216	264	270	344	352	221	221	264	266	278	278	298	298	374	386	370	370	270	282
CP	CP231	264	270	352	356	221	223	262	264	278	280	292	294	374	374	370	370	270	270
CP	CP246	264	264	344	356	221	223	262	264	278	278	284	284	386	386	366	366	270	270
CP	CP254	268	270	352	356	221	221	262	262	280	280	284	292	386	386	370	370	270	270
CP	CP255	264	268	352	356	221	223	262	264	280	280	290	294	374	376	370	370	270	272
CP	CP276	264	270	344	344	221	223	262	266	280	280	288	298	386	386	366	370	270	270
CP	CP277	264	270	344	356	221	221	262	266	280	280	298	298	376	386	370	370	272	272
CP	CP280	264	268	344	354	221	221	264	264	278	280	296	298	374	386	370	370	270	270
CP	CP297	268	270	344	356	221	223	262	262	278	278	284	284	386	386	366	366	270	272
CP	CP304	264	264	344	344	221	221	264	266	278	278	290	298	0	0	366	370	270	270
CP	CP335	264	270	354	356	221	221	252	264	278	282	292	292	386	386	370	370	0	0
CP	CP340	264	264	344	346	219	221	264	266	280	280	290	298	374	378	370	370	270	274
CP	CP350	268	270	352	352	221	221	262	262	278	280	282	290	386	386	370	370	270	282
CP	CP361	264	268	344	352	221	221	262	264	280	280	292	294	386	386	366	370	270	282
GB	GB012	264	270	344	352	221	221	262	262	278	280	292	294	374	386	366	370	270	270
GB	GB013	270	270	346	352	221	225	262	262	278	280	290	292	386	386	366	370	282	282
GB	GB027	264	268	348	348	221	221	250	262	280	280	270	282	376	386	370	370	272	282
GB	GB035	264	268	344	354	221	225	262	262	280	280	282	282	376	386	366	370	272	282
GB	GB055	264	268	344	352	221	221	240	248	280	280	292	298	386	386	366	366	282	282
GB	GB056	270	274	348	354	221	221	262	262	280	280	294	294	374	386	366	370	270	282
GB	GB074	268	268	344	352	221	221	262	262	278	280	282	298	386	386	366	366	270	282
GB	GB076	264	264	344	352	221	221	254	264	278	278	294	294	374	386	364	364	270	282
GB	GB083	264	268	348	352	221	221	254	260	278	278	282	298	386	386	370	370	282	282
GB	GB096	268	274	344	354	221	221	262	262	278	278	282	298	386	386	366	366	282	282
GB	GB126	268	268	346	352	221	221	262	264	280	280	270	298	374	386	366	366	270	270
GB	GB128	270	274	344	354	221	225	264	264	280	280	284	290	386	386	364	364	282	282
GB	GB133	264	268	340	352	221	221	260	260	280	280	0	0	0	0	370	370	282	282
GB	GB137	270	270	344	348	225	225	260	262	280	280	292	298	386	386	366	366	282	282
GB	GB139	268	268	352	352	221	225	262	264	252	280	290	292	386	386	370	370	282	282
GB	GB162	268	268	352	352	221	221	264	264	0	0	0	0	0	0	370	370	270	282
GB	GB167	264	270	344	352	221	221	262	262	280	280	284	288	386	386	366	370	282	282
GB	GB201	264	268	352	354	221	221	264	290	278	280	282	298	378	386	366	366	274	282
GB	GB203	264	268	344	354	221	221	240	258	276	276	282	298	376	386	366	366	272	282
GB	GB219	270	274	346	352	221	221	262	264	278	278	282	292	386	386	366	366	282	282
GB	GB220	264	268	340	360	221	221	0	0	0	0	0	0	0	0	370	370	0	0
GB	GB226	268	268	352	352	221	221	262	264	278	280	270	298	376	386	370	370	272	282
GB	GB249	268	270	344	354	221	225	262	262	278	278	0	0	376	386	370	370	272	282
GB	GB274	268	274	344	346	221	221	264	266	278	278	288	298	386	386	366	366	282	282
GB	GB276	268	270	354	360	221	225	264	264	280	280	290	290	374	386	366	366	270	282
GB	GB318	268	268	354	354	221	221	262	264	278	278	0	0	0	0	368	370	0	0
GB	GB324	268	268	0	0	221	221	262	262	276	276	0	0	0	0	370	370	0	0

GB	GB350	264	264	348	352	221	221	242	266	278	280	282	288	386	386	370	370	282	282
PL	PL067	264	264	346	352	221	221	262	262	278	282	290	292	376	376	366	366	272	272
PL	PL068	264	268	352	352	221	221	262	266	278	282	298	298	386	386	364	364	282	282
PL	PL069	264	268	336	354	221	221	262	266	280	282	290	298	386	386	366	370	270	282
PL	PL070	264	268	348	352	221	221	262	262	280	280	292	292	374	376	364	364	270	272
PL	PL071	264	268	344	344	221	221	252	266	280	280	292	292	376	386	366	370	272	282
PL	PL072	264	270	346	348	221	221	244	266	0	0	0	0	0	0	366	366	0	0
PL	PL073	264	270	344	352	221	221	252	262	278	282	300	300	376	378	364	370	272	274
PL	PL074	264	264	348	352	221	221	262	262	252	278	298	298	386	386	366	370	282	282
PL	PL075	268	268	352	352	221	221	264	266	278	280	288	298	376	376	366	370	272	272
PL	PL076	264	274	352	352	221	221	262	266	280	280	270	290	386	386	364	364	272	272
PL	PL077	264	268	348	352	221	221	252	254	252	278	298	298	376	378	370	370	272	274
PL	PL078	264	264	344	354	221	221	254	266	280	280	298	298	376	386	366	370	272	282
PL	PL079	264	268	344	346	221	221	252	262	280	280	290	298	386	386	364	364	282	282
PL	PL080	268	270	346	346	221	221	262	262	0	0	0	0	0	0	366	366	0	0
PL	PL081	268	274	346	352	221	225	266	266	278	280	298	300	376	386	364	364	272	282
PL	PL082	264	268	344	346	221	221	262	264	278	278	298	298	386	386	364	364	282	282
PL	PL083	264	268	340	352	221	221	244	262	278	280	290	298	386	386	366	370	282	282
PL	PL084	264	268	346	352	221	225	254	258	278	278	298	300	374	376	366	366	270	272
PL	PL085	268	268	340	348	221	221	264	266	278	280	290	290	386	386	366	366	282	282
PL	PL086	264	268	342	348	221	221	262	264	278	282	292	300	376	386	366	370	272	272
PL	PL087	268	270	352	354	221	221	254	258	278	278	298	298	386	386	366	370	282	282
PL	PL088	270	272	352	354	221	221	256	256	280	280	282	298	386	386	366	370	282	282
PL	PL089	268	272	338	344	221	221	254	266	278	278	298	298	376	376	366	370	272	272
PL	PL090	264	264	344	344	221	225	262	266	280	280	270	290	374	374	366	366	270	270
PL	PL091	264	268	336	354	221	221	260	262	278	280	298	298	378	386	366	370	274	282
PL	PL092	264	268	354	354	221	221	240	262	280	280	270	298	386	386	364	364	272	282
PL	PL093	264	274	336	352	221	221	240	250	278	278	288	290	386	386	366	370	272	282
PL	PL094	264	264	0	0	221	221	260	260	0	0	0	0	0	0	370	370	0	0
PL	PL095	264	272	352	356	221	221	240	262	280	282	298	298	376	386	366	366	272	282
PL	PL096	268	272	346	354	221	221	240	262	280	282	298	298	386	386	366	366	282	282
RC	RC001	268	268	344	354	221	225	256	264	278	278	282	290	376	386	366	366	272	282
RC	RC004	268	268	344	354	221	221	254	260	278	278	270	298	376	386	366	366	272	282
RC	RC015	264	268	352	354	221	221	256	262	278	278	290	298	374	386	364	364	270	282
RC	RC020	264	268	342	362	221	221	264	264	280	280	292	298	386	386	366	370	282	282
RC	RC085	268	268	346	354	221	221	256	264	280	280	298	298	374	374	364	364	270	272
RC	RC111	264	264	352	362	221	221	254	256	280	280	292	292	374	376	364	364	270	272
RC	RC113	268	270	350	352	221	221	240	254	278	278	298	298	376	386	366	366	272	282
RC	RC125	264	268	350	354	221	221	262	266	278	280	292	298	386	386	366	366	272	282
RC	RC138	264	268	352	352	221	225	256	264	278	278	298	298	374	386	366	370	270	282
RC	RC142	270	270	338	352	221	221	262	264	278	278	298	298	376	376	364	364	272	272
RC	RC144	268	268	352	354	221	225	260	264	278	278	292	298	376	386	364	364	272	282
RC	RC154	268	274	338	354	221	221	240	262	280	282	282	290	386	386	366	366	282	282
RC	RC158	264	270	354	354	221	221	254	254	278	280	292	292	386	386	366	370	272	282
RC	RC186	268	268	352	352	221	221	262	264	280	280	282	298	376	376	366	366	272	272
RC	RC200	268	268	346	352	221	221	240	266	280	282	270	296	374	376	366	366	270	272
RC	RC203	264	268	340	348	221	221	262	262	280	280	282	298	376	386	366	366	272	282
RC	RC216	264	268	344	352	221	221	254	254	278	282	292	298	376	386	366	370	272	282
RC	RC217	268	270	338	352	221	221	262	264	280	280	292	298	376	386	366	370	272	272
RC	RC222	268	268	344	352	221	221	262	262	278	280	282	292	376	386	364	364	272	282
RC	RC233	264	268	352	352	221	221	254	266	280	280	298	298	376	376	370	370	272	272
RC	RC235	268	268	352	356	221	221	240	262	280	280	290	298	386	386	366	370	282	282

RC	RC252	264	268	352	354	221	221	262	262	278	280	282	302	376	386	364	364	272	282
RC	RC255	264	264	344	352	221	221	254	256	278	278	292	292	376	386	364	364	272	282
RC	RC264	264	264	352	354	221	221	266	266	278	278	292	298	376	376	366	366	272	272
RC	RC270	268	268	354	354	221	221	254	262	280	280	298	298	386	386	366	370	282	282
RC	RC280	268	268	340	352	221	221	264	264	280	280	280	298	376	386	366	370	272	282
RC	RC295	264	268	352	352	221	221	264	264	278	280	292	298	376	376	364	364	272	272
RC	RC301	270	270	338	352	221	221	262	264	278	278	292	298	376	386	370	380	272	282
RC	RC317	264	268	338	342	221	221	256	266	280	282	270	292	376	386	366	366	272	282
RC	RC365	264	264	340	352	221	221	262	262	278	280	270	298	376	386	370	370	272	282
SH	SH001	264	268	350	350	221	221	262	262	278	280	288	300	386	386	364	364	272	272
SH	SH002	268	268	344	350	221	221	240	240	278	280	296	296	386	386	364	364	272	272
SH	SH003	264	264	350	350	221	221	240	262	278	280	288	300	386	386	366	366	274	282
SH	SH004	264	268	352	354	221	221	248	262	280	280	270	282	0	0	370	370	270	270
SH	SH005	264	268	344	354	221	221	254	262	280	280	270	300	386	386	366	366	270	270
SH	SH006	264	270	354	354	221	221	254	264	280	280	282	298	374	374	370	370	270	272
SH	SH007	264	268	344	346	221	221	258	262	278	282	290	290	386	386	366	366	270	270
SH	SH008	268	268	352	354	221	223	254	254	280	280	288	290	386	386	364	364	0	0
SH	SH009	264	264	354	354	221	221	258	264	282	282	296	298	0	0	364	364	270	272
SH	SH010	268	270	352	352	221	221	258	262	280	280	290	296	386	386	364	364	270	282
SH	SH011	264	274	354	354	221	225	262	264	280	280	290	298	386	386	364	364	272	282
SH	SH012	264	268	344	352	221	223	252	262	280	280	288	296	386	386	364	364	272	272
SH	SH013	268	268	346	354	221	221	260	264	280	280	0	0	374	376	366	366	270	272
SH	SH014	264	270	346	356	221	223	258	262	280	280	282	290	386	386	364	364	0	0
SH	SH015	264	264	362	362	221	223	262	262	280	282	288	290	386	386	364	364	270	270
SH	SH016	264	268	352	354	221	221	262	262	280	280	290	300	374	374	366	370	270	270
SH	SH017	264	268	344	352	221	221	252	262	280	280	278	288	386	386	366	366	282	282
SH	SH018	264	264	354	354	221	221	264	270	280	282	270	300	386	386	366	366	282	282
SH	SH019	268	274	352	354	221	221	248	264	280	280	276	296	386	386	364	364	280	280
SH	SH020	268	268	344	350	221	221	240	240	278	280	296	296	386	386	364	364	272	280
SH	SH021	264	268	352	352	221	223	260	270	280	280	290	290	386	386	364	364	272	272
SH	SH022	264	264	354	354	221	221	252	260	280	280	288	298	386	386	364	364	272	280
SH	SH023	268	274	338	352	221	223	260	262	282	282	298	298	386	386	364	364	270	270
SH	SH024	268	274	338	354	221	221	262	262	280	280	288	290	386	386	370	370	0	0
SH	SH025	268	268	344	344	221	221	260	270	278	280	296	296	386	386	370	370	270	280
SH	SH026	264	272	352	362	221	221	264	270	280	280	296	298	386	386	366	368	282	282
SH	SH027	264	264	338	354	221	221	262	262	280	280	300	300	386	386	364	364	270	270
SH	SH028	264	264	344	354	221	221	262	264	280	280	270	296	374	374	364	364	270	282
SH	SH029	264	268	352	356	221	221	262	262	282	282	270	300	0	0	366	370	270	272
SH	SH030	264	264	352	354	221	221	240	262	278	280	296	300	386	386	364	364	272	282
SJ	SJ001	264	274	354	354	221	221	264	264	252	280	288	290	386	386	370	370	282	282
SJ	SJ002	262	264	346	364	221	221	264	264	278	280	282	290	386	386	366	370	272	282
SJ	SJ003	262	264	346	364	221	221	264	264	278	280	282	290	386	386	366	366	272	282
SJ	SJ004	264	264	352	352	221	221	264	264	276	282	270	292	386	386	366	366	272	282
SJ	SJ005	264	270	342	342	221	221	248	262	278	280	290	298	374	376	364	364	270	272
SJ	SJ006	0	0	0	0	221	221	262	262	280	280	290	290	376	386	366	366	272	282
SJ	SJ007	264	264	348	352	221	221	250	250	276	282	290	294	376	376	366	370	272	272
SJ	SJ008	264	264	342	346	221	221	254	254	280	280	290	290	374	376	364	364	270	272
SJ	SJ009	262	264	364	364	221	221	262	266	280	280	290	292	386	386	366	366	272	282
SJ	SJ010	264	274	342	364	221	221	266	266	276	280	288	292	386	386	366	366	272	282
SJ	SJ011	264	266	346	364	221	221	244	262	280	280	282	292	376	386	364	364	272	282
SJ	SJ012	270	270	352	354	221	221	244	262	278	282	282	292	376	376	364	364	272	272
SJ	SJ013	262	270	342	364	221	221	266	266	278	280	290	292	376	386	370	370	272	282

SJ	SJ014	264	264	346	364	221	221	250	250	276	282	288	290	386	386	364	364	282	282
SJ	SJ015	268	268	350	364	221	221	266	266	278	280	282	288	386	386	370	370	272	282
SJ	SJ016	264	268	342	346	221	221	244	244	276	282	290	292	376	386	364	364	272	282
SJ	SJ017	264	270	352	352	221	221	244	262	276	280	290	292	376	376	366	370	272	272
SJ	SJ018	268	268	342	364	221	221	262	262	278	280	284	292	376	386	366	366	272	282
SJ	SJ019	262	264	354	356	221	221	262	266	280	280	288	298	376	386	364	364	272	282
SJ	SJ020	264	270	348	364	221	221	262	266	252	278	282	292	386	386	364	364	270	282
SJ	SJ021	264	274	348	354	221	221	264	264	280	282	290	292	386	386	370	370	272	282
SJ	SJ022	264	264	342	356	221	223	252	262	280	280	282	290	374	376	366	370	270	272
SJ	SJ023	264	268	342	352	221	221	262	264	276	280	290	292	376	386	364	364	272	282
SJ	SJ024	264	264	354	364	221	221	266	266	276	280	288	290	376	386	370	370	272	282
SJ	SJ025	264	264	344	362	221	221	266	266	278	280	282	282	374	376	370	370	270	272
SJ	SJ026	264	264	348	364	221	221	262	264	280	280	282	292	386	386	366	366	282	282
SJ	SJ027	264	274	348	354	221	221	266	266	276	280	290	290	376	376	370	370	272	272
SJ	SJ028	262	264	354	354	221	221	266	266	278	280	284	290	376	386	366	370	272	282
SJ	SJ029	264	264	342	362	221	221	248	262	276	280	288	294	376	386	364	364	272	282
SJ	SJ030	264	264	344	362	221	221	266	266	278	280	282	282	374	376	370	370	270	272
ST	ST006	264	268	360	362	221	221	248	262	280	280	0	0	386	386	364	364	282	282
ST	ST007	264	264	348	348	221	221	262	264	252	278	296	298	386	386	366	366	282	282
ST	ST008	262	264	344	346	221	221	248	248	252	278	296	298	386	386	366	370	272	282
ST	ST009	262	268	346	352	221	221	256	264	280	282	290	296	386	386	364	364	272	282
ST	ST010	262	262	338	354	221	221	240	240	280	280	294	296	386	386	366	366	272	282
ST	ST011	264	268	346	360	221	221	248	264	278	278	298	302	386	386	364	364	272	282
ST	ST012	264	264	346	348	221	221	248	264	280	280	292	296	386	386	364	364	282	282
ST	ST013	268	268	348	352	221	221	262	264	278	280	296	296	386	386	366	370	270	282
ST	ST014	264	264	344	348	221	221	256	256	278	282	290	302	386	386	366	366	270	282
ST	ST015	262	262	348	356	221	221	252	260	278	280	290	298	374	376	366	370	270	272
ST	ST016	264	264	348	360	221	221	264	264	280	280	292	296	374	374	366	370	270	270
ST	ST017	264	268	348	352	221	221	262	262	278	282	290	296	386	386	366	370	282	282
ST	ST018	264	264	348	348	221	221	260	260	278	280	290	316	386	386	366	370	282	282
ST	ST019	264	268	346	356	221	221	260	262	278	278	296	298	386	386	366	366	282	282
ST	ST020	264	268	352	356	221	221	264	264	280	280	292	292	376	376	366	366	272	272
ST	ST021	262	264	354	362	221	221	262	262	278	278	0	0	0	0	366	366	0	0
ST	ST022	264	264	346	346	221	221	248	248	282	282	290	296	386	386	366	366	282	282
ST	ST023	264	264	346	352	221	221	262	262	280	280	296	298	386	386	366	370	282	282
ST	ST024	264	264	346	352	221	221	256	264	280	280	290	292	386	386	366	366	282	282
ST	ST025	264	264	346	348	221	221	262	264	280	282	290	296	386	386	366	370	272	282
ST	ST026	262	264	344	346	221	221	248	264	282	282	302	302	386	386	366	370	282	282
ST	ST027	262	264	344	348	221	221	256	262	278	278	290	296	386	386	366	370	272	272
ST	ST028	264	268	348	352	221	221	248	264	280	280	298	298	386	386	364	364	282	282
ST	ST029	264	264	344	354	221	221	248	262	280	280	282	296	386	386	366	366	282	282
ST	ST030	268	272	354	354	221	221	248	260	278	278	296	316	386	386	366	366	272	282
BC	BC001	264	270	352	352	221	221	262	266	282	282	270	290	386	386	366	366	282	282
BC	BC002	264	268	344	348	221	221	262	262	280	280	288	296	386	386	366	366	282	282
BC	BC003	264	268	344	344	221	221	264	264	278	282	288	296	378	386	366	366	274	282
BC	BC004	264	272	348	350	221	221	262	262	278	282	282	296	374	374	364	366	270	270
BC	BC005	264	264	344	350	221	225	256	262	278	280	282	296	374	386	366	366	270	282
BC	BC006	264	264	354	360	221	223	248	262	278	280	282	282	374	378	366	366	272	272
BC	BC007	264	264	352	354	221	221	258	262	278	282	292	292	374	386	366	366	272	282
BC	BC008	264	270	342	344	221	221	262	262	252	280	270	282	376	386	366	366	272	272
BC	BC009	264	264	350	360	221	221	262	262	278	280	282	292	378	386	366	366	270	270
BC	BC010	264	264	348	348	221	225	262	262	278	280	290	296	374	374	366	366	270	270

BC	BC011	270	270	344	352	221	221	262	262	278	282	270	296	386	386	364	366	282	282
BC	BC012	264	264	344	350	221	221	244	262	276	282	290	290	376	386	366	366	272	282
BC	BC013	270	270	344	350	221	221	244	262	280	282	282	300	374	374	366	366	270	270
BC	BC014	264	268	344	348	221	223	260	262	278	282	288	296	376	376	366	366	272	272
BC	BC015	264	264	346	352	221	221	260	262	280	282	270	292	374	386	366	366	270	282
BC	BC016	264	264	352	352	221	221	262	264	278	282	282	296	374	386	366	366	270	282
BC	BC017	264	264	344	348	221	221	262	262	278	280	282	290	386	386	366	366	282	282
BC	BC018	264	268	348	352	221	223	258	258	282	282	282	292	386	386	366	366	282	282
BC	BC019	264	264	348	350	221	221	256	262	278	278	290	296	374	386	366	366	270	282
BC	BC020	264	264	344	348	221	221	256	266	278	280	288	296	374	374	366	366	272	272
BC	BC021	264	264	342	354	221	221	258	266	280	282	290	290	376	386	366	366	272	282
BC	BC022	264	268	342	344	221	221	240	266	278	280	282	290	376	386	366	366	282	282
BC	BC023	264	268	350	352	221	221	262	262	278	278	296	296	376	386	366	366	272	282
BC	BC024	268	268	350	360	221	223	262	262	278	282	296	296	376	386	366	366	272	282
BC	BC025	264	268	344	354	221	223	248	262	278	282	270	290	386	386	366	366	270	282
BC	BC026	264	270	352	352	221	223	262	262	278	282	270	282	374	386	366	366	270	282
BC	BC027	264	270	342	348	221	221	262	262	280	282	282	290	374	374	366	366	270	270
BC	BC028	264	264	342	344	221	221	262	262	280	282	290	296	386	386	366	366	270	282
BC	BC029	264	268	348	354	221	223	248	260	252	282	282	296	374	386	366	366	282	282
BC	BC030	264	264	342	354	221	221	258	266	280	282	288	292	386	386	366	366	282	282

doi:10.1371/journal.pone.0156694.s004

S1 Table. Predicted lifespan for *M. affinis*. DBH_{max} refers the largest DBH found for a tree in each of the five intensively surveyed populations. Predicted lifespan was calculated dividing DBH_{max} by the mean annual growth of *M. affinis* in the 50-ha BCI plot (i.e., 1.75 mm per year).

Population	DBH_{max} (mm)	Predicted lifespan (years)
CP	117.7	67.3
GB	125.5	71.7
PL	109.3	62.3
RC	95.3	54.5
AG	115.3	65.9
mean	*112.6*	*64.3*

S2 Table. Model averaged coefficients (β) and their standard errors (SE) calculated from the candidate model set using allelic richness as the response variable (i.e. models with ΔAIC < 5).

Spatial isolation refers the mean geographic distance of each population to the other populations. Forest cover represents the percentage of forest cover within a 1, 2 and 3 km radius. Forest cover considers all vegetation taller than 5 meters in height. Model averaged coefficients not overlapping with zero are shown in bold.

	Allelic richness											
	Forest cover (1 km)				Forest cover (2 km)				Forest cover (3 km)			
	β	SE	Z value	P value	β	SE	Z value	P value	β	SE	Z value	P value
Spatial isolation	-0.00296	0.01392	0.186	0.85	-0.00295	0.01393	0.186	0.85	-0.00294	0.01391	0.185	0.85
Elevation	-0.00044	0.00035	1.106	0.27	-0.00044	0.00035	1.105	0.27	-0.00044	0.00035	1.104	0.27
Forest cover	0.00004	0.00546	0.006	0.99	-0.00068	0.00931	0.063	0.95	-0.00110	0.00803	0.119	0.91

S3 Table. Model averaged coefficients (β) and their standard errors (SE) calculated from the candidate model set using allelic private richness as the response variable (i.e. models with ΔAIC < 5).

Spatial isolation refers the mean geographic distance of each population to the other populations. Forest cover represents the percentage of forest cover within a 1, 2, and 3 km radius. Forest cover considers all vegetation taller than 5 meters in height. Model averaged coefficients not overlapping with zero are indicated with asterisks.

	Allelic private richness											
	Forest cover (1 km)				Forest cover (2 km)				Forest cover (3 km)			
	β	SE	Z value	P value	β	SE	Z value	P value	β	SE	Z value	P value
Spatial isolation	0.00224	0.00232	0.833	0.40	0.00222	0.00232	0.826	0.41	0.00269	0.00125	1.863	0.06*
Elevation	0.00006	0.00007	0.702	0.48	0.00150	0.00171	0.757	0.45	0.00206	0.00218	0.814	0.42
Forest cover	0.00049	0.00103	0.414	0.68	0.00006	0.00007	0.697	0.49	0.00005	0.00006	0.683	0.49

S4 Table. Model averaged coefficients (β) and their standard errors (SE) calculated from the candidate model set using unbiased Nei's gene diversity as the response variable (i.e. models with ΔAIC < 5).

Spatial isolation refers the mean geographic distance of each population to the other populations. Forest cover represents the percentage of forest cover within a 1, 2, and 3 km radius. Forest cover considers all vegetation taller than 5 meters in height.

	Nei's gene diversity											
	Forest cover (1 km)				Forest cover (2 km)				Forest cover (3 km)			
	β	SE	Z value	P value	β	SE	Z value	P value	β	SE	Z value	P value
Spatial isolation	-0.00008	0.00004	1.513	0.13	-0.00008	0.00004	1.510	0.13	-0.00008	0.00004	1.508	0.13
Elevation	-0.00212	0.00170	1.082	0.28	-0.00211	0.00170	1.081	0.28	-0.00211	0.00170	1.079	0.28
Forest cover	0.00006	0.00075	0.067	0.95	-0.00009	0.00128	0.064	0.95	-0.00051	0.00109	0.402	0.69

S5 Table. Ranked models explaining landscape effects on genetic differentiation (i.e., G_{ST}) among *Miconia affinis*' populations in Panama.

Models are ranked based upon the difference between Akaike's Information Criterion (AIC) in each individual model and the lowest AIC model (ΔAIC). Akaike's weights (ω_i) for each model are listed. The genetic differentiation between populations was calculated as $[G_{ST} / (1-G_{ST})]$. Geographic refers to log-transformed null resistance distance.

Model statement	AIC	ΔAIC	ω_i
Geographic + Elevation	-315.21	0.00	0.584
Geographic + Elevation + Deforestation	-313.43	1.78	0.240
Geographic	-312.14	3.07	0.126
Geographic + Deforestation	-310.17	5.05	0.047
Elevation	-304.24	10.97	0.002
Deforestation + Elevation	-302.66	12.56	0.001
Deforestation	-294.97	20.25	0.000

S6 Table. Model averaged coefficients (β) and their standard errors (SE) calculated from the candidate model set (i.e. models with ΔAIC < 5).

Model averaged coefficients not overlapping with zero are indicated with asterisks. The genetic differentiation between populations was calculated as $[G_{ST} / (1-G_{ST})]$. Geographic refers to log-transformed null resistance distance.

	β	SE	Z value	P value
Geographic	0.17780	0.04660	3.732	0.001*
Elevation	0.00006	0.00003	2.369	0.018*
Deforestation	-0.00059	0.00130	0.446	0.656

S7 Table. Ranked models explaining landscape effects on genetic differentiation (i.e., D_{ST}) among *Miconia affinis*' populations in Panama.

Models are ranked based upon the difference between Akaike's Information Criterion (AIC) in each individual model and the lowest AIC model (ΔAIC). Akaike's weights (ω_i) for each model are listed. The genetic differentiation between populations was calculated as $[D_{ST} / (1-D_{ST})]$. Geographic refers to log-transformed null resistance distance.

Model statement	AIC	ΔAIC	ω_i
Geographic + Elevation	-169.51	0.00	0.432
Geographic + Elevation + Deforestation	-168.40	1.10	0.249
Geographic	-168.13	1.37	0.217
Geographic + Deforestation	-166.54	2.97	0.098
Elevation	-159.52	9.98	0.003
Elevation + Deforestation	-157.52	11.98	0.001
Deforestation	-150.46	19.04	0.000

S8 Table. Model averaged coefficients (β) and their standard errors (SE) calculated from the candidate model set (i.e. models with ΔAIC < 5).

Model averaged coefficients not overlapping with zero are indicated with asterisks. The genetic differentiation between populations was calculated as $[D_{ST} / (1-D_{ST})]$. Geographic refers to log-transformed null resistance distance.

	β	SE	Z value	P value
Geographic	0.6890	0.1800	3.748	0.001*
Elevation	0.0002	0.0001	1.927	0.053*
Deforestation	0.0038	0.0045	0.818	0.414

S9 Table. Ranked models explaining landscape effects on F_{ST} **between** *Miconia affinis*' among populations in Panama.

Models are ranked based upon the difference between Akaike's Information Criterion (AIC) in each individual model and the lowest AIC model (ΔAIC). Akaike's weights (ω_i) provide the weight for each model. Elevation was computed as the Euclidean elevation distance matrix. Deforestation represents the percentage of deforested cover and was coded as a resistance variable. Geographic refers to null resistance distance.

Model statement	AIC	ΔAIC	ω_i
Geographic + Elevation	-241.03	0.00	0.622
Geographic + Elevation + Deforestation	-239.05	1.98	0.231
Geographic	-237.49	3.55	0.106
Geographic + Deforestation	-235.51	5.52	0.039
Elevation	-228.13	12.91	0.001
Elevation + Deforestation	-227.93	13.10	0.001
Deforestation	-216.16	24.87	0.000

S10 Table. Model averaged coefficients (β) and their standard errors (SE) calculated from the candidate model set (i.e. models with ΔAIC < 5).

Model averaged coefficients not overlapping with zero are indicated with asterisks. Population pairwise F_{ST} were used as estimates of genetic differentiation between *M. affinis*' populations. Elevation was computed as the Euclidean elevation distance matrix. Deforestation refers to the percentage of deforested cover and was coded as a resistance variable. Geographic refers to null resistance distance. All explanatory variables are centered.

	β	SE	Z value	P value
Geographic	0.37960	0.09664	3.845	<0.001*
Elevation	0.00005	0.00002	2.29	0.022*
Forest cover	-0.00044	0.00322	0.135	0.893

ACKNOWLEDGMENTS

We thank Herbario Universidad de Panama, J. Wright, R. Condit, J. Hall, A. Espinosa, N. Gottdenker, and R. A. Perez for their invaluable help in locating the *M. affinis* populations. We thank J. A. Ramirez, E. Sanchez, M. Ureña, A. G. Castillo, and numerous volunteers for their help in the collection of leaf material. The BCI forest dynamics research project was founded by S.P. Hubbell and R.B. Foster and is now managed by R. Condit, S. Lao and R. Perez under the Center for Tropical Forest Science and the Smithsonian Tropical Research Institute in Panama. Numerous organizations have provided funding, principally the U.S. National Science Foundation, and hundreds of field workers have contributed to the project. Finally, we would also like to thank the Jha lab, Edgardo Ortiz, and three anonymous reviewers for helpful feedback on the analyses and manuscript.

AUTHOR CONTRIBUTIONS

Conceived and designed the experiments: ARC SJ. Performed the experiments: ARC. Analyzed the data: ARC NP. Contributed reagents/materials/analysis tools: NP SJ. Wrote the paper: ARC NP RJ SJ.

REFERENCES

1. Ricketts TH, Daily GC, Ehrlich PR, Michener CD. Economic value of tropical forest to coffee production. Proc Natl Acad Sci U S A. 2004;101: 12579–12582. doi: 10. 1073/pnas. 0405147101. pmid:15306689
2. Laurance WF, Sayer J, Cassman KG. Agricultural expansion and its impacts on tropical nature. Trends in Ecology and Evolution. 2014. pp. 107–116. doi: 10. 1016/j. tree. 2013. 12. 001. pmid:24388286
3. Blignaut J, Aronson J, de Groot R. Restoration of natural capital: A key strategy on the path to sustainability. Ecol Eng. 2014;65: 54–61. doi: 10. 1016/j. ecoleng. 2013. 09. 003.
4. Chazdon RL. Beyond Deforestation: Restoring Degraded Lands. Communities. 2008;1458: 1458–1460. doi: 10. 1126/science. 1155365
5. Wishnie MH, Dent DH, Mariscal E, Deago J, Cedeño N, Ibarra D, et al. Initial performance and reforestation potential of 24 tropical tree species planted across a precipitation gradient in the Republic of Panama. For Ecol Manage. 2007;243: 39–49. doi: 10. 1016/j. foreco. 2007. 02. 001.
6. Garen EJ, Saltonstall K, Ashton MS, Slusser JL, Mathias S, Hall JS. The tree planting and protecting culture of cattle ranchers and small-scale agriculturalists in rural Panama: Opportunities for reforestation and land restoration. For Ecol Manage. 2011;261: 1684–1695. doi: 10. 1016/j. foreco. 2010. 10. 011.
7. Schröder R, Prasse R. Cultivation and hybridization alter the germination behavior of native plants used in revegetation and restoration. Restor Ecol. 2013;21: 793–800. doi: 10. 1111/rec. 12018.
8. Schröder R, Prasse R. Do cultivated varieties of native plants have the ability to outperform their wild relatives? PLoS One. Public Library of Science; 2013;8: e71066. doi: 10. 1371/journal. pone. 0071066. pmid:23951081
9. Gallagher MK, Wagenius S. Seed source impacts germination and early establishment of dominant grasses in prairie restorations. Wilsey B, editor. J Appl Ecol. 2016;53: 251–263. doi: 10. 1111/1365-2664. 12564.
10. Laikre L, Schwartz MK, Waples RS, Ryman N. Compromising genetic diversity in the wild: Unmonitored large-scale release of plants and animals. Trends in Ecology and Evolution. 2010. pp. 520–529. doi: 10. 1016/j. tree. 2010. 06. 013. pmid:20688414
11. Byrne M, Stone L, Millar MA. Assessing genetic risk in revegetation. J Appl Ecol. Blackwell Publishing Ltd; 2011;48: 1365–1373. doi: 10. 1111/j. 1365-2664. 2011. 02045. x.
12. Banks SC, Cary GJ, Smith AL, Davies ID, Driscoll DA, Gill AM, et al. How does ecological disturbance influence genetic diversity? Trends Ecol Evol. 2013;28: 670–9. doi: 10. 1016/j. tree. 2013. 08. 005. pmid:24054910
13. Wright S. Isolation by Distance. Genetics. 1943;28: 114–138. Article pmid:17247074
14. Jenkins DG, Carey M, Czerniewska J, Fletcher J, Hether T, Jones A, et al. A meta-analysis of isolation by distance: Relic or reference standard for landscape genetics? Ecography (Cop). 2010;33: 315–320. doi: 10. 1111/j. 1600-0587. 2010. 06285. x.
15. Primmer CR, Veselov AJ, Zubchenko A, Poututkin A, Bakhmet I, Koskinen MT. Isolation by distance within a river system: Genetic population structuring of Atlantic salmon, Salmo salar, in tributaries of the Varzuga River in northwest Russia. Mol Ecol. 2006;15: 653–666. doi: 10. 1111/j. 1365-294X. 2005. 02844. x. pmid:16499692
16. Petit RJ, Mousadik AEL, Pons O. Identifying populations for conservation on the basis of genetic markers. Conserv Biol. 1998;12: 844–855. doi: 10. 1046/j. 1523-1739. 1998. 96489. x.
17. Lemes MR, Gribel R, Proctor J, Grattapaglia D. Population genetic structure of mahogany (Swietenia macrophylla King, Meliaceae) across the Brazilian Amazon, based on variation at microsatellite loci: Implications for conservation. Mol Ecol. 2003;12: 2875–2883. doi: 10. 1046/j. 1365-294X. 2003. 01950. x. pmid:14629369
18. Collevatti RG, Grattapaglia D, Hay JD. Population genetic structure of the endangered tropical tree species Caryocar brasiliense, based on variability at microsatellite loci. Mol Ecol. 2001;10: 349–356. doi: 10. 1046/j. 1365-294X. 2001. 01226. x. pmid:11298950
19. Hardesty BD, Dick CW, Hamrick JL, Degen B, Hubbell SP, Bermingham E. Geographic influence on genetic structure in the widespread Neotropical tree Simarouba amara (Simaroubaceae). Trop Plant Biol. 2010;3: 28–39. doi: 10. 1007/s12042-010-9044-3.
20. Beals EW. Vegetational Change Along Altitudinal Gradients. Science (80-). American Association for the Advancement of Science; 1969;165: 981–985. Available: http://www. jstor. org/ stable/1727058 pmid:17791022
21. Herrera CM, Bazaga P. Adding a third dimension to the

edge of a species' range: altitude and genetic structuring in mountainous landscapes. Heredity (Edinb). 2008;100: 275–285. doi: 10. 1038/sj. hdy. 6801072.

22. Pellmyr O. Three pollination morphs in Cimicifuga simplex; incipient speciation due to inferiority in competition. Oecologia. 1986;68: 304–307. doi: 10. 1007/BF00384804.

23. Castilla AR, Alonso C, Herrera CM. Exploring local borders of distribution in the shrub Daphne laureola: Individual and populations traits. Acta Oecologica. 2011;37: 269–276. doi: 10. 1016/j. actao. 2011. 02. 011.

24. Lomolino M V. Elevation gradients of species-density: Historical and prospective views. Glob Ecol Biogeogr. 2001;10: 3–13. doi: 10. 1046/j. 1466-822x. 2001. 00229. x.

25. Dalsgaard B, Martín González AM, Olesen JM, Ollerton J, Timmermann A, Andersen LH, et al. Plant-hummingbird interactions in the West Indies: Floral specialisation gradients associated with environment and hummingbird size. Oecologia. 2009;159: 757–766. doi: 10. 1007/s00442-008-1255-z. pmid:19132403

26. Stiles FG. Ecological and evolutionary implications of bird pollination. Integr Comp Biol. 1978;18: 715–727. doi: 10. 1093/icb/18. 4. 715.

27. Cruden RW. Pollinators in high-elevation ecosystems: relative effectiveness of birds and bees. Science (80-). 1972;176: 1439–1440. doi: 10. 2307/1734592. pmid:17834653

28. Jordano P, Godoy JA. RAPD variation and population genetic structure in Prunus mahaleb (Rosaceae), an animal-dispersed tree. Mol Ecol. 2000;9: 1293–1305. doi: 10. 1046/j. 1365-294X. 2000. 01009. x. pmid:10972769

29. Schuster WS, Alles DL, Mitton JB. Gene Flow in Limber Pine: Evidence from Pollination Phenology and Genetic Differentiation Along an Elevational Transect. Am J Bot. 1989;76: 1395. doi: 10. 2307/2444563.

30. Gauzere J, Klein EK, Oddou-Muratorio S. Ecological determinants of mating system within and between three Fagus sylvatica populations along an elevational gradient. Mol Ecol. 2013;22: 5001–5015. doi: 10. 1111/ mec. 12435. pmid:23952125

31. Ohsawa T, Ide Y. Global patterns of genetic variation in plant species along vertical and horizontal gradients on mountains. Global Ecology and Biogeography. 2008. pp. 152–163. doi: 10. 1111/j. 1466-8238. 2007. 00357. x.

32. Young A, Boyle T, Brown T. The population genetic consequences of habitat fragmentation for plants. Trends Ecol Evol. 1996;11: 413–418. doi: 10. 1016/0169-5347(96)10045-8. pmid:21237900

33. Lowe A, Harris S, Ashton P. Ecological genetics: design, analysis and application. Anal Appl Oxford Blackwell Publ. 2004; doi: 10. 1017/s0030605304260860

34. Kramer AT, Ison JL, Ashley M V. , Howe HF. The paradox of forest fragmentation genetics. Conservation Biology. 2008. pp. 878–885. doi: 10. 1111/j. 1523-1739. 2008. 00944. x. pmid:18544089

35. Lowe a J, Cavers S, Boshier D, Breed MF, Hollingsworth PM. The resilience of forest fragmentation genetics—no longer a paradox—we were just looking in the wrong place. Heredity (Edinb). 2015;115: 97–99. doi: 10. 1038/ hdy. 2015. 40.

36. Montoya D, Zavala MA, Rodríguez MA, Purves DW. Animal versus wind dispersal and the robustness of tree species to deforestation. Science. American Association for the Advancement of Science; 2008;320: 1502–4. doi: 10. 1126/science. 1158404. pmid:18535208

37. Markl JS, Schleuning M, Forget PM, Jordano P, Lambert JE, Traveset A, et al. Meta-analysis of the effects of human disturbance on seed dispersal by animals. Conserv Biol. 2012;26: 1072–81. doi: 10. 1111/j. 1523-1739. 2012. 01927. x. pmid:22971077

38. Breed MF, Ottewell KM, Gardner MG, Marklund MHK, Dormontt EE, Lowe AJ. Mating patterns and pollinator mobility are critical traits in forest fragmentation genetics. Heredity (Edinb). The Genetics Society; 2015;115: 108–114. Available:http://dx. doi. org/10. 1038/hdy. 2013. 48

39. Davies SJ, Cavers S, Finegan B, White A, Breed MF, Lowe AJ. Pollen flow in fragmented landscapes maintains genetic diversity following stand-replacing disturbance in a neotropical pioneer tree, Vochysia ferruginea Mart. Heredity (Edinb). The Genetics Society; 2015;115: 125–129. Available:http://dx. doi. org/10. 1038/hdy. 2013. 95

40. Breed MF, Ottewell KM, Gardner MG, Marklund MHK, Stead MG, Harris JBC, et al. Mating system and early viability resistance to habitat fragmentation in a bird-pollinated eucalypt. Heredity (Edinb). The Genetics Society; 2015;115: 100–107. Available: http://dx. doi. org/10. 1038/hdy. 2012. 72

41. Malhi Y, Silman M, Salinas N, Bush M, Meir P, Saatchi S. Introduction: Elevation gradients in the tropics: Laboratories for ecosystem ecology and global change research. Global Change Biology. 201016: 3171–3175. doi: 10. 1111/j. 1365-2486. 2010. 02323. x.

42. Ollerton J, Winfree R, Tarrant S. How many flowering plants are pollinated by animals? Oikos. 2011;120: 321–326. doi: 10. 1111/j. 1600-0706. 2010. 18644. x.

43. Hansen MC, Potapov P V, Moore R, Hancher M, Turubanova S a, Tyukavina a, et al. High-resolution global maps of 21st-century forest cover change. Science. 2013;342: 850–3. doi: 10. 1126/science. 1244693. pmid:24233722

44. Montagnini F, Sancho F. Impacts of Native Trees on Tropical Soils: A Study in the Atlantic Lowlands of Costa Rica. Ambio. 1990;19: 386–390. doi: 10. 2307/4313746.

45. Condit R, Robinson WD, Ibáñez R, Aguilar S, Sanjur A, Martínez R, et al. The Status of the Panama Canal Watershed and Its Biodiversity at the Beginning of the 21st Century: Long-term ecological studies reveal a diverse flora and fauna near the Panama Canal, harbored within a corridor of forest stretching from the Caribbean to. Biosci. 2001;51: 389–398. doi: 10. 1641/0006-3568(2001)051[0389:TSOTPC]2. 0. CO;2.

46. Jha S, Dick CW. Shade coffee farms promote genetic diversity of native trees. Current Biology. 2008. doi: 10. 1016/j. cub. 2008. 11. 017.

47. Renner SS. Phylogeny and classification of the

Melastomataceae and Memecylaceae. Nord J Bot. 1993;13: 519–540. doi: 10. 1111/j. 1756-1051. 1993. tb00096. x.

48. Goldenberg R, Penneys D, Almeda F, Judd W, Michelangeli F. Phylogeny of Miconia (Melastomataceae): Patterns of Stamen Diversification in a Megadiverse Neotropical Genus. Int J Plant Sci. University of Chicago Press; 2008;169: 963–979. doi: 10. 1086/589697.

49. Jha S, Dick CW. Native bees mediate long-distance pollen dispersal in a shade coffee landscape mosaic. Proc Natl Acad Sci U S A. 2010;107: 13760–13764. doi: 10. 1073/pnas. 1002490107. pmid:20660738

50. Croat T. Flora of Barro Colorado Island. Stanford Calif. : Stanford Univ. Pr. ; 1978.

51. Meyer J-Y. Observations on the Reproductive Biology of Miconia calvescens DC (Melastomataceae), an Alien lnvasive Tree on the Island of Tahiti (South Pacific Ocean)'. Biotropica. 1998;30: 609–624. doi: 10. 1111/j. 1744-7429. 1998. tb00101. x.

52. Le Roux JJ, Wieczorek AM, Meyer JY. Genetic diversity and structure of the invasive tree Miconia calvescens in Pacific islands. Divers Distrib. 2008;14: 935–948. doi: 10. 1111/j. 1472-4642. 2008. 00504. x.

53. Parrotta JA, Knowles OH. Restoring tropical forests on lands mined for bauxite: Examples from the Brazilian Amazon. Ecol Eng. 2001;17: 219–239. doi: 10. 1016/ S0925-8574(00)00141-5.

54. Condit R. Tropical Forest Census Plots: Methods and Results from Barro Colorado Island, Panama and a Comparison with Other Plots [Internet]. 1998. Available:https://books. google. com/books?hl=es&lr= &id=i59sdrERVwIC&pgis=1

55. Hubbell SP, Foster RB, O'Brien ST, Harms KE, R. C, Wechsler B, et al. Light gap disturbances, recruitment limitation, and tree diversity in a neotropical forest. Science (80-). 1999;283: 554–557. pmid:9915706 doi: 10. 1126/science. 283. 5401. 554

56. Hubbell S, Condit R, Foster R. Barro Colorado forest census plot data. URL http://ctfs si/edu/datasets/bci. 2005; Available: https://scholar. google. com/scholar?q =Barro+Colorado+Forest+Census+Plot+Data&btnG=& hl=es&as_sdt=0%2C44#0

57. Condit R, Hubbell SP, Foster RB. Mortality rates of 205 neotropical tree and shrub species and the impact of a severe drought. Ecol Monogr. 1995;65: 419–439. doi: 10. 2307/2963497.

58. Condit R, Hubbell SP, Foster RB. Assessing the response of plant functional types to climatic change in tropical forests. J Veg Sci. 1996;7: 405–416. doi: 10. 2307/3236284.

59. Condit R, Ashton PS, Manokaran N, LaFrankie J V, Hubbell SP, Foster RB. Dynamics of the forest communities at Pasoh and Barro Colorado: comparing two 50–ha plots. Philos Trans R Soc London B Biol Sci. 1999;354: 1739–1748. Available:http://rstb. royalsocietypublishing. org/content/354/1391/1739. abstractpmid:11605618

60. Brokaw NVL. Gap-phase regeneration of three pioneer tree species in a tropical forest. J Ecol. 1987;75: 9–19. doi: 10. 2307/2260533.

61. Castilla AR, Pope N, Jha S. Positive density-dependent reproduction regulated by local kinship and size in an understorey tropical tree. Ann Bot. 2016;117: 319–329. doi: 10. 1093/aob/mcv170. pmid:26602288

62. Doyle JJ, Doyle JL. A rapid DNA isolation procedure for small quantities of fresh leaf tissue. Phytochem Bull. 1987;19: 11–15. doi: 10. 2307/4119796.

63. Jha S, Dick CW. Isolation and characterization of nine microsatellite loci for the tropical understory tree Miconia affinis Wurdack (Melastomataceae). Mol Ecol Resour. 2009;9: 344–345. doi: 10. 1111/j. 1755-0998. 2008. 02428. x. pmid:21564645

64. Le Roux JJ, Wieczorek AM. Isolation and characterization of polymorphic microsatellite markers from the velvet tree, Miconia calvescens DC. (Melastomataceae). Mol Ecol Resour. 2008;8: 961–964. doi: 10. 1111/j. 1755-0998. 2008. 02102. x. pmid:21585941

65. Van Oosterhout C, Hutchinson WF, Wills DPM, Shipley P. MICRO-CHECKER: Software for identifying and correcting genotyping errors in microsatellite data. Mol Ecol Notes. 2004;4: 535–538. doi: 10. 1111/j. 1471-8286. 2004. 00684. x.

66. Raymond M, Rousset F. GENEPOP (version 1. 2): population genetics software for exact tests and ecumenicism. J Hered. 1995;86: 248–249. doi: 10. 1111/j. 0021-8790. 2004. 00839. x.

67. Kalinowski ST. HP-RARE 1. 0: A computer program for performing rarefaction on measures of allelic richness. Mol Ecol Notes. 2005;5: 187–189. doi: 10. 1111/j. 1471-8286. 2004. 00845. x.

68. Peakall R, Smouse PE. GenALEx 6. 5: Genetic analysis in Excel. Population genetic software for teaching and research-an update. Bioinformatics. 2012;28: 2537–2539. doi: 10. 1093/bioinformatics/bts460. pmid:22820204

69. Barton K. MuMIn: Multi-model inference. R package version 1. 9. 13. Version. 2015;1: 18. citeulike:11961261

70. Excoffier L, Smouse PE, Quattro JM. Analysis of molecular variance inferred from metric distances among DNA haplotypes: Application to human mitochondrial DNA restriction data. Genetics. 1992;131: 479–491. doi: 10. 1007/s00424-009-0730-7 pmid:1644282

71. Pritchard JK, Stephens M, Donnelly P. Inference of population structure using multilocus genotype data. Genetics. 2000;155: 945–59. doi: 10. 1111/j. 1471-8286. 2007. 01758. x

72. Earl DA, vonHoldt BM. STRUCTURE HARVESTER: A website and program for visualizing STRUCTURE output and implementing the Evanno method. Conserv Genet Resour. 2012;4: 359–361. doi: 10. 1007/s12686-011-9548-7.

73. Evanno G, Regnaut S, Goudet J. Detecting the number of clusters of individuals using the software STRUCTURE: A simulation study. Mol Ecol. 2005;14: 2611–2620. doi: 10. 1111/j. 1365-294X. 2005. 02553. x. pmid:15969739

74. McRae BH. Isolation by resistance. Evolution. 2006;60:

1551–1561. doi: 10. 1554/05-321. 1. pmid:17017056

75. Clarke RT, Rothery P, Raybould AF. Confidence limits for regression relationships between distance matrices: estimating gene flow with distance. J Agric Biol Environ Stat. 2002;7: 361–372. doi: 10. 1198/108571102320.

76. Fox J, Weisberg S. An R Companion to Applied Regression: Appendices. In: Robust Regression in R [Internet]. 2014 pp. 1–17. doi: 10. 1177/0049124105277200.

77. Pinheiro J, Bates D, DebRoy S, Sarkar D, Team TRC. nlme: Linear and Nonlinear Mixed Effects Models. October. 2013. pp. 1–3.

78. Van Strien MJ, Keller D, Holderegger R. A new analytical approach to landscape genetic modelling: Least-cost transect analysis and linear mixed models. Mol Ecol. 2012;21: 4010–4023. doi: 10. 1111/j. 1365-294X. 2012. 05687. x. pmid:22738667

79. Jha S. Contemporary human-altered landscapes and oceanic barriers reduce bumble bee gene flow. Mol Ecol. 2015;24: 993–1006. doi: 10. 1111/mec. 13090. pmid:25626470

80. Jaffé R, Castilla A, Pope N, Imperatriz-Fonseca VL, Metzger JP, Arias MC, et al. Landscape genetics of a tropical rescue pollinator. Conservation Genetics. 2015. doi: 10. 1007/s10592-015-0779-0.

81. Laurance WF, Nascimento HEM, Laurance SG, Andrade AC, Fearnside PM, Ribeiro JEL, et al. Rain forest fragmentation and the proliferation of successional trees. Ecology. 2006;87: 469–482. doi: 10. 1890/05-0064. pmid:16637371

82. Wilson GA, Rannala B. Bayesian inference of recent migration rates using multilocus genotypes. Genetics. 2003;163: 1177–1191. Article pmid:12663554

83. Beerli P, Felsenstein J. Maximum likelihood estimation of a migration matrix and effective population sizes in n subpopulations by using a coalescent approach. Proc Natl Acad Sci. 2001;98: 4563–4568. doi: 10. 1073/pnas. 081068098. pmid:11287657

84. Garza JC, Williamson EG. Detection of reduction in population size using data from microsatellite loci. Mol Ecol. 2001;10: 305–318. doi: 10. 1046/j. 1365-294x. 2001. 01190. x. pmid:11298947

85. Goslee SC, Urban DL. The ecodist Package for Dissimilarity-based Analysis of Ecological Data. J Stat Softw. 2007;22: 1–19. citeulike-article-id:12008924 doi: 10. 18637/jss. v022. i07

86. Ciofi C, Bruford MW. Genetic structure and gene flow among Komodo dragon populations inferred by microsatellite loci analysis. Mol Ecol. 1999;8: S17–S30. doi: 10. 1046/j. 1365-294X. 1999. 00734. x. pmid:10703549

87. Nybom H. Comparison of different nuclear DNA markers for estimating intraspecific genetic diversity in plants. Mol Ecol. 2004;13: 1143–1155. doi: 10. 1111/j. 1365-294X. 2004. 02141. x. pmid:15078452

88. Andrew Jones F, Cerón-Souza I, Hardesty BD, Dick CW. Genetic evidence of Quaternary demographic changes in four rain forest tree species sampled across the Isthmus of Panama. J Biogeogr. 2013;40: 720–731. doi: 10. 1111/jbi. 12037.

89. Dick CW, Hardy OJ, Jones FA, Petit RJ. Spatial Scales of Pollen and Seed-Mediated Gene Flow in Tropical Rain Forest Trees. Trop Plant Biol. 2008;1: 20–33. doi: 10. 1007/s12042-007-9006-6.

90. Diamond JM, Gilpin ME, Mayr E. Species-distance relation for birds of the Solomon Archipelago, and the paradox of the great speciators. Proc Natl Acad Sci U S A. 1976;73: 2160–2164. doi: 10. 1073/pnas. 73. 6. 2160. pmid:16592328

91. Jordano P, García C, Godoy J a, García-Castaño JL. Differential contribution of frugivores to complex seed dispersal patterns. Proc Natl Acad Sci U S A. 2007;104: 3278–3282. doi: 10. 1073/pnas. 0606793104. pmid:17360638

92. Luck GW, Daily GC. Tropical countryside bird assemblages: Richness, composition, and foraging differ by landscape context. Ecol Appl. 2003;13: 235–247. doi: 10. 1890/1051-0761(2003)013[0235:TCBARC]2. 0. CO;2.

93. Shi MM, Michalski SG, Chen XY, Durka W. Isolation by elevation: Genetic structure at neutral and putatively non-neutral loci in a dominant tree of subtropical forests, castanopsis eyrei. PLoS One. 2011;6. doi: 10. 1371/journal. pone. 0021302.

94. Brito VLG, Sazima M. Tibouchina pulchra (Melastomataceae): Reproductive biology of a tree species at two sites of an elevational gradient in the Atlantic rainforest in Brazil. Plant Syst Evol. 2012;298: 1271–1279. doi: 10. 1007/s00606-012-0633-5.

95. Aragão D V, Fortini LB, Mulkey SS, Zarin DJ, Araujo MM, de Carvalho CJR. Correlation but no causation between leaf nitrogen and maximum assimilation: the role of drought and reproduction in gas exchange in an understory tropical plant Miconia ciliata (Melastomataceae). Am J Bot. 2005;92: 456–461. doi: 10. 3732/ajb. 92. 3. 456. pmid:21652422

96. Loiselle BA, Blake JG. Temporal variation in birds and fruits along an elevational gradient in Costa Rica. Ecology. 1991. pp. 180–193. doi: 10. 2307/1938913.

97. Tabarelli M, Lopes A V. , Peres CA. Edge-effects drive tropical forest fragments towards an early-successional system. Biotropica. 2008. pp. 657–661. doi: 10. 1111/j. 1744-7429. 2008. 00454. x.

98. Hall JS, Ashton MS, Garen EJ, Jose S. The ecology and ecosystem services of native trees: Implications for reforestation and land restoration in Mesoamerica. For Ecol Manage. 2011;261: 1553–1557. doi: 10. 1016/j. foreco. 2010. 12. 011.

99. Breugel M van, Hall JS, Craven DJ, Gregoire TG, Park A, Dent DH, et al. Early growth and survival of 49 tropical tree species across sites differing in soil fertility and rainfall in Panama. For Ecol Manage. 2011;261: 1580–1589. doi: 10. 1016/j. foreco. 2010. 08. 019.

SECTION III: GENETIC DIVERSITY IN MICROBES: IMPACTS ON DISEASES AND THERAPIES

CHAPTER 14

ECOLOGY AND EVOLUTION OF THE HUMAN MICROBIOTA: FIRE, FARMING AND ANTIBIOTICS

Michael R. Gillings, Ian T. Paulsen and Sasha G. Tetu

Keywords: Integrin, evolution, mercury; disinfectant, resistance, antibiotic, Anthropocene, dysbiosis

INTRODUCTION

Humans have become the most important evolutionary force on the planet [1]. Our activities have significant effects on the climate, atmosphere, hydrosphere, and on global biogeochemistry [2]. These changes are leading to intense rounds of natural selection, adding to the pressures already imposed by extinction, homogenization of species distributions, and human technologies that direct evolution [3,4]. The influence of human activities is pervasive, even extending to the microbial world, where we have impacts on dispersal, population structure, and the contributions of microorganisms to biogeochemical cycles [5,6].

We have significantly influenced the biosphere over the last 10,000 years. Across the same time period we have also influenced the composition, properties and diversity of our internal biosphere, the microorganisms that comprise the human microbiota [7]. These microbial cells outnumber our human cells by ten to one, exhibit diversity between body sites and between individuals, and contain two orders of magnitude more genes than the human genome [8,9].

Perturbations to the composition and functions of the microbiota may have significant effects on health [10]. Understanding and dealing with the negative consequences of such dysbiosis requires an understanding of what a "normal" microbiota should look like [11]. It also requires an understanding of the dynamics of our internal microbial communities. In this regard, our microbiota can be examined in the same way as more conventional ecosystems, by analyzing colonization, competition and selection [12].

There is no doubt that the human microbiota has been subjected to various selection pressures over the last 10,000 years. Selective sweeps probably accompanied changes in diet resulting from the spread of agriculture. Direct selection pressures were undoubtedly imposed by increasing exposure to environmental contaminants after the industrial revolution, and by heavy metals, disinfectants and antimicrobial agents in the modern era. Here we review the ancestral microbiota, and examine changes in diversity and composition wrought by various selective agents, including those most potent agents of selection, the antibiotics. To begin this analysis, we concentrate on the genes encoded by the human microbiota: the human microbiome.

THE HISTORY OF THE HUMAN MICROBIOME

The term "microbiome" has been used in a number of different ways over the past decade. Here we will refer to the human microbiome strictly as the sum total of the genes carried by micro-organisms commensal with humans. This is to distinguish this concept from the human microbiota, which are those species of micro-organisms commensal with humans [13]. It is also important to distinguish between an individual's microbiome/microbiota, and those of populations, or those of the whole human species.

Does the microbiome have a "steady state" and has the human microbiome changed over time? While the microbiota of modern, westernized populations differs from individual to individual, the functions encoded by their microbiome are much more constant. In particular, the presence and abundance of genes assigned to conserved orthologous groups (COGs) remain similar [10]. In other words, the core functional microbiome remains relatively constant, despite the species housing these functions varying considerably between individuals.

While the core functions of the microbiome might be well conserved, some elements of the microbiome can confer important advantages, while being a comparatively rare component of the total microbiome. These "accessory" genes are not part of the core genome of any particular species, and may only find temporary residence within the microbiome.

Citation: Michael R. Gillings, Ian T. Paulsen and Sasha G. Tetu, Ecology and Evolution of the Human Microbiota: Fire, Farming and Antibiotics, doi:10.3390/genes6030841.

Bacteria exchange such accessory genes via the process of lateral gene transfer (LGT), itself mediated by the processes of conjugation, transformation and transduction [14].

The human microbiome interacts with the global pangenome (the sum of all genes in the biosphere) through environmental exposure and LGT. Potentially any gene in the biosphere can find its way into the human microbiome by lateral transfer [15]. There are some routes of LGT that are well trodden. Gene exchange is easier to accomplish between organisms in the same Phylum, those with similar GC content, and those that inhabit similar niches [16,17]. In some cases, there are communities where LGT is common, creating networks of gene exchange [18,19]. High cell density and biofilms improve the efficiency of transfer [20]. This makes the high densities and biofilm communities that comprise the human microbiota a potential hotspot for gene exchange, and accumulation of accessory genes [21,22].

To persist in a population, including within the microbiota, a newly acquired accessory gene usually has to be: (i) stably inherited, by incorporation into a replicon; and (ii) able to confer some advantage to the host cell, such that it is not lost by drift or purifying selection [14]. Consequently, we might envisage the human microbiome as a clearing house, where diverse genes constantly arrive by LGT, and are then eliminated by failure to replicate, or are subjected to fixation or elimination by natural selection.

This must have been the state of the human microbiome during our early evolution. Because diet strongly affects the composition of the microbiota, it also affects the microbiome [13], and the likelihood of successful LGT. For instance, distantly related members of the phylum *Bacteroidetes* can readily share genes for degradation of refractory high molecular weight polysaccharides [23], providing an advantage to human populations that rely on fibrous plant foods [24]. Subsequent changes to diet, such as the increasing consumption of cereals after the adoption of agriculture, would then select for altered components of the microbiota [25,26], and simultaneously select for new accessory genes in the microbiome. Adoption of unique dietary components would also promote selection of appropriate accessory genes. For example, a gene for the enzyme porphyranase appears to have made a successful lateral transfer from a marine *Bacteroides* into *B. plebeius* in the microbiota of the Japanese population [27]. This enzyme helps digest seaweed, a common item in Japanese diet. Another example might be the significant change in oral microbiota following the increased availability of processed foods and refined sugar after the industrial revolution [28].

There are two main differences between the dynamics of ancestral human microbiomes and those of humans living in the developed world. The first of these is that ancient humans and early agrarian populations were exposed to the microbiota and microbiome from a comparatively small geographic area (Figure 1). The diversity of microbiota and microbiome components to which individuals could potentially be exposed increased during the age of exploration, as the movement of humans, plants and animals around the world had an increasing impact on microbial biogeography [6]. In the developed world, an individual has the potential to be exposed to microbiota and microbiome components from anywhere in the world (Figure 1), as demonstrated by the ability of antibiotic resistant enterobacteria to rapidly spread via international travel [29].

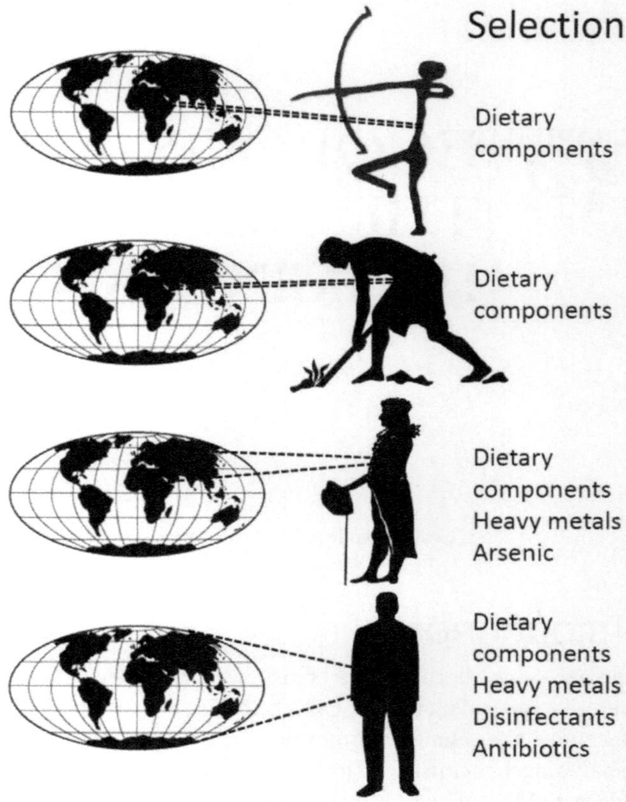

Figure 1. Schematic history of selective forces acting on the human microbiome and microbiota. As human populations moved from hunter-gatherer, to agrarian communities, to the industrial revolution, to modern civilization (vertical dimension), the selective forces on the microbiota and microbiome have changed. Early selection was mainly driven by ability to metabolize dietary components, while more recently, strong selective agents such as heavy metals and antimicrobial agents have become increasingly important. In addition, the potential sources of both microbiota and microbiome components (species and genes) have broadened as mass movement of materials and organisms became a feature of a globalized world.

The second difference in microbiome dynamics is driven by the mode and strength of selection on microbiome components. In hunter-gatherers and agrarian communities, the microbiota and its component microbiome are subject to selection mainly through diet. This selection is essentially "soft", in that it arises through competition between individual cells and species in their ability to metabolize food sources. Cells might gain an incremental metabolic advantage by acquiring a new microbiome component via LGT, and thus increase in relative abundance. In contrast, after the industrial revolution, humans have been increasingly exposed to many more agents of "hard" selection (Figure 1). These agents, such as heavy metals, disinfectants, biocides and antibiotics act directly, killing or severely debilitating many cells, while positively selecting those rare cells that carry resistance determinants (serendipitously acquired by LGT). The speed and intensity of this selection has a second consequence, the fixation of particular clonal lineages from within the microbiota. Repeated rounds of selection by different agents can also then result in the sequential assembly of multiple resistance determinants, each with different phylogenetic origins, into complex, mosaic DNA elements [30,31].

ASSEMBLY OF MULTI-RESISTANCE DNA ELEMENTS WITHIN THE MICROBIOME

Genes that can confer resistance phenotypes are collectively termed the resistome [32,33,34]. The resistome is a useful organizing concept from a human and medical perspective, but it is not a natural or functional grouping, as it comprises diverse genes whose functions may have simply been co-opted to confer resistance [15,35]. Nevertheless, the resistome concept helps to highlight the point that resistance genes in pathogens often originate from unrelated environmental organisms, and that the resistance genes of medical concern are just a small proportion of the resistome [36,37].

It is important to distinguish between sequences that are assigned to the resistome based on bioinformatic prediction of function, as opposed to genes that confer a demonstrable resistance phenotype. Many resistome elements do not confer resistance in their native context because they are not expressed, or require critical mutations to confer resistance [35,38]. The ancient human microbiota probably did contain diverse elements that could be assigned to the resistome, but these did not confer resistance in the clinical sense [11]. Culture independent sampling of human microbiota reveals a vast reservoir of potential resistance genes. However, the majority of these genes have not been identified in clinical contexts, and are evolutionarily distant from known resistance determinants [39]. In contrast, if aerobic proteobacteria are cultured from fecal samples, over half of the resistome elements are identical to known resistance genes. These resistance determinants are readily exchanged with pathogens, and spread rapidly between individuals [40]. This suggests that proteobacteria from environmental sources might act as a conduit for clinically relevant resistance genes to make their way into the human microbiome and thence into pathogens [14].

Before the Industrial Revolution, it is unlikely that clinically relevant resistance genes were at high frequencies in the human microbiome. Resistome elements would have occasionally found their way into the microbiome via LGT, but in the absence of any selective advantage would be lost by drift or purifying selection. This view is supported by characterization of plasmids collected before the antibiotic era. These plasmids have similar backbones to modern plasmids, but lack resistance genes [41].

The first rounds of hard selection to retain resistome elements in the microbiome may have occurred some 120–180 years ago, when there was a significant shift in abundance and diversity of mercury resistance genes in lake sediments. This shift was presumably driven by increased mercury pollution as a consequence of industrialization [42]. Certainly by the late 1800s, both mercury and arsenic were being used prophylactically, and this would have increased the abundance of relevant resistance genes in the microbiome, in particular transposons carrying *mer* operons [43] (Figure 2).

The subsequent development and use of diverse antimicrobial agents during the 20th Century introduced sequential selection pressures, each of which favored cells that happened to have recently acquired an appropriate resistance gene. There is a strong stochastic nature to these events, because two criteria must be met: the acquisition of a relevant element of the resistome; and the coincident application of a selective agent. Consequently, there would be a strong founder effect. Once a particular resistance gene was driven to high abundance by selection, there would be less chance that a second gene could achieve similar success, even if it conferred the same phenotype. This may explain why only certain genes from within the resistome have become clinically important [32,37,44,45].

Each selection event helped raise the abundance and distribution of resistome elements and their mobile DNA vectors such as transposons and plasmids. Recruitment of new resistome elements into existing transposons or plasmids increased the potential for co-selection based on the possession of multiple resistance determinants [46,47,48]. In particular, the acquisition of gene accumulation platforms such as the class 1 integron [49], increased the flexibility and rate at which new resistance determinants could be added to existing multi-resistance elements.

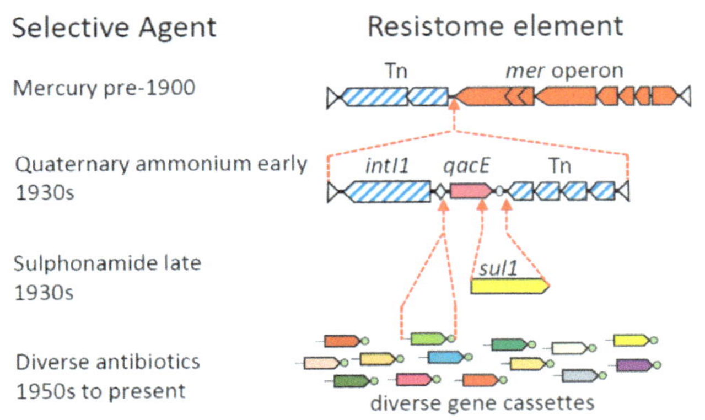

Figure 2. Schematic illustrating the sequential assembly of complex, mosaic DNA elements under the influence of human selection. The proposed history of transposon Tn*21* is used as an example [43,49,50]. Open reading frames conferring resistance phenotypes are represented by solid color; genes for DNA recombination, such as the integron-integrase *intI1* and transposition machinery, have diagonal hatching. The sequential series of selection events that is thought to have fixed each additional resistance gene to form larger and more complex elements are shown. These resistance determinants include: a *mer* operon (mercury resistance); *qacE* (resistance to quaternary ammonium compounds); *sul1* (resistance to sulphonamides); and integron gene cassettes (resistance to diverse antibiotics). Only one of the many acquisitions of different integron gene cassettes is shown. For details of integron diversity and activity see [51].

Each extant plasmid, integrative conjugative element or compound transposon has its own unique evolutionary history, which, given the promiscuous nature of lateral gene transfer, might never be precisely reconstructed. Nevertheless, a general illustration of the processes involved can be given by outlining the most likely explanation for the origins and subsequent evolution of the transposons and clinical class 1 integrons that have played such a central role in the dissemination of antibiotic resistance (Figure 2).

Assembly of composite, mosaic resistance elements probably began with exposure to mercury in the 19th and early 20th Century. Environmental or prophylactic mercury exposure may have fixed a *mer* operon and attendant transposon in the human microbiome. The introduction of quaternary ammonium compounds as disinfectants [52] then provided a second selective force. A class 1 integron that carried a *qacE* gene cassette was captured from a betaproteobacterial chromosome by a Tn*402* transposon [49,51], and this hybrid element was then inserted into a *mer* transposon [43]. The resultant mosaic transposon conferred resistance to mercury, and also to disinfectants via the *qacE* integron gene cassette (Figure 2). Both the*mer* operon and the class 1 integron were probably derived from environmental microorganisms, while the Tn*402*transposon might have already been resident in the human microbiome [49]. The fusion of these three elements resulted in the compound transposon Tn*21*, whose subsequent derivatives have become widely distributed on different plasmids in the human microbiome [43,53].

Introduction of sulphonamides, the first true antibiotic, in the late 1930s [54] provided the selective pressure to fix an insertion of the *sul1* gene into the class 1 integron, deleting the terminal portion of the *qacE* gene cassette [50]. Introduction of further antibiotics over the next 60 years selected, at each time point, for integron variants that had acquired the appropriate resistance genes (Figure 2). Over this time period, class 1 integrons captured over 130 distinct antibiotic resistance gene cassettes [55]. The transposons in which the class 1 integrons are embedded have themselves moved between different plasmids, and have also moved by LGT to diverse species, such that class 1 integrons are found in 40%–70% of Gram-negative clinical pathogens [56,57]. They are also commonly carried in the microbiota of humans, their livestock, and domestic animals [14,58], and have become a permanent feature of the microbiome.

There are a number of points to be made here about the influence of antimicrobials on the human microbiome. Antimicrobial compounds provide the motive force to fix appropriate resistance genes when these arrive in the microbiome via LGT. Each additional antimicrobial treatment promotes insertion of more resistance determinants into compound elements of growing complexity. The individual components of these mosaic, multi-resistance elements are derived from diverse, unrelated organisms [31]. The fusion of genes with such disparate origins over such a short time frame would not have occurred but for the sequential application of different selective agents in the modern era. In this sense, the multi-resistance elements now circulating in the human microbiome are xenogenetic, that is, foreign to the natural world, and fixing only through human activity [15,59].

Further, the linkage of diverse resistance genes on single genetic elements promotes their retention, since application of any one selective force co-selects for all the linked resistance genes and for the compound element itself [47,48,60]. Additionally, any gene that can embed itself within a complex xenogenetic element can take advantage of co-selection. This particularly applies to mobile DNA elements, whose presence can both expand the phylogenetic distribution of the parent element, and increase the number of possible interactions with other elements, thus generating enormous complexity through combinatorial assembly [61]. As a consequence, multi-resistance plasmids and other resistance loci have become home to an extraordinary array of insertion elements and transposons [62,63,64].

HISTORY OF THE HUMAN MICROBIOTA

The human microbiota is a complex ecosystem, consisting of multiple species that interact with each other, and with the host, to generate a dynamic community. The final microbial assemblage at any particular time, and at any one body site, is determined by ecological and evolutionary forces, including: (i) colonization processes, themselves subject to dispersal, ecology and opportunity; (ii) recruitment to particular body sites, driven by interactions between microbial and host genotypes; and (iii) competition between cells for resources, coupled with natural selection [65]. Within such a framework, each individual human can be regarded as an island of habitat, ready to be colonized by dispersing microbiota [12].

Just as humans have had significant effects on the world's ecosystems, our activities have also had adverse effects on our internal ecosystems, perturbing our microbiota. These perturbations have roughly corresponded with the key transitions that define the Anthropocene: the origin of farming; the industrial revolution; and the "Great Acceleration" since the 1950's [7,66]. To fully understand the changes that might have occurred over human evolutionary history, it would be useful to have a better idea of what an ancient human microbiota looked like.

There are three possible ways to reconstruct our ancestral microbiota. These are: to examine the microbiota in our closest relatives, the great apes [67]; to examine microbial fossils in human coprolites, fossil dental calculus or other preserved material [68,69]; or to examine the microbiota of extant human cultures that have lifestyles similar to our hunter-gatherer ancestors [25,26]. These three approaches all yield results that are in broad agreement.

Changes to the composition of gut microbiota in the great apes are clock-like, and consistent with the dates of their evolutionary divergence [67]. In contrast, the human microbiota has evolved at an accelerated rate since diverging from chimpanzees, and has lost ancestral diversity. This loss of diversity was accompanied by an increase in taxa associated with diets containing animal products [67]. In this context, it would be interesting to compare the human microbiota with that of baboons, which might have a diet similar to early hominines.

It is also clear that the habitual use of fire for preparing food from about 350,000 years ago [70] may have relaxed the selection for retaining some microbial species, in particular, for taxa capable of dealing with the more refractory components of plant foods. A shift in allocation of resources, from the gut to supporting a larger brain, results in humans having a relatively small gut for our body size [71]. Further, the small intestine accounts for over half the gut volume in humans, while in the other great apes it occupies only 14%–29% [72]. Both these factors mean that there is comparatively less volume for the high microbial population density typical of the large intestine, and this may in turn have lowered overall microbial diversity during human evolution.

In some cases, the microbiota of ancient humans has been preserved as fossils. Given the caveats of bacterial preservation and DNA recovery, these fossils can shed light on ancient microbiota. Coprolites from 8000 to 1400 years B.P. show that early humans had microbiota that was similar to rural, rather than cosmopolitan communities [73]. Calcified dental plaque documents compositional shifts and declining diversity in the oral microbiota during the transition from hunter-gatherers to the Neolithic, and again around the industrial revolution [28].

Studies on human populations with lifestyles similar to our pre-agricultural ancestors confirm these general conclusions. The Hadza of Tanzania [24] and the Amerindian Yanomami [11] both maintain hunter-gatherer lifestyles, and have much higher levels of microbial diversity than other human populations. The Yanomami, in particular, have had little or no outside contact, and represent a human population that may have been isolated for 11,000 years. They have unprecedented diversity of both microbial species and gene functions [11]. The Asaro and Sausi of Papua New Guinea are traditional societies that rely on subsistence agriculture, and also have very high diversity in their microbiota. A significant proportion of their microbial taxa are not found in US populations [74].

In all the traditional populations, there is a high microbial α-diversity (diversity of species within the microbiota of an individual), attributed variously to retention of diverse taxa that deal with refractory foods, frequent meals, food seasonality or increased environmental exposure [11,74]. In comparison, there is less α-diversity in subsistence agriculturalists from Venezuela and Malawi, and even less in Italian and US metropolitan populations [11,24,26,74]. In contrast, microbial β-diversity (differences in species composition between individuals), is much higher in westernized societies [11,74]. Microbial assemblages in westernized populations may be sufficiently different to reliably identify individuals [75].

These data suggest that ancestral humans had high microbial diversity, but individuals carried roughly the same diverse flora. Why then do westernized humans differ so much in their microbial carriage, when there are apparently less species for them to acquire? One explanation is that the greater dietary ranges of individuals in modern societies drive the selection of a greater range of microbial species. Another explanation, often cited in the literature, is that there are decreased dispersal rates amongst westernized populations [74]. On the face of it, improvements in sanitation and hygiene would appear to support this view. However, this does not sit easily with the vastly increased opportunities for exposure to organisms from virtually any continent, brought about by globalization (Figure 1). The rapid spread of infectious disease agents [29,76] and newly emerging antibiotic resistance genes around the world [77,78] demonstrate that dispersal of microorganisms is perhaps more efficient than it has ever been.

We propose an alternate view, that the elevated β-diversity in modern humans is a result of interplay between ecological and evolutionary factors. These include interruption to colonization opportunities, stochastic sampling from a depauperate pool of microbial species, and the sequential imposition of multiple bottlenecks that affect both species and genetic diversity. Under this hypothesis, each individual in modern society is a microbial island, upon which the potent effects of diverse antimicrobial agents are unleashed, at different intervals, and in different relative orders, unique to each person.

What are the effects of antibiotic administration on the human microbiota? Antibiotic therapy leads to rapid shifts in composition, and a decline in microbial diversity. Susceptible microorganisms are depleted while resistant cells increase in relative abundance, with the consequent fixation and persistence of resistance determinants [79]. Although communities do recover after cessation of antibiotic therapy, they may never return to their original diversity, and often still carry the relevant resistance genes, even four years later [80]. In many populations, the average time period between different courses of antibiotics is much less than four years. On any given day, between 1% and 3% of Europeans are on antibiotic therapy [81].

The response of the gut microbiota to antibiotics differs between individuals [82], consequently α-diversity is decreased, while simultaneously increasing β-diversity. The same phenomenon is likely to occur even within species, because resistant clonal lines are favored by antibiotic selection. During antibiotic therapy, declines in intra-species genetic variation have been observed, caused by significant shifts in clonal population structure. These changes persisted over the entire course of several studies (2–4 years), and have been observed in genera as diverse as *Bacteroides*, *Enterococcus* and *Staphylococcus*[79,83,84]. The consequences of repeated antibiotic therapy would then be a series of genetic bottlenecks [85], each of which takes the microbial assemblages of individuals on increasingly divergent paths. High coverage sequencing of metagenomic DNA will be needed to resolve microbial strains within the microbiota of individuals [86].

CONCLUSIONS

The clear conclusion emerging from the literature is that the diversity of the human microbiota has been declining, initially during our divergence from the great apes, but progressively more so during the Neolithic, Industrial Revolution and Modern Eras (Table 1). The reasons for this decline are multifactorial, and involve intersecting ecological and evolutionary drivers. Changes in diet through the use of fire in food preparation, and the invention of agriculture, probably both reduced α-diversity, resulting in a loss of species [87], and a loss of gene functions in microbial populations [11,24].

Table 1. A proposed history of ecological and evolutionary effects on the human microbiota.

Timeframe	Ecological or Evolutionary driver	Possible effect on microbiota	Possible effect on microbiome
Pre-350,000 bp [A]	fire & diet gut *vs.* brain investment	↓ α-diversity	speeds molecular clock
Neolithic 10,000 bp	farming & diet	↓ α-diversity zoonotic exchanges	↓ gene functions
Industrial revolution 18th century	processed foods pollutants, heavy metals globalization	↓ α-diversity selection ↑ dispersal	selection for *mer* operon
Modern era 1930's	disinfectants sanitation	↓ α-diversity ↑ β-diversity?	selection for *qacE*, spread of class 1 integron
First antibiotic 1930's	sulphonamides	↓ α-diversity selection	selection for *sul1*
Antibiotic era 1950's on	diverse antibiotics caesarean section bottle feeding globalization sanitation	↓ α-diversity ↓ transmission ↓ transmission ↑ dispersal ↑ β-diversity	selection for resistance genes ↓ intra-species genetic diversity

A = before present.

However, it is during the last 300 years that the most significant changes are likely to have occurred. Exposure to pollutants and the increasing use of antimicrobial compounds has imposed strong selection directly on all components of the microbiota, in a manner that was unknown in the past. Because of the essentially stochastic nature of antibiotic therapy, and the variation in individual responses [82], this leads to a series of population, species and genetic bottlenecks within the microbiota of individuals [85], potentially leading to unique microbial assemblages in every person. Of course, many other factors might be at play, including the consumption of increasingly hygienic and processed food [23], and reduced transmission between individuals due to sanitation, caesarian section and bottle feeding [7,74,87]. Further, declining α-diversity in each generation reduces the diversity available for transmission to the next generation, and does so in a step-wise manner [87].

It is useful to think of each individual human as an island, colonized and populated by microorganisms. From this perspective, many of the principles of island biogeography become useful and informative. The human body is originally colonized by microorganisms from relatives, and from the environment, and these assemblages undergo progressive changes to reach a final, climax community as an adult [88,89]. In the modern world, declining microbial exposure and decreased α-diversity in our relatives limits the diversity of potential colonizers, and thus makes the eventual climax community less predictable. Each island of microorganisms is also then exposed to different agents of selection, which preferentially remove some elements of the assemblage, and lower α-diversity at both species and genetic levels [83,85,90]. Each island experiences these perturbations at different times, with different agents of selection, and thus each human microbial island has a different ecological and evolutionary history. This sequential series of independent bottlenecks [85,91] then generates the elevated β-diversity that characterizes modern populations [11,74].

Depauperate ecosystems are less resilient than richly diverse ecosystems, and this appears to be true of our internal microbial ecosystems as well. Perturbation to the human microbiota is now being associated with an increasing list of diseases. These include many diseases of uncertain etiology that have rapidly increased in frequency in the last 50 years, such as allergic disorders, chronic inflammatory conditions, and autoimmune diseases [10,92]. The microbiota is also thought to influence behavior, via feedbacks on brain development [93,94], suggesting that microbiome perturbations might contribute to the increasing frequency of anxiety and depression in the developed world [95].

The interactions between the microbiota and host tissues are essential for good health, yet antibiotics alter microbial physiology and gene expression [96], and if administered during a critical window, can have permanent effects on host metabolism [97]. We are only just beginning to understand the complexities of our internal ecosystems, and with some notable exceptions [98], do not have effective therapies for restoring a "healthy" microbiota. In the quest for such solutions, it would be good to remember that while human genetic diversity consists of allelic diversity in the nuclear and mitochondrial genomes, the potential genetic diversity of the microbiome outstrips this by several orders of magnitude. Managing and maintaining microbial diversity may turn out to be a central part of human health into the future.

ACKNOWLEDGMENTS

Michael R. Gillings, Ian T. Paulsen and Sasha G. Tetu are supported by the Australian Research Council. Thanks to Elizabeth Hagan-Lawson for critical discussions on the comparative evolution of digestive systems in the great apes, and to David Wilkinson for insightful comments.

Author Contributions

Michael R. Gillings, Ian T. Paulsen and Sasha G. Tetu planned the manuscript. Michael R. Gillings wrote the first draft, and

Ian T. Paulsen and Sasha G. Tetu provided critical feedback and editing.

CONFLICTS OF INTEREST

The authors declare no conflict of interest.

REFERENCES

1. Palumbi, S.R. Humans as the world's greatest evolutionary force. *Science* **2001**, *293*, 1786–1790.
2. Rockstrom, J.; Steffen, W.; Noone, K.; Persson, A.; Chapin, F.S.; Lambin, E.F.; Lenton, T.M.; Scheffer, M.; Folke, C.; Schellnhuber, H.J.; *et al.* A safe operating space for humanity. *Nature* **2009**, *461*, 472–475.
3. Williams, M.; Zalasiewicz, J.; Haff, P.; Schwagerl, C.; Barnosky, A.D.; Ellis, E.C. The Anthropocene biosphere. *Anthr. Rev.* **2015**.
4. Gillings, M.R.; Westoby, M. DNA technology and evolution of the central dogma. *Trends Ecol. Evol.* **2014**, *29*, 1–2.
5. Falkowski, P.G.; Fenchel, T.; Delong, E.F. The microbial engines that drive Earth's biogeochemical cycles. *Science* **2008**,*320*, 1034–1039.
6. Wilkinson, D.M. Have we underestimated the importance of humans in the biogeography of free-living terrestrial microorganisms? *J. Biogeogr.* **2010**, *37*, 393–397.
7. Gillings, M.R.; Paulsen, I.T. Microbiology of the Anthropocene. *Anthropocene* **2014**, *5*, 1–8.
8. Morgan, X.C.; Segata, N.; Huttenhower, C. Biodiversity and functional genomics in the human microbiome. *Trends Genet.* **2013**, *29*, 51–58.
9. Gill, S.R.; Pop, M.; DeBoy, R.T.; Eckburg, P.B.; Turnbaugh, P.J.; Samuel, B.S.; Gordon, J.I.; Relman, D.A.; Fraser-Liggett, C.M.; Nelson, K.E. Metagenomic analysis of the human distal gut microbiome. *Science* **2006**, *312*, 1355–1359.
10. Cho, I.; Blaser, M.J. The human microbiome: At the interface of health and disease. *Nat. Rev. Genet.* **2012**, *13*, 260–270.
11. Clemente, J.C.; Pehrsson, E.C.; Blaser, M.J.; Sandhu, K.; Gao, Z.; Wang, B.; Magris, M.; Hidalgo, G.; Contreras, M.; Noya-Alarcón, Ó. The microbiome of uncontacted Amerindians. *Sci. Adv.* **2015**.
12. Costello, E.K.; Stagaman, K.; Dethlefsen, L.; Bohannan, B.J.; Relman, D.A. The application of ecological theory toward an understanding of the human microbiome. *Science* **2012**, *336*, 1255–1262.
13. Xu, Z.; Knight, R. Dietary effects on human gut microbiome diversity. *Br. J. Nutr.* **2015**, *113*, S1–S5.
14. Stokes, H.W.; Gillings, M.R. Gene flow, mobile genetic elements and the recruitment of antibiotic resistance genes into Gram-negative pathogens. *FEMS Microbiol. Rev.* **2011**, *35*, 790–819.
15. Gillings, M.R. Evolutionary consequences of antibiotic use for the resistome, mobilome and microbial pangenome.*Front. Microbiol.* **2013**, *4*.
16. Popa, O.; Dagan, T. Trends and barriers to lateral gene transfer in prokaryotes. *Curr. Opin. Microbiol.* **2011**, *14*, 615–623.
17. Thomas, C.M.; Nielsen, K.M. Mechanisms of, and barriers to, horizontal gene transfer between bacteria. *Nat. Rev. Microbiol.* **2005**, *3*, 711–721.
18. Kloesges, T.; Popa, O.; Martin, W.; Dagan, T. Networks of gene sharing among 329 proteobacterial genomes reveal differences in lateral gene transfer frequency at different phylogenetic depths. *Mol. Biol. Evol.* **2010**, *28*, 1057–1074.
19. Skippington, E.; Ragan, M.A. Lateral genetic transfer and the construction of genetic exchange communities. *FEMS Microbiol. Rev.* **2011**, *35*, 707–735.
20. Sorensen, S.J.; Bailey, M.; Hansen, L.H.; Kroer, N.; Wuertz, S. Studying plasmid horizontal transfer *in situ*: A critical review. *Nat. Rev. Microbiol.* **2005**, *3*, 700–710.
21. Smillie, C.S.; Smith, M.B.; Friedman, J.; Cordero, O.X.; David, L.A.; Alm, E.J. Ecology drives a global network of gene exchange connecting the human microbiome. *Nature* **2011**, *480*, 241–244.
22. Kurokawa, K.; Itoh, T.; Kuwahara, T.; Oshima, K.; Toh, H.; Toyoda, A.; Takami, H.; Morita, H.; Sharma, V.K.; Srivastava, T.P. Comparative metagenomics revealed commonly enriched gene sets in human gut microbiomes. *DNA Res.* **2007**, *14*, 169–181.
23. Thomas, F.; Hehemann, J.-H.; Rebuffet, E.; Czjzek, M.; Michel, G. Environmental and gut *Bacteroidetes*: The food connection. *Front. Microbiol.* **2011**, *2*.
24. Schnorr, S.L.; Candela, M.; Rampelli, S.; Centanni, M.; Consolandi, C.; Basaglia, G.; Turroni, S.; Biagi, E.; Peano, C.; Severgnini, M. Gut microbiome of the Hadza hunter-gatherers. *Nat. Commun.* **2014**.
25. De Filippo, C.; Cavalieri, D.; Di Paola, M.; Ramazzotti, M.; Poullet, J.B.; Massart, S.; Collini, S.; Pieraccini, G.; Lionetti, P. Impact of diet in shaping gut microbiota revealed by a comparative study in children from Europe and rural Africa. *Proc. Natl. Acad. Sci. USA* **2010**, *107*, 14691–14696.
26. Yatsunenko, T.; Rey, F.E.; Manary, M.J.; Trehan, I.; Dominguez-Bello, M.G.; Contreras, M.; Magris, M.; Hidalgo, G.; Baldassano, R.N.; Anokhin, A.P.; *et al.* Human gut microbiome viewed across age and geography. *Nature* **2012**, *486*, 222–227.
27. Hehemann, J.-H.; Correc, G.; Barbeyron, T.; Helbert, W.; Czjzek, M.; Michel, G. Transfer of carbohydrate-active enzymes from marine bacteria to Japanese gut microbiota. *Nature* **2010**, *464*, 908–912.
28. Adler, C.J.; Dobney, K.; Weyrich, L.S.; Kaidonis, J.; Walker, A.W.; Haak, W.; Bradshaw, C.J.A.; Townsend, G.; Soltysiak, A.; Alt, K.W.; *et al.* Sequencing ancient calcified dental plaque shows changes in oral microbiota with dietary shifts of the Neolithic and Industrial Revolutions. *Nat. Genet.* **2013**, *45*, 450–455.
29. Van der Bij, A.K.; Pitout, J.D. The role of international travel in the worldwide spread of multiresistant*Enterobacteriaceae. J. Antimicrob. Chemother.* **2012**, *67*, 2090–2100.
30. Davies, J.; Davies, D. Origins and evolution of antibiotic resistance. *Microbiol. Mol. Biol. Rev.* **2010**, *74*, 417–433.

31. Toussaint, A.; Chandler, M. Prokaryote genome fluidity: Toward a system approach of the mobilome. In *Bacterial Molecular Networks*; Helden, J., Toussaint, A., Thieffry, D., Eds.; Springer: New York, NY, USA, 2012; Volume 804, pp. 57–80.

32. D'Costa, V.M.; McGrann, K.M.; Hughes, D.W.; Wright, G.D. Sampling the antibiotic resistome. *Science* **2006**, *311*, 374–377.

33. Forsberg, K.J.; Reyes, A.; Wang, B.; Selleck, E.M.; Sommer, M.O.A.; Dantas, G. The shared antibiotic resistome of soil bacteria and human pathogens. *Science* **2012**, *337*, 1107–1111.

34. Wright, G.D. The antibiotic resistome. *Expert Opin. Drug Discov.* **2010**, *5*, 779–788.

35. Dantas, G.; Sommer, M.O.A. Context matters—The complex interplay between resistome genotypes and resistance phenotypes. *Curr. Opin. Microbiol.* **2012**, *15*, 577–582.

36. Allen, H.K.; Donato, J.; Wang, H.H.; Cloud-Hansen, K.A.; Davies, J.; Handelsman, J. Call of the wild: Antibiotic resistance genes in natural environments. *Nat. Rev. Microbiol.* **2010**, *8*, 251–259.

37. Riesenfeld, C.S.; Goodman, R.M.; Handelsman, J. Uncultured soil bacteria are a reservoir of new antibiotic resistance genes. *Environ. Microbiol.* **2004**, *6*, 981–989.

38. Perry, J.A.; Westman, E.L.; Wright, G.D. The antibiotic resistome: What's new? *Curr. Opin. Microbiol.* **2014**, *21*, 45–50.

39. Sommer, M.O.A.; Dantas, G.; Church, G.M. Functional characterization of the antibiotic resistance reservoir in the human microflora. *Science* **2009**, *325*, 1128–1131.

40. Sommer, M.O.; Church, G.M.; Dantas, G. The human microbiome harbors a diverse reservoir of antibiotic resistance genes. *Virulence* **2010**, *1*, 299–303.

41. Datta, N.; Hughes, V.M. Plasmids of the same Inc groups in enterobacteria before and after the medical use of antibiotics. *Nature* **1983**, *306*, 616–617.

42. Poulain, A.J.; Aris-Brosou, S.; Blais, J.M.; Brazeau, M.; Keller, W.B.; Paterson, A.M. Microbial DNA records historical delivery of anthropogenic mercury. *ISME J.* **2015**.

43. Liebert, C.A.; Hall, R.M.; Summers, A.O. Transposon Tn*21*, flagship of the floating genome. *Microbiol. Mol. Biol. Rev.***1999**, *63*, 507–522.

44. Allen, H.K.; Moe, L.A.; Rodbumrer, J.; Gaarder, A.; Handelsman, J. Functional metagenomics reveals diverse β-lactamases in a remote Alaskan soil. *ISME J.* **2008**, *3*, 243–251.

45. D'Costa, V.M.; King, C.E.; Kalan, L.; Morar, M.; Sung, W.W.L.; Schwarz, C.; Froese, D.; Zazula, G.; Calmels, F.; Debruyne, R.; *et al*. Antibiotic resistance is ancient. *Nature* **2011**, *477*, 457–461.

46. Gaze, W.H.; Abdouslam, N.; Hawkey, P.M.; Wellington, E.M.H. Incidence of class 1 integrons in a quaternary ammonium compound-polluted environment. *Antimicrob. Agents Chemother.* **2005**, *49*, 1802–1807.

47. Rosewarne, C.P.; Pettigrove, V.; Stokes, H.W.; Parsons, Y.M. Class 1 integrons in benthic bacterial communities: Abundance, association with Tn*402*-like transposition modules and evidence for coselection with heavy-metal resistance. *FEMS Microbiol. Ecol.* **2010**, *72*, 35–46.

48. Wright, M.S.; Baker-Austin, C.; Lindell, A.H.; Stepanauskas, R.; Stokes, H.W.; McArthur, J.V. Influence of industrial contamination on mobile genetic elements: Class 1 integron abundance and gene cassette structure in aquatic bacterial communities. *ISME J.* **2008**, *2*, 417–428.

49. Gillings, M.; Boucher, Y.; Labbate, M.; Holmes, A.; Krishnan, S.; Holley, M.; Stokes, H.W. The evolution of class 1 integrons and the rise of antibiotic resistance. *J. Bacteriol.* **2008**, *190*, 5095–5100.

50. Kholodii, G.Y.; Mindlin, S.; Bass, I.; Yurieva, O.; Minakhina, S.; Nikiforov, V. Four genes, two ends, and a res region are involved in transposition of Tn*5053*: A paradigm for a novel family of transposons carrying either a *mer* operon or an integron. *Mol. Microbiol.* **1995**, *17*, 1189–1200.

51. Gillings, M.R. Integrons: Past, present, and future. *Microbiol. Mol. Biol. Rev.* **2014**, *78*, 257–277.

52. Russell, A. Introduction of biocides into clinical practice and the impact on antibiotic-resistant bacteria. *J. Appl. Microbiol.* **2002**, *92*, S121–S135.

53. Partridge, S.R.; Brown, H.J.; Stokes, H.; Hall, R.M. Transposons Tn*1696* and Tn*21* and their integrons In4 and In2 have independent origins. *Antimicrob. Agents Chemother.* **2001**, *45*, 1263–1270.

54. Sköld, O. Sulfonamide resistance: Mechanisms and trends. *Drug Resist. Updates* **2000**, *3*, 155–160.

55. Partridge, S.R.; Tsafnat, G.; Coiera, E.; Iredell, J.R. Gene cassettes and cassette arrays in mobile resistance integrons.*FEMS Microbiol. Rev.* **2009**, *33*, 757–784.

56. Martinez-Freijo, P.; Fluit, A.; Schmitz, F.; Grek, V.; Verhoef, J.; Jones, M. Class 1 integrons in gram-negative isolates from different European hospitals and association with decreased susceptibility to multiple antibiotic compounds. *J. Antimicrob. Chemother.* **1998**, *42*, 689–696.

57. Van Essen-Zandbergen, A.; Smith, H.; Veldman, K.; Mevius, D. Occurrence and characteristics of class 1, 2 and 3 integrons in *Escherichia coli*, *Salmonella* and *Campylobacter* spp. in the Netherlands. *J. Antimicrob. Chemother.* **2007**, *59*, 746–750.

58. Goldstein, C.; Lee, M.D.; Sanchez, S.; Hudson, C.; Phillips, B.; Register, B.; Grady, M.; Liebert, C.; Summers, A.O.; White, D.G. Incidence of class 1 and 2 integrases in clinical and commensal bacteria from livestock, companion animals, and exotics. *Antimicrob. Agents Chemother.* **2001**, *45*, 723–726.

59. Gillings, M.R.; Stokes, H.W. Are humans increasing bacterial evolvability? *Trends Ecol. Evol.* **2012**, *27*, 346–352.

60. Gillings, M.R.; Xuejun, D.; Hardwick, S.A.; Holley, M.P.; Stokes, H.W. Gene cassettes encoding resistance to

quaternary ammonium compounds: A role in the origin of clinical class 1 integrons? *ISME J.* **2009**, *3*, 209–215.

61. Garriss, G.; Waldor, M.K.; Burrus, V. Mobile antibiotic resistance encoding elements promote their own diversity.*PLoS Genet.* **2009**, *5*, e1000775.
62. Gillings, M.R.; Labbate, M.; Sajjad, A.; Giguere, N.J.; Holley, M.P.; Stokes, H.W. Mobilization of a Tn*402*-like class 1 integron with a novel cassette array via flanking miniature inverted-repeat transposable element-like structures. *Appl. Environ. Microbiol.* **2009**, *75*, 6002–6004.
63. Harmer, C.J.; Moran, R.A.; Hall, R.M. Movement of IS*26*-associated antibiotic resistance genes occurs via a translocatable unit that includes a single IS*26* and preferentially inserts adjacent to another IS*26*. *MBio* **2014**, *5*, e01801–e01814.
64. Reid, C.; Chowdhury, P.R.; Djordjevic, S.P. Tn*6026* and Tn*6029* are found in complex resistance regions mobilised by diverse plasmids and chromosomal islands in multiple antibiotic resistant *Enterobacteriaceae*. *Plasmid* **2015**.
65. Dethlefsen, L.; McFall-Ngai, M.; Relman, D.A. An ecological and evolutionary perspective on human–microbe mutualism and disease. *Nature* **2007**, *449*, 811–818.
66. Lewis, S.L.; Maslin, M.A. Defining the Anthropocene. *Nature* **2015**, *519*, 171–180.
67. Moeller, A.H.; Li, Y.; Ngole, E.M.; Ahuka-Mundeke, S.; Lonsdorf, E.V.; Pusey, A.E.; Peeters, M.; Hahn, B.H.; Ochman, H. Rapid changes in the gut microbiome during human evolution. *Proc. Natl. Acad. Sci. USA* **2014**, *111*, 16431–16435.
68. Santiago-Rodriguez, T.M.; Narganes-Storde, Y.M.; Chanlatte, L.; Crespo-Torres, E.; Toranzos, G.A.; Jimenez-Flores, R.; Hamrick, A.; Cano, R.J. Microbial communities in pre-Columbian coprolites. *PLoS ONE* **2013**, *8*.
69. Warinner, C.; Speller, C.; Collins, M.J.; Lewis, C.M. Ancient human microbiomes. *J. Hum. Evol.* **2015**, *79*, 125–136.
70. Shimelmitz, R.; Kuhn, S.L.; Jelinek, A.J.; Ronen, A.; Clark, A.E.; Weinstein-Evron, M. "Fire at will": The emergence of habitual fire use 350,000 years ago. *J. Hum. Evol.* **2014**, *77*, 196–203.
71. Aiello, L.C.; Wheeler, P. The expensive-tissue hypothesis: The brain and the digestive system in human and primate evolution. *Curr. Anthropol.* **1995**, *36*, 199–221.
72. Milton, K. A hypothesis to explain the role of meat-eating in human evolution. *Evol. Anthropol. Issues News Rev.* **1999**,*8*, 11–21.
73. Tito, R.Y.; Knights, D.; Metcalf, J.; Obregon-Tito, A.J.; Cleeland, L.; Najar, F.; Roe, B.; Reinhard, K.; Sobolik, K.; Belknap, S. Insights from characterizing extinct human gut microbiomes. *PLoS ONE* **2012**, *7*, e51146.
74. Martínez, I.; Stegen, J.C.; Maldonado-Gómez, M.X.; Eren, A.M.; Siba, P.M.; Greenhill, A.R.; Walter, J. The gut microbiota of rural Papua New Guineans: Composition, diversity patterns, and ecological processes. *Cell Rep.* **2015**,*11*, 527–538.
75. Franzosa, E.A.; Huang, K.; Meadow, J.F.; Gevers, D.; Lemon, K.P.; Bohannan, B.J.; Huttenhower, C. Identifying personal microbiomes using metagenomic codes. *Proc. Natl. Acad. Sci. USA* **2015**, *112*, E2930–E2938.
76. Molton, J.S.; Tambyah, P.A.; Ang, B.S.P.; Ling, M.L.; Fisher, D.A. The global spread of healthcare-associated multidrug-resistant bacteria: A perspective from Asia. *Clin. Infect. Dis.* **2013**, *56*, 1310–1318.
77. Cantón, R.; Akóva, M.; Carmeli, Y.; Giske, C.G.; Glupczynski, Y.; Gniadkowski, M.; Livermore, D.M.; Miriagou, V.; Naas, T.; Rossolini, G. Rapid evolution and spread of carbapenemases among *Enterobacteriaceae* in Europe. *Clin. Microbiol. Infect.* **2012**, *18*, 413–431.
78. Johnson, A.P.; Woodford, N. Global spread of antibiotic resistance: The example of New Delhi metallo-β-lactamase (Ndm)-mediated carbapenem resistance. *J. Med. Microbiol.* **2013**, *62*, 499–513.
79. Jernberg, C.; Löfmark, S.; Edlund, C.; Jansson, J.K. Long-term ecological impacts of antibiotic administration on the human intestinal microbiota. *ISME J.* **2007**, *1*, 56–66.
80. Jakobsson, H.E.; Jernberg, C.; Andersson, A.F.; Sjölund-Karlsson, M.; Jansson, J.K.; Engstrand, L. Short-term antibiotic treatment has differing long-term impacts on the human throat and gut microbiome. *PLoS ONE* **2010**, *5*, e9836.
81. Goossens, H.; Ferech, M.; Vander Stichele, R.; Elseviers, M.; Group, E.P. Outpatient antibiotic use in Europe and association with resistance: A cross-national database study. *Lancet* **2005**, *365*, 579–587.
82. Dethlefsen, L.; Relman, D.A. Incomplete recovery and individualized responses of the human distal gut microbiota to repeated antibiotic perturbation. *Proc. Natl. Acad. Sci. USA* **2011**, *108*, 4554–4561.
83. Sjölund, M.; Tano, E.; Blaser, M.J.; Andersson, D.I.; Engstrand, L. Persistence of resistant *Staphylococcus epidermidis*after single course of clarithromycin. *Emerg. Infect. Dis.* **2005**, *11*, 1389–1392.
84. Sjölund, M.; Wreiber, K.; Andersson, D.I.; Blaser, M.J.; Engstrand, L. Long-term persistence of resistant *Enterococcus*species after antibiotics to eradicate *Helicobacter pylori*. *Ann. Intern. Med.* **2003**, *139*, 483–487.
85. Schloissnig, S.; Arumugam, M.; Sunagawa, S.; Mitreva, M.; Tap, J.; Zhu, A.; Waller, A.; Mende, D.R.; Kultima, J.R.; Martin, J. Genomic variation landscape of the human gut microbiome. *Nature* **2013**, *493*, 45–50.
86. Morowitz, M.J.; Denef, V.J.; Costello, E.K.; Thomas, B.C.; Poroyko, V.; Relman, D.A.; Banfield, J.F. Strain-resolved community genomic analysis of gut microbial colonization in a premature infant. *Proc. Natl. Acad. Sci. USA* **2011**, *108*, 1128–1133.
87. Blaser, M.J.; Falkow, S. What are the consequences of the disappearing human microbiota? *Nat. Rev. Microbiol.* **2009**, *7*, 887–894.

88. Koenig, J.E.; Spor, A.; Scalfone, N.; Fricker, A.D.; Stombaugh, J.; Knight, R.; Angenent, L.T.; Ley, R.E. Succession of microbial consortia in the developing infant gut microbiome. *Proc. Natl. Acad. Sci. USA* **2011**, *108*, 4578–4585.

89. Koren, O.; Knights, D.; Gonzalez, A.; Waldron, L.; Segata, N.; Knight, R.; Huttenhower, C.; Ley, R.E. A guide to enterotypes across the human body: Meta-analysis of microbial community structures in human microbiome datasets.*PLoS Comput. Biol.* **2013**, *9*, e1002863.

90. Jernberg, C.; Löfmark, S.; Edlund, C.; Jansson, J.K. Long-term impacts of antibiotic exposure on the human intestinal microbiota. *Microbiology* **2010**, *156*, 3216–3223.

91. Lahti, L.; Salojärvi, J.; Salonen, A.; Scheffer, M.; de Vos, W.M. Tipping elements in the human intestinal ecosystem.*Nat. Commun.* **2014**.

92. Ehlers, S.; Kaufmann, S.H. Infection, inflammation, and chronic diseases: Consequences of a modern lifestyle. *Trends Immunol.* **2010**, *31*, 184–190.

93. Bravo, J.A.; Forsythe, P.; Chew, M.V.; Escaravage, E.; Savignac, H.M.; Dinan, T.G.; Bienenstock, J.; Cryan, J.F. Ingestion of *Lactobacillus* strain regulates emotional behavior and central GABA receptor expression in a mouse via the vagus nerve. *Proc. Natl. Acad. Sci. USA* **2011**, *108*, 16050–16055.

94. Heijtz, R.D.; Wang, S.; Anuar, F.; Qian, Y.; Björkholm, B.; Samuelsson, A.; Hibberd, M.L.; Forssberg, H.; Pettersson, S. Normal gut microbiota modulates brain development and behavior. *Proc. Natl. Acad. Sci. USA* **2011**, *108*, 3047–3052.

95. Foster, J.A.; McVey Neufeld, K.-A. Gut–brain axis: How the microbiome influences anxiety and depression. *Trends Neurosci.* **2013**, *36*, 305–312.

96. Maurice, C.F.; Haiser, H.J.; Turnbaugh, P.J. Xenobiotics shape the physiology and gene expression of the active human gut microbiome. *Cell* **2013**, *152*, 39–50.

97. Cox, L.M.; Yamanishi, S.; Sohn, J.; Alekseyenko, A.V.; Leung, J.M.; Cho, I.; Kim, S.G.; Li, H.; Gao, Z.; Mahana, D. Altering the intestinal microbiota during a critical developmental window has lasting metabolic consequences. *Cell***2014**, *158*, 705–721.

98. Van Nood, E.; Vrieze, A.; Nieuwdorp, M.; Fuentes, S.; Zoetendal, E.G.; de Vos, W.M.; Visser, C.E.; Kuijper, E.J.; Bartelsman, J.F.; Tijssen, J.G. Duodenal infusion of donor feces for recurrent *Clostridium difficile*. *N. Engl. J. Med.* **2013**,*368*, 407–415.

CHAPTER 15

HELICOBACTER PYLORI GENETIC DIVERSITY AND GASTRO-DUODENAL DISEASES

Selva Perumal Gunaletchumy, Indran Seevasant, Mun Hua Tan, Laurence J. Croft, Hazel M. Mitchell, Khean Lee Goh, Mun Fai Loke & Jamuna Vadivelu

INTRODUCTION

H*elicobacter pylori* contributes world-wide to various gastro-duodenal diseases ranging from chronic gastritis to the development of peptic ulcer disease, gastric cancer (GC) and mucosa-associated lymphoid tissue (MALT) lymphoma[1,2]. Significant variation exist in the prevalence and incidence of peptic ulcer disease and gastric cancer due to infection with *H. pylori* among the different multiethnic populations in Malaysia (Malays, Chinese and Indians)[3,4]. The *H. pylori* prevalence amongst Indians is 35.6%, but the incidence of peptic ulcer disease and gastric cancer is relatively low. Despite lower *H. pylori* prevalence of 28.6% among the Chinese, incidence of peptic ulcer disease and gastric cancer are relatively high[3,5].

Phylogenetic analysis of the *H. pylori* genomes was carried out to determine the ancestry relationship among the overall international isolates. Analysis of seven housekeeping genes of *H. pylori* by multi-locus sequence typing (MLST) suggest that the bacteria originated in Africa and later split into seven distinct population groups (hpEurope, hpNEAfrica, hpAfrica1, hpAfrica2, hpAsia2, hpSahul and hpEastAsia) and subpopulations that are strongly associated with geographical localization. As for the subpopulations, AE1 and AE2 are within hpEurope, hspWafrica and hspSAfrica within hpAfrica1, hspIndia within hpAsia2 and hspEAsia, hspAmerind and hspMaori within the population of hpEastAsia[6,7,8]. In Malaysia, *H. pylori* hpAsia2/hspIndia mainly colonises Indian and Malay subjects while hpEastAsia/hspEAsia predominantly colonises Chinese subjects. These two groups (hpEastAsia/hspEAsia and hpAsia2/hspIndia) accounted for 41.5% and 39.0% of all *H. pylori* isolates respectively[6].

H. pylori genetic factors may play a role in influencing the disease outcome as Chinese strains from Malaysia that are similar to strains from other areas of high gastric cancer incidence (Japan, Korea and China) predominantly belongs to hspEAsia subpopulation. Meanwhile, local Malay/Indian strains together with their counterparts from India (high *H. pylori* prevalence but low incidence of gastric cancer) belong to the hspIndia subpopulation[6]. Besides *H. pylori* genetic factors, host and environmental factors may also influence disease outcome.

In Western countries, the *H. pylori* protein encoded by cytotoxin-associated gene A, CagA, has been shown to be strongly associated with peptic ulcer disease, atrophic gastritis and gastric cancer[1,9]. However,*cagA*, carried by almost all Asian *H. pylori* strains, does not predict disease outcome in this region of the world[1,10,11]. Previous studies demonstrated that *cagA* containing the EPIYA-D tend to be more virulent than those carrying the EPIYA-C motif due to the higher level of IL-8 secretion produced by EPIYA-D strains as compared with EPIYA-C strains[12,13]. Studies have proven that Cag-A containing the EPIYA-D motif possessed among the East Asian isolates has a stronger affinity for SHP-2 binding activity in contrast with Western Cag-A. Thus, East Asian isolates are found to be greater in countries with a high prevalence of gastric cancer[12,14,15,16,17,18,19]. However, Schmidt *et al.*(2009) and other studies have demonstrated that no association exists between EPIYA motifs and gastro-duodenal disease progression[11,20,21]. Thus, there is a need to discover new virulence factors by studying the genomic makeup of Asian strains.

H. pylori genetic diversity found in Malaysia, as demonstrated by MLST typing, provides ideal conditions for studying the interaction of co-existing *H. pylori* populations at the genomic level in a multiethnic society. In this study, we analyzed *H. pylori* strains from Asians (Malaysian Chinese, Malay, Indian), Japanese and mainland Chinese subjects presenting with different disease status. The aim of the study was to undertake whole genome comparative analysis of these Asian strains to define a subset of *H. pylori* disease-associated genes that may contribute to gastro-duodenal diseases.

Citation: Gunaletchumy, S.P. et al. Helicobacter pylori Genetic Diversity and Gastro-duodenal Diseases. Sci. Rep. 4, 7431; DOI:10.1038/srep07431 (2014).

RESULTS

In this study, we carried out comparative genome analysis of 27 *H. pylori* strains (4 gastric cancer, 10 peptic ulcer and 13 non-ulcer dyspepsia strains). Strains were selected based on the phylogenetic analysis construction using the full genome sequences (Figure 1). Strains isolated from local Chinese subjects clustered closely with hspEAsia strains from Japan (F30, F32 and F57), China (XZ274 and HLJ271) and Korea (HP51). Most strains from the local Malay and Indian subjects (except UM037) were found clustered together to form the hspIndia branch. Out of 27 genomes selected for comparative analysis, 21 were isolated from Malaysia, three from Japan, two from China and one from Korea (Table 1). The genomes of non-Malaysian isolates were obtained from GenBank.

Comparative genomic analysis revealed six genes that showed significant association with peptic ulcer disease and/or gastric cancer (Table 2). Three genes was associated with gastric cancer and the remaining 3 were associated with both peptic ulcer disease and gastric cancer. All 6 genes were statistically significantly correlated with disease with P-value <0.05 when the percentage of identity were compared using Student's t-test (Table 3). Gene ontology and prediction of protein families, domains and functional sites is presented in Table 4. Only genes with predicted structure and functions were selected for further analysis. The Pearson's correlation coefficient test was performed to evaluate the correlation between these genes and disease status, as well as the correlation between different genes (Table 5).

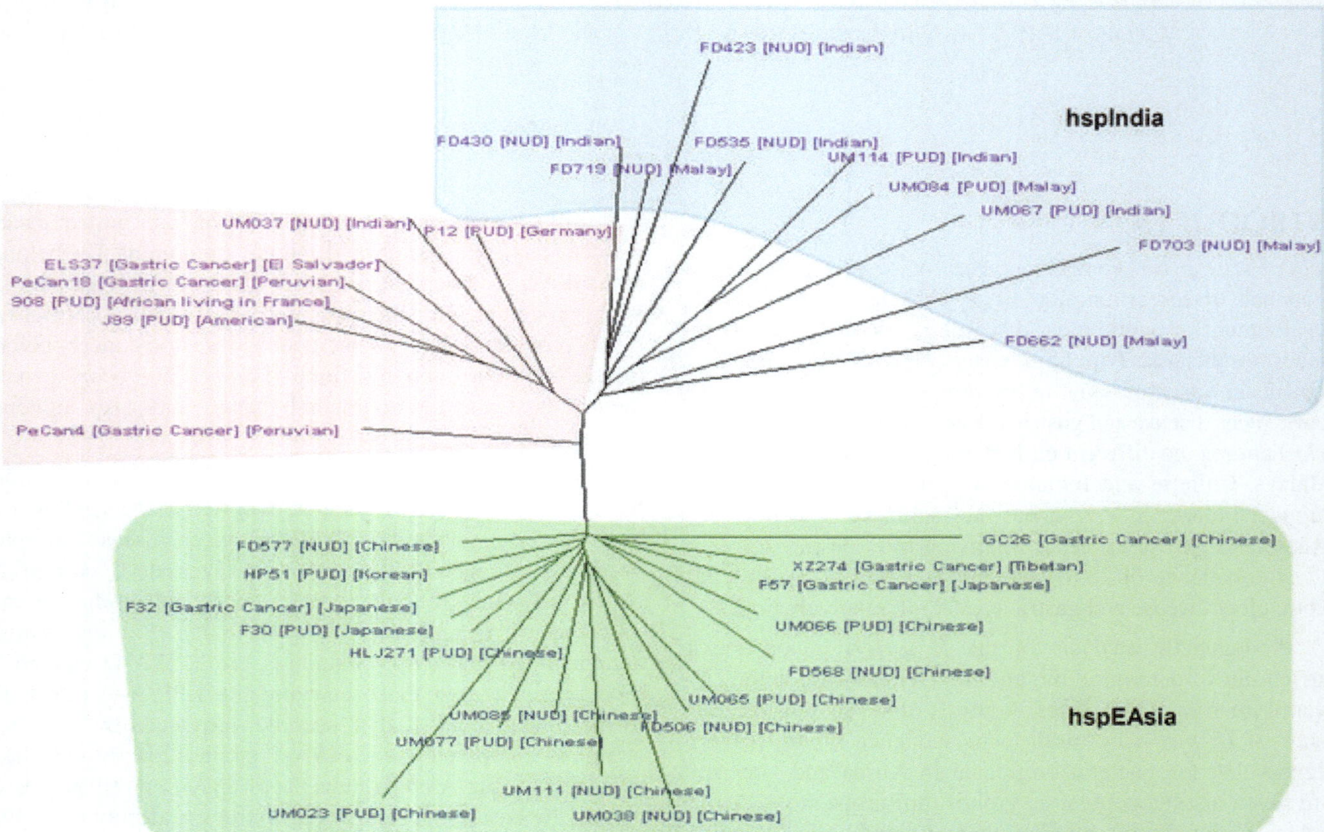

Figure 1: Phylogenetic tree constructed on a whole genome of 33*H. pylori* **strains.** Asian strains used in this study mostly clustered together with hspIndia and hspEAsia strains according to host ethnicity (except UM037).

Table 1: Details of strains used in this study

Strain	Ethnicity	Age	Gender	Year of Isolation	Geographical	Disease	Reference
GC26	Chinese	70	Female	2005	Malaysia	Gastric cancer	30
F57	Japanese	-	-	-	Japan	Gastric cancer	32
F32	Japanese	-	-	-	Japan	Gastric cancer	32
XZ274	Tibetan	-	-	-	China	Gastric cancer	33
UM023	Chinese	71	Female	2011	Malaysia	Peptic ulcer disease	30
UM065	Chinese	57	Female	2011	Malaysia	Peptic ulcer disease	30
UM066	Chinese	30	Female	2011	Malaysia	Peptic ulcer disease	30
UM077	Chinese	59	Female	2011	Malaysia	Peptic ulcer disease	30
UM067	Indian	64	Male	2011	Malaysia	Peptic ulcer disease	30
UM084	Malay	60	Male	2011	Malaysia	Peptic ulcer disease	30
UM114	Indian	42	Female	2012	Malaysia	Peptic ulcer disease	30
F30	Japanese	-	-	-	Japan	Peptic ulcer disease	32
HP51	Korean	-	-	-	Korea	Peptic ulcer disease	-
HLJ271	-	-	-	-	China	Peptic ulcer disease	34
FD506	Chinese	56	Female	2005	Malaysia	Non-ulcer dyspepsia	30
FD568	Chinese	50	Female	2005	Malaysia	Non-ulcer dyspepsia	30
FD577	Chinese	67	Female	2005	Malaysia	Non-ulcer dyspepsia	30
UM038	Chinese	73	Male	2011	Malaysia	Non-ulcer dyspepsia	30
UM085	Chinese	68	Female	2011	Malaysia	Non-ulcer dyspepsia	30
UM111	Chinese	76	Male	2012	Malaysia	Non-ulcer dyspepsia	30
FD662	Malay	47	Female	2006	Malaysia	Non-ulcer dyspepsia	30
FD719	Malay	60	Female	2006	Malaysia	Non-ulcer dyspepsia	30
FD703	Malay	51	Male	2006	Malaysia	Non-ulcer dyspepsia	30
FD423	Indian	39	Male	2005	Malaysia	Non-ulcer dyspepsia	30
FD430	Indian	34	Male	2005	Malaysia	Non-ulcer dyspepsia	30
FD535	Indian	44	Male	2005	Malaysia	Non-ulcer dyspepsia	30
UM037	Indian	67	Female	2011	Malaysia	Non-ulcer dyspepsia	30

Table 2: Candidate genes identified as associated with GC, PUD and NUD with percentage of identity. Percentage of identity generated by RAST >80% was considered similar

	GC				PUD				
	Chinese	Japanese		Tibetan	Chinese				
Gene/Function	GC26	F57	F32	XZ274	UM023	UM065	UM066	UM077	HLJ271
Membrane protein GC26_77	100	78.7	86.6	88.67	0	0	0	0	0
Hypothetical ATPase protein GC26_69	100	93.8	98.6	95.65	0	0	0	0	76.88
Hypothetical protein GC26_73	100	92.1	97.4	88.24	0	0	74.1	0	0
Outer membrane protein GC26_66 (type 1)	100	96.2	98.0	97.6	0	98.08	96.63	96.46	98.02
Outer membrane protein GC26_66 (type 2)	0	0	0	0	0	0	0	0	0
Phospho-2-Dehydro-3-Deoxyheptonate Aldolase	100	99.6	99.0	98.66	99.33	98.44	99.33	98.66	99.33
Hypothetical protein GC26_33	100	84.1	100	98.18	100	96.36	90.91	96.36	0

	PUD					**NUD**			
	Japanese	**Korean**	**Malay**	**Indian**		**Chinese**			
	F30	**HP51**	**UM084**	**UM067**	**UM114**	**FD506**	**FD568**	**FD577**	**UM038**
Membrane Protein GC26_77	0	0	0	0	0	89.3	95.2	100	96.0
Hypothetical ATPase Protein GC26_69	82.55	92.22	0	0	0	0	0	0	0
Hypothetical protein GC26_73	89.19	0	0	0	0	0	0	0	0
Outer membrane protein GC26_66 (type 1)	97.12	98.56				0	0	0	0
Outer membrane protein GC26_66 (type 2)	0	0	95.43	93.91	94.42	0	0	0	0
Phospho-2-Dehydro-3-Deoxyheptonate Aldolase	98.66	97.09	97.55	0	98.22	0	0	0	0
Hypothetical protein GC26_33	74.55	0	0	0	81.82	0	0	0	0

	NUD								
	Chinese		**Malay**			**Indian**			
	UM085	**UM111**	**FD662**	**FD703**	**FD719**	**FD423**	**FD430**	**FD535**	**UM037**
Membrane Protein GC26_77	87.3	92.0	93.0	96.7	90.0	92.2	94.0	93.2	89.3
Hypothetical ATPase Protein GC26_69	0	0	0	0	0	0	0	0	0
Hypothetical protein GC26_73	0	94.7	0	0	0	0	0	0	0
Outer membrane protein GC26_66 (type 1)	0	0	0	0	0	0	0	0	0
Outer membrane protein GC26_66 (type 2)	0	0	94.92	95.94	96.45	95.43	93.40	96.45	94.44
Phospho-2-Dehydro-3-Deoxyheptonate Aldolase	99.33	99.0	0	0	0	0	0	0	0
Hypothetical protein GC26_33	100	0	0	0	0	0	0	0	0

Percentage of similarity >80% is considered significa.

Table 3: Significance of association of candidate genes with disease was examined by Student's t-test

	Student's t-test (P-value)	
Genes/Function	GC vs. PUD	GC vs. NUD
Membrane Protein GC26_77	0.000**	0.390
Hypothetical ATPase Protein GC26_69	0.000**	0.000**
Hypothetical protein GC26_73	0.000**	0.000**
Outer membrane protein GC26_66 (type 1)	0.282	0.005**
Phospho-2-Dehydro-3-Deoxyheptonate Aldolase	0.308	0.000**
Hypothetical protein GC26_33	0.022*	0.000**

*. P-value <0.05 is considered significant.
**. P-value <0.01 is considered highly significant.

Table 4: Structural and domain prediction of candidate genes by Blast2Go

Gene/Function	Accession No.	Gene Length (bp)	Hits	Min. e-value	Mean Similarity	Go ID	InterPro
Membrane Protein GC26_77	**KJ716798**	456	20	1.11E-70	92.2%	-	i. Transmembrane protein
Hypothetical ATPase Protein GC26_69	**KJ721220**	1683	20	0	86.35%	Fatty acid biosynthetic process: GTP catabolic process; GTPase activity; GTP binding	i. Dynamin, GTPase domain ii.P-loop containing nucleoside triphosphate hydrolase
Hypothetical protein GC26_73	**KJ721221**	150	6	5.58E-17	93.5%	-	i. Signal peptide
Outer membrane protein GC26_66	**KJ721222**	2394	20	5.99E-140	98.65%	-	i. *Helicobacter pylori* outer membrane protein
Phospho-2-Dehydro-3-Deoxyheptonate Aldolase	**KJ721223**	1353	20	0	99.05%	3-deoxy-7-phosphoheptulonate synthase activity; aromatic amino acid family biosynthetic process	DAHP synthetase, class II
Hypothetical protein GC26_33	**KJ721224**	168	20	4.57E-28	93.55%	-	i. Transmembrane protein

Table 5: Pearson's correlation coefficient test on relationship between candidate gene and disease and between different genes

		GC	PUD	NUD	Race	Membrane protein GC26_77	Hypothetical ATPase protein GC26_69	Hypothetical protein GC26_73	Outer membrane protein GC26_66 (type 1)	Phospho-2-dehydro-3-deoxyheptonate aldolase	Hypothetical protein GC26_33
GC	Pearson Correlation	1	−0.320	−0.402*	−0.346	0.287	0.760**	0.739**	0.546**	0.377	0.523**
	Sig. (2-tailed)		0.104	0.038	0.077	0.147	0.000	0.000	0.003	0.052	0.005
	N	27	27	27	27	27	27	27	27	27	27
PUD	Pearson Correlation	−0.320	1	−0.739**	−0.168	−0.996**	0.028	−0.138	0.363	0.527**	0.269
	Sig. (2-tailed)	0.104		0.000	0.403	0.000	0.890	0.491	0.063	0.005	0.175
	N	27	27	27	27	27	27	27	27	27	27
NUD	Pearson Correlation	−0.402*	−0.739**	1	0.408*	0.759**	−0.567**	−0.392*	−0.739**	−0.778**	−0.632**
	Sig. (2-tailed)	0.038	0.000		0.035	0.000	.002	0.043	0.000	0.000	0.000
	N	27	27	27	27	27	27	27	27	27	27
Race	Pearson Correlation	−0.346	−0.168	0.408*	1	0.176	−0.488**	−0.310	−0.636**	−0.474*	−0.549**
	Sig. (2-tailed)	0.077	0.403	0.035		0.380	0.010	0.116	0.000	0.012	0.003
	N	27	27	27	27	27	27	27	27	27	27
Membrane protein GC26_77	Pearson Correlation	0.287	−0.996**	0.759**	0.176	1	−0.054	0.113	−0.384*	−0.556**	−0.297
	Sig. (2-tailed)	0.147	0.000	0.000	0.380		0.789	0.575	0.048	0.003	0.132
	N	27	27	27	27	27	27	27	27	27	27
Hypothetical ATPase protein GC26_69	Pearson Correlation	0.760**	0.028	−0.567**	−0.488**	−0.054	1	0.667**	0.771**	0.528**	0.382*
	Sig. (2-tailed)	0.000	0.890	0.002	0.010	0.789		0.000	0.000	0.005	0.049
	N	27	27	27	27	27	27	27	27	27	27
Hypothetical protein GC26_73	Pearson Correlation	0.739**	−0.138	−0.392*	−0.310	0.113	0.667**	1	0.586**	0.531**	0.513**
	Sig. (2-tailed)	0.000	0.491	0.043	0.116	0.575	0.000		0.001	0.004	0.006
	N	27	27	27	27	27	27	27	27	27	27
Outer membrane protein GC26_66	Pearson Correlation	0.546**	0.363	−0.739**	−0.636**	−0.384*	0.771**	0.586**	1	0.687**	0.602**
	Sig. (2-tailed)	0.003	0.063	0.000	0.000	0.048	0.000	0.001		0.000	0.001
	N	27	27	27	27	27	27	27	27	27	27
Phospho-2-dehydro-3-deoxyheptonate aldolase	Pearson Correlation	0.377	0.527**	−0.778**	−0.474*	−0.556**	0.528**	0.531**	0.687**	1	0.740**
	Sig. (2-tailed)	0.052	0.005	0.000	0.012	0.003	0.005	0.004	0.000		0.000
	N	27	27	27	27	27	27	27	27	27	27
Hypothetical protein GC26_33	Pearson Correlation	0.523**	0.269	−0.632**	−0.549**	−0.297	0.382*	0.513**	0.602**	0.740**	1
	Sig. (2-tailed)	0.005	0.175	0.000	0.003	0.132	0.049	0.006	0.001	0.000	
	N	27	27	27	27	27	27	27	27	27	27

*. Correlation is significant at the 0.05 level (2-tailed).
**. Correlation is significant at the 0.01 level (2-tailed).

Gastric Cancer

A 456 bp gene encoding for a *H. pylori* membrane protein GC26_77 was absent in all (0/10) peptic ulcer disease (PUD) strains analyzed. This gene was present in all gastric cancer (GC) (4/4) and non-ulcer dyspepsia (NUD) (13/13) strains (Table 2). While no significant statistical significance (P-value ≥ 0.05) was observed between presence of this gene in strains isolated from GC and NUD patients, the gene was absent in PUD patients (P-value < 0.001) (Table 3). Pearson's correlation also revealed a strong negative correlation with PUD (Table 5).

Further, the presence of hypothetical ATPase protein GC26_69 was shown to correlate with strains isolated from GC patients. The ATPase gene was detected in 100% (4/4) GC strains as compared with 30% (3/10) of PUD strains and 0% (0/13) of isolates from NUD. Interestingly, the 3 isolates in which the hypothetical ATPase protein GC26_69 gene was detected originated from PUD patients from China, Japan and Korea. None of the PUD strains isolated from Malaysia carried this gene (Table 2). Highly significant associations were observed in the prevalence of this gene in isolates from GC patients as compared with isolates from PUD and NUD (P-value <0.001) (Table 3). The association of hypothetical ATPase protein GC26_69 with GC was confirmed by a strong positive correlation by Pearson's analysis (Table 5).

In addition a hypothetical protein GC26_73 was detected in all (4/4) GC strains but only 20% (2/10) and 7.7% (1/13) of PUD and NUD strains respectively (Table 2). Comparison of the prevalence of this gene in strains isolated from GC patients with that in both PUD and NUD patients showed a highly significant difference to exist (P-value <0.001) (Table 3). The association of hypothetical protein GC26_73 with GC was confirmed by a strong Pearson's positive correlation (Table 5).

Peptic Ulcer Disease and Gastric Cancer

Genes encoding for the *H. pylori* outer membrane protein GC26_66, phospho-2-dehydro-3-deoxyheptonate aldolase, and hypothetical protein GC26_33 were all highly associated with both GC and PUD strains. The gene encoding for outer membrane protein GC26_66 being present in all GC strains (4/4) and 90% (9/10) of the PUD strains. The outer membrane protein gene was not detected in any of the six Chinese NUD strains (Table 2). A highly statistically significant difference was observed between the prevalence of outer membrane protein in GC strains and Chinese NUD strains (P-value <0.001) (Table 3). While outer membrane protein GC26_66 was also present in NUD strains from local Malay and Indians, alignment of the translated protein sequence revealed that these local Malay and Indian strains from NUD patients differed from GC/PUD strains in the 2nd to 5th amino acid positions (Figure 2). For the purpose of discussion, we will differentiate the former as type 1 and the later as type 2. Outer membrane protein GC26_66 (type 1) was found to have strong negative Pearson's correlation with NUD (Table 5).

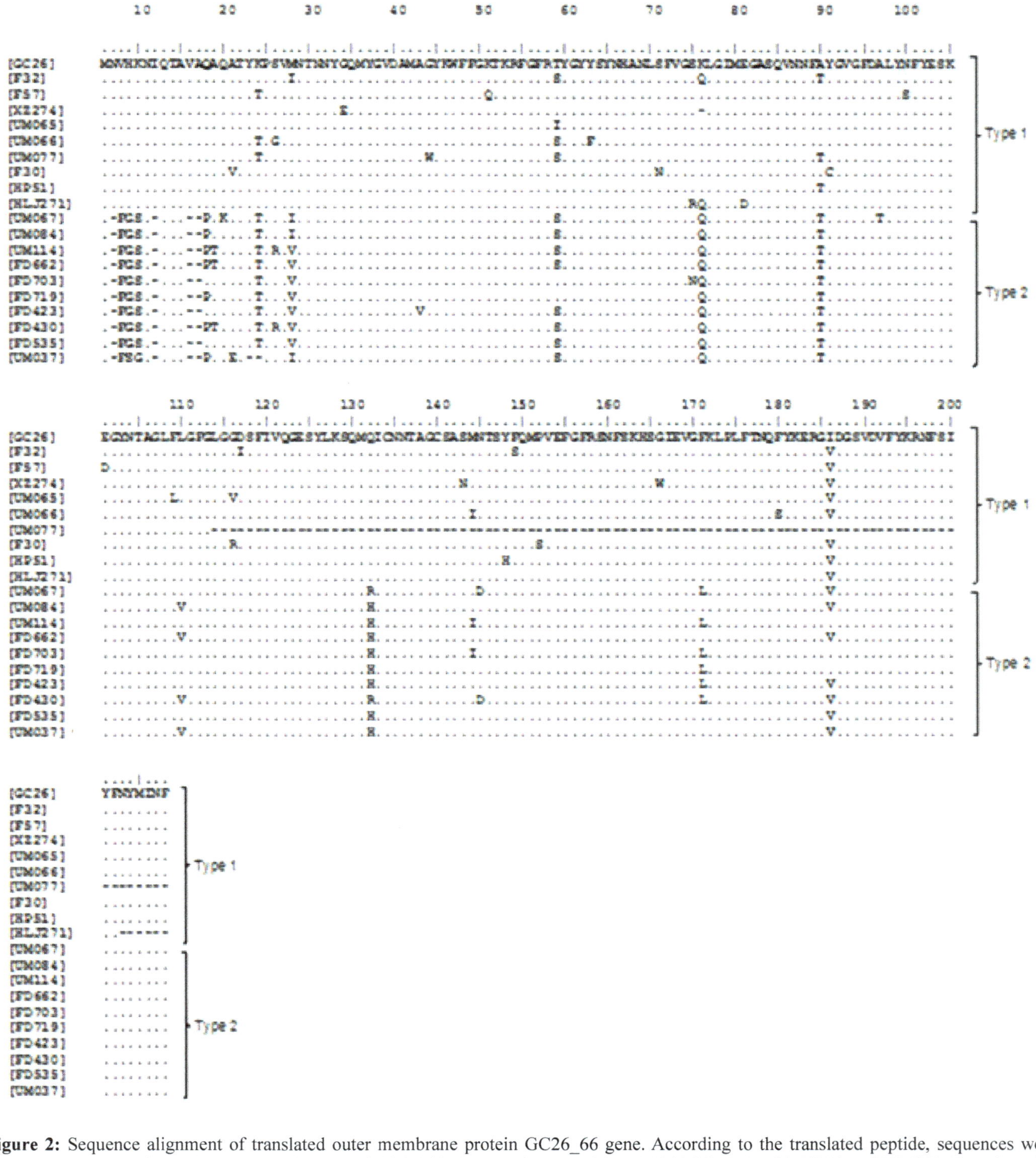

Figure 2: Sequence alignment of translated outer membrane protein GC26_66 gene. According to the translated peptide, sequences were grouped as type 1 and type 2. Type 1 strains were GC26, F32, F57, XZ274, UM065, UM066, UM077, F30, HP51 and HLJ271. Type 2 strains were UM067, UM084, UM114, FD662, FD703, FD719, FD423, FD430, FD535 and UM037. There was close correlation with host ethnicity.

The phospho-2-dehydro-3-deoxyheptonate aldolase gene was present in all of the GC strains (4/4) and 90% (9/10) of the PUD strains. In contrast, this gene was only found in 15.4% (2/13) of the NUD strains (Table 2). A statistically increased prevalence of this gene was observed in isolates from GC patients as compared with isolates from NUD (P-value <0.001) (Table 3). Phospho-2-dehydro-3-deoxyheptonate aldolase was also found to have strong negative Pearson's correlation with NUD (Table 5).

The gene encoding for hypothetical protein GC26_33 was present in all GC strains (4/4). 60% (6/10) of the PUD strains and only 7.7% (1/13) of the NUD strains (Table 2). Comparison of the prevalence of this gene in*H. pylori* strains isolated from GC and PUD patients showed a significantly increased prevalence (P-value <0.05) in those patients with GC. Comparison of strains isolated from GC and NUD patients showed a significantly increased prevalence in isolates from GC patients (P-value <0.001) (Table 3). Hypothetical protein GC26_33 was also found to have strong negative Pearson's correlation with NUD (Table 5).

DISCUSSION

In this study, we identified six genes that were associated with *H. pylori*-related disease status. In agreement with previous studies, no single *H. pylori* genetic marker on its own was shown to be associated with any specific disease group22·23. However, it is possible that these genes, working individually or collaborating with other genetic elements, are potential risk factors influencing the severity of disease or disease progression. Based on Pearson's analysis of the correlation between disease status and genes, as well as the correlation between different genes presented in Table 4, a correlation map was prepared to summarize the relationship between disease and genes (Figure 3). This map illustrates the complexity of *H. pylori* factors that may affect disease outcome. From the map, the number of genes that potentially influence disease status appears to increase with severity of disease. However, it should be noted that the correlation map only illustrates the co-existence of genes but does not imply that these genes actually interacts or work together to have an impact on disease development.

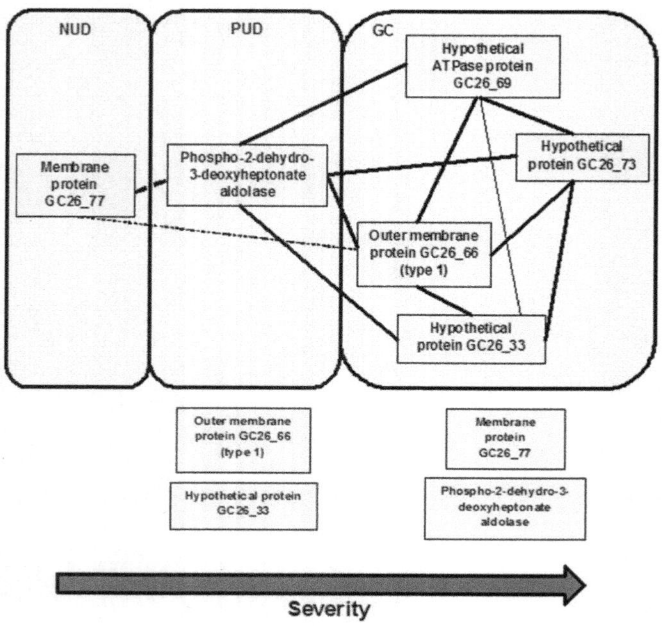

Figure 3: Correlation map of candidate genes with disease states and between genes. The correlation was determined by Pearson's correlation (Table 5). Dotted line represents negative correlation and full line represents positive correlation. Bold line represents strong correlation with significance <0.01 and fine line represents correlation with significance <0.05. Genes outside the boxes represent genes present in the respective disease strains (Student's t-test) but no significant correlation with respective disease status (Pearson's).

Based on our comparative genomic analysis, the presence of *H. pylori*phospho-2-dehydro-3-deoxyheptonate aldolase in the absence of membrane protein GC-26_77 was demonstrated to be a risk factor for the development of peptic ulcer disease. In contrast, outer membrane protein GC26_66 (type 1), hypothetical ATPase protein GC26_69, hypothetical protein GC26_73 and hypothetical protein GC26_33 were shown to be risk factors for gastric cancer development. Outer membrane GC26_77 may have a protective effect against peptic ulcer disease. However, these data were based on bioinformatic analysis of strains from Malaysia and few strains from other parts of Asia. There is a need to validate these results experimentally in the laboratory. Furthermore, it will be important to be able screen and estimate the carriage rate of these genes in a larger number of strains with different disease presentations from other parts of Asia as well as other parts of the world.

H. pylori is an ancient and permanent resident of the human stomach and has likely been part of the gastric microbiome since the origin of human species. Given this, it is not surprising that more than 80% of those infected with *H. pylori* remain asymptomatic. It has been estimated that *H. pylori*-positive individuals have a 10 to 20% lifetime risk of developing peptic ulcer disease and a 1 to 2% risk of developing gastric cancer24. While there is strong evidence to support the role of*H. pylori* in gastro-duodenal disease, there is some evidence that the decline in *H. pylori* infection among human populations has been suggested to have contributed to the increase in other diseases (e.g., esophageal adenocarcinoma [EAC], allergic asthma, rhinitis and atrophy)25. Furthermore, it has been reported that in adult males *H. pylori* colonization is associated with reduced circulating leptin levels, a finding that may explain the observation that significant weight gain may occur with *H. pylori* eradication26. Furthermore, Blaser, M.J. and Atherton, J.C. have suggested that the fall in *H. pylori* prevalence in developed countries may also contribute to high risk of metabolic syndrome, type II diabetes and metabolic obesity27. Current literature would suggest that infection with *H. pylori* is rapidly disappearing in developed countries28.

These risk factors can be potential biomarkers to identify those strains that present with higher risk of developing severe gastro-duodenal complications. The use of an approach where selective eradication therapy is given to individuals with these risk genes instead of all those infected with *H. pylori* will also help to preserve the usual gut microbiota. The understanding of *H. pylori* genetic factors and its association with severe gastro-duodenal diseases is necessary to decide on the optimal management of *H. pylori* infections.

A similar study has examined the genomic characteristics among 84 *H. pylori* isolates frm China with differing clinical status using microarray29. In this study, regions associated with genes involved in bacterial R-M systems and type IV secretion system were identified to be linked to disease status. However, these genes were not identified in this study, instead, a different set of genes were identified in this study. This inconsistency could be due to the different approach adopted and the strains selected for these studies. The microarray was designed based on 6 sequenced strains of European and American origins. Thus, genomic variations present only in Asian strains but not in Western strains will not be identified using the microarray approach. Although the next-generation sequencing approach is not limited by probe design, it can be limited by the quality of sequencing data. Also, the higher cost by the next-generation sequencing approach will also complicate mass screening of large number of strains.

In summary, we have identified a number of *H. pylori* genetic factors that may enable the identification of those at risk of peptic ulcer disease strains and gastric cancer using a comparative genomic approach. However, screening of a larger number of *H. pylori* strains from different disease groups is also required. Furthermore, it would be of interest to investigate the functions of these largely hypothetical proteins and how they interact to contribute to *H. pylori*-induced pathogenesis.

METHODS

Sample Background

Twenty-one *H. pylori* from Malaysia (GC26, UM023, UM065, UM066, UM077, UM067, UM084, UM114, FD506, FD568, FD577, UM038, UM085, UM111, FD662, FD719, FD703, FD423, FD430, FD535 and UM037) were isolated from symptomatic patients undergoing endoscopy procedure at University of Malaya Medical Centre (UMMC, Kuala Lumpur, Malaysia) between the years 2007 and 2012 (Table 1). Based on endoscopic and histological examinations, patients were diagnosed as having gastric cancer (GC), peptic ulcer disease (PUD) or non-ulcer dyspepsia/functional dyspepsia (NUD/FD). All biopsies were obtained with the informed consent of patients and approval of the Human Ethics Committees of UMMC and UNSW. The genomes of the twenty-one Malaysian *H. pylori* strains were sequenced and assembled *de novo*as previously described30. For comparison, we downloaded the genome sequences of another 6 Asian *H. pylori* strains from the National Center for Biotechnology Information (NCBI) GenBank. Further information of these strains are provided in Table 1. Only F30, F32, F57 and HP51 are complete genomes, the rest are available as drafts.

Sequence Alignment and Phylogenetic Analysis

In addition to the 27 genomes analyzed in this study, an additional of six strains regardless of East Asian origin was downloaded from NCBI database resulting in a total of 33 strains that were used to align with MAUVE (version 2.3.1) progressive alignment software and the output data was viewed using the SplitsTree program (version 4.12.8) as Super Network.

Downstream analysis

Annotation of all twenty-seven *H. pylori* strains were performed using the Rapid Annotation using Subsystem Technology (RAST) version 4.031. Comparative genomic analysis was performed using the sequence comparison function available on the SEED viewer version 2.0. A List of core genes present among all strains belonging to the same group was generated by SEED for each disease groups (GC, PUD and NUD). FASTA file containing the sequences these of core genes were generated for the respective disease groups. Further, RAST analysis was performed to this data for annotation and comparing against other *H. pylori* strains inTable 1 to obtain the percentage of similarity. The Student's unpaired two-tailed t-test was performed and the genes with *P-values* of < 0.05 and < 0.01were considered statistically significant and highly significant respectively. Two-tailed Pearsons' correlation analysis was adopted to examine for correlation between candidate genes and disease, as well as between different genes.

Accession Numbers

The accession numbers of the *H. pylori* genome sequences reported in this paper are: GC26 (AKHV00000000), UM023 (AUSK00000000), UM065 (AUSM00000000), UM066 (AUSJ00000000 and CP005493), UM077 (AUSQ00000000), UM067 (AUSN00000000), UM084 (AUSO00000000), UM114 (AUSS00000000), FD506 (AKHO00000000), FD568 (AKHQ00000000), FD577 (AKHR00000000), UM038 (AUSL00000000), UM111 (AUSR00000000), FD662 (AKHT00000000), FD719 (AKHU00000000), FD703 (AKHS00000000), FD423 (AKHM00000000), FD430 (AKHN00000000), FD535 (AKHP00000000), UM037 (AUSI00000000 and CP005492), UM085 (AUSP00000000), F30 (AP011941), F32 (AP011943), F57 (AP011945), HP51 (CP000012), XZ274 (CP003419) and HLJ271 (ALKB00000000).

REFERENCES

1. Israel, D. A. et al. Helicobacter pylori strain-specific differences in genetic content, identified by microarray, influence host inflammatory responses. J Clin Invest. 107, 611–620 (2001).
2. Chalker, A. F. et al. Systematic identification of selective essential genes in Helicobacter pylori by genome prioritization and allelic replacement mutagenesis. J Bacteriol. 183, 1259–1268 (2001).
3. Goh, K. L. & Parasakthi, N. The racial cohort phenomenon: seroepidemiology of Helicobacter pylori infection in a multiracial South-East Asian country. Eur J Gastroenterol Hepatol 13, 177–183 (2001).
4. Goh, K. L. Epidemiology of Helicobacter pylori infection in Malaysia-- observations in a multiracial Asian population. Med J Malaysia. 64, 187–192 (2009).
5. Musa, A. F., Yunos, M. N. M., Rahman, S. A. & Nordin, R. B. The seroprevalence and eradication success of Helicobacter pylori in indigenous people of Seletar in Southern Malaysia. Br J Med Med Res. 4, 1854–1863 (2014).
6. Tay, C. Y. et al. Population structure of Helicobacter pylori among ethnic groups in Malaysia: recent acquisition of the bacterium by the Malay population. BMC Microbiol 19, 126 (2009).
7. Falush, D. et al. Traces of human migrations in Helicobacter pylori populations. Science. 299, 1582–1585 (2003).
8. Breurec, S.et al. Evolutionary history of Helicobacter pylori sequences reflect past human migrations in Southeast Asia. PLoS One. 6, e22058 (2011).
9. Fock, K. M. & Ang, T. L. Epidemiology of Helicobacter pylori infection and gastric cancer in Asia. J Gastroenterol Hepatol. 25, 479–486 (2010).
10. Fock, K. M. et al. Asia-Pacific consensus guidelines on gastric cancer prevention. J Gastroenterol Hepatol 23, 351–365 (2008).
11. Conteduca, V.et al. H. pylori infection and gastric cancer: state of the art (review). Int J Oncol. 42, 5–18 (2013).
12. Argent, R. H., Hale, J. L., El-Omar, E. M. & Atherton, J. C. Differences in Helicobacter pylori CagA tyrosine

phosphorylation motif patterns between western and East Asian strains, and influences on interleukin-8 secretion. J Med Microbiol. 57, 1062–1067 (2008).

13. Schmidt, H. M. et al. The cag PAI is intact and functional but HP0521 varies significantly in Helicobacter pylori isolates from Malaysia and Singapore. Eur J Clin Microbiol Infect Dis. 29, 439–451 (2010).
14. Kim, S. Y., Lee, Y. C., Kim, H. K. & Blaser, M. J. Helicobacter pylori CagA transfection of gastric epithelial cells induces interleukin-8. Cell Microbiol. 8, 97–106 (2006).
15. Sahara, S.et al. Role of Helicobacter pylori cagA EPIYA motif and vacA genotypes for the development of gastrointestinal diseases in Southeast Asian countries: a meta-analysis. BMC Infect Dis. 12, 223 (2012).
16. Jones, K. R. et al. Polymorphism in the CagA EPIYA motif impacts development of gastric cancer. J Clin Microbiol. 47, 959–968 (2009).
17. Azuma, T. et al. Association between diversity in the Src homology 2 domain-- containing tyrosine phosphatase binding siteof Helicobacter pylori CagA protein and gastric atrophy and cancer. J Infect Dis. 189, 820–827 (2004).
18. Higashi, H.et al. Biological activity of the Helicobacter pylori virulence factor CagA is determined by variation in the tyrosine phosphorylation sites. Proc Natl Acad Sci USA. 99, 14428–14433 (2002)
19. Shiota, S., Suzuki, R. & Yamaoka, Y. The significance of virulence factors in Helicobacter pylori. J Dig Dis. 14, 341–349 (2013).
20. Choi, K. D.et al. Analysis of the 3' variable region of the cagA gene of Helicobacter pylori isolated in Koreans. Dig Dis Sci. 52, 960–966 (2007).
21. Zhu, Y. L., Zheng, S., Du, Q., Qian, K. D. & Fang, P. C. Characterization of CagA variable region of Helicobacter pylori isolates from Chinese patients. World J Gastroenterol. 11, 880–884 (2005).
22. Tan, H. J., Rizal, A. M., Rosmadi, M. Y. & Goh, K. L. Distribution of Helicobacter pylori cagA, cagE and vacA in different ethnic groups in Kuala Lumpur, Malaysia. J Gastroenterol Hepatol. 20, 589–594 (2005).
23. Ahmed, N., Tenguria, S. & Nandanwar, N. Helicobacter pylori - a seasoned pathogen by any other name. Gut Pathog. 1, 24 (2009).
24. Kusters, J. G., van Vliet, A. H. M. & Kuipers, E. J. Pathogenesis of Helicobacter pylori infection. Clin Microbiol Rev. 19, 449–90 (2006).
25. Blaser, M. J. Who are we? Indigenous microbes and the ecology of human diseases. EMBO Rep. 7, 956–960 (2006).
26. Roper, J. et al. Leptin and ghrelin in relation to Helicobacter pylori status in adult males. J Clin Endocrinol Metab. 93, 2350–2357 (2008).
27. Blaser, M. J. & Atherton, J. C. Helicobacter pylori persistence: biology and disease. J Clin Inves 113, 321–333 (2004).
28. Hadley, C. The infection connection. Helicobacter pylori is more than just the cause of gastric ulcers--it offers an unprecedented opportunity to study changes in human microecology and the nature of chronic disease. EMBO Rep. 7, 470–473 (2006).
29. You, Y. et al. Comparative Genomics of Helicobacter pylori Strains of China Associated with Different Clinical Outcome. PLoS One. 7, e38528 (2012).
30. Rehvathy, V. et al. Multiple Genome Sequences of Helicobacter pylori Strains of Diverse Disease and Antibiotic Resistance Backgrounds from Malaysia. Genome Announc. 1 (2013).
31. Aziz, R. K. et al. The RAST Server: rapid annotations using subsystems technology. BMC Genomics. 9, 75 (2008).
32. Kawai, M. et al. Evolution in an oncogenic bacterial species with extreme genome plasticity: Helicobacter pylori East Asian genomes. BMC Microbiol. 11, 104 (2011)
33. Guo, Y.et al. Genome of Helicobacter pyloristrain XZ274, an isolate from a Tibetan patient with gastric cancer in China. J Bacteriol 194, 4146–4147 (2012).
34. You, Y. et al. Genome sequences of three Helicobacter pylori strains isolated from atrophic gastritis and gastric ulcer patients in China. J Bacteriol. 194, 6314–6315 (2012).

CHAPTER 16

PSEUDOMONAS AERUGINOSA EVOLUTIONARY ADAPTATION AND DIVERSIFICATION IN CYSTIC FIBROSIS CHRONIC LUNG INFECTIONS

Craig Winstanley, Siobhan O'Brien, and Michael A. Brockhurst

PSEUDOMONAS AERUGINOSA INFECTION IN CYSTIC FIBROSIS

Cystic fibrosis (**CF**; see Glossary) is a debilitating, genetically inherited disease characterised by defects in a transport protein (the cystic fibrosis transmembrane regulator), resulting in sticky mucus, most notably in the respiratory tract [1. CF patients are susceptible to chronic lung infections, the predominant cause of the morbidity and mortality associated with the disease. The most common pathogen in this respect is *Pseudomonas aeruginosa*, a highly versatile bacterium capable of causing a wide range of mostly opportunistic infections, as well as occupying a variety of environmental niches [2. The ecological flexibility of *P. aeruginosa* can be attributed to its large genome (typically >6 Mb), which contains a particularly high proportion of regulatory genes, as well as a large number of genes involved in the catabolism, transport, and efflux of organic compounds [3, 4]. In CF, there has been some progress with development of aggressive early eradication therapies, whereby treatment is initiated as soon as the pathogen is detected, which delays the onset of chronic infection [5. However, once a chronic infection is established by*P. aeruginosa*, it is apparently impossible to eradicate.

During the course of chronic infection, CF patients produce samples (most commonly sputum) that are subjected to microbiological analysis for diagnostic purposes (identification of pathogens and antimicrobial susceptibility testing) and have proven to be a rich resource for researchers interested in analyses of the evolution of bacteria during chronic infection. As a discipline, microbiology has depended heavily on analysis of cultured organisms [6, and microbiologists are ingrained with the importance of obtaining single pure colonies. Hence, the study of bacterial pathogens during infection, including chronic CF infections, has relied heavily upon an assumption that bacterial populations at any given time are genetically uniform, at least at the level of 'strains', and that therefore it is justifiable to study and diagnose infections on the basis of single isolates. As a result, for many years researchers have assumed that it is possible to draw conclusions about the whole infecting bacterial population from the traits of single bacterial colonies. It is, however, increasingly clear that this assumption may not always be true.

In this review we discuss the phenotypic and genomic studies that have advanced our understanding of the adaptation and evolution of *P. aeruginosa* populations during chronic infections in the CF lung, highlighting the evidence demonstrating that infecting *P. aeruginosa* populations are highly diverse both genetically and phenotypically. We further discuss the causes and consequences of this diversity with respect to the underlying evolutionary processes and the clinical implications.

Citation: Craig Winstanley,@ Siobhan O'Brien, and Michael A. Brockhurst, Pseudomonas aeruginosa Evolutionary Adaptation and Diversification in Cystic Fibrosis Chronic Lung Infections, http://dx.doi.org/10.1016/j.tim.2016.01.008.

P. aeruginosa PHENOTYPIC ADAPTATIONS COMMONLY ASSOCIATED WITH CF INFECTIONS

TheCFlungisaheterogeneous,hostile,andstressfulenvironment for invading bacteria, and *P. aeruginosa* populations must overcome these challenges to persist and survive. Postulated stressors in the CF lung include osmotic stress [7 due to the viscous mucus, oxidative [8 and nitrosative [9 stresses due to host responses, sublethal concentrations of antibiotics [10, and the presence of other microorganisms [11, 12]. It has been recognised for many years that *P. aeruginosa* undergoes evolutionary changes in response to these selective forces during the chronic infection process. Phenotypic analysis of isolates show the emergence of mucoid colonies [13, caused by overproduction of the polysaccharide alginate, which is widely considered to be a marker for the transition to chronic infection. Alginate is one of three exopolysaccharides (along with Pel and Psl) that play important roles in the development and structural maintenance of a biofilm matrix that can offer *P. aeruginosa* protection from antibiotics and host responses [14. Other adaptations include the accumulation of auxotrophic mutations in the amino acid-rich lung environment [15, loss of motility [16, and the emergence of hypermutators [17, which display elevated mutation rates due to defects in DNA repair mechanisms. Given that CF patients are subjected to prolonged and often intensive therapy with antibiotics [18, the evolution of antibiotic resistances is also a common adaptation [19.

A recurring theme in the phenotypic analysis of CF isolates is the tendency for*P. aeruginosa* to become defective in terms of some of its key virulence factors (Box 1), such as type III secretion and the **quorum sensing (QS)** system (Figure 1). In CF, the accumulation of mutations in the gene encoding the key QS regulator, LasR, causing loss of QS regulation, is especially common [20. Other mutations often found associated with CF isolates of *P. aeruginosa* include mutations in *gacS* and *retS*, genes implicated in the switch between acute and chronic virulence (Box 1, Figure 1). Other regulators, such as AmpR [21, have also been implicated in this switch to chronicity, and it seems likely that a number of global regulatory systems may actually be involved. The accumulation of virulence factor mutations has been interpreted as *P. aeruginosa* adapting to lose its acute virulence during chronic infections [22. However, it is clear that many virulence factors (for example, QS-regulated factors and QS signal molecules) can still be detected in patient sputum samples during chronic infections [23.

Box 1. ***Pseudomonas aeruginosa*** **Virulence Factors**

P. aeruginosa virulence has largely been defined in terms of acute infections, for which models are readily available and loss of function mutations can be easily demonstrated [71–73]. The expression of virulence genes in P. aeruginosa is controlled by extremely complex, interweaving regulatory circuits [35] and multiple signalling systems [74]. For an opportunistic pathogen, P. aeruginosa produces an impressive array of particularly secreted virulence factors, utilising its type III secretion system to secrete various exotoxins (ExoS, ExoU, ExoT, and ExoY) [75], and quorum sensing (QS) systems (cell-density-dependent regulation) to control numerous important secreted virulence factors [23,76], including secreted pyocyanin, elastase, cyanide, and rhamnolipid. The secretion of toxins, coupled with phenotypes such as motility (swimming, swarming, and twitching), is considered to be important for acute infections. In contrast, apart from biofilm formation, virulence factors that are important in chronic infections are less well understood. It has been proposed that the GAC system network (incorporating the two-component regulatory system GacA/GacS, two other sensor kinases RetS/LadS, the small regulatory protein RsmA and the small RNAs RsmZ/RsmY) controls the reversible transition from acute to chronic infections [74]. This is based on a switch between acute virulence factors (swarming motility, lipase, rhamnolipids and type III secretion) and chronic virulence factors, such as biofilm formation, but also type VI secretion, and QS-regulated factors more traditionally thought of as involved in acute infections, namely pyocyanin and hydrogen cyanide. Although a role in acute infections has been demonstrated in animal models [77], because of the limitations of the available animal models of chronic infection, the role of the GAC system in chronic infection is less clear. However, an improved natural inhalation model that does not require the implantation of agar beads, and during which the bacteria adapt and exhibit features of chronic infections, has been developed [78]. This offers considerable potential to better characterise key chronic infection virulence factors.

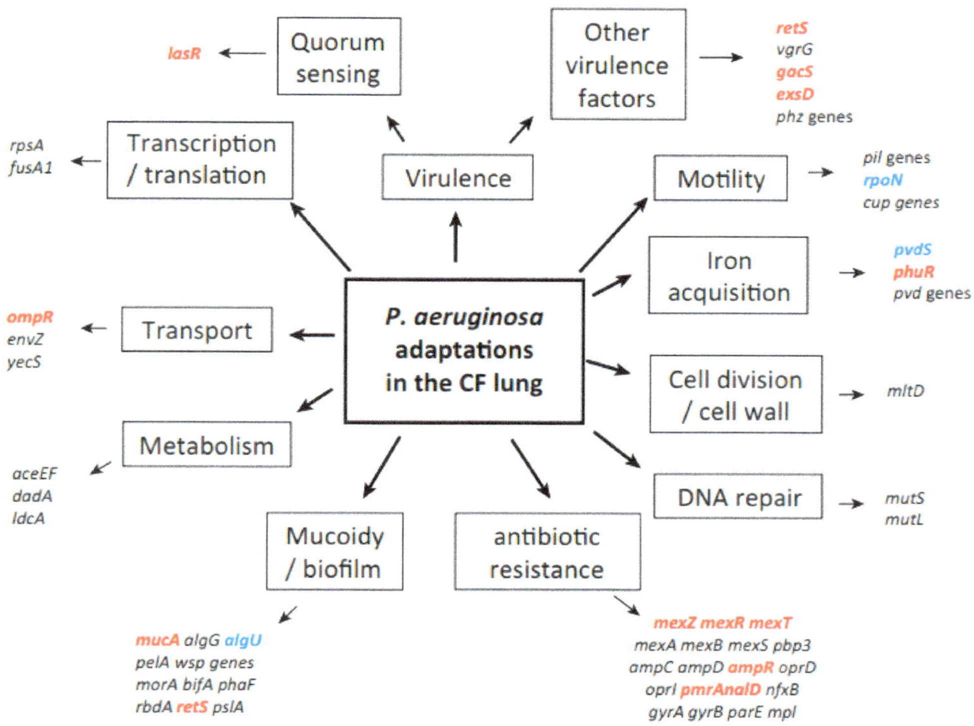

Figure 1. Pathoadaptive Mutations in Pseudomonas aeruginosa. Genes encoding regulatory proteins are highlighted in red. Genes encoding sigma factors are highlighted in blue.

GENOMIC ANALYSIS OF ADAPTATION USING SEQUENTIAL ISOLATES

The advent of affordable whole-genome sequencing technologies triggered considerable interest in using genomics to define the genetic basis of adaptations that occur during infections in the CF lung environment. In particular, there have been a number of studies reporting comparisons of clonally related longitudinal isolates, mostly contrasting the mutated genes in isolates from early and late in the infection process [24, 25, 26, 27, 28, 29]. These studies have revealed common mutations falling into various functional categories such as virulence (including QS and mucoidy), motility, transport, antibiotic resistance, iron acquisition, DNA replication or repair, transcription/translation, cell division or metabolism. In particular, mutations in genes encoding key global regulators are common (for example, *lasR*, *rpoN*, *mucA*, *mexT*, *retS*, *exsD*, and *ampR*; seeFigure 1). Together, the suite of traits affected by these mutations have been termed 'pathoadaptive' traits.

An analysis of an extensive retrospective collection of isolates, many of which represented a transmissible lineage (DK2), demonstrated that after initial transmission, sublineages evolved independently in patients, accumulating pathoadaptive mutations [25. It was further demonstrated that hypermutator lineages can coexist with nonhypermutators, developing distinct evolutionary pathways [30. Expanding this work to a study of 474 longitudinal isolates from 34 CF children and young adults, representing 36 different lineages of *P. aeruginosa*, the same group were able to show parallel evolution at 52 genes [31, suggesting common adaptations and constraints during the process of adaptation.

Each of these studies suggests evidence for adaptive evolution, with selection for mutations that are beneficial in the CF lung environment. For example, there is evidence for adaptation towards iron acquisition from haemoglobin repeatably and independently across multiple patients [32. Notably, mutations of the complex *P. aeruginosa* regulatory and metabolic networks in the lung environment are likely to extensively modify gene expression levels and alter metabolic fluxes. Mutations occuring early in the infection, and therefore presumably the most beneficial, are located in global regulatory network control hubs [33, with other mutations occurring later and leading to fine-tuning [33. However, it is also notable that the pathoadaptive mutations are not consistent between studies, suggesting the existence of multiple evolutionary trajectories to pathoadaptation [34. Observations in CF suggest that this may be a result of the inherent complexity of *P. aeruginosa* regulatory networks [35: similarly beneficial effects can result from different mutations, or combinations of mutations in regulators, leading to changes in multiple processes. As a result, different mutations may converge upon similar phenotypes and levels of increased fitness. Likewise, there may be epistatic interactions among mutations such that particular combinations of mutations are required, making the evolutionary trajectory within a given patient highly contingent upon which early mutations arise and reach fixation [34.

The tendency for *P. aeruginosa* to acquire loss-of-function mutations during adaptation to the CF lung could suggest that it is travelling towards an evolutionary 'dead-end'. However, the existence of transmissible strains, such as the DK2 lineage and the **Liverpool epidemic strain** (LES) [36, 37], argues that this is not always the case. Anecdotally, the LES is most likely to infect patients already infected with another *P. aeruginosa* strain. It is conceivable, therefore, that transmissible strains have acquired mutations that not only favour transmission but also enhance competitive ability in the lung.

WHAT ARE THE DRIVERS OF PATHOADAPTATION?

Parallel evolution of particular traits or genes independently in multiple patients is strongly suggestive of positive selection at these loci, leading to the identification of the suite of pathoadaptive traits. Thus, we now have detailed phenotypic and genetic descriptions of how natural selection targets *P. aeruginosa* populations in the CF lung, but we still lack a full understanding of why these particular traits and genes are experiencing selection. This is in part due to the fact that the CF host environment is highly complex, and in part a reflection of our incomplete understanding of the physiology of the bacteria. Perhaps the clearest case where the selective force can be linked to the evolutionary response is for antibiotic resistance evolution: for example, in a recent genomics study of *P. aeruginosa* adapting to the CF lung, the fitness of particular alleles at the penicillin-binding protein 3 could be linked to the use of particular antibiotics [38. Similarly, in *in vitro* evolution experiments, stereotypical resistance mutations become enriched under antibiotic selection, clearly establishing causality [39. Other traits are less easily associated to particular selective causes. For example, it has been suggested that the loss of virulence-associated and motility traits is a response to immune selection [16; however, there have been few direct tests of this hypothesis. Indeed, a recent study using nematode hosts showed no effect of host immunity on the trajectory of *P. aeruginosa* adaptation: virulence traits were lost with and without immune selection [40. Similarly, adaptation to the CF-like conditions of artificial sputum medium selects for mutations that cause a switch to an immotile biofilm lifestyle [39, suggesting that the sputum environment itself is sufficient to select against motility. Moreover, growth of *P. aeruginosa* in flow cells selects for mutations causing mucoidy and loss of pilus-dependent motility, suggesting that simply dwelling in a biofilm is sufficient to cause the evolution of these characteristic CF-associated phenotypes [41. Other common adaptations include changes in metabolism, DNA repair, and iron acquisition (Figure 1). However, there is a clear need for careful experiments testing evolutionary hypotheses about the drivers of selection within the host which disentangle this complex multifaceted environment. Box 2 outlines our current knowledge with respect to adaptations driven by social interactions, demonstrating how experimental evolution can help us to both generate and test hypotheses.

Box 2. Adaptation in the Context of Social Evolution

The transition from acute to chronic infection can radically alter costs and benefits associated with the secretion of extracellular metabolites. Some metabolites are costly to produce when not needed, which may explain the redundancy of virulence-associated genes observed during cystic fibrosis (CF) infections. However, secreted metabolites are often costly to produce, so even when they are crucial for growth their loss might be selectively advantageous – providing there remain enough producing individuals to compensate for this small reduction in metabolite concentration. Hence, nonproducers ('cheats') can still reap the benefits of metabolites produced by others ('cooperators'), while paying little or no cost [79]. For instance, individuals that produce the costly iron-scavenging siderophore pyoverdine can be exploited by pyoverdine-negative mutants, who benefit from increased iron-acquisition without paying a cost. Experimental evolution has revealed that in spite of cheats exhibiting reduced relative fitness in an iron-limited medium, they can spread through a population of cooperators, over a matter of days [80,81]. A recent significant paper shows that social selection may be similarly resulting in pyoverdine mutants in the CF lung [82]. Loss-of-function mutations occur in pyoverdine synthesis genes, but not in the receptor gene (at least as long as producers and nonproducers coexist together), indicating that the loss of pyoverdine in vivo is not driven by disuse, but by loss-due-to-cheating. Cheats can only exploit producers as long as they maintain a functional receptor. If loss were a result of disuse, mutations would be predicted to accumulate in both receptor and synthesis genes. However, with the notable exception of pyoverdine, we know very little in practice about the role of social evolution in shaping trait loss in the context of the CF lung. It could be tentatively suggested that virulence factors and secreted toxins may benefit bacteria other than the producer, and hence may also be vulnerable to nonproducing cheats. If this is the case, the dominance of the 'loss-due-to-disuse' framework for explaining loss of secreted traits in CF lung infections may be open to uncertainty

COMPLEXITY OF *P. AERUGINOSA* POPULATIONS IN THE CF LUNG

It has been known anecdotally for many years that *P. aeruginosa* populations in CF can be diverse in terms of phenotypes such as colony morphology (e.g., coexistence of mucoid and nonmucoid colonies). Recent detailed analyses have revealed that there is extensive phenotypic heterogeneity within populations of *P. aeruginosa* in the CF lung beyond their colony morphology [42, 43, 44, 45, 46, 47, 48, 49] (Figure 2, Key Figure). Importantly, *P. aeruginosa* populations exhibit within-population diversity in many of the phenotypes observed to be altered during evolutionary adaptation, including motility, virulence factor production, siderophore production, antibiotic resistance, auxotrophy, and hypermutability. Hence, although some members of the population have acquired mutations affecting these phenotypes, these mutants coexist in patients alongside other genotypes that have not (Figure 3). The most practical clinically relevant consequence of this phenotypic diversity is in relation to antimicrobial susceptibility testing in diagnostic laboratories, which is typically carried out on either single isolates, or two colonies acting as representatives of mucoid and nonmucoid colony morphotypes. However, there is almost always considerable diversity in the antimicrobial susceptibilities within the population isolated from an individual sputum sample (and within colony morphotypes) [43, 45] (Figure 3). Hence, it is perhaps not surprising that therapy based on antimicrobial testing is a poor predictor of clinical outcome [50, 51].

This closer scrutiny of *P. aeruginosa* within-population diversity has been extended to genomic analysis, with reports that contemporary within-patient genomic diversity between isolates can be comparable to the variation reported between sequential isolates [52. Further studies have built upon this by analysing larger numbers of isolates per sample, to characterise the diversity present within populations [44, 53]. By sequencing *P. aeruginosa* populations from a group of patients infected with the same strain (the LES), it was possible to demonstrate the coexistence of divergent sublineages within individual patients, strongly suggesting the likelihood of ongoing transmission between patients [53. Data from this study also demonstrated that, although mutations in some genes are common, they are neither always present in all patients, nor carried by all members of the *P. aeruginosa* population within individual patients (Figure 4). Further evidence for the merits of studying fine-scale evolutionary dynamics was provided in another recent paper, which reported deep analysis of 12 sputum samples isolated from one patient over the course of a year [38, with the evolutionary emergence of two clonal sublineages within the patient.

This observed diversification is consistent with the idea that genetically diverged isolates coexist and interact within an ecologically cohesive population in the lung. However, there is likely to be a role for spatial structure and environmental heterogeneity within the lung environment in the origin and maintenance of the observed genetic diversity. Different regions of lung tissue are likely to vary along a range of environmental axes, including variations in mixtures of nutrients, concentrations of penetrating antibiotics, other microorganisms, or factors such as oxygen availability, potentially leading to spatially variable selection for different ecotypes. At a macro-scale this is supported by the coexistence of distinct sublineages associated with different parts of the airway in patients (paranasal sinus vs lung) [54. Moreover, in a recent landmark paper, different regions of explanted lungs from chronically infected CF patients were analysed to study the variations in clonally related isolates in great detail [55. The *P. aeruginosa* isolates occupying different regions of the lung had evolved independently, and they differed in phenotypic characteristics such as nutritional requirements, antibiotic resistance, and virulence. The study, however, reported that there was limited intermixing between these separate communities, suggesting that regional isolation mediated by spatial structure promotes both the origin and the maintenance of this diversity [55. It is important to note that spatial divergence of this kind could arise simply due to random drift rather than any adaptive divergence.

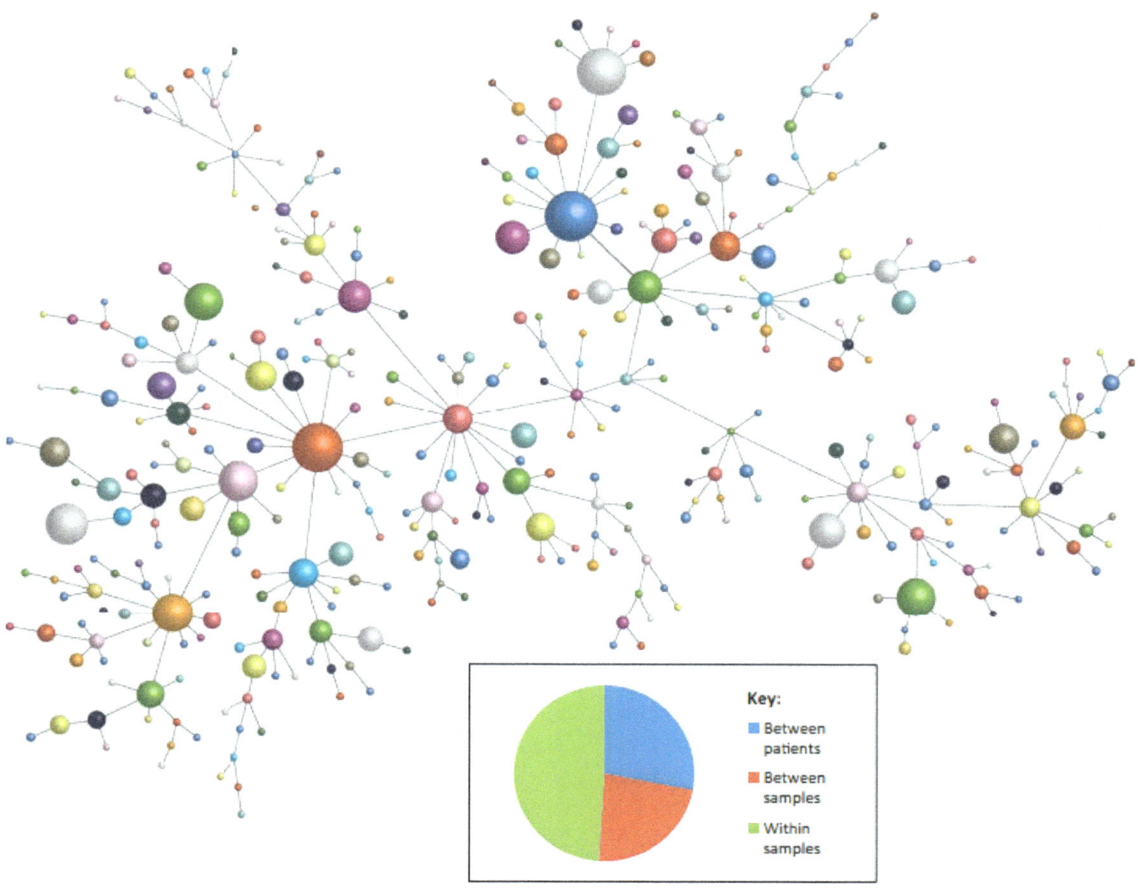

Figure 2. The figure shows a population structure based on 15 variable traits using the eBURST algorithm [83]. From ten patients infected with the Liverpool epidemic strain (LES) of P. aeruginosa, 1720 isolates from 43 different sputum samples were analysed, giving rise to 398 unique 'subtypes' of the LES. Each sphere represents a different subtype. The size reflects the relative abundance of each subtype. Two subtypes connected by a single line differ in only one characteristic. The pie chart inset indicates the percentage contribution to diversity of variation between patients, between samples or within samples, demonstrating the major contribution of the latter. Adapted from [43].

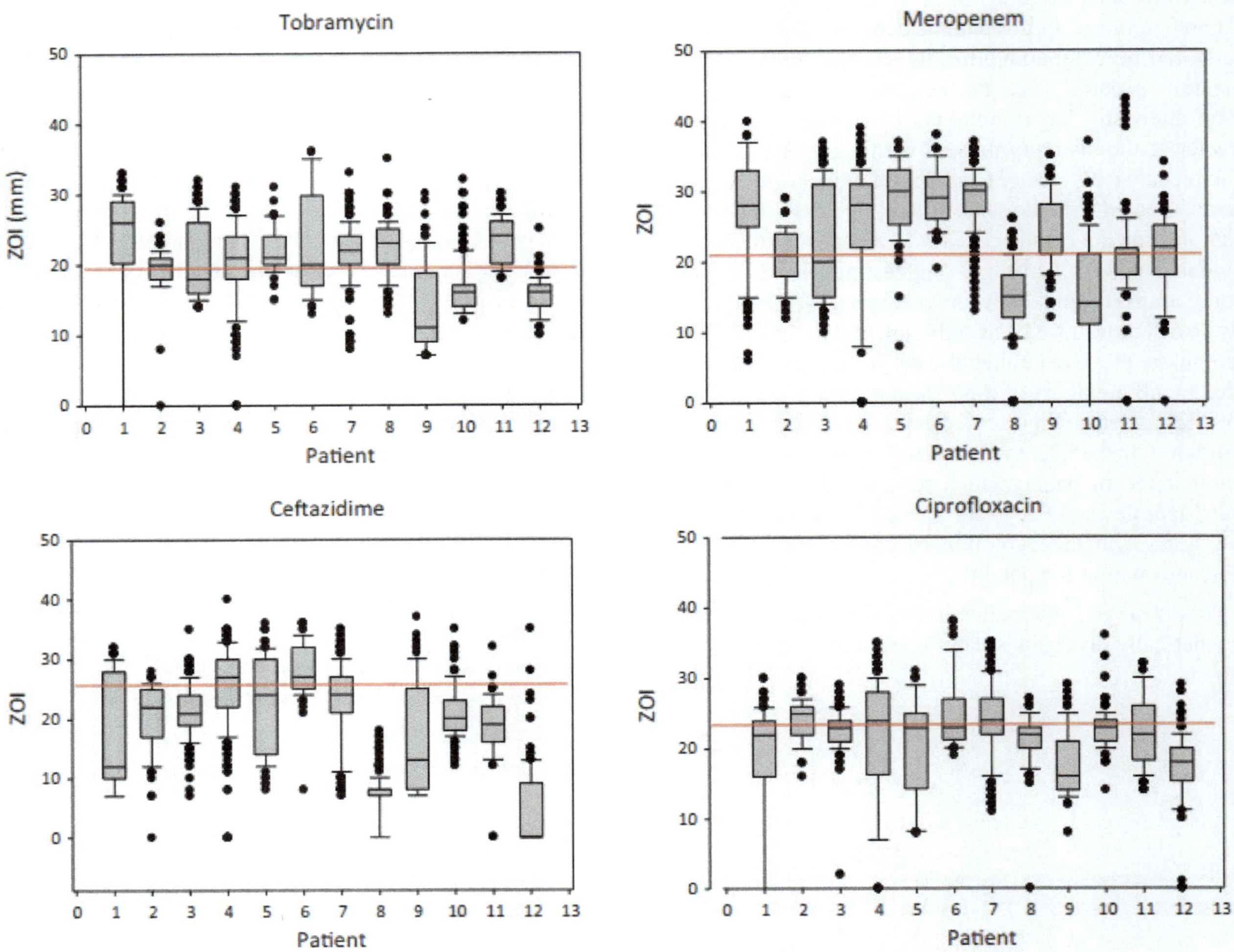

Figure 3. Variations in the Antimicrobial Susceptibilities of Pseudomonas aeruginosa within Individual Patients. The figure summarises, for four antibiotics, the spread of zone of inhibition data (ZOI) in mm for multiple isolates taken from 13 patients infected with the Liverpool epidemic strain (LES). For each patient, a minimum of 80 isolates was analysed, taken at multiple sampling points (40 isolates per sample point). The red line indicates the recognised cut-off points as defined by the British Society for Antimicrobial Susceptibility [84]. Note the tendency for isolates from the same patient to occur both above (susceptible) and below (resistant) the red line. Data adapted from two studies [43,46].

Therefore, upon current evidence, it appears that the observed diversity in CF lungs is the result of both adaptive (spatially heterogeneous selection driving the evolution of different ecotypes in different regions of the lung) and nonadaptive (genetic drift due to spatial structure combined with limited mixing) processes that promote the origin and coexistence of genotypes. Although recombination can also play a role in the diversification process [44, 52, 53, 56], there is disagreement about the extent of its contribution. One study suggested that recombination is a key driver of genomic and phenotypic diversity [44. However, refined analysis of these data [57, and the levels reported in other studies, suggest that recombination rates are low. Although experimental tests are limited, it has been shown in several studies that genetic diversity readily evolves in spatially structured biofilm populations [41, 58, 59, 60]. Moreover, several studies have demonstrated a role for stressors likely to occur in the CF lung in selecting for the evolution of diversity in *P. aeruginosa* biofilms (oxidative stress [59) and in populations growing in a CF-like artificial sputum medium environment (subinhibitory concentrations of antibiotics [61). The complexity of the system is emphasised by the fact that diversity can occur both within a biofilm population in a ‘microcompartment’ and in the wider CF airway environment, where physically separate biofilm populations can be exposed to different environmental conditions, including variable antibiotic concentrations [62. Furthermore, as the lung tissue deteriorates over time during CF disease progression, it is likely that the selective forces operating on the pathogen population vary, such that mutations beneficial earlier in the infection may be less favoured later.

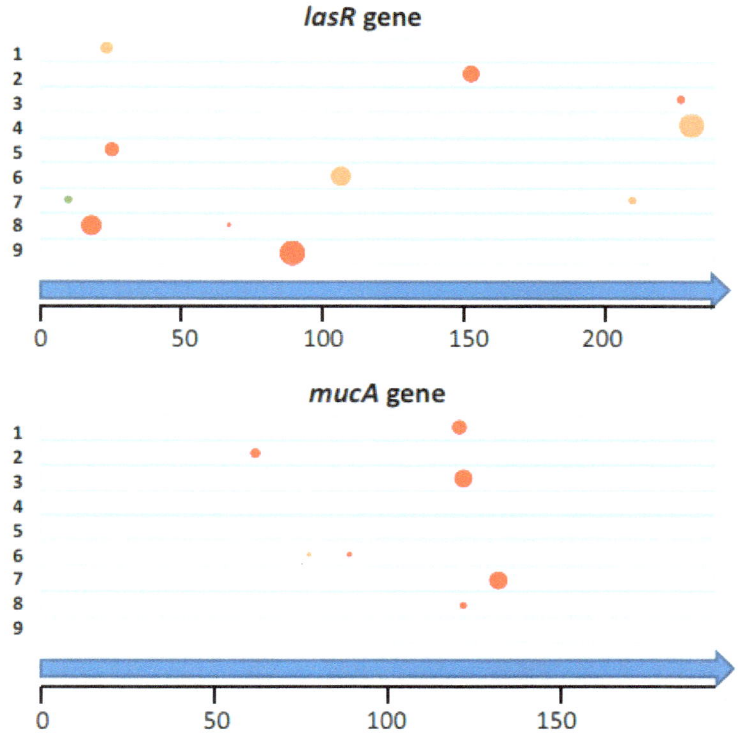

Figure 4. Within-patient Variations in the Prevalence and Location within a Gene of Common Pathoadaptive Mutations. Based on sets of 40 isolates from a single sputum sample, for each of nine patients infected with the Liverpool epidemic strain, the prevalence and location of mutations is indicated for the lasR and mucA genes (with location relative to amino acid position on the predicted protein sequence indicatedon the scale).Each patient is represented by the space between lines. Each circle represents the location of a mutation that is either severe (red; e.g., a frame-shift), a nonsynonymous single nucleotide polymorphism (orange; e.g., a single amino acid change), or a change that would not impact on the protein sequence (green). The size of the circle reflects the relative abundance in each set of 40 isolates (e.g., the severe lasR mutation in patient 9 was present in all 40 isolates tested). Analysis of data from a published study [53].

The effects of the many selective forces likely to operate in the lung remain to be experimentally tested. These include the other members of the complex multispecies microbial communities (the microbiome) [63, the host immune system, and phage (which can occur at high densities in CF patient samples [64). Experimental evolution approaches are likely to be a powerful tool for shedding light on these issues.

CONCLUDING REMARKS

Driven by the advent of affordable high-throughput genome sequencing, there has been rapid progress in our understanding of how *P. aeruginosa* adapts and evolves in the context of chronic CF lung infections. More recently, this has extended to fine-scale analysis of evolutionary dynamics within infecting populations, revealing high levels of coexisting genetic and phenotypic diversity, including at clinically important traits. The clinical consequences are not fully understood, though, given the extensive phenotypic diversity, there are clear implications for false diagnoses based upon antimicrobial susceptibility testing using single/pairs of isolates. Given the limited efficacy of current antibiotics in these chronically infected patients, it is important that we improve our understanding of the evolution of bacterial populations during chronic infections in order to design better strategies for clinical intervention (see Outstanding Questions). For example, the issue of whether the loss of social traits in *P. aeruginosa* populations is a result of disuse or cheating has relevance that extends beyond the field of evolutionary biology *per se.* If cheating can indeed accelerate the loss of virulence-associated metabolites [65, 66], meddling in the social lives of microbes could present a novel strategy for the development of virulence-attenuating therapeutics, as well as controlling the spread of infectious diseases [67, 68]. In particular, there has been particular emphasis on developing novel antivirulence therapeutics, especially strategies aimed at inhibition of the *P. aeruginosa* QS system [69, 70].

Given the tools at our disposal, we are now well placed to undertake a detailed characterisation of the structure and dynamics of bacterial populations during infections, and it is likely that a huge amount of data will be published over the coming years. However, it is also important that we accelerate our understanding of the fundamental biology underlying the complexity of pathogen physiology and evolution. One key approach to bridging the gap between *in vivo* observational data and an understanding of the biological mechanisms is to use data from clinical samples to generate hypotheses that can be tested subsequently using experimental evolution. By using an iterative cycle of these kinds of approaches we have the potential to unpick the, often daunting, complexity of real microbial populations during the infection process.

ACKNOWLEDGMENTS

C.W. and M.A.B. would like to acknowledge the support of the Wellcome Trust (Project Grants 089215/Z/09/Z and 093306/Z/10). S.O.B. is supported by the Wellcome Trust (ref: 105624) through the Centre for Chronic Diseases and Disorders (C2D2) at the University of York.

REFERENCES

1. Cohen, T.S. and Prince, A. (2012) Cystic fibrosis: a mucosal immunodeficiency syndrome. Nat. Med. 18, 509–519
2. Lyczak, J.B. et al. (2000) Establishment of Pseudomonas aeruginosa infection: lessons from a versatile opportunist. Microbes Infect. 2, 1051–1060
3. Stover, C.K. et al. (2000) Complete genome sequence of Pseudomonas aeruginosa PAO1, an opportunistic pathogen. Nature 406, 959–964
4. Silby, M.W. et al. (2011) Pseudomonas genomes: diverse and adaptable. FEMS Microbiol. Rev. 35, 652–680
5. Langton Hewer, S.C. and Smyth, A.R. (2014) Antibiotic strategies for eradicating Pseudomonas aeruginosa in people with cystic fibrosis. Cochrane Database Syst. Rev. 11, CD004197
6. Burns, J.L. and Rolain, J.M. (2014) Culture-based diagnostic microbiology in cystic fibrosis: can we simplify the complexity? J. Cyst. Fibrosis 13, 1–9
7. Brocker, C. et al. (2012) The role of hyperosmotic stress in inflammation and disease. Biomol. Concepts 3, 345–364
8. Hector,A. et al. (2014)Oxidative stress in cystic fibrosis lung disease: an early event, but worth targeting? Eur. Respir. J. 44, 17–19
9. Wood, S.R. et al. (2007) Nitrosative stress inhibits production of the virulence factor alginate in mucoid Pseudomonas aeruginosa. Free Radic. Res. 41, 208–215
10. Andersson, D.I. and Hughes, D. (2014) Microbiological effects of sublethal levels of antibiotics. Nat. Rev. Microbiol. 12, 465–478
11. Fodor, A.A. et al. (2012) The adult cystic fibrosis airway microbiota is stable over time and infection type, and highly resilient to antibiotic treatment of exacerbations. PLoS ONE 7, e45001
12. Lopes, S.P. et al. (2014) Microbiome in cystic fibrosis: shaping polymicrobial interactions for advances in antibiotic therapy. Crit. Rev. Microbiol. 41, 353–565
13. Govan, J.R. and Deretic, V. (1996) Microbial pathogenesis in cystic fibrosis: mucoid Pseudomonas aeruginosa and Burkholderia cepacia. Microbiol. Rev. 60, 539–574
14. Bjarnsholt, T. et al. (2009) Pseudomonas aeruginosa biofilms in the respiratory tract of cystic fibrosis patients. Pediatr. Pulmonol. 44, 547–558
15. Barth, A.L. and Pitt, T.L. (1996) The high amino-acid content of sputum from cystic fibrosis patients promotes growth of auxotrophic Pseudomonas aeruginosa. J. Med. Microbiol. 45, 110–119
16. Mahenthiralingam, E. et al. (1994) Nonmotility and phagocytic resistance of Pseudomonas aeruginosa isolates from chronically colonized patients with cystic fibrosis. Infect. Immun. 62, 596–605
17. Oliver, A. et al. (2000) High frequency of hypermutable Pseudomonas aeruginosa in cystic fibrosis lung infection. Science 288, 1251–1254
18. Doring, G. et al. (2000) Antibiotic therapy against Pseudomonas aeruginosa in cystic fibrosis: a European consensus. Eur. Respir. J. 16, 749–767
19. Breidenstein, E.B. et al. (2011) Pseudomonas aeruginosa: all roads lead to resistance. Trends Microbiol. 19, 419–426
20. Hoffman, L.R. et al. (2009) Pseudomonas aeruginosa lasR mutants are associated with cystic fibrosis lung disease progression. J. Cyst. Fibrosis 8, 66–70
21. Balasubramanian, D. et al. (2015) Pseudomonas aeruginosa AmpR: an acute–chronic switch regulator. Pathog. Dis. 73, 1–14
22. Nguyen, D. and Singh, P.K. (2006) Evolving stealth: genetic adaptation of Pseudomonas aeruginosa during cystic fibrosis infections. Proc. Natl. Acad. Sci. U.S.A. 103, 8305–8306
23. Winstanley, C. and Fothergill, J.L. (2009) The role of quorum sensing in chronic cystic fibrosis Pseudomonas aeruginosa infections. FEMS Microbiol. Lett. 290, 1–9
24. Smith, E.E. et al. (2006) Genetic adaptation by Pseudomonas aeruginosa to the airways of cystic fibrosis patients. Proc. Natl. Acad. Sci. U.S.A. 103, 8487–8492
25. Yang, L. et al. (2011) Evolutionary dynamics of bacteria in a human host environment. Proc. Natl. Acad. Sci. U.S.A. 108, 7481–7486
26. Bragonzi, A. et al. (2009) Pseudomonas aeruginosa microevolution during cystic fibrosis lung infection establishes clones with adapted virulence. Am. J. Respir. Crit. Care Med. 180, 138–145
27. Cramer, N. et al. (2011) Microevolution of the major common Pseudomonas aeruginosa clones C and PA14 in cystic fibrosis lungs. Environ. Microbiol. 13, 1690–1704
28. Hoboth, C. et al. (2009) Dynamics of adaptive microevolution of hypermutable Pseudomonas aeruginosa during chronic pulmonary infection in patients with cystic fibrosis. J. Infect. Dis. 200, 118–130
29. Marvig, R.L. et al. (2015) Within-host microevolution of Pseudomonas aeruginosa in Italian cystic fibrosis patients. BMC Microbiol. 15, 218
30. Marvig, R.L. et al. (2013) Genome analysis of a transmissible lineage of Pseudomonas aeruginosa reveals pathoadaptive mutations and distinct evolutionary paths of hypermutators. PLoS Genet. 9, e1003741
31. Marvig, R.L. et al. (2015) Convergent evolution and adaptation of Pseudomonas aeruginosa within patients with cystic fibrosis. Nat. Genet. 47, 57–64
32. Marvig, R.L. et al. (2014) Within-host evolution of Pseudomonas aeruginosa reveals adaptation toward iron acquisition from hemoglobin. mBio 5, e00966-14
33. Folkesson, A. et al. (2012) Adaptation of Pseudomonas aeruginosa to the cystic fibrosis airway: an evolutionary perspective. Nat. Rev. Microbiol. 10, 841–851
34. Marvig, R.L. et al. (2015) Evolutionary insight from whole-genome sequencing of Pseudomonas aeruginosa from cystic fibrosis patients. Future Microbiol. 10, 599–611

35. Balasubramanian, D. et al. (2013) A dynamic and intricate regulatory network determines Pseudomonas aeruginosa virulence. Nucleic Acids Res. 41, 1–20

36. Fothergill, J.L. et al. (2012) Transmissible strains of Pseudomonas aeruginosa in cystic fibrosis lung infections. Eur. Respir. J. 40, 227–238

37. Winstanley, C. et al. (2009) Newly introduced genomic prophage islands are critical determinants of in vivo competitiveness in the Liverpool Epidemic Strain of Pseudomonas aeruginosa. Genome Res. 19, 12–23

38. Diaz Caballero, J. et al. (2015) Selective sweeps and parallel pathoadaptation drive Pseudomonas aeruginosa evolution in the cystic fibrosis lung. mBio 6, e00981-15

39. Wong, A. et al. (2012) Genomics of adaptation during experimental evolution of the opportunistic pathogen Pseudomonas aeruginosa. PLoS Genet. 8, e1002928

40. Jansen, G. et al. (2015) Evolutionary transition from pathogenicity to commensalism: global regulator mutations mediate fitness gains through virulence attenuation. Mol. Biol. Evol. 32, 2883– 2896

41. McElroy, K.E. et al. (2014) Strain-specific parallel evolution drives short-term diversification during Pseudomonas aeruginosa biofilm formation. Proc. Natl. Acad. Sci. U.S.A. 111, E1419–E1427

42. Foweraker, J.E. et al. (2005) Phenotypic variability of Pseudomonas aeruginosa in sputa from patients with acute infective exacerbation of cystic fibrosis and its impact on the validity of antimicrobial susceptibility testing. J. Antimicrob. Chemother. 55, 921–927

43. Mowat, E. et al. (2011) Pseudomonas aeruginosa population diversity and turnover in cystic fibrosis chronic infections. Am. J. Respir. Crit. Care Med. 183, 1674–1679

44. Darch, S.E. et al. (2015) Recombination is a key driver of genomic and phenotypic diversity in a Pseudomonas aeruginosa population during cystic fibrosis infection. Sci. Rep. 5, 7649

45. Workentine, M.L. et al. (2013) Phenotypic heterogeneity of Pseudomonas aeruginosa populations in a cystic fibrosis patient. PLoS ONE 8, e60225

46. Fothergill, J.L. et al. (2010) Fluctuations in phenotypes and genotypes within populations of Pseudomonas aeruginosa in the cystic fibrosis lung during pulmonary exacerbations. J. Med. Microbiol. 59, 472–481

47. Ashish, A. et al. (2013) Extensive diversification is a common feature of Pseudomonas aeruginosa populations during respiratory infections in cystic fibrosis. J. Cyst. Fibrosis 12, 790–793

48. Clark, S.T. et al. (2015) Phenotypic diversity within a Pseudomonas aeruginosa population infecting an adult with cystic fibrosis. Sci. Rep. 5, 10932

49. Wilder, C.N. et al. (2009) Instantaneous within-patient diversity of Pseudomonas aeruginosa quorum-sensing populations from cystic fibrosis lung infections. Infect. Immun. 77, 5631–5639

50. Hurley, M.N. et al. (2012) Results of antibiotic susceptibility testing do not influence clinical outcome in children with cystic fibrosis. J. Cyst. Fibrosis 11, 288–292

51. Smith, A.L. et al. (2003) Susceptibility testing of Pseudomonas aeruginosa isolates and clinical response to parenteral antibiotic administration: lack of association in cystic fibrosis. Chest 123, 1495–1502

52. Chung, J.C. et al. (2012) Genomic variation among contemporary Pseudomonas aeruginosa isolates from chronically infected cystic fibrosis patients. J. Bacteriol. 194, 4857–4866

53. Williams, D. et al. (2015) Divergent, coexisting Pseudomonas aeruginosa lineages in chronic cystic fibrosis lung infections. Am. J. Resp. Crit. Care Med. 191, 775–785

54. Markussen, T. et al. (2014) Environmental heterogeneity drives within-host diversification and evolution of Pseudomonas aeruginosa. mBio 5, e01592-14

55. Jorth, P. et al. (2015) Regional isolation drives bacterial diversifi- cation within cystic fibrosis lungs. Cell Host Microbe 18, 307–319

56. Dettman, J.R. et al. (2013) Evolutionary genomics of epidemic and nonepidemic strains of Pseudomonas aeruginosa. Proc. Natl. Acad. Sci. U.S.A. 110, 21065–21070

57. Williams, D. et al. (2016) Refined analyses suggest that recombination is a minor source of genomic diversity in Pseudomonas aeruginosa chronic cystic fibrosis infections. Microb. Genom. Published online January 18, 2016. http://dx.doi.org/10.1099/ mgen.0.000051

58. Boles, B.R. et al. (2004) Self-generated diversity produces "insurance effects" in biofilm communities. Proc. Natl. Acad. Sci. U.S.A. 101, 16630–16635

59. Boles, B.R. and Singh, P.K. (2008) Endogenous oxidative stress produces diversity and adaptability in biofilm communities. Proc. Natl. Acad. Sci. U.S.A. 105, 12503–12508

60. Traverse, C.C. et al. (2013) Tangled bank of experimentally evolved Burkholderia biofilms reflects selection during chronic infections. Proc. Natl. Acad. Sci. U.S.A. 110, E250–E259

61. Wright, E.A. et al. (2013) Sub-inhibitory concentrations of some antibiotics can drive diversification of Pseudomonas aeruginosa populations in artificial sputum medium. BMC Microbiol. 13, 170

CHAPTER 17 EXPLOITING GENETIC INTERFERENCE FOR ANTIVIRAL THERAPY

Elizabeth J. Tanner, Karla A. Kirkegaard, Leor S. Weinberger

INTRODUCTION

RNA viruses, such as influenza virus, HIV, and dengue virus, are dynamic, rapidly evolving pathogens that evade most therapeutic interventions. The error-prone viral RNA polymerase generates genetically heterogeneous viral populations that, in combination with large population sizes, enable these viruses to overcome static (chemical) drug therapies. Drug-resistant viral infections often emerge within days after treatment starts, the natural result of a strong selection pressure (the drug) acting on a genetically heterogeneous viral population [1–5]. Since RNA viruses are responsible for many of the world's ongoing and emerging epidemics (e.g., HIV, dengue virus, Zika virus), strategies that withstand viral dynamics and evolution are required. Below, we discuss two such strategies that aim to therapeutically target viruses not by blocking viral enzymes or receptors but by interfering with the fundamental viral structure or assembly processes. These two interventions—based upon the relatively arcane viral phenomena of "phenotypic masking" and defective interfering particles—have the potential to greatly diminish, and possibly overcome, problems associated with drug resistance and access to therapy.

Target for interference: virion capsid assembly

All virions are assembled from a few essential components: the viral genome, a protective protein shell called the capsid, and often an outer lipid envelope. Capsids are self-assembling oligomers composed of hundreds to thousands of individual proteins (i.e., they are homo-oligomers) (Fig 1). Assembled capsids conform to a variety of shapes. For example, many RNA virus capsids are icosahedral, while the HIV capsid is conical, and the flaviviruses, such as dengue virus, have a disordered capsid often referred to as the core [6–10]. Regardless of their topology, capsids are high-order oligomers composed of identical subunits that self-assemble with high apparent cooperativity [11].

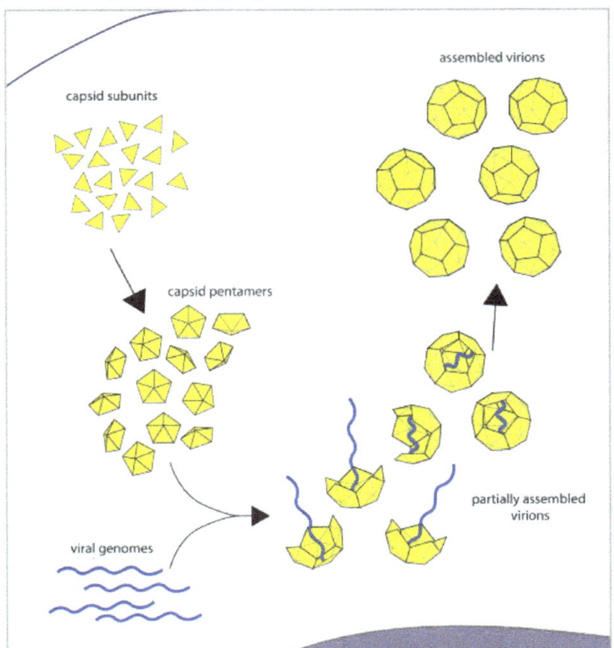

Figure 1. Schematic of targets of genetic interference: capsid assembly and genome encapsidation. Individual capsid subunits (yellow triangles) assemble into pentamers, which assemble to form the full icosahedral capsid. The viral genome (blue) is packaged into the capsid.

Citation: Tanner EJ, Kirkegaard KA, Weinberger LS (2016) Exploiting Genetic Interference for Antiviral Therapy. PLoS Genet 12(5): e1005986. doi:10.1371/journal.pgen.1005986.

Capsid Heterogeneity and "Phenotypic masking"

The homo-oligomeric nature of capsid formation makes capsid assemblies a rich target for genetic interference. Capsid assembly proceeds from a common pool of proteins, and mature capsids are often chimeras assembled from the proteins of multiple viral variants within a cell. In some cases, one variant may be susceptible to interference by toxic capsid subunits generated by another viral variant (i.e., dominant-negative interactions). For example, when antibody-resistant variants replicate in cells alongside antibody-sensitive variants, chimeric capsids can form. The antibody-sensitive capsid subunits in the structure can be bound by the antibody, leading to neutralization of the entire virion, despite the presence of escape-variant capsid subunits within the virion [12].

This dominant-negative interaction between antibody-sensitive and antibody-resistant capsid subunits is referred to as phenotypic masking. As described below, the formation of chimeric capsids can be exploited as a therapeutic strategy to suppress drug-resistant variants.

Target for Interference: Encapsidation of Viral Genetic Material

In most cases, the viral genome is packaged through direct interactions with viral packaging proteins, such as the capsid. For some RNA viruses, much of the specificity of packaging likely derives from direct coupling of packaging to RNA synthesis; only newly synthesized genomic RNAs are packaged. In many other cases, the packaging proteins recognize specific nucleotide sequences embedded in the viral RNA, regardless of the age or origin of the RNA. For example, the HIV-1 genome encodes *cis*-acting packaging signals, including the psi packaging element [13–16], reviewed in [17]. In contrast to the packaging signal, many viral proteins, including capsid and envelope proteins, act in *trans*. Hence, any RNA that encodes the requisite *cis*-packaging sequences can be packaged into the HIV-1 capsid, including reporter constructs, deletion mutants, and even RNAs of unrelated viruses [13]. This ability to encapsidate orthogonal genetic elements into a given viral capsid enables viral vectors to be used for gene delivery and gene therapy but also provides an opportunity for genetic interference.

Parasitic Genome Encapsidation by Interfering Particles

Genetic interference via encapsidation is feasible because virion assembly often entails mass-action processes [11,18,19] that proceed regardless of the source of the raw materials provided. In lentiviruses, for example, orthogonal capsid proteins, envelope proteins, and even genomes can be "spiked" into a cell from various sources, processed, and integrated into virion assembly in essentially the same manner as raw materials from any other source (standard complementation assays between viral mutants also suggest that viral variants within a cell share a common pool of viral products). Of course, not all viral processes necessarily proceed from a well-mixed pool of viral proteins (e.g., certain poliovirus proteins required for genome synthesis are not efficiently complemented in *trans* [20]). Nevertheless, in viral processes where mass-action mixing readily occurs, if two viral variants are present in a cell, their genomes inevitably compete for shared viral products, such as capsid and envelope proteins [21,22].

This competition between viral genomes can generate a fitness cost, as illustrated by a phenomenon described in the 1940s by both Henle and von Magnus. During serial passaging of influenza virus at high multiplicities of infection (MOI), they observed the formation of "incomplete" viruses that decreased the overall yield of infectious virus [23,24]. In 1970, the term defective interfering particle (DIP) was coined to describe these incomplete genomes, which are replication-incompetent and depend on a complete helper virus for replication. For example, influenza DIPs are typically composed of genomes in which one or more segments contain internal deletions [25], often in the polymerase genes [26], and thus rely on a helper virus to provide the missing polymerase.

Many other single-stranded RNA viruses, including dengue virus, poliovirus, and vesicular stomatitis virus, were subsequently found to produce DIPs [27–29]. However, it took decades to determine the mechanism of interference: deletion variants acting as molecular parasites of the wild-type virus [30]. Importantly, DIP genomes tend to be more abundant in the cell because of a replication advantage provided by their smaller size or increased affinity for the polymerase [31–35]. Since genome encapsidation is typically a mass-action process, the more abundant DIP genomes tend to stoichiometrically outcompete the wild-type genomes for packaging proteins [30,36–38]. Overall, these cheater variants (i.e., DIPs) amplify themselves at the expense of the wild-type virus [39,40]. This phenomenon is exploited for the therapeutic interfering particle (TIP) strategy discussed below.

HARNESSING GENETIC INTERFERENCE FOR ANTIVIRAL THERAPY

Exploiting Phenotypic Masking for Dominant Drug Targeting: Experimental Validation

Since the replication processes of RNA viruses are highly error-prone, genetic variants are unavoidable and rapidly arise. These variants generate resistance to virtually any monotherapy. However, the existence of drug-resistant variants does not guarantee their selection, and genetic interference can ensure that these mutants are not selected. For example, as explained below, targeting capsid or self-assembly processes with an inhibitor enables wild-type, drug-susceptible genomes to naturally suppress the outgrowth of drug-resistant variants [41,42]. The term dominant drug target was coined to highlight the dominant-negative role of the drug-susceptible variant.

Dominant drug targeting is based on the fact that when a drug-resistant variant first arises within a cell (through the error-prone replication of the viral genome), it is most often a minority variant in a population of wild-type, drug-susceptible genomes generated during genome amplification in that cell (Fig 2A). When an oligomeric structure such as the capsid is the drug target, both drug-susceptible and drug-resistant genomes contribute products to the common pool of proteins, leading to the assembly of capsids that contain both drug-resistant and drug-susceptible subunits. Thus, chimeric capsids—susceptible to drug inhibition despite the presence of resistant subunits—are likely to form in each cell in which drug-resistant genomes arise, and the outgrowth of the drug-resistant mutants is likely halted.

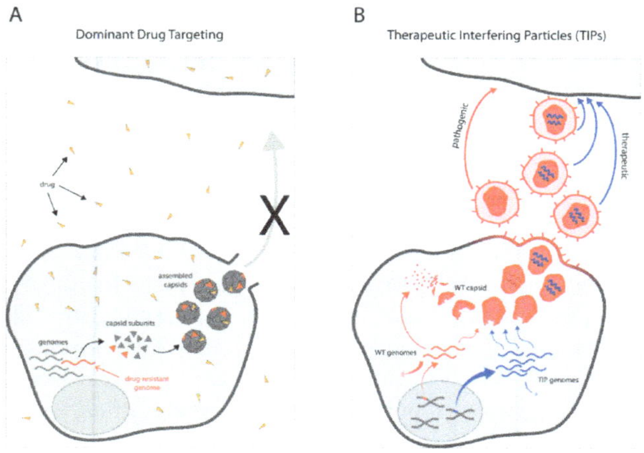

Figure 2. Exploiting genetic interference in capsid assembly and encapsidation for antiviral therapy. (A) Dominant drug targeting. Drug-resistant genomes (red curves) arise from wild-type genomes (gray curves) during replication within an infected cell. Drug-resistant and drug-sensitive genomes generate capsid proteins (red and gray triangles, respectively), which assemble into chimeric capsids. The drug (orange triangle) interferes with the function of both wild-type and chimeric capsids. (B) Therapeutic interfering particles (TIPs). After reverse transcription and integration, the HIV-1 proviral genome (red line) is integrated within the host genome and generates cytoplasmic mRNAs (red curves) by using its *trans* elements Tat and Rev as well as host cofactors. These mRNAs translate viral proteins, including capsid (red), and some of the mRNAs also serve as genomic RNA and are encapsidated in diploid form. TIP genomes (blue line) carry large deletions but are also reverse transcribed and integrated into the host genome. TIPs also utilize wild-type HIV-1 *trans* elements (e.g., Tat and Rev) to express genomic RNAs in the cytoplasm (blue curves). Since they have extensive deletions, these RNAs encode few if any proteins but are transcribed in higher proportions than wild-type RNAs. Because they are present at higher concentrations, TIP RNAs can outcompete wild-type virus (red) for packaging proteins and encapsidate their own diploid genomes in place of HIV-1 genomes.

The concept of dominant drug targeting has been demonstrated for both poliovirus and dengue virus using capsid and core inhibitors, respectively [41,42]. Targeting these oligomeric structures with inhibitors suppressed the emergence of drug resistance in infected mice. In striking contrast, targeting proteins involved in genome replication resulted in the emergence of drug resistance within days of initiating treatment. The efficacy of the capsid and core inhibitors could not be attributed to a lack of resistance mutations; resistant viral variants arise frequently during replication, and a single nucleotide change confers full resistance to either drug [41,43–45]. For both of these dominant drug targets (capsid and core), biochemical analyses confirmed the formation of oligomeric complexes susceptible to drug inhibition (e.g., Tanner et al. [41] directly show formation of chimeric, drug-inhibitable poliovirus capsids). These studies were the first to show that dominant drug targeting can suppress the emergence of drug resistance.

Expanding on Dominant Drug Targets

Given that the dominant drug target strategy has been validated for two different virus families (one enveloped and one nonenveloped), the strategy is likely to be broadly applicable to other viruses and targets. Two key conditions must be met for dominant drug targeting to be successful: (i) the virus must exhibit intracellular heterogeneity, and (ii) the target must be susceptible to dominant-negative interactions. Condition (i) is met by most single-stranded RNA viruses. These viruses rely on error-prone polymerases that introduce new mutations into each progeny genome, thereby generating genetic heterogeneity within an individual infected cell. Condition (ii) can be empirically tested by screening for viral factors susceptible to dominant-negative interactions.

A screen in poliovirus identified viral factors other than the capsid that are susceptible to dominant-negative interactions [46]. These targets included elements involved in genome replication, such as the viral polymerase and an RNA structure that acts as a template for genome synthesis. Curiously, only some polymerase mutants were able to act as dominant negatives, suggesting that only some functions of the polymerase are susceptible to genetic interference. While it remains unclear how genome synthesis can be susceptible to dominant-negative interactions, one clue may lie in the fact that, in many viruses, the polymerase forms an oligomer [47–50]. These oligomers are functional, and RNA replication is cooperative, suggesting that multiple polymerase molecules may replicate the same template simultaneously [51–54]. Perhaps defective subunits within the polymerase oligomer act as toxic subunits. Similarly, an intramolecularly cleaving protease identified in the screen may generate a toxic fusion protein by failing to cleave itself from the viral polyprotein precursor. Given that the poliovirus screen identified multiple factors susceptible to dominant-negative interference, similar screens in other viruses may illuminate additional dominant drug targets.

Moving forward, targeting inhibitors to dominant drug targets could reduce the number of antivirals required to treat infections and substantially lower the chance of resistance for the targeted viral pathogen. These factors could lower the cost and potential side effects of antiviral therapy regimens.

Potential Pitfalls of Dominant Drug Targeting

Dominant drug targeting relies on inherent viral interactions that must be sufficiently strong to suppress drug resistance over the course of treatment. If between-variant interactions are rare (perhaps because of insufficient mixing of proteins), the drug-resistant genome may escape suppression. Any drug-resistant variant that seeds a new round of infection in the absence of drug-susceptible genomes will no longer be controlled by genetic interference. Given that drug resistance did not emerge from the treated mice [41,42], mixing appears to be common in the capsid and core of poliovirus and dengue virus, respectively, and interference within these processes is quite effective. Nevertheless, these mouse studies were carried out on the order of days, so the long-term efficacy of dominant drug targeting has not been assessed. Modeling the probability that an infection escapes suppression could help determine how susceptible each target will be to drug failure and the potential consequences for population-level infection control.

GENERATING DE NOVO INTERFERENCE FOR THERAPY

Theory: Therapeutic Interfering Particles

Overcoming drug resistance remains a critical challenge. However, for many communicable diseases, two other challenges are at least as problematic: behavioral barriers, such as poor adherence to therapy and risk disinhibition in the target populations, and unequal deployment of therapies, which results from differences in access to therapy in at-risk populations. These three challenges—resistance, adherence, and access—arise from three core characteristics inherent to viruses but absent in antiviral interventions: evolution (viruses mutate, but pharmaceuticals do not, invariably leading to resistant variants), persistence (viruses evolved mechanisms to remain in the host, but pharmaceuticals are cleared, necessitating sustained adherence), and transmission (viruses are communicable, but pharmaceuticals are not, so deployment of therapies always lags).

These challenges are exemplified by the HIV-1/AIDS pandemic. HIV-1 rapidly evolves resistance [55,56], often necessitating cocktails of three antiviral drugs to effectively suppress viral replication. HIV-1 also establishes lifelong persistence in infected persons, typically requiring lifelong adherence to therapeutic regimens [57]. Finally, HIV-1 transmission patterns are highly heterogeneous, with hard-to-reach superspreaders (e.g., commercial sex workers and their clients) who engage in high-risk behaviors, contributing disproportionately to disease spread [58–60]. Hence, despite the development of highly suppressive antiretroviral drugs, effective pandemic control has been unattainable, and some 2 million people are infected each year, primarily in resource-limited settings, such as sub-Saharan Africa [61].

A proposed solution to overcome these challenges is to utilize the concept of DIPs by engineering optimized TIPs [62] that molecularly parasitize wild-type viruses across multiple biological scales, from the cellular to the epidemiological scale. While proposals to use DIPs for therapy are not new [63–68], the hypothesis that DIPs engineered to stoichiometrically outcompete wild-type viruses at the cellular scale could lessen disease at the epidemiological scale [62] is a unique therapeutic proposition.

Provided that TIPs can be engineered, theoretical models predict that for HIV/AIDS, TIPs could piggyback on wild-type HIV-1 replication (Fig 2B) to produce substantially greater reductions in disease prevalence compared to conventional interventions. Specifically, these data-driven models linked molecular, patient, and epidemiological scales and assessed the feasibility of TIPs as population-level HIV-1 therapeutic interventions compared to conventional vaccines and drug therapies [62,69–71]. Strikingly, the epidemiological analyses predicted that introducing TIPs into a small percentage of infected individuals could significantly lower the HIV/AIDS prevalence, particularly because TIPs would automatically concentrate in the superspreader populations that are so challenging to reach with conventional therapies and vaccines [62].

The fundamental reason that TIPs outperformed conventional interventions in silico was based on the ability of DIPs to transmit through human populations [72]. Consequently, TIPs could disseminate through natural HIV transmission networks, despite the model accounting for a lower transmission efficacy of TIPs than HIV-1. Being lentiviral vectors, TIPs integrate into the host genome and can lie latent [73,74] for the lifetime of the cell until subsequent wild-type virus infection. Hence, despite DIP transmission being less successful when viral loads are low (i.e., during initial infection), TIPs have the potential to reactivate upon subsequent wild-type virus infection. Accounting for this long-lived latent state increases effective TIP transmission and reduces HIV/AIDS prevalence in epidemiological models [62], but in high-risk populations, TIPs retain efficacy even absent latency [62,71].

Remarkably, the models also showed that HIV-1 is unlikely to escape TIP parasitism through adaptation because the two populations experience competing selection pressures. For example, HIV-1 could theoretically escape TIP interactions by altering the specificity of packaging. Such a functional change would require at least two independent mutations, one in the packaging signal and one in the capsid. However, TIPs are under pressure to maintain their interaction with HIV-1 packaging proteins and will only have to make the single mutation in the packaging signal to bind an altered capsid protein. Being dependent on the same error-prone replication machinery as HIV-1, TIPs would adapt at the same evolutionary rate as HIV-1 and would be able to maintain the parasitic interaction. Extensive modeling of various escape scenarios suggests that HIV-1 cannot escape the TIP interaction without severely limiting its own growth [69–71]. Hence, TIPs may constitute "resistance-proof" antivirals.

Constructing TIPs for HIV-1 and Beyond

While a bona-fide TIP has not yet been experimentally validated, a number of HIV-based vectors have been engineered to interfere with viral replication [75–80]. These engineered vectors partially fulfill the criteria of a TIP.

The models [62,69–71] indicate that, for therapeutic effectiveness and halting resistance, TIPs must exhibit a basic reproductive number (R_0) > 1 (specifically, a within-patient $R_0 > 1$). This criterion means that TIP-infected cells must, on average, stably and continually generate more than one new TIP-infected cell. Theoretically, TIPs that have an $R_0 > 1$ necessarily reduce the HIV-1 R_0, thereby reducing HIV-1 viremia in patients and ultimately reducing HIV-1 transmission and prevalence.

Molecularly, the design criteria that enable $R_0 > 1$ are: (i) that excess TIP genomic RNA (gRNA) be generated compared to wild-type HIV-1 gRNA within an HIV-1 infected cell and (ii) that the excess TIP gRNAs encode all relevant *cis* elements required for efficient replication and packaging. These two design criteria ensure that the TIP genome will spread from one cell to the next more efficiently than its helper (HIV-1) virus. The excess (or asymmetry) of TIP gRNA synthesis can be achieved by enhanced production or diminished degradation of TIP gRNA, compared to wild-type HIV-1 gRNA (Fig 2B).

For HIV, abrogating mRNA splicing can cause more RNA to remain as gRNA [81] and could thus satisfy the requirement for TIP gRNA production asymmetry. The resulting excess of TIP gRNA would enable TIPs to outcompete HIV-1 for encapsidation resources in infected cells.

Unfortunately, the previously constructed therapeutic vectors [75–80] encode genetic elements that directly antagonize HIV replication, likely precluding $R_0 > 1$ because production of essential *trans* elements is inhibited [82]. For example, the Dropulic group designed an HIV-based vector that expresses antisense RNA to the virus envelope transcript [83,84]. This vector mobilizes in patients to some extent, but viral suppression was temporary, and the vector was lost over time [85,86]. Nevertheless, the fact that these vectors transiently mobilized in patients provides a rationale that TIPs may be engineered against HIV-1. Moreover, HIV-1 deletion mutants lacking HIV-inhibitory elements also inhibit the spread of HIV-1 in cell culture and mobilize in the presence of a wild-type virus [76,78], but these mutants were selected based on inhibition (not transmission), and so it is unclear if $R_0 > 1$ was satisfied.

Beyond HIV-1, interventions that address the universal barriers to disease control (i.e., resistance, adherence, and access) will likely be critical to control future epidemics of other viruses (e.g., retro- or lentiviruses that zoonotically cross over from nonhuman primates). Retroviruses related to HIV-1 are endemic to primate species in Africa, and 12 cases of retroviral zoonotic transmissions from nonhuman primates to humans have been documented in the last century [87–92]. Given that one of these transmissions is responsible for the HIV-1 pandemic, concern over future retroviral pandemics is not unfounded; another pandemic virus may emerge from the primate reservoir and overwhelm conventional disease-control approaches, as did HIV-1. Hence, development of TIP-like interventions capable of overcoming the barriers to effective disease control would seem prudent.

Potential Pitfalls of the TIP Strategy

The safety and ethical implications of transmissible, integrating therapies are important considerations that have been addressed extensively by Notton et al., [93], but we briefly summarize the main points here.

The chief concerns center around: (i) safety risks associated with insertional mutagenesis of TIPs into the host genome and potential oncogenic transformation of the cell, (ii) viral evolution and recombination, leading to more virulent HIV-1 strains, and (iii) the ethical implications of treating individuals without consent.

While insertional mutagenesis leading to oncogenesis was problematic with early gamma retroviral gene therapy in hematopoietic stem cells [94], this genotoxicity appears limited to the specific combination of gamma retrovirus and hematopoietic stem cells. Carl June and colleagues found no evidence of oncogenesis or clonal dominance in a decade-long trial in human patients [95] when retroviral vectors were used on terminally differentiated T cells. For lentiviral gene therapies, no examples of oncogenesis have been reported when either differentiated or undifferentiated cell types have been transduced in the clinic [85,86,95–98]. Multiyear follow-up studies of patients transduced with lentiviral vectors targeting hematopoietic stem cells—highly transformable cell types—showed no genotoxicity or preferential expansion of transduced cells in patients [97,98]. The genotoxicity of the original gamma retroviral vectors in hematopoietic cells may arise from their differential promoter architecture compared to lentiviruses and the proclivity of retroviruses to integrate near the 5′ transcription start sites as well as in proto-oncogenes, while the lentiviral patient trials noted no increased integration events near the 5′ ends of genes or near proto-oncogenes (the integration patterns were quasi-random in genic regions). Given that TIPs are lentiviruses and would infect the same cells as HIV-1 (primarily differentiated T cells), the patient trials above suggest that potential therapeutic benefits of TIPs may outweigh risks associated with insertional oncogenesis.

The risk that TIPs may lead to the evolution of increased virulence in HIV-1 is a theoretical possibility that has been investigated using quantitative evolutionary dynamics models [62,69–71]. These models indicate that it is unlikely that HIV-1 could evolve away from TIPs, at either the patient [69] or epidemiological scale [71]. Intuitively, the modeling results can be understood from the fact that TIPs parasitize and replicate more efficiently than HIV-1; while HIV-1 needs to mutate both *cis* and *trans* factors simultaneously to escape TIPs, TIPs only mutate *cis* factors (they lack *trans* factors). This two-to-one mutation rate advantage largely precludes HIV-1 escape variants at the individual-patient scale. Moreover, the models argue that recombination with wild-type HIV, while certainly possible, is selected against by both HIV-1 and TIPs [69]. In the event that recombination occurs, it will be limited to already infected individuals, and even if it occurs in every infected cell, it will ultimately reconstitute the already present wild-type virus (i.e., the equivalent of an HIV mutation that results in loss of efficacy for a therapy).

While the notion of a transmissible therapy—even a therapy such as TIPs that only transmit in a limited fashion between already infected individuals—may raise ethical concerns regarding consent, many of these ethical issues have already been addressed for live-attenuated vaccines. For example, the oral poliovirus vaccine (OPV) is a live-attenuated vaccine chosen by the World Health Organization (WHO) for the worldwide polio eradication effort [99] (most recently, OPV successfully eradicated polio in India [100]). Like many live-attenuated vaccines, OPV is known to transmit between individuals—vaccinated individuals shed OPV virus for days, inoculating other individuals against the poliovirus—and this transmission of vaccination is widely viewed as a benefit of OPV. While there are safety risks associated with live-attenuated vaccines such as OPV (i.e., reversion to virulence), TIPs exhibit far greater safety in this regard, as they can only recombine to become virulent if they have mobilized (i.e., in individuals already infected with wild-type virus). Hence, for TIPs, the consent and reversion issues are circumscribed to the infected populations as opposed to the general populace.

All medical interventions carry inherent risk. Interventions that enter into clinical practice are those in which the benefits have been determined to outweigh the risks. Risk–benefit analysis is thus an essential aspect of any medical innovation. Overall, while the safety risks and ethical issues associated with TIPs cannot be overlooked, given the potential benefits TIPs may provide in overcoming universal disease control barriers and the precedents above, not pursuing research on these potential therapeutic interventions may ultimately be more ethically problematic.

Conclusion

The dominant drug targeting and TIPs strategies employ different mechanisms, but both rely on genetic interference to counter viral population heterogeneity, lower the viral burden, and overcome the obstacles that face current antiviral therapies. Dominant drug targeting operates by clearing an infection with chemical inhibitors targeting highly cooperative processes, and TIPs control infection at multiple biological scales (molecular to epidemiological) through a transmissible molecular parasite particle. The dominant drug targeting strategy exploits naturally occurring dominant-negative interaction to suppress the key target—drug-resistant genomes. Such dominant-negative interactions require intracellular heterogeneity. The TIP strategy, on the other hand, introduces a molecular parasite into the population and harnesses genetic interference through parasitic genome encapsidation, so TIPs can target viruses that generate intercellular genetic diversity even if these viruses do not exhibit intracellular diversity. These therapeutic strategies based on genetic interference have the potential to overcome major barriers in disease control, including drug resistance and population-level barriers to effective deployment of therapy. Going forward, quantitative viral genetics approaches may expose new targets and strategies to exploit genetic interference for therapy.

REFERENCES

1. Kimberlin DW, Acosta EP, Prichard MN, Sanchez PJ, Ampofo K, Lang D, et al. Oseltamivir pharmacokinetics, dosing, and resistance among children aged <2 years with influenza. J Infect Dis. 2013;207(5):709–20. doi: 10. 1093/infdis/jis765. pmid:23230059
2. Kiso M, Mitamura K, Sakai-Tagawa Y, Shiraishi K, Kawakami C, Kimura K, et al. Resistant influenza A viruses in children treated with oseltamivir: descriptive study. Lancet. 2004;364(9436):759–65. pmid:15337401 doi: 10. 1016/s0140-6736(04)16934-1
3. Larder BA, Darby G, Richman DD. HIV with reduced sensitivity to zidovudine (AZT) isolated during prolonged therapy. Science. 1989;243(4899):1731–4. pmid:2467383 doi: 10. 1126/science. 2467383
4. Pevear DC, Hayden FG, Demenczuk TM, Barone LR, McKinlay MA, Collett MS. Relationship of pleconaril susceptibility and clinical outcomes in treatment of common colds caused by rhinoviruses. Antimicrob Agents Chemother. 2005;49(11):4492–9. pmid:16251287 doi: 10. 1128/aac. 49. 11. 4492-4499. 2005
5. Young KC, Lindsay KL, Lee KJ, Liu WC, He JW, Milstein SL, et al. Identification of a ribavirin-resistant NS5B mutation of hepatitis C virus during ribavirin monotherapy. Hepatology. 2003;38(4):869–78. pmid:14512874 doi: 10. 1002/hep. 1840380413
6. Caspar DL, Klug A. Physical principles in the construction of regular viruses. Cold Spring Harb Symp Quant Biol. 1962;27:1–24. pmid:14019094 doi: 10. 1101/sqb. 1962. 027. 001. 005
7. Ganser BK, Li S, Klishko VY, Finch JT, Sundquist WI. Assembly and analysis of conical models for the HIV-1 core. Science. 1999;283(5398):80–3. pmid:9872746 doi: 10. 1126/science. 283. 5398. 80
8. Li S, Hill CP, Sundquist WI, Finch JT. Image reconstructions of helical assemblies of the HIV-1 CA protein. Nature. 2000;407(6802):409–13. pmid:11014200 doi: 10. 1038/35030177
9. Ganser-Pornillos BK, Yeager M, Sundquist WI. The structural biology of HIV assembly. Curr Opin Struct Biol. 2008;18(2):203–17. doi: 10. 1016/j. sbi. 2008. 02. 001. pmid:18406133
10. Kuhn RJ, Zhang W, Rossmann MG, Pletnev SV, Corver J, Lenches E, et al. Structure of dengue virus: implications for flavivirus organization, maturation, and fusion. Cell. 2002;108(5):717–25. pmid:11893341 doi: 10. 1142/9789814513357_0037
11. Zlotnick A. To build a virus capsid. An equilibrium model of the self assembly of polyhedral protein complexes. J Mol Biol. 1994;241(1):59–67. pmid:8051707
12. Holland JJ, de la Torre JC, Steinhauer DA, Clarke D, Duarte E, Domingo E. Virus mutation frequencies can be greatly underestimated by monoclonal antibody neutralization of virions. J Virol. 1989;63(12):5030–6. pmid:2479770
13. Aldovini A, Young RA. Mutations of RNA and protein sequences involved in human immunodeficiency virus type 1 packaging result in production of noninfectious virus. J Virol. 1990;64(5):1920–6. pmid:2109098
14. Clever J, Sassetti C, Parslow TG. RNA secondary structure and binding sites for gag gene products in the 5' packaging signal of human immunodeficiency virus type 1. J Virol. 1995;69(4):2101–9. pmid:7884856
15. McBride MS, Schwartz MD, Panganiban AT. Efficient encapsidation of human immunodeficiency virus type 1 vectors and further characterization of cis elements required for encapsidation. J Virol. 1997;71(6):4544–54. pmid:9151848
16. De Guzman RN, Wu ZR, Stalling CC, Pappalardo L, Borer PN, Summers MF. Structure of the HIV-1 nucleocapsid protein bound to the SL3 psi-RNA recognition element. Science. 1998;279(5349):384–8. pmid:9430589 doi: 10. 2210/pdb1a1t/pdb
17. Kuzembayeva M, Dilley K, Sardo L, Hu WS. Life of psi: how full-length HIV-1 RNAs become packaged genomes in the viral particles. Virology. 2014;454–455:362–70. doi: 10. 1016/j. virol. 2014. 01. 019. pmid:24530126
18. Endres D, Zlotnick A. Model-based analysis of assembly kinetics for virus capsids or other spherical polymers. Biophys J. 2002;83(2):1217–30. pmid:12124301 doi: 10. 1016/s0006-3495(02)75245-4
19. Olson AJ, Hu YH, Keinan E. Chemical mimicry of viral capsid self-assembly. Proc Natl Acad Sci U S A. 2007;104(52):20731–6. pmid:18093950 doi: 10. 1073/pnas. 0709489104
20. Teterina NL, Zhou WD, Cho MW, Ehrenfeld E. Inefficient complementation activity of poliovirus 2C and 3D proteins for rescue of lethal mutations. J Virol.

1995;69(7):4245–54. pmid:7769684

21. Chen J, Nikolaitchik O, Singh J, Wright A, Bencsics CE, Coffin JM, et al. High efficiency of HIV-1 genomic RNA packaging and heterozygote formation revealed by single virion analysis. Proc Natl Acad Sci U S A. 2009;106(32):13535–40. doi: 10. 1073/pnas. 0906822106. pmid:19628694
22. Dilley KA, Ni N, Nikolaitchik OA, Chen J, Galli A, Hu WS. Determining the frequency and mechanisms of HIV-1 and HIV-2 RNA copackaging by single-virion analysis. J Virol. 2011;85(20):10499–508. doi: 10. 1128/ JVI. 05147-11. pmid:21849448
23. Henle W, Henle G. Interference of Inactive Virus with the Propagation of Virus of Influenza. Science. 1943;98(2534):87–9. pmid:17749157 doi: 10. 1126/ science. 98. 2534. 87
24. Von Magnus P. Incomplete forms of influenza virus. Adv Virus Res. 1954;2:59–79. pmid:13228257 doi: 10. 1016/ s0065-3527(08)60529-1
25. Nayak DP, Chambers TM, Akkina RK. Defective-interfering (DI) RNAs of influenza viruses: origin, structure, expression, and interference. Curr Top Microbiol Immunol. 1985;114:103–51. pmid:3888540 doi: 10. 1007/978-3-642-70227-3_3
26. Davis AR, Nayak DP. Sequence relationships among defective interfering influenza viral RNAs. Proc Natl Acad Sci U S A. 1979;76(7):3092–6. pmid:290988 doi: 10. 1073/pnas. 76. 7. 3092
27. Li D, Lott WB, Lowry K, Jones A, Thu HM, Aaskov J. Defective interfering viral particles in acute dengue infections. PLoS One. 2011;6(4):e19447 doi: 10. 1371/ journal. pone. 0019447
28. Cole CN, Smoler D, Wimmer E, Baltimore D. Defective interfering particles of poliovirus. I. Isolation and physical properties. J Virol. 1971;7(4):478–85. pmid:4329564
29. Cooper PD, Bellett AJ. A transmissible interfering component of vesicular stomatitis virus preparations. J Gen Microbiol. 1959;21:485–97. pmid:13811805 doi: 10. 1099/00221287-21-3-485
30. Huang AS, Baltimore D. Defective viral particles and viral disease processes. Nature. 1970;226(5243):325–7. pmid:5439728 doi: 10. 1038/226325a0
31. Sekellick MJ, Marcus PI. Viral interference by defective particles of vesicular stomatitis virus measured in individual cells. Virology. 1980;104(1):247–52. pmid:6249028 doi: 10. 1016/0042-6822(80)90385-2
32. Giachetti C, Holland JJ. Altered replicase specificity is responsible for resistance to defective interfering particle interference of an Sdi- mutant of vesicular stomatitis virus. J Virol. 1988;62(10):3614–21. pmid:2843664
33. Von Laer DM, Mack D, Kruppa J. Delayed formation of defective interfering particles in vesicular stomatitis virus-infected cells: kinetic studies of viral protein and RNA synthesis during autointerference. J Virol. 1988;62(4):1323–9. pmid:2831393
34. Akpinar F, Timm A, Yin J. High-Throughput Single-Cell Kinetics of Virus Infections in the Presence of Defective Interfering Particles. J Virol. 2015;90(3):1599–612. doi: 10. 1128/JVI. 02190-15. pmid:26608322
35. Laske T, Heldt FS, Hoffmann H, Frensing T, Reichl U. Modeling the intracellular replication of influenza A virus in the presence of defective interfering RNAs. Virus Res. 2015;213:90–9. doi: 10. 1016/j. virusres. 2015. 11. 016. pmid:26592173
36. Giachetti C, Holland JJ. Vesicular stomatitis virus and its defective interfering particles exhibit in vitro transcriptional and replicative competition for purified L-NS polymerase molecules. Virology. 1989;170(1):264–7. pmid:2541540 doi: 10. 1016/0042-6822(89)90375-9
37. Odagiri T, Tominaga K, Tobita K, Ohta S. An amino acid change in the non-structural NS2 protein of an influenza A virus mutant is responsible for the generation of defective interfering (DI) particles by amplifying DI RNAs and suppressing complementary RNA synthesis. J Gen Virol. 1994;75 (Pt 1):43–53. pmid:8113739 doi: 10. 1099/0022-1317-75-1-43
38. Odagiri T, Tashiro M. Segment-specific noncoding sequences of the influenza virus genome RNA are involved in the specific competition between defective interfering RNA and its progenitor RNA segment at the virion assembly step. J Virol. 1997;71(3):2138–45. pmid:9032347
39. Thompson KA, Yin J. Population dynamics of an RNA virus and its defective interfering particles in passage cultures. Virol J. 2010;7:257. doi: 10. 1186/1743-422X-7-257. pmid:20920247
40. Dennehy JJ, Turner PE. Reduced fecundity is the cost of cheating in RNA virus phi6. Proc Biol Sci. 2004;271(1554):2275–82. pmid:15539353 doi: 10. 1098/rspb. 2004. 2833
41. Tanner EJ, Liu HM, Oberste MS, Pallansch M, Collett MS, Kirkegaard K. Dominant drug targets suppress the emergence of antiviral resistance. Elife. 2014;3: e03830. doi: 10. 7554/elife. 03830
42. Mateo R, Nagamine CM, Kirkegaard K. Suppression of Drug Resistance in Dengue Virus. MBio. 2015;6(6):e01960–15. doi: 10. 1128/mBio. 01960-15. pmid:26670386
43. Kouiavskaia DV, Dragunsky EM, Liu HM, Oberste MS, Collett MS, Chumakov KM. Immunological and pathogenic properties of poliovirus variants selected for resistance to antiviral drug V-073. Antivir Ther. 2011;16(7):999–1004. doi: 10. 3851/IMP1838. pmid:22024515
44. Liu HM, Roberts JA, Moore D, Anderson B, Pallansch MA, Pevear DC, et al. Characterization of poliovirus variants selected for resistance to the antiviral compound V-073. Antimicrob Agents Chemother. 2012;56(11):5568–74. doi: 10. 1128/AAC. 00539-12. pmid:22890765
45. Byrd CM, Dai D, Grosenbach DW, Berhanu A, Jones KF, Cardwell KB, et al. A novel inhibitor of dengue virus replication that targets the capsid protein. Antimicrob Agents Chemother. 2013;57(1):15–25. doi: 10. 1128/ AAC. 01429-12. pmid:23070172

46. Crowder S, Kirkegaard K. Trans-dominant inhibition of RNA viral replication can slow growth of drug-resistant viruses. Nat Genet. 2005;37(7):701–9. pmid:15965477 doi: 10. 1038/ng1583

47. Lyle JM, Bullitt E, Bienz K, Kirkegaard K. Visualization and functional analysis of RNA-dependent RNA polymerase lattices. Science. 2002;296(5576):2218–22. pmid:12077417 doi: 10. 1126/science. 1070585

48. Qin W, Luo H, Nomura T, Hayashi N, Yamashita T, Murakami S. Oligomeric interaction of hepatitis C virus NS5B is critical for catalytic activity of RNA-dependent RNA polymerase. J Biol Chem. 2002;277(3):2132–7. pmid:11673460 doi: 10. 1074/jbc. m106880200

49. Wang QM, Hockman MA, Staschke K, Johnson RB, Case KA, Lu J, et al. Oligomerization and cooperative RNA synthesis activity of hepatitis C virus RNA-dependent RNA polymerase. J Virol. 2002;76(8):3865–72. pmid:11907226 doi: 10. 1128/jvi. 76. 8. 3865-3872. 2002

50. Goregaoker SP, Culver JN. Oligomerization and activity of the helicase domain of the tobacco mosaic virus 126- and 183-kilodalton replicase proteins. J Virol. 2003;77(6):3549–56. pmid:12610130 doi: 10. 1128/jvi. 77. 6. 3549-3556. 2003

51. Pata JD, Schultz SC, Kirkegaard K. Functional oligomerization of poliovirus RNA-dependent RNA polymerase. RNA. 1995;1(5):466–77. pmid:7489508

52. Beckman MT, Kirkegaard K. Site size of cooperative single-stranded RNA binding by poliovirus RNA-dependent RNA polymerase. J Biol Chem. 1998;273(12):6724–30. pmid:9506971 doi: 10. 1074/jbc. 273. 12. 6724

53. Hobson SD, Rosenblum ES, Richards OC, Richmond K, Kirkegaard K, Schultz SC. Oligomeric structures of poliovirus polymerase are important for function. EMBO J. 2001;20(5):1153–63. pmid:11230138 doi: 10. 1093/emboj/20. 5. 1153

54. Spagnolo JF, Rossignol E, Bullitt E, Kirkegaard K. Enzymatic and nonenzymatic functions of viral RNA-dependent RNA polymerases within oligomeric arrays. RNA. 2010;16(2):382–93. doi: 10. 1261/rna. 1955410. pmid:20051491

55. Rhee SY, Gonzales MJ, Kantor R, Betts BJ, Ravela J, Shafer RW. Human immunodeficiency virus reverse transcriptase and protease sequence database. Nucleic Acids Res. 2003;31(1):298–303. pmid:12520007 doi: 10. 1093/nar/gkg100

56. Shafer RW. Rationale and uses of a public HIV drug-resistance database. J Infect Dis. 2006;194 Suppl 1:S51–8. pmid:16921473 doi: 10. 1086/505356

57. Finzi D, Blankson J, Siliciano JD, Margolick JB, Chadwick K, Pierson T, et al. Latent infection of CD4+ T cells provides a mechanism for lifelong persistence of HIV-1, even in patients on effective combination therapy. Nat Med. 1999;5(5):512–7. pmid:10229227

58. Woodhouse DE, Rothenberg RB, Potterat JJ, Darrow WW, Muth SQ, Klovdahl AS, et al. Mapping a social network of heterosexuals at high risk for HIV infection. AIDS. 1994;8(9):1331–6. pmid:7802989 doi: 10. 1097/00002030-199409000-00018

59. Lloyd-Smith JO, Schreiber SJ, Kopp PE, Getz WM. Superspreading and the effect of individual variation on disease emergence. Nature. 2005;438(7066):355–9. pmid:16292310 doi: 10. 1038/nature04153

60. Woolhouse ME, Dye C, Etard JF, Smith T, Charlwood JD, Garnett GP, et al. Heterogeneities in the transmission of infectious agents: implications for the design of control programs. Proc Natl Acad Sci U S A. 1997;94(1):338–42. pmid:8990210 doi: 10. 1073/pnas. 94. 1. 338

61. WHO. Global update on the health sector response to HIV, 2014. 2014. p. 16.

62. Metzger VT, Lloyd-Smith JO, Weinberger LS. Autonomous targeting of infectious superspreaders using engineered transmissible therapies. PLoS Comput Biol. 2011;7(3):e1002015. doi: 10. 1371/journal. pcbi. 1002015. pmid:21483468

63. Doyle M, Holland JJ. Prophylaxis and immunization in mice by use of virus-free defective T particles to protect against intracerebral infection by vesicular stomatitis virus. Proc Natl Acad Sci U S A. 1973;70(7):2105–8. pmid:4352972 doi: 10. 1073/pnas. 70. 7. 2105

64. Chambers TM, Webster RG. Protection of chickens from lethal influenza virus infection by influenza A/chicken/Pennsylvania/1/83 virus: characterization of the protective effect. Virology. 1991;183(1):427–32. pmid:2053293 doi: 10. 1016/0042-6822(91)90160-d

65. Nelson GW, Perelson AS. Modeling defective interfering virus therapy for AIDS: conditions for DIV survival. Math Biosci. 1995;125(2):127–53. pmid:7881191 doi: 10. 1016/0025-5564(94)00021-q

66. Madsen MS, Snelling DF, Heaphy S, Norris V. Antiviruses as therapeutic agents: a mathematical analysis of their potential. J Theor Biol. 1997;184(2):111–6. pmid:9059590 doi: 10. 1006/jtbi. 1996. 0238

67. Kim H, Yin J. Quantitative analysis of a parasitic antiviral strategy. Antimicrob Agents Chemother. 2004;48(3):1017–20. pmid:14982798 doi: 10. 1128/aac. 48. 3. 1017-1020. 2004

68. Dimmock NJ, Easton AJ. Defective interfering influenza virus RNAs: time to reevaluate their clinical potential as broad-spectrum antivirals? J Virol. 2014;88(10):5217–27. doi: 10. 1128/JVI. 03193-13. pmid:24574404

69. Rouzine IM, Weinberger LS. Design requirements for interfering particles to maintain coadaptive stability with HIV-1. J Virol. 2013;87(4):2081–93. doi: 10. 1128/JVI. 02741-12. pmid:23221552

70. Rouzine IM, Weinberger LS. Reply to "Coadaptive stability of interfering particles with HIV-1 when there is an evolutionary conflict". J Virol. 2013;87(17):9960–2. doi: 10. 1128/JVI. 00932-13. pmid:23943740

71. Rast LI, Rouzine IM, Rozhnova G, Bishop L, Weinberger AD, Weinberger LS. Conflicting Selection Pressures will Constrain Viral Escape from Interfering Particles: Principles for Designing Resistance-Proof Antivirals. PLoS Comput Biol. 2016. In press.

72. Aaskov J, Buzacott K, Thu HM, Lowry K, Holmes EC. Long-term transmission of defective RNA

viruses in humans and Aedes mosquitoes. Science. 2006;311(5758):236–8. pmid:16410525 doi: 10. 1126/science. 1115030

73. Weinberger LS, Burnett JC, Toettcher JE, Arkin AP, Schaffer DV. Stochastic gene expression in a lentiviral positive-feedback loop: HIV-1 Tat fluctuations drive phenotypic diversity. Cell. 2005;122(2):169–82. pmid:16051143 doi: 10. 1016/j. cell. 2005. 06. 006

74. Jordan A, Defechereux P, Verdin E. The site of HIV-1 integration in the human genome determines basal transcriptional activity and response to Tat transactivation. EMBO J. 2001;20(7):1726–38. pmid:11285236 doi: 10. 1093/emboj/20. 7. 1726

75. Dropulic B, Hermankova M, Pitha PM. A conditionally replicating HIV-1 vector interferes with wild-type HIV-1 replication and spread. Proc Natl Acad Sci U S A. 1996;93(20):11103–8. pmid:8855316 doi: 10. 1073/pnas. 93. 20. 11103

76. An DS, Morizono K, Li QX, Mao SH, Lu S, Chen IS. An inducible human immunodeficiency virus type 1 (HIV-1) vector which effectively suppresses HIV-1 replication. J Virol. 1999;73(9):7671–7. pmid:10438857

77. Mautino MR, Ramsey WJ, Reiser J, Morgan RA. Modified human immunodeficiency virus-based lentiviral vectors display decreased sensitivity to trans-dominant Rev. Hum Gene Ther. 2000;11(6):895–908. pmid:10779166 doi: 10. 1089/10430340050015509

78. Klimatcheva E, Planelles V, Day SL, Fulreader F, Renda MJ, Rosenblatt J. Defective lentiviral vectors are efficiently trafficked by HIV-1 and inhibit its replication. Mol Ther. 2001;3(6):928–39. pmid:11407907 doi: 10. 1006/mthe. 2001. 0344

79. Morris KV, Rossi JJ. Lentivirus-mediated RNA interference therapy for human immunodeficiency virus type 1 infection. Hum Gene Ther. 2006;17(5):479–86. pmid:16716105 doi: 10. 1089/hum. 2006. 17. 479

80. Turner AM, De La Cruz J, Morris KV. Mobilization-competent Lentiviral Vector-mediated Sustained Transcriptional Modulation of HIV-1 Expression. Mol Ther. 2009;17(2):360–8. doi: 10. 1038/mt. 2008. 268. pmid:19066594

81. D'Costa J, Brown H, Kundra P, Davis-Warren A, Arya S. Human immunodeficiency virus type 2 lentiviral vectors: packaging signal and splice donor in expression and encapsidation. J Gen Virol. 2001;82(Pt 2):425–34. pmid:11161282 doi: 10. 1099/0022-1317-82-2-425

82. Weinberger LS, Schaffer DV, Arkin AP. Theoretical design of a gene therapy to prevent AIDS but not human immunodeficiency virus type 1 infection. J Virol. 2003;77(18):10028–36. pmid:12941913 doi: 10. 1128/jvi. 77. 18. 10028-10036. 2003

83. Lu X, Humeau L, Slepushkin V, Binder G, Yu Q, Slepushkina T, et al. Safe two-plasmid production for the first clinical lentivirus vector that achieves >99% transduction in primary cells using a one-step protocol. J Gene Med. 2004;6(9):963–73. pmid:15352069 doi: 10. 1002/jgm. 593

84. Lu X, Yu Q, Binder GK, Chen Z, Slepushkina T, Rossi J, et al. Antisense-mediated inhibition of human immunodeficiency virus (HIV) replication by use of an HIV type 1-based vector results in severely attenuated mutants incapable of developing resistance. J Virol. 2004;78(13):7079–88. pmid:15194784 doi: 10. 1128/jvi. 78. 13. 7079-7088. 2004

85. Levine BL, Humeau LM, Boyer J, MacGregor RR, Rebello T, Lu X, et al. Gene transfer in humans using a conditionally replicating lentiviral vector. Proc Natl Acad Sci U S A. 2006;103(46):17372–7. pmid:17090675 doi: 10. 1073/pnas. 0608138103

86. Tebas P, Stein D, Binder-Scholl G, Mukherjee R, Brady T, Rebello T, et al. Antiviral effects of autologous CD4 T cells genetically modified with a conditionally replicating lentiviral vector expressing long antisense to HIV. Blood. 2013;121(9):1524–33. doi: 10. 1182/blood-2012-07-447250. pmid:23264589

87. Hirsch VM, Olmsted RA, Murphey-Corb M, Purcell RH, Johnson PR. An African primate lentivirus (SIVsm) closely related to HIV-2. Nature. 1989;339(6223):389–92. pmid:2786147 doi: 10. 1038/339389a0

88. Gao F, Yue L, White AT, Pappas PG, Barchue J, Hanson AP, et al. Human infection by genetically diverse SIVSM-related HIV-2 in west Africa. Nature. 1992;358(6386):495–9. pmid:1641038 doi: 10. 1038/358495a0

89. Hahn BH, Shaw GM, De Cock KM, Sharp PM. AIDS as a zoonosis: scientific and public health implications. Science. 2000;287(5453):607–14. pmid:10649986 doi: 10. 1126/science. 287. 5453. 607

90. Keele BF, Van Heuverswyn F, Li Y, Bailes E, Takehisa J, Santiago ML, et al. Chimpanzee reservoirs of pandemic and nonpandemic HIV-1. Science. 2006;313(5786):523–6. pmid:16728595 doi: 10. 1126/science. 1126531

91. Van Heuverswyn F, Li Y, Neel C, Bailes E, Keele BF, Liu W, et al. Human immunodeficiency viruses: SIV infection in wild gorillas. Nature. 2006;444(7116):164. pmid:17093443 doi: 10. 1038/444164a

92. Plantier JC, Leoz M, Dickerson JE, De Oliveira F, Cordonnier F, Lemee V, et al. A new human immunodeficiency virus derived from gorillas. Nat Med. 2009;15(8):871–2. doi: 10. 1038/nm. 2016. pmid:19648927

93. Notton T, Sardanyes J, Weinberger AD, Weinberger LS. The case for transmissible antivirals to control population-wide infectious disease. Trends Biotechnol. 2014;32(8):400–5. doi: 10. 1016/j. tibtech. 2014. 06. 006. pmid:25017994

94. Pike-Overzet K, van der Burg M, Wagemaker G, van Dongen JJ, Staal FJ. New insights and unresolved issues regarding insertional mutagenesis in X-linked SCID gene therapy. Mol Ther. 2007;15(11):1910–6. pmid:17726455 doi: 10. 1038/sj. mt. 6300297

95. Scholler J, Brady TL, Binder-Scholl G, Hwang WT, Plesa G, Hege KM, et al. Decade-long safety and function of retroviral-modified chimeric antigen receptor T cells. Sci Transl Med. 2012;4(132):132ra53. doi: 10. 1126/scitranslmed. 3003761. pmid:22553251

96. McGarrity GJ, Hoyah G, Winemiller A, Andre K, Stein D, Blick G, et al. Patient monitoring and follow-up in

lentiviral clinical trials. J Gene Med. 2013;15(2):78–82. doi: 10. 1002/jgm. 2691. pmid:23322669

97. Aiuti A, Biasco L, Scaramuzza S, Ferrua F, Cicalese MP, Baricordi C, et al. Lentiviral hematopoietic stem cell gene therapy in patients with Wiskott-Aldrich syndrome. Science. 2013;341(6148):1233151. doi: 10. 1126/science. 1233151. pmid:23845947

98. Biffi A, Montini E, Lorioli L, Cesani M, Fumagalli F, Plati T, et al. Lentiviral hematopoietic stem cell gene therapy benefits metachromatic leukodystrophy. Science. 2013;341(6148):1233158. doi: 10. 1126/ science. 1233158. pmid:23845948

99. Fine PE, Carneiro IA. Transmissibility and persistence of oral polio vaccine viruses: implications for the global poliomyelitis eradication initiative. American journal of epidemiology. 1999;150(10):1001–21. pmid:10568615 doi: 10. 1093/oxfordjournals. aje. a009924

100. O'Connor PM, Allison R, Thapa A, Bahl S, Chunsuittiwat S, Hasan M, et al. Update on polio eradication in the World Health Organization South-East Asia Region, 2013. J Infect Dis. 2014;210 Suppl 1:S216–24. doi: 10. 1093/infdis/jit585. pmid:25316838

INDEX

M

N

O

P

Q

R

S

T

W

Z